W9-BIF-340

Calculus for Management, Social, and Life Sciences

Calculus for Management,

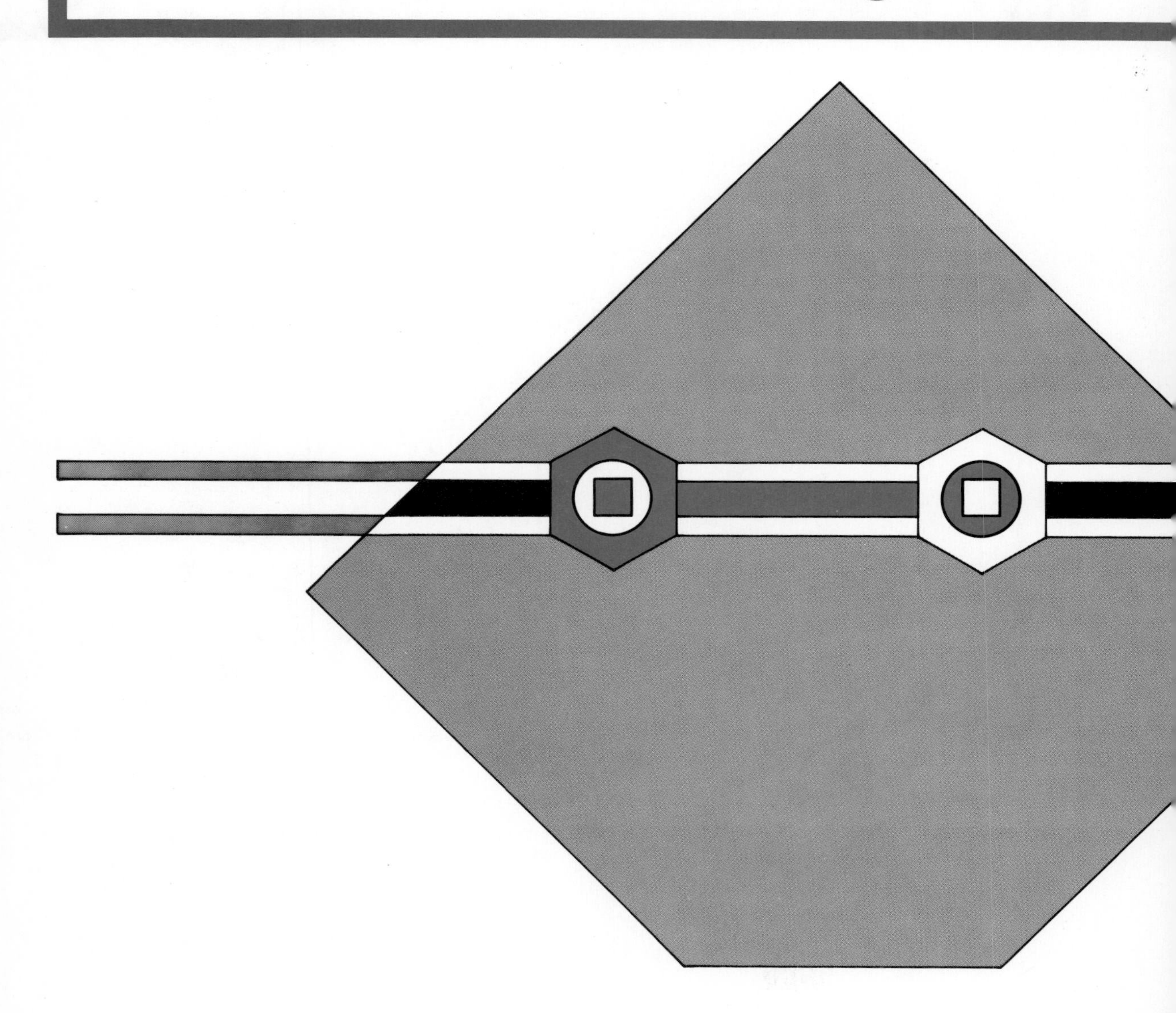

Social, and Life Sciences

Raymond J. Cannon
Mathematics Department,
Baylor University

Gareth Williams
Mathematics Department,
Stetson University

wcb

WM. C. BROWN PUBLISHERS

DUBUQUE, IOWA

Library of Congress Catalog Card Number: 87-19561

ISBN 0-697-06762-9

Printed in the United States of America.
10 9 8 7 6 5 4 3 2

To our wives,
Josephine Jolley Cannon
Donna Williams

Contents

7

Techniques of Integration 419

8

Calculus of Several Variables 457

7 Techniques of Integration [illegible]

8 Calculus of Several Variables [illegible]

Preface

For most of the three hundred years since Newton and Leibniz, calculus has been studied by people interested in mathematics either for its own sake or for its applications to physics and the other natural sciences. In the latter half of this century, however, the use of calculus has spread to other disciplines that have become more quantitatively oriented. This text is intended for students whose main interest is in the social, managerial, or life sciences. Generally speaking, these students may not have been previously attracted to mathematics, so we have kept the prerequisites to a minimum, assuming about three semesters of high school algebra. Chapter 0 provides a brief review of this material for those who may need to refresh their memory.

Features

We have concentrated on maintaining a style that is clear, friendly, and considerate of the student. The applications of calculus are stressed, but in a way that does not require extensive knowledge of other fields.

We have found that frequently students view their mathematics text solely as the source for their homework exercises. Our text is structured to encourage students to read the explanations and study the examples. We have cross-referenced certain exercises and examples so that after working through an example, the student can turn immediately to a similar exercise. Likewise, in certain exercises we indicate similar examples the student can turn to for help. Furthermore, we have tried to include enough detail in the computations so that the student can follow successive steps.

We have also included special warnings of common errors and reminders of previous concepts; these and other special comments are flagged with a bold Warning, Reminder, or some similar heading.

Although a good understanding of most calculus topics depends on a good understanding of preceding topics, we have tried to keep the text as flexible as possible. Different instructors will have different tastes, some emphasizing one topic that others may omit; we have attempted to structure the text with this fact in mind. Further comments along this line are included in the instructor's manual.

Acknowledgments

A number of people contributed to the writing of this book. We would like to thank our students who have, with good spirit, helped us keep the writing clear and the examples helpful. We are most grateful to the following reviewers who provided constructive criticism that helped to clarify ideas and to improve the presentation of the concepts:

Professor Bill Bompart
Augusta College
Department of Mathematics
Augusta, GA 30910

Professor Robert Bumcrot
Department of Mathematics
Hofstra University
Hempstead, NY 11550

Professor Donald Cathcart
Department of Math Sciences
Salisbury State College
Salisbury, MD 21801

Professor Philip Crooke
Vanderbilt University
Box 6205 Station B
Nashville, TN 37235

Professor Bruce Edwards
University of Florida
Department of Mathematics
201 Walker Hall
Gainesville, FL 32611

Professor Garrett Etgen
University of Houston—
University Park
Department of Mathematics
Houston, TX 77004

Professor Joe S. Evans
Department of Mathematics
and Computer Science
Middle Tennessee State University
Murfreesboro, TN 37132

Professor Roseanne Hofmann
C. E. Hofmann Associates
1709 Bantry Drive
Dresher, PA 19025

Dr. Lloyd Koontz
Eastern Illinois University
Department of Mathematics
Charleston, IL 61920

Professor Robert Piziak
Department of Mathematics
Baylor University
Waco, TX 76798

Professor Ann Thorne
Department of Natural Sciences
College of DuPage
22nd & Lambert Rd.
Glen Ellyn, IL 60137

Professor H. G. Williams, Jr.
Virginia Military Institute
Department of Mathematics
Lexington, VA 24450

The monumental task of transforming terrible handwriting into a readable manuscript fell upon Dee Nieman, Karen LaPoint, and Joan Maxwell. Thanks for your patience and a job well done.

Calculus for Management, Social, and Life Sciences

This chapter contains basic algebra topics that are necessary for the materials in this book. You are encouraged to study those topics for which you need review and skip those topics with which you are familiar.

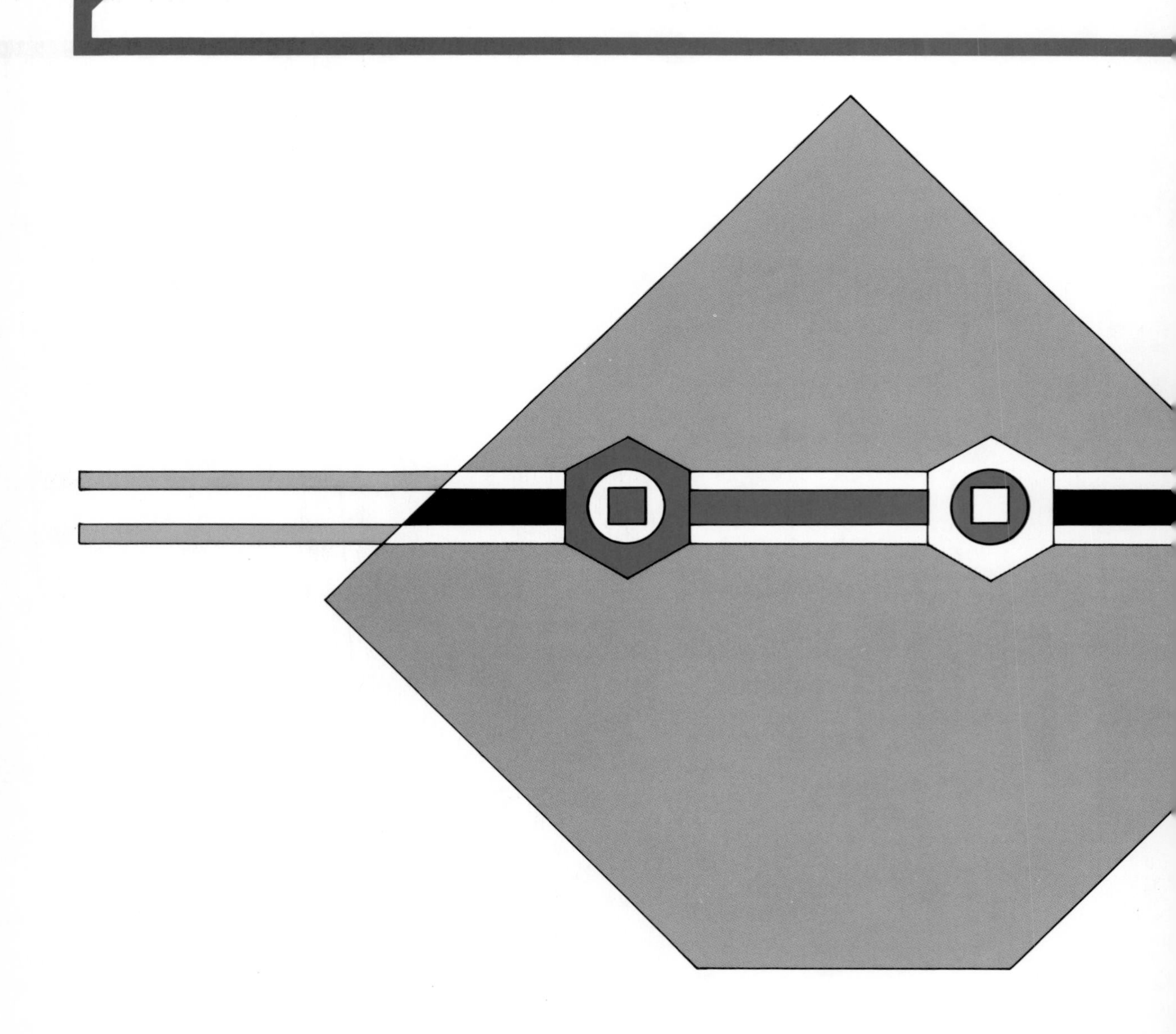

Review Topics

- Properties of Real Numbers
- Solving Linear Equations
- Coordinate Systems
- Linear Inequalities and Interval Notation
- Polynomial Arithmetic
- Exponents

0–1 Properties of Real Numbers

The Real Number Line
The Arithmetic of Real Numbers

The Real Number Line

The most basic numbers in our study of mathematics are the **natural** numbers;

$$1, 2, 3, 4, \ldots$$

These are, in fact, the numbers that we use to count. They can be represented graphically as points on a line, as in Figure 0–1.

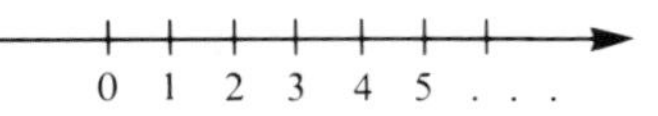

Figure 0–1

Begin with a point 0 on the line and a convenient unit scale. Starting at 0, mark off equal lengths to the right. These marks represent $1, 2, 3, \ldots$.

We can mark off lengths in the opposite direction from 0 as well to get the negative numbers $-1, -2, -3, \ldots$ (Figure 0–2). This collection of numbers

$$\ldots, -4, -3, -2, -1, 0, 1, 2, 3, 4, \ldots$$

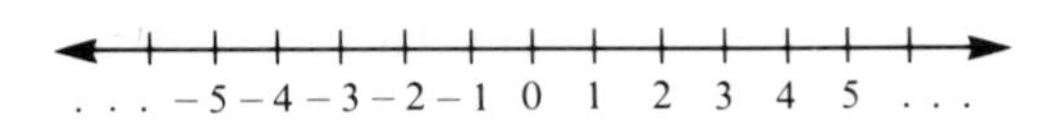

Figure 0–2

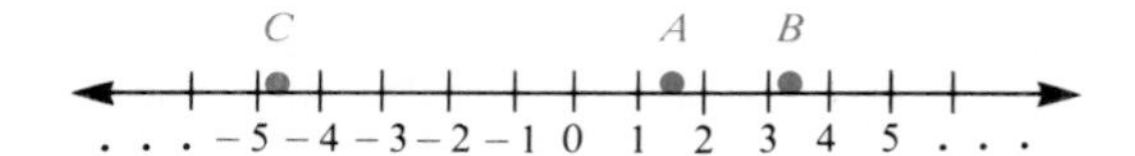

Figure 0–3

is called the set of **integers**. We call the number represented by the symbol 0, "zero." It is neither positive nor negative.

Other numbers, the **fractions**, can be represented with points between the integers. In Figure 0–3, point A, halfway between 1 and 2, represents $1\frac{1}{2}$; point B, one quarter of the way between 3 and 4, represents $3\frac{1}{4}$. Point C is three quarters of the way between -4 and -5; so it represents $-4\frac{3}{4}$. These numbers together with the integers are called **rational numbers**.

All these points can be expressed in terms of finite or infinite **decimals**. Point A is the point 1.5, B the point 3.25, and C the point -4.75. Every rational number can be expressed either as a decimal that terminates, such as A, B, and C above, or as a decimal that repeats infinitely. For example, $5\frac{1}{3}$ is a rational number that can be written in decimal form as 5.333...; the 3 repeats endlessly.

There are, however, certain numbers called **irrational** numbers that do not have any pattern of repetition in their decimal form. One such number is $\sqrt{2}$. Its decimal form is 1.414213....

The set of all rational and irrational numbers is called the set of **real numbers**. One way to visualize the set of real numbers is to think of each point on the line as a real number.

The Arithmetic of Real Numbers

There are four useful operations on the set of real numbers: addition, subtraction, multiplication, and division. Table 0–1 summarizes some rules that govern these operations.

Table 0–1.

Rule		Example
Division by zero is not allowed.		$\frac{5}{0}$ and $\frac{0}{0}$ have no meaning.
	Rules of Operations	
$a + b = b + a$	Numbers can be added in either order.	$3 + 7 = 7 + 3$
$ab = ba$	Numbers can be multiplied in either order.	$5 \times 8 = 8 \times 5$
$a(b + c) = ab + ac$	A common number can be factored from each term in a sum.	$3(4a + 5b) = 12a + 15b$

Table 0–1. (*continued*)

Rule		Example
Rules of Signs (In this section, *a* and *b* are positive real numbers.)		
$-a = (-1)a$	The negative of a number a is the product $(-1)a$.	$-12 = (-1)12$
$-(-a) = a$		$-(-7) = 7$
$(-a)b = a(-b) = -ab$	The product of a positive and a negative is negative.	$(-4)8 = -32$ $5(-3) = -15$
$(-a)(-b) = ab$	The product of two negative numbers is positive.	$(-2)(-7) = 14$
$(-a) + (-b) = -(a + b)$	The sum of two negative numbers is negative.	$(-2) + (-3) = -5$
$-(a - b) = (-a) + b$		$-(8 - 3) = -8 + 3$
$-a/b = a/(-b)$ $= -(a/b)$ $(b \neq 0)$	Division using a positive and a negative number is negative.	$-18/3 = -6$ $22/(-11) = -2$
Arithmetic of Fractions		
$\frac{ac}{bc} = \frac{a}{b}$ $(b, c \neq 0)$	The value of a fraction is unchanged if both numerator and denominator are multiplied, or divided, by the same number.	$\frac{12}{8} = \frac{3}{2}$ $\frac{4}{3} = \frac{20}{15}$
$\frac{a}{d} + \frac{b}{d} = \frac{a + b}{d}$ $(d \neq 0)$	To add two fractions with the same denominators, add the numerators and keep the same denominator.	$\frac{2}{7} + \frac{3}{7} = \frac{5}{7}$
$\frac{a}{d} + \frac{b}{c} = \frac{ac}{dc} + \frac{bd}{dc}$ $= \frac{ac + bd}{dc}$ $(c, d \neq 0)$	To add two fractions with different denominators, convert them to fractions with the same denominators by multiplying numerator and denominator of each by the denominator of the other.	$\frac{2}{5} + \frac{3}{4} = \frac{2(4)}{5(4)} + \frac{3(5)}{5(4)}$ $= \frac{2(4) + 3(5)}{5(4)}$ $= \frac{8 + 15}{20}$ $= \frac{23}{20}$

Table 0–1. *(continued)*

Rule		Example
Arithmetic of Fractions		
$\frac{a}{d} \times \frac{b}{c} = \frac{ab}{dc}$ $(c, d \neq 0)$	To multiply two fractions, multiply their numerators and multiply their denominators.	$\frac{3}{8} \times \frac{2}{7} = \frac{6}{56} = \frac{3}{28}$
$\frac{a}{d} \div \frac{b}{c} = \frac{a}{d} \times \frac{c}{b} = \frac{ac}{db}$ $(b, c, d \neq 0)$	To divide by a fraction, invert the divisor and multiply.	$\frac{3}{4} \div \frac{2}{5} = \frac{3}{4} \times \frac{5}{2}$ $= \frac{15}{8}$
$\frac{a/d}{b/c} = \frac{a}{d} \div \frac{b}{c}$ $= \frac{a}{d} \times \frac{c}{b} = \frac{ac}{db}$ $(b, c, d \neq 0)$		$\frac{4/5}{8/9} = \frac{4}{5} \div \frac{8}{9}$ $= \frac{4}{5} \times \frac{9}{8}$ $= \frac{1}{5} \times \frac{9}{2}$ $= \frac{9}{10}$

0–1 Exercises

Evaluate each of the following:

1. $(-1)13$
2. $(-1)(-7)$
3. $-(-23)$
4. $(-10)(-4)$
5. $(-5)(6)$
6. $(-2)(-4)$
7. $5(-7)$
8. $(-6) + (-11)$
9. $-(7 - 2)$
10. $(-10)/5$
11. $21/(-3)$
12. $(5 \times 3)/(7 \times 3)$
13. $(-4) + (-6)$
14. $(-3)(-2)$
15. $(-4)2$
16. $\frac{4}{9} + \frac{2}{9}$
17. $\frac{5}{3} + \frac{4}{3}$
18. $\frac{4}{11} - \frac{2}{11}$
19. $\frac{12}{5} - \frac{3}{5}$
20. $\frac{6}{10} - \frac{13}{10}$
21. $\frac{2}{3} + \frac{3}{4}$
22. $\frac{5}{8} - \frac{1}{3}$
23. $\frac{5}{6} - \frac{7}{4}$
24. $\frac{5}{12} - \frac{1}{6}$
25. $\frac{2}{5} + \frac{1}{4}$
26. $(-3) + 6$
27. $\frac{4}{7} - \frac{3}{5}$
28. $\frac{2}{3} \times \frac{4}{5}$

29. $(\frac{3}{4})/(\frac{9}{8})$ **30.** $\frac{3}{8} + \frac{2}{5}$ **31.** $(\frac{2}{7})/(\frac{4}{5})$ **32.** $(\frac{4}{3})(\frac{6}{7})$
33. $(\frac{1}{3})(\frac{1}{5})$ **34.** $6 \times (-3)$ **35.** $\frac{2}{5} \times \frac{4}{3}$ **36.** $(-\frac{2}{3})(\frac{1}{9})$
37. $(-\frac{3}{5})(-\frac{4}{7})$ **38.** $\frac{4}{5} \div \frac{2}{15}$ **39.** $\frac{3}{11} + \frac{1}{3}$ **40.** $(-\frac{4}{9})/(\frac{5}{2})$
41. $\frac{5}{7} \div \frac{15}{28}$ **42.** $(\frac{4}{9})/(\frac{16}{3})$ **43.** $(\frac{5}{8}) \div (\frac{1}{3})$ **44.** $(\frac{1}{2} - \frac{1}{3})(\frac{5}{7})$
45. $(\frac{3}{4} + \frac{1}{5}) \div (\frac{2}{9})$ **46.** $5(4a + 2b)$ **47.** $-2(3a + 11b)$ **48.** $2(a - 3b)$
49. $-5(2a + 10b)$

0–2 Solving Linear Equations

Numerous disciplines including science, technology, social sciences, business, manufacturing, and government find mathematical techniques essential in day-to-day operations. They depend heavily on mathematical equations that describe conditions or relationships between quantities.

In an equation such as

$$4x - 5 = 7$$

the symbol x, called a **variable**, represents an arbitrary, an unspecified, or an unknown number just as John Doe and Jane Doe often denote an arbitrary, unspecified, or unknown person.

The equation $4x - 5 = 7$ may be true or false depending on the choice of the number x. If we substitute the number 3 for x in

$$4x - 5 = 7$$

both sides become equal and we say that $x = 3$ is a **solution** of the equation. If 5 is substituted for x, then both sides are not equal so $x = 5$ is not a solution.

It may help a businessman to have an equation describing the relationship between sales and profits, for example. However, he may also need to find a solution to the equation in order to make some decision.

One basic procedure for solving an equation is to obtain a sequence of equivalent equations with the goal of isolating the variable on one side of the equation and the appropriate number on the other side.

The following two operations help to isolate the variable and find the solution.

1. The same number may be added to or subtracted from both sides of an equation.
2. Both sides of an equation may be multiplied or divided by a nonzero number.

Either of these operations yields another equation that is equivalent to the first, that is, a second equation having the same solution as the first.

Example 1 (*Compare Exercise 3*)
Solve the equation $3x + 4 = 19$.

Solution We begin to isolate x by subtracting 4 from both sides.

$$3x + 4 - 4 = 19 - 4$$
$$3x = 15$$

Next divide both sides by 3

$$\frac{3x}{3} = \frac{15}{3}$$

$x = 5$ is the solution.

We can check our answer by substituting $x = 5$ into the original equation.

$3(5) + 4 = 15 + 4 = 19$ so the solution checks.

Example 2 (*Compare Exercise 6*)
Solve $4x - 2 = 2x + 12$.

Solution

$$4x - 2 = 2x + 12 \quad \text{First, add 2 to both sides}$$
$$4x - 2 + 2 = 2x + 12 + 2$$
$$4x = 2x + 14 \quad \text{Next, subtract } 2x \text{ from both sides}$$
$$4x - 2x = 2x + 14 - 2x$$
$$2x = 14 \quad \text{Now, divide both sides by 2}$$
$$x = 7$$

CHECK $4(7) - 2 = 28 - 2 = 26$ (left side) and $2(7) + 12 = 14 + 12 = 26$ (right side) so it checks.

Example 3 (*Compare Exercise 9*)
Solve $7x + 13 = 0$.

Solution

$$7x + 13 = 0 \quad \text{Subtract 13 from both sides}$$
$$7x = -13 \quad \text{Divide both sides by 7}$$
$$x = -\frac{13}{7}$$

The above examples all use **linear equations**.

Definition

A linear equation in one variable is an equation that can be written in the form

$$ax + b = 0 \quad \text{where} \quad a \neq 0$$

A linear equation in two variables is an equation that can be written in the form

$$y = ax + b \quad \text{where} \quad a \neq 0$$

Example 4 (*Compare Exercise 13*)

Solve $\frac{3x - 5}{2} + \frac{x + 7}{3} = 8$.

Solution We show two ways to solve this. First, use rules of fractions to combine the terms on the left side:

$$\frac{3x - 5}{2} + \frac{x + 7}{3} = 8 \qquad \text{Convert fractions to the same denominator}$$

$$\frac{3(3x - 5)}{6} + \frac{2(x + 7)}{6} = 8 \qquad \text{Now add the fractions}$$

$$\frac{3(3x - 5) + 2(x + 7)}{6} = 8$$

$$\frac{9x - 15 + 2x + 14}{6} = 8$$

$$\frac{11x - 1}{6} = 8 \qquad \text{Now multiply both sides by 6}$$

$$11x - 1 = 48$$

$$11x = 49$$

$$x = \frac{49}{11}$$

An alternate, and simpler, method is the following:

$$\frac{3x - 5}{2} + \frac{x + 7}{3} = 8$$

Multiply through by 6.

$$3(3x - 5) + 2(x + 7) = 48$$

$$9x - 15 + 2x + 14 = 48$$

$$11x - 1 = 48$$

$$11x = 49$$

$$x = \frac{49}{11}$$

0–2 Exercises

Determine which of the following values of x are solutions to the equations in Exercises 1 and 2. Use $x = 1, 2, -3, 0, 4$, and -2.

1. $2x - 4 = -10$
2. $3x + 1 = x + 5$

Solve the following equations:

3. (*See Example 1*) $2x - 3 = 5$

4. $-4x + 2 = 6$

5. $4x - 3 = 5$

6. (*See Example 2*) $7x + 2 = 3x + 4$

7. $5 - x = 8 + 3x$

8. $2x - 4 = -5x + 2$

9. (*See Example 3*) $12x + 21 = 0$

10. $7x - 4 = 0$

11. $3(x - 5) + 4(2x + 1) = 9$

12. $6(4x + 5) + 7 = 2$

13. (*See Example 4*) $\dfrac{2x + 3}{3} + \dfrac{5x - 1}{4} = 2$

14. $\dfrac{4x + 7}{6} + \dfrac{2 - 3x}{5} = 5$

15. $\dfrac{12x + 4}{2x + 7} = 4$

16. $\dfrac{x + 1}{x - 1} = \dfrac{3}{4}$

17. The U-Drive-It Rental charges \$0.20 per mile plus \$112 per week for car rental. The weekly rental fee for a car is represented by a linear equation

$$y = 0.20x + 112$$

where x is the number of miles driven and y is the weekly rental charge.

(a) Determine the rental fee if the car is driven 650 miles during the week.
(b) Determine the rental fee if the car is driven 1500 miles.
(c) The weekly rental fee is \$302. How many miles were driven?

18. Joe Cool has a summer job of selling real estate in a subdivision development. He receives a base pay of \$100 per week plus \$50 for each lot sold. Thus, the equation

$$y = 50x + 100$$

represents his weekly income where x is the number of lots sold.

(a) What is his weekly income if he sells 7 lots?
(b) What is his weekly income if he sells 15 lots?
(c) If he receives \$550 one week, how many lots did he sell?

19. A Boy Scout troop collects aluminium cans for a project. The recycling center weighs the cans in a container that weighs 8 pounds, so the Boy Scouts are paid according to the equation

$$y = 0.21(x - 8)$$

where x is the weight in pounds given by the scale and y is the payment in dollars.

(a) How much money do the Boy Scouts receive if the scale reads 42 pounds?
(b) How much do they receive if the scale reads 113 pounds?
(c) They received \$11.13 for one weekend's collection. What was the reading on the scale?

20. The tuition and fees paid by students at a local junior college is given by the equation

$$y = 27x + 85$$

where x is the number of hours enrolled and y is the total tuition-fees cost in dollars.

(a) How much does a student pay who is enrolled in 13 hours?

(b) A student who pays $517 is enrolled in how many hours?

0–3 Coordinate Systems

We have all seen a map, a house plan, or a wiring diagram that shows information recorded on a flat surface. Each of these uses some notation unique to the subject to convey the desired information. In mathematics, we often use a flat surface called a **plane** to draw figures and locate points. A reference system in the plane helps to record and communicate information accurately. The standard mathematical reference system consists of a horizontal and a vertical line (called **axes**). These two perpendicular axes form a *Cartesian*, or **rectangular, coordinate system.** They intersect at a point called the **origin.**

We name the horizontal axis the **x-axis**, and we name the vertical axis the **y-axis**. The origin is labeled O.

Two numbers are used to describe the location of a point in the plane and they are recorded in the form (x, y). For example, $x = 3$ and $y = 2$ for the point $(3, 2)$. The first number, 3, called the **x-coordinate** or **abscissa**, represents the horizontal distance from the y-axis to the point. The second number, 2, called the **y-coordinate** or **ordinate**, represents the vertical distance measured from the x-axis to the point. The point $(3, 2)$ is shown as point P in Figure 0–4. Points located to the right of the y-axis have positive x-coordinates; those to the left have negative x-coordinates. The y-coordinate is positive for points located above the x-axis and negative for those located below.

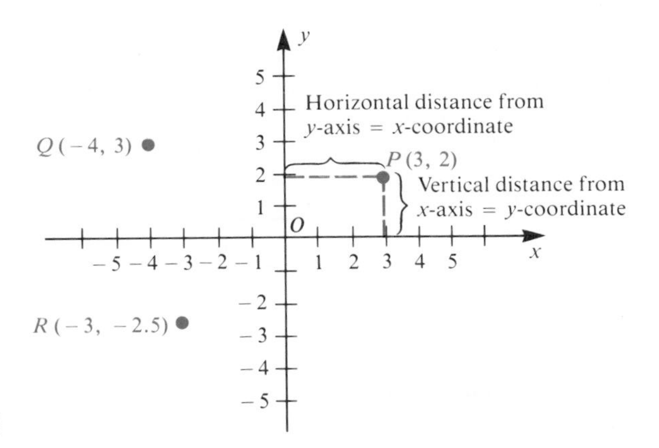

Figure 0–4

Figure 0–4 shows other examples of points in this coordinate system: Q is the point $(-4, 3)$ and R is the point $(-3, -2.5)$. The origin O has coordinates $(0, 0)$.

Figure 0–5 shows the points $(-3, 2)$, $(-4, -2)$, $(1, 1)$, and $(1, -2)$ plotted on the Cartesian coordinate system.

The coordinate axes divide the plane into four parts called **quadrants**. The quadrants are labeled I, II, III, and IV as shown in Figure 0–6. Point A is in the first quadrant, B in the second quadrant, C in the third, and D in the fourth. Points A and E are in the same quadrant.

René Descartes (1596–1650), a French philosopher-mathematician, invented the Cartesian coordinate system. His invention of the coordinate system is one of the outstanding events in the history of mathematics because it combined algebra and geometry in a way that enables us to use algebra to solve geometry problems and to use geometry to clarify algebraic concepts.

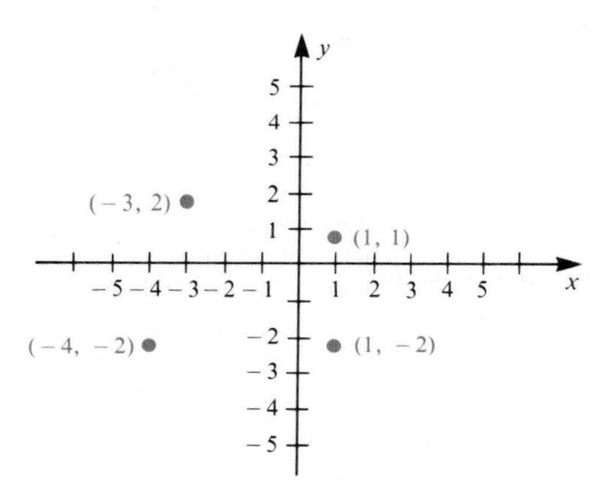

Figure 0–5

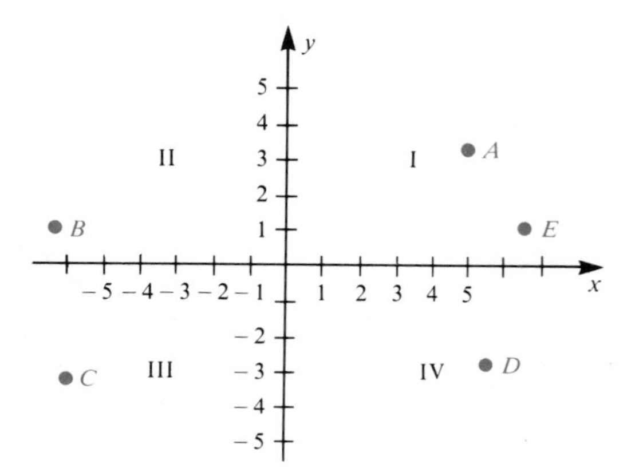

Figure 0–6

0–3 Exercises

1. The following are the coordinates of points in a rectangular Cartesian coordinate system. Plot these points.

$$(-5, 4), (-2, -3), (-2, 4), (1, 5), (2, -5)$$

2. What are the coordinates of the points P, Q, R, and S in the coordinate system in Figure 0–7?

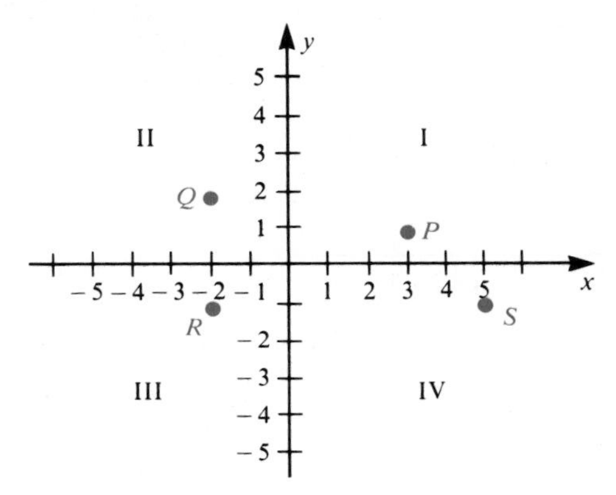

Figure 0–7

3. Locate the following points on a Cartesian coordinate system:

$$(-2, 5), (3, -2), (0, 4), (-2, 0), \left(\frac{7}{2}, 2\right)$$

$$\left(\frac{2}{3}, \frac{9}{4}\right), (-4, -2), (0, -5), (0, -2), (-6, -3)$$

4. Give the coordinates of A, B, C, D, E, and F in the coordinate system shown in Figure 0–8.

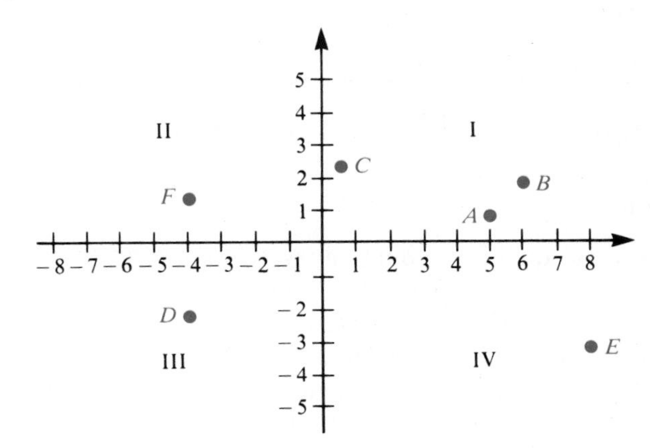

Figure 0–8

5. Note that all points in the first quadrant have positive x-coordinates and positive y-coordinates. What are the characteristics of the points in

   (a) the second quadrant?
   (b) the third quadrant?
   (c) the fourth quadrant?

6. For each case shown in Figure 0–9, find the property the points have in common.

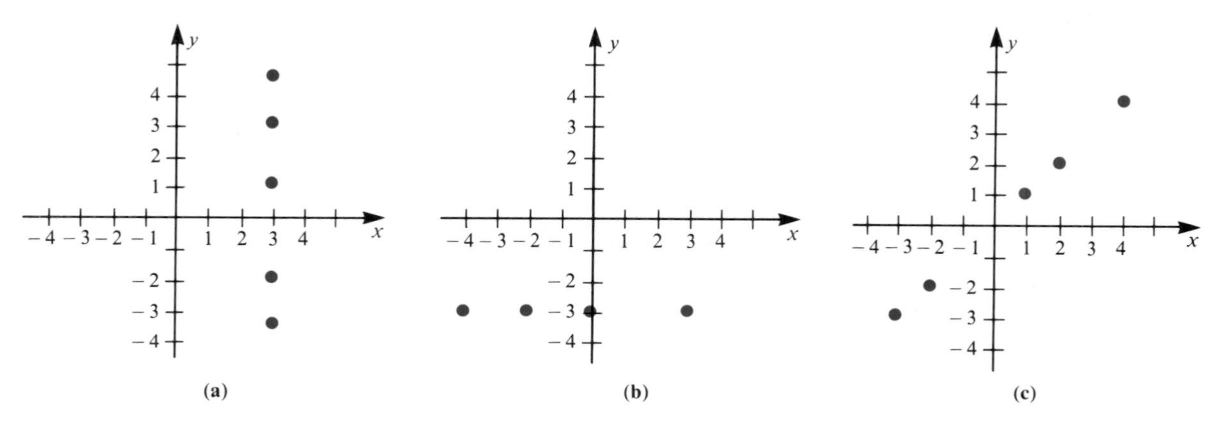

Figure 0–9

7. An old map gives these instructions to find a buried treasure:

   Start at giant oak tree. Go north 15 paces, then east 22 paces to a half buried rock. The key to the treasure chest is buried at the spot that is 17 paces west and 13 paces north of the rock.

   From the place where the key is buried, go 32 paces west and 16 paces south to the location of the buried treasure. Use a coordinate system to represent the location of the oak tree, the rock, the key, and the treasure.

0–4 Linear Inequalities and Interval Notation

Solving Inequalities
Interval Notation

We frequently use inequalities in our daily conversation. They may take the form "Which store has the lower price?" "Did you make a higher grade?" "Our team scored more points." "My expenses are greater than my income." Statements such as these basically state that one quantity is greater than another. Statements using the terms "greater than" or "less than" are called **inequalities**. Our goal is to solve inequalities. First, we give some terminology and notation.

The symbol < means "less than" and > means "greater than." Just remember that each of these symbols points to the smaller quantity. The notation $a > b$ and $b < a$ have exactly the same meaning. We may state the definition of $a > b$ in three ways. At times one may be more useful than the other, so choose the most appropriate one.

If a and b are real numbers, the following statements have the same meaning.

(a) $a > b$ means that a lies to the right of b on a number line.

(b) $a > b$ means that there is a positive number p such that $a = b + p$.

(c) $a > b$ means that $a - b$ is a positive number p.

The positive numbers lie to the right of zero on a number line and the negative numbers lie to the left.

Example 1 (*Compare Exercise 1*)
The numbers 5, 8, 17, -2, -3, and -15 are plotted on a number line in Figure 0–10. Notice the following.

1. **(a)** 17 lies to the right of 5.
   (b) $17 = 5 + 12$
   (c) $17 - 5 = 12$

Each of these three statements is equivalent to saying that $17 > 5$.

2. $8 > -3$ because $8 - (-3) = 8 + 3 = 11$. (By part c)
3. $-2 > -15$ because $-2 = -15 + 13$. (By part b where $p = 13$)

Figure 0–10

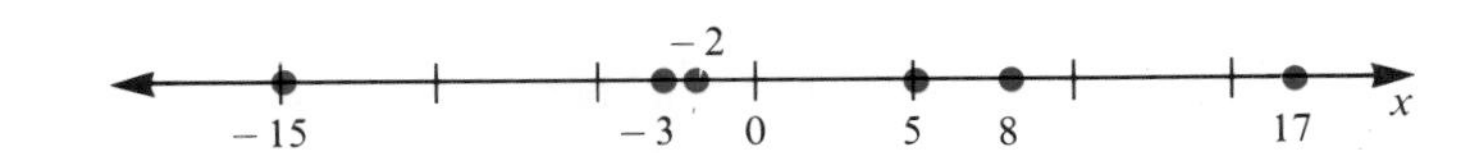

Solving Inequalities

By the **solution** of an inequality like

$$3x + 5 > 23$$

we mean the value, or values, of x that make the statement true. The method for solving inequalities is similar to that for solving equations. We want to operate on an inequality in a way that gives an equivalent inequality, but which enables us to determine the solution.

Here are some simple examples of useful properties of inequalities.

Example 2

1. Since $18 > 4$, $18 + 6 > 4 + 6$, that is, $24 > 10$. (6 added to both sides)
2. Since $23 > -1$, $23 - 7 > -1 - 7$, that is, $16 > -8$. (7 subtracted from both sides)

3. Since $6 > 2$, $4(6) > 4(2)$, that is, $24 > 8$. (Both sides multiplied by 4)

4. Since $10 > 3$, $-2(10) < -2(3)$, that is, $-20 < -6$. (Both sides multiplied by -2)

Warning! The inequality symbol reverses when we multiply each side by a negative number.

5. Since $-15 > -21$, $\frac{-15}{3} > \frac{-21}{3}$, that is, $-5 > -7$. (Divide both sides by 3)

6. Since $20 > 6$, $\frac{20}{-2} < \frac{6}{-2}$, that is, $-10 < -3$. (Divide both sides by -2)

Warning! The inequality symbol reverses when dividing each side by a negative number.

These examples illustrate basic properties that are useful in solving inequalities.

Properties of Inequalities

For real numbers a, b, and c, the following are true.

1. If $a > b$, then $a + c > b + c$.
2. If $a > b$, then $a - c > b - c$.
3. If $a > b$ and c is positive, then $ca > cb$.
4. If $a > b$ and c is negative, then $ca < cb$. Notice the change from $>$ to $<$.
5. If $a > b$ and c is positive, then $\frac{a}{c} > \frac{b}{c}$.
6. If $a > b$ and c is negative, then $\frac{a}{c} < \frac{b}{c}$. Notice the change from $>$ to $<$.

Similar properties hold if each inequality symbol is reversed.

We use these properties to solve an inequality; that is, to find the values of x that make the inequality true. In general, we proceed by finding equivalent inequalities that will eventually isolate x on one side of the inequality and the appropriate number on the other side.

Example 3 (*Compare Exercise 2*)
Solve the inequality $3x + 5 > 14$.

Solution Begin with the given inequality.

$$3x + 5 > 14 \qquad \text{Next, subtract 5 from each side (Property 2)}$$
$$3x > 9 \qquad \text{Now divide each side by 3 (Property 5)}$$
$$x > 3$$

Thus, all x greater than 3 make the inequality true. This solution can be graphed on a number line as shown in Figure 0–11. The empty circle indicates that $x = 3$ is omitted from the solution and the heavy line indicates the values of x included in the solution.

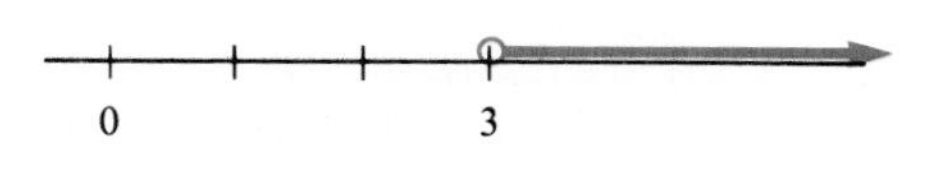

Figure 0–11 $x > 3$

Example 4 (*Compare Exercise 10*)
Solve the inequality $5x - 17 > 8x + 14$ and indicate the solution on a graph.

Solution Start with the given inequality.

$5x - 17 > 8x + 14$ Now add 17 to both sides (Property 1)

$5x > 8x + 31$ Now subtract $8x$ from both sides (Property 2)

$-3x > 31$ Now divide both sides by -3 (Property 6)

$x < \dfrac{-31}{3}$ This reverses the inequality symbol

Thus, the solution consists of all x to the left of $-31/3$. See Figure 0–12.

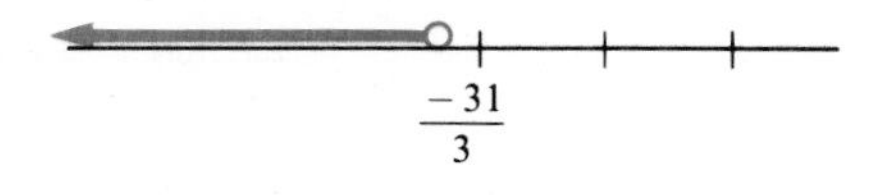

Figure 0–12 $x < \dfrac{-31}{3}$

We will also use the symbols $\geq$ (greater than or equal to) and $\leq$ (less than or equal to). All the properties of inequalities hold if $<$ is replaced with $\leq$ and $>$ is replaced with $\geq$. An inequality with $\geq$ simply includes the possibility of an equality as well as greater than.

Example 5 (*Compare Exercise 7*)
Solve and graph $2(x - 3) \leq 3(x + 5) - 7$

Solution

$2(x - 3) \leq 3(x + 5) - 7$ First perform the indicated multiplications

$2x - 6 \leq 3x + 15 - 7$

$2x - 6 \leq 3x + 8$ Now add 6 to both sides (Property 1)

$2x \leq 3x + 14$ Subtract $3x$ from both sides (Property 2)

$-x \leq 14$ Multiply both sides by -1 (Property 4)

$x \geq -14$

Since the solution includes -14 and all numbers greater, the graph shows a solid circle at -14 (see Figure 0–13).

Figure 0–13 $x \geq -14$

The next example illustrates a problem that involves two inequalities.

Example 6 (*Compare Exercise 16*)
Solve and graph $3 < 2x + 5 \leq 13$.

Solution This inequality means both $3 < 2x + 5$ *and* $2x + 5 \leq 13$. Solve it in a manner similar to the preceding examples except that you try to isolate the x in the middle.

Begin with the given inequality.

$3 < 2x + 5 \leq 13$	Subtract 5 from all parts of the inequality
$-2 < 2x \leq 8$	Divide each part by 2
$-1 < x \leq 4$	The solution consists of all numbers between -1 and 4, including 4 but not including -1

The graph of the solution (see Figure 0–14) shows an empty circle at -1 because -1 is not a part of the solution. It shows a solid circle at 4 because 4 is a part of the solution. The solid line between -1 and 4 indicates that all numbers between -1 and 4 are included in the solution.

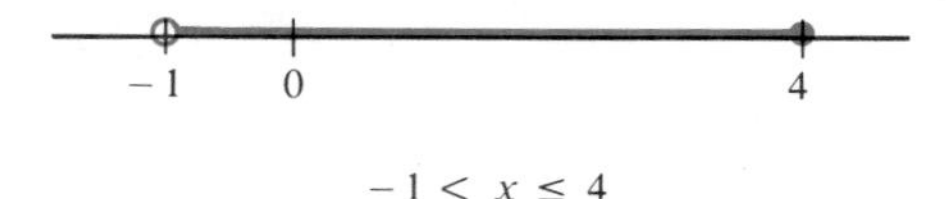

Figure 0–14 $-1 < x \leq 4$

Interval Notation

There is yet another notation for indicating the solution of an inequality. It is the **interval notation**. The portion of the number line that represents the solution of an inequality is identified by its endpoints; brackets or parentheses indicate whether or not the endpoint is included in the solution. A parenthesis indicates that the endpoint is not included and a bracket indicates that the endpoint is included. The notation $(-1, 4]$ indicates the set of all numbers between -1 and 4 with -1 excluded and 4 included in the set. The notation $(-1, 4)$ indicates that both -1 and 4 are excluded.

The notation $(-1, \infty)$ means all numbers greater than -1. The symbol ∞ denotes infinity and indicates that there is no upper bound to the interval.

Table 0–2 shows the variations of the interval notation.

Example 7 (*Compare Exercise 22*)
Solve $1 \leq 2(x - 5) + 3 < 5$.

Solution

$1 \leq 2(x - 5) + 3 < 5$	Multiply to remove the parentheses
$1 \leq 2x - 10 + 3 < 5$	
$1 \leq 2x - 7 < 5$	Add 7 throughout
$8 \leq 2x < 12$	Divide through by 2
$4 \leq x < 6$	

The solution consists of all values of x in the interval $[4, 6)$ and the graph is shown in Figure 0–15.

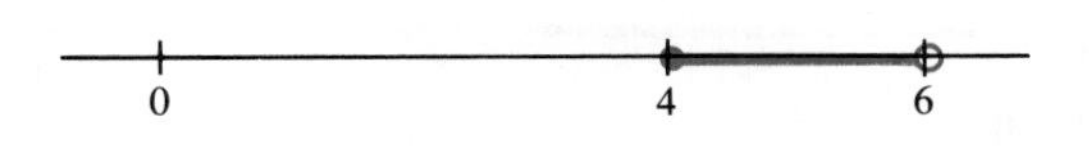

Figure 0–15 $4 \le x < 6$

Example 8 (*Compare Exercise 29*)
The total points on an exam given by Professor Passmore are determined by: 20 points plus 2.5 points for each correct answer. A total score in [70, 80) is a C. If Scott made a C on the exam, how many questions did he answer correctly?

Solution The score on an exam is given by $20 + 2.5x$, where x is the number of correct answers. Thus, the condition for a C is

$$70 \le 20 + 2.5x < 80$$

Solve for x to obtain the number of correct answers.

$$70 \le 20 + 2.5x < 80$$
$$50 \le 2.5x < 60$$
$$\frac{50}{2.5} \le x < \frac{60}{2.5}$$
$$20 \le x < 24$$

In this case only whole numbers make sense so, Scott got 20, 21, 22, or 23 correct answers.

Table 0–2.

Inequality notation		Interval notation		
General	**Example**	**General**	**Example**	**Graph of interval**
$a < x < b$	$-1 < x < 4$	(a, b)	$(-1, 4)$	
$a \le x < b$	$-1 \le x < 4$	$[a, b)$	$[-1, 4)$	
$a < x \le b$	$-1 < x \le 4$	$(a, b]$	$(-1, 4]$	
$a \le x \le b$	$-1 \le x \le 4$	$[a, b]$	$[-1, 4]$	
$x < b$	$x < 4$	$(-\infty, b)$	$(-\infty, 4)$	
$x \le b$	$x \le 4$	$(-\infty, b]$	$(-\infty, 4]$	
$a < x$	$-1 < x$	(a, ∞)	$(-1, \infty)$	
$a \le x$	$-1 \le x$	$[a, \infty)$	$[-1, \infty)$	

0–4 Exercises

I.

1. (*See Example 1*) The following inequalities are of the form $a > b$. Verify the truth or falsity of each one by using the property $a > b$ means $a - b$ is a positive number.

   (a) $9 > 3$ **(b)** $4 > 0$ **(c)** $-5 > 0$
   (d) $-3 > -15$ **(e)** $\frac{5}{6} > \frac{2}{3}$

Solve the inequalities in Exercises 2 through 9. State your solution using inequalities.

2. (*See Example 3*) $2x + 5 > 17$
3. $3x - 5 < x + 4$
4. $12 > 1 - 5x$
5. $5x - 22 \leq 7x + 10$
6. $13x - 5 \leq 7 - 4x$
7. (*See Example 5*) $3(x + 4) < 2(x - 3) + 14$
8. $14 < 3x + 8 < 32$
9. $-9 \leq 3(x + 2) - 15 < 27$

Solve the inequalities in Exercises 10 through 17. Graph the solution.

10. (*See Example 4*) $3x + 2 \leq 4x - 3$
11. $6x + 5 < 5x - 4$
12. $3x + 2 < 2x - 3$
13. $78 < 6 - 3x$
14. $4(x - 2) > 5(2x + 1)$
15. $3(2x + 1) < -1(3x - 10)$
16. (*See Example 6*) $-16 < 3x + 5 < 22$
17. $124 > 5 - 2x \geq 68$

Solve the inequalities in Exercises 18 through 23. Give the solution in interval form.

18. $5x - 7 > 3$
19. $3x + 4 \leq 1$
20. $-3x + 4 < 2x - 6$
21. $-7x + 4 \geq 2x + 3$
22. (*See Example 7*) $-45 < 4x + 7 \leq -10$
23. $16 > 2x - 10 \geq 4$

II.

Solve the following inequalities.

24. $\dfrac{2x - 5}{3} < \dfrac{x + 7}{4}$
25. $\dfrac{6x + 5}{-2} \geq \dfrac{4x - 3}{5}$
26. $\dfrac{3}{4} < \dfrac{7x + 1}{6} < \dfrac{5}{2}$
27. $\dfrac{2}{3} < \dfrac{x + 5}{-4} \leq \dfrac{3}{2}$

III.

28. A sporting goods store runs a special on jogging shoes. The manager expects to make a profit if the number of shoes sold, x, satisfies $32x - 4230 > 2x + 480$. How many shoes must be sold in order to make a profit?
29. (*See Example 8*) Professor Tuff computes a grade on a test by $35 + 5x$, where

x is the number of correct answers. A grade in [75, 90) is a B. If a student receives a B, how many correct answers were given?

30. On a final exam any grade in [85, 100] was an A. The professor gave 3 points for each correct answer and then adjusted the grades by adding 25 points. If a student made an A, how many correct answers were given?

0–5 Polynomial Arithmetic

Addition of Polynomials

A polynomial is an expression of the form

$$a_n x^n + a_{n-1}x^{n-1} + \cdots + a_1 x + a_0$$

where n is a positive integer and $a_n, a_{n-1}, \ldots, a_1, a_0$ are real numbers. The following are examples of polynomials.

$$2x^2 + 8x + 4$$
$$5x^4 + 3x^2 + 2x + 2$$

Polynomials are added by adding the coefficients of corresponding powers of x.

Example 1 *(Compare Exercise 3)*
Add the polynomials $2x^2 + 8x + 4$ and $5x^4 + 3x^2 + 2x + 2$.

Solution We group terms involving corresponding powers of x together, starting with the highest power.

$$\begin{aligned}(2x^2 + 8x + 4) + (5x^4 + 3x^2 + 2x + 2) &= 5x^4 + 2x^2 + 3x^2 + 8x + 2x + 4 + 2\\ &= 5x^4 + 5x^2 + 10x + 6\end{aligned}$$

Example 2 Simplify $2(4x^4 + 3x^2 + 2x - 1) - 5(2x^4 - 3x^2 + 2x + 5)$

Solution Multiply each term in parentheses by the number outside. We get

$$\begin{aligned}&2(4x^4 + 3x^2 + 2x - 1) - 5(2x^4 - 3x^2 + 2x + 5)\\ &= 8x^4 + 6x^2 + 4x - 2 - 10x^4 + 15x^2 - 10x - 25\\ &= 8x^4 - 10x^4 + 6x^2 + 15x^2 + 4x - 10x - 2 - 25\\ &= -2x^4 + 21x^2 - 6x - 27\end{aligned}$$

Multiplication of Polynomials

Example 3 (*Compare Exercise 4*)
Perform the multiplication $(4x + 3)(2x^2 + 3x - 4)$.

Solution The second polynomial is multiplied by every term in the first polynomial.

$$\begin{aligned}(4x + 3)(2x^2 + 3x - 4) &= 4x(2x^2 + 3x - 4) + 3(2x^2 + 3x - 4)\\ &= 8x^3 + 12x^2 - 16x + 6x^2 + 9x - 12\\ &= 8x^3 + 12x^2 + 6x^2 - 16x + 9x - 12\\ &= 8x^3 + 18x^2 - 7x - 12\end{aligned}$$

Example 4 (*Compare Exercise 4*)
Perform the following multiplication:

$$(3x^2 - 4x + 2)(2x^2 + 5x - 3)$$

Solution By multiplying the second polynomial by every term in the first polynomial we get

$$\begin{aligned}&(3x^2 - 4x + 2)(2x^2 + 5x - 3)\\ &\quad = 3x^2(2x^2 + 5x - 3) - 4x(2x^2 + 5x - 3) + 2(2x^2 + 5x - 3)\\ &\quad = 6x^4 + 15x^3 - 9x^2 - 8x^3 - 20x^2 + 12x + 4x^2 + 10x - 6\\ &\quad = 6x^4 + 15x^3 - 8x^3 - 9x^2 - 20x^2 + 4x^2 + 12x + 10x - 6\\ &\quad = 6x^4 + 7x^3 - 25x^2 + 22x - 6\end{aligned}$$

Factoring Polynomials

The reverse of multiplying polynomials is factoring polynomials. You use factoring when you start with a polynomial and want to express it as a product of shorter polynomials. In some circumstances, you may need to multiply polynomials, while at other times you may need to go in the opposite direction, to factor polynomials.

Example 5 (*Compare Exercise 5*)
Factor $2x^2 + 4x$.

Solution First we look for a common factor in the coefficients. We see that 2 is a common factor, and we can write

$$2x^2 + 4x = 2(x^2 + 2x)$$

Next, look to see if some power of x is a common factor. Here x is a factor, and we can write

$$2x^2 + 4x = 2x(x + 2)$$

Example 6 Factor $8x^3 - 16x^2$.

Solution

$$8x^3 - 16x^2 = 8(x^3 - 2x^2)$$

Next, x^2 can also be factored out.

$$8x^3 - 16x^2 = 8x^2(x - 2)$$

Factoring Polynomials of Degree Two

Given a polynomial of the type $ax^2 + bx + c$ we want, if possible, to express it in terms of factors, $(px + q)(rx + s)$. The constants p, q, r, and s must satisfy

$$\begin{aligned} pr &= a \\ qs &= c \\ ps + qr &= b \end{aligned}$$

Use trial and error to find p, q, r, and s.

Example 7 *(Compare Exercise 5)*
Factor $x^2 + 5x + 6$.

Solution We want to find p, q, r, and s such that

$$x^2 + 5x + 6 = (px + q)(rx + s)$$

On comparison with the preceding notation we see that $a = 1$, $b = 5$, and $c = 6$. The three conditions on p, q, r, and s become

$$\begin{aligned} pr &= 1 \\ qs &= 6 \\ ps + qr &= 5 \end{aligned}$$

The first condition suggests $p = r = 1$. The second condition suggests $q = 3$, $s = 2$. We see that these values satisfy the third condition, $ps + qr = (1)(2) + (3)(1) = 5$.

Thus,

$$x^2 + 5x + 6 = (x + 3)(x + 2)$$

Example 8 Factor $12x^2 + 4x - 5$.

Solution Let $12x^2 + 4x - 5 = (px + q)(rx + s)$
The conditions on p, q, r, and s are

$$\begin{aligned} pr &= 12 \\ qs &= -5 \\ ps + qr &= 4 \end{aligned}$$

If we try $p = 4$ and $r = 3$ in the first equation and $q = 5$ and $s = -1$ in the second equation, we find that $ps + qr = -4 + 15 = 11$. The third equation is not satisfied. This combination will not do.

However, $p = 6$ and $r = 2$ in the first equation and $q = 5$ and $s = -1$ in the second lead to $ps + qr = -6 + 10 = 4$. The third equation is satisfied. This is a correct set of values. We get

$$12x^2 + 4x - 5 = (6x + 5)(2x - 1)$$

You should check the answers to Examples 7 and 8 by multiplying the factors. This will help you see how the conditions on p, q, r, and s are obtained.

Solving Quadratic Equations

A quadratic equation is an equation of the form

$$ax^2 + bx + c = 0$$

where $a \neq 0$. Suppose the polynomial $ax^2 + bx + c$ can be factored,

$$ax^2 + bx + c = (px + q)(rx + s)$$

The equation then becomes

$$(px + q)(rx + s) = 0$$

This gives

$$px + q = 0 \qquad \text{or} \qquad rx + s = 0$$

Thus,

$$x = -\frac{q}{p} \qquad \text{or} \qquad -\frac{s}{r}$$

Example 9 (*Compare Exercise 6*)
Solve the equation $12x^2 + 4x - 5 = 0$.

Solution We saw in the previous example that $12x^2 + 4x - 5$ can be factored

$$12x^2 + 4x - 5 = (6x + 5)(2x - 1)$$

The solutions are given by

$$(6x + 5)(2x - 1) = 0$$

$$6x + 5 = 0 \qquad \text{or} \qquad 2x - 1 = 0$$

$$x = -\frac{5}{6} \qquad \text{or} \qquad x = \frac{1}{2}$$

The Quadratic Formula

We now introduce a method that can always be used to solve a quadratic equation, whether or not you can successfully factor the polynomial.

The solutions to the quadratic equation $ax^2 + bx + c = 0$ are given by

$$x = \frac{-b \pm \sqrt{b^2 - 4ac}}{2a}$$

Example 10 (*Compare Exercise 7*)
Solve the quadratic equation $2x^2 + 7x + 3 = 0$

Solution We have that $a = 2$, $b = 7$, $c = 3$.

Using the above formula,

$$x = \frac{-7 \pm \sqrt{7^2 - 4(2)(3)}}{2(2)} = \frac{-7 \pm \sqrt{49 - 24}}{4}$$

$$= \frac{-7 \pm \sqrt{25}}{4} = \frac{-7 \pm 5}{4}$$

The two solutions are $\dfrac{-7+5}{4}$ and $\dfrac{-7-5}{4}$

$$= \frac{-2}{4} \text{ and } \frac{-12}{4}$$

$$= \frac{-1}{2} \text{ and } -3$$

Example 11 Solve the quadratic equation

$$3x^2 - 2x + 4 = 0$$

Solution For this equation $a = 3$, $b = -2$, $c = 4$.

The quadratic formula gives

$$x = \frac{-(-2) \pm \sqrt{(-2)^2 - 4(3)(4)}}{2(3)} = \frac{2 \pm \sqrt{4 - 48}}{6}$$

$$= \frac{2 \pm \sqrt{-44}}{6}$$

But there is no real number that corresponds to $\sqrt{-44}$. We say that a solution does not exist. Whenever the number under the radical sign in the quadratic formula is negative, a solution will not exist.

Rational Expressions

An algebraic fraction, one whose numerator and denominator are polynomials, is called a rational expression. Examples of rational expressions are

$$\frac{1}{x} \qquad \frac{2x-3}{4x^2+2x+1} \qquad \frac{x^4+2x}{4x+3}$$

The rules for working with these expressions are similar to those that apply to ordinary fractions. We summarize some of the more important rules. Let P, Q, R, and S be polynomials.

Addition $\dfrac{P}{Q} + \dfrac{R}{Q} = \dfrac{P+R}{Q} \quad (Q \neq 0)$

Subtraction $\dfrac{P}{Q} - \dfrac{R}{Q} = \dfrac{P-R}{Q} \quad (Q \neq 0)$

Multiplication $\dfrac{P}{Q} \cdot \dfrac{R}{S} = \dfrac{PR}{QS} \quad (Q \neq 0, S \neq 0)$

Division $\dfrac{P}{Q} \div \dfrac{R}{S} = \dfrac{PS}{QR} \quad (Q \neq 0, S \neq 0, R \neq 0)$

In addition to the above rules the numerator and denominator of a rational expression can both be multiplied by a polynomial. This rule proves to be useful in simplifying sums of rational expressions.

$$\frac{P}{Q} = \frac{PR}{QR} \quad (Q \neq 0, R \neq 0)$$

Example 12 (*Compare Exercise 8*)
Simplify

$$\frac{x^2 + 3x}{4x + 2} + \frac{3x^3 - 2x + 1}{4x + 2}$$

Solution Using the rule for addition, we get

$$\begin{aligned}\frac{x^2 + 3x}{4x + 2} + \frac{3x^3 - 2x + 1}{4x + 2} &= \frac{x^2 + 3x + 3x^3 - 2x + 1}{4x + 2} \\ &= \frac{3x^3 + x^2 + 3x - 2x + 1}{4x + 2} \\ &= \frac{3x^3 + x^2 + x + 1}{4x + 2}\end{aligned}$$

Example 13 (*Compare Exercise 8*)
Simplify

$$\frac{(x + 1)(x + 2)}{x^2 + 1} \div \frac{3(x + 2)}{4x}$$

given that $x + 2 \neq 0$ and $x \neq 0$.

Solution Apply the rule for division

$$\begin{aligned}\frac{(x + 1)(x + 2)}{x^2 + 1} \div \frac{3(x + 2)}{4x} &= \frac{(x + 1)(x + 2)}{x^2 + 1} \cdot \frac{4x}{3(x + 2)} \\ &= \frac{(x + 1)(x + 2)4x}{3(x^2 + 1)(x + 2)}\end{aligned}$$

At this stage we can simplify by dividing both the numerator and denominator by the $(x + 2)$ factor as in the case of ordinary fractions. We get

$$\frac{(x + 1)(x + 2)}{x^2 + 1} \div \frac{3(x + 2)}{4x} = \frac{4x(x + 1)}{3(x^2 + 1)}$$

Example 14 Simplify

$$\frac{x^2 + 3}{x} + \frac{x - 1}{x + 1}$$

Solution The common denominator is $x(x + 1)$. Multiply both the numerator and denominator of the first expression by $(x + 1)$ and the numerator and denominator of the second expression by x. We get

$$\frac{x^2+3}{x}+\frac{x-1}{x+1}=\frac{(x^2+3)(x+1)}{x(x+1)}+\frac{x(x-1)}{x(x+1)}$$
$$=\frac{(x^2+3)(x+1)+x(x-1)}{x(x+1)}$$
$$=\frac{x^2(x+1)+3(x+1)+x^2-x}{x(x+1)}$$
$$=\frac{x^3+x^2+3x+3+x^2-x}{x(x+1)}$$
$$=\frac{x^3+2x^2+2x+3}{x(x+1)}$$

0–5 Exercises

Answers are given for problems with an asterisk.

In Exercise 1, solve the given equations.

1. *(a) $2x-3=5$ (b) $-4x+2=6$
   *(c) $4x-3=5$ (d) $3x+2=7x+4$
   *(e) $5-x=8+3x$ (f) $2x-4=-5x+2$

In Exercise 2, solve the given algebraic equations involving parentheses.

2. *(a) $4(x+2)-3=5$ (b) $4x-2(x-1)=6$
   *(c) $5x-2+3(2-x)=5$ (d) $3(x-2)+4(2x+3)=2x-1$
   *(e) $-2(3x+2)+5=6x-4(x+3)$ (f) $2(x+3)-3(5-x)=4x+3$

In Exercise 3, add the given polynomials.

3. (*See Example 1*)
   *(a) $2x^2+3x-1$ and $5x^2-3x+2$ (b) $3x^3-2x+3$ and $5x^3+2x+1$
   *(c) $4x^3-2x^2+3$ and $2x^2-5x+2$ (d) $7x^4-2x+3$ and $-4x^3-5x+2$
   *(e) $5x^3-2x^2+3x+2$ and $-4x^3-5x^2+5$ (f) $7x^4-2x^2-3x+6$ and $-3x^3-4x^2-x+2$

In Exercise 4, perform the multiplication of the polynomials.

4. *(a) $(x+2)(x-3)$ (b) $(2x-1)(x+3)$
   *(c) $(3x^2+2x-1)(x+3)$ (d) (*See Example 3*) $(-x^2+2x+4)(x^2-2x+1)$
   *(e) (*See Example 4*) $(2x^2-x+1)(2x^3+2x+1)$ (f) $(x^4+3x)(x^2-2)$

In Exercise 5, factor the given polynomials.

5. *(a) $x^2 + 3x$
(b) $2x^3 - 4x$
*(c) (*See Example 5*) $5x^4 - 25x^3 + 10x^2$
(d) $x^2 + x - 6$
*(e) $x^2 + 6x + 5$
(f) $x^2 - 7x + 12$
*(g) (*See Example 7*) $2x^2 + x - 6$
(h) $3x^2 + 2x - 5$
*(i) $4x^2 + 4x - 3$
(j) $6x^2 - 7x - 5$
*(k) $4x^2 - 11x + 6$

In Exercise 6, solve the given quadratic equations by first factoring.

6. *(a) $x^2 + 2x - 3 = 0$
(b) (*See Example 9*) $x^2 + 7x + 10 = 0$
*(c) $2x^2 - x - 3 = 0$
(d) $2x^2 + 9x + 10 = 0$
*(e) $6x^2 - 5x - 4 = 0$

In Exercise 7, solve the given quadratic equations by using the quadratic formula, if solutions exist.

7. *(a) (*See Example 10*) $2x^2 - x - 3 = 0$
(b) $6x^2 - x - 1 = 0$
*(c) $4x^2 - 11x - 3 = 0$
(d) $10x^2 - x - 2 = 0$
*(e) $9x^2 - 9x - 4 = 0$
(f) $2x^2 - x + 6 = 0$
*(g) $2x^2 - x - 1 = 0$
(h) $x^2 - 2x - 8 = 0$

In Exercise 8, perform the given operations.

8. (*See Examples 12 and 13*)

*(a) $\dfrac{x^2 + 2}{x - 1} + \dfrac{x - 3}{x - 1}$
(b) $\dfrac{x + 2}{x^2 - 3} + \dfrac{2x + 4}{x^2 - 3}$
*(c) $\dfrac{2x - 3}{x + 2} - \dfrac{3x - 3}{x + 2}$
(d) $\dfrac{4x + 1}{x - 1} \cdot \dfrac{x}{3x + 2}$
*(e) $\dfrac{5x - 1}{2x + 1} \cdot \dfrac{3x + 2}{4x - 1}$
(f) $\dfrac{4x + 3}{2x + 3} \cdot \dfrac{x}{3x^2 - x + 2}$
*(g) $\dfrac{2x + 3}{4x - 1} \cdot \dfrac{5}{2x + 3}$
(h) $\dfrac{5x - 1}{2x} \div \dfrac{3x}{2}$
*(i) $\dfrac{4x + 3}{5x - 1} \div \dfrac{4x}{2x + 1}$
(j) $\dfrac{4x}{3x + 2} \div \dfrac{5x^2 + 2}{4x^2 - 3}$
*(k) $\dfrac{2x - 3}{4} \cdot \dfrac{2x}{7}$
(l) $\dfrac{3x - 1}{4} \div \dfrac{3x - 2}{2x}$
*(m) $\dfrac{2}{x} + \dfrac{4}{x - 1}$
(n) $\dfrac{5}{x + 2} + \dfrac{7x + 3}{x}$

*(o) $\frac{5x-1}{x-1}+\frac{4x+2}{x+1}$

(p) $\frac{5x}{2}-\frac{7}{x-1}$

*(q) $\frac{4x+3}{5x-1}-\frac{2x+1}{x}$

(r) $\frac{3x-1}{x+1}+\frac{x+2}{2x+5}$

*(s) $\frac{5}{x}+\frac{7}{x^2}$

(t) $\frac{5x}{2x^2-3}+\frac{2x-1}{x}$

0–6 Exponents

Positive Integer Exponents
Other Integer Exponents
Fractional Exponents
Special Remark on b when the Exponent Is a Fraction

Most people seem to remember that exponents stand for "repeated multiplication," and they have little problem computing 2^3; $2^3 = 2 \cdot 2 \cdot 2 = 8$. Thinking of exponents this way, however, causes problems when trying to understand what is meant by 2^0, 2^{-3} and $2^{1/5}$. You can't multiply two times itself $\frac{1}{5}$ times, for example. There are also rules for manipulating exponents—rules like $2^3 \cdot 2^2 = 2^5$ and $(2^3)^2 = 2^6$ that cause some trouble, especially when the exponents are not positive integers. We will start with the notion of "repeated multiplication" and show what the rules for combining exponents must be when the exponents are positive integers. Then the meaning we attach to, say 2^{-3}, is motivated by a desire to keep these exponent rules the same no matter what the exponent is. If you have trouble handling exponents, we hope to encourage you to think about the meaning of the rules in a very concrete setting.

Positive Integer Exponents

In this *one* case, we do think of the exponent as a symbol meaning "repeated multiplication." Thus, $3^4 = 3 \cdot 3 \cdot 3 \cdot 3 = 81$; $b^n = b \cdot b \cdot \cdots \cdot b$ (n times). Now we establish the two main rules for positive integers, and then introduce other types of numbers as exponents so that *the rules stay the same.*

How can we simplify $b^n \cdot b^k$? Think concretely—how can we simplify $4^3 \cdot 4^2$? $4^3 = 4 \cdot 4 \cdot 4$, and $4^2 = 4 \cdot 4$, so $4^3 \cdot 4^2 = (4 \cdot 4 \cdot 4)(4 \cdot 4) = 4 \cdot 4 \cdot 4 \cdot 4 \cdot 4 = 4^5$. $5 = 3 + 2$, so $4^3 \cdot 4^2 = 4^5$. This gives us the rule for simplifying $b^n \cdot b^k$.

Rule 1 $b^n \cdot b^k = b^{n+k}$

Example 1 (*Compare Exercise 34*)
Simplify $7^5 \cdot 7^3$

Solution $7^5 \cdot 7^3 = 7^{5+3} = 7^8$

Next, how can we simplify $(b^n)^k$? Again, let's look at a concrete example with $b = 4$, $n = 3$, and $k = 2$. $(4^3)^2 = 4^3 \cdot 4^3 = (4 \cdot 4 \cdot 4)(4 \cdot 4 \cdot 4) = 4^6$, and $6 = 2 \cdot 3$, so we know what Rule 2 must be.

Rule 2

$$(b^n)^k = b^{n \cdot k}$$

Example 2 (*Compare Exercise 28*)
Simplify $(7^5)^3$

Solution $(7^5)^3 = 7^{5 \cdot 3} = 7^{15}$

Other Integer Exponents

We want Rule 1 and Rule 2 to be true for all possible exponents n and k, so we extend the meaning of exponents to numbers that are not positive integers. But when we do, we can no longer explain the exponent as "repeated multiplication." Furthermore, as we allow more kinds of exponents, we must introduce some restrictions on b.

First, let's try to define 2^0. We want Rule 1 to hold, so we want $2^3 \cdot 2^0 = 2^{3+0} = 2^3$. Multiplying by 2^0 doesn't change the value of 2^3; 2^0 must equal 1.

Definition

If $b \neq 0$, $b^0 = 1$

With this definition we know Rule 1 is always satisfied. What about Rule 2? We compute $(b^n)^0$ and b^{n0} separately. If we let $c = b^n$, then $(b^n)^0 = c^0 = 1$. Also, $b^{(n \cdot 0)} = b^0 = 1$. Thus, $(b^n)^0 = 1 = b^{n \cdot 0}$, and Rule 2 holds.

Next we want to define b^n when n is a negative integer. What do we want 2^{-3} to mean? If Rule 1 is to still hold we must have $2^3 \cdot 2^{-3} = 2^{3+-3} = 2^0 = 1$. So 2^{-3} is the reciprocal of 2^3.

Definition

If n is a positive integer and $b \neq 0$, $b^{-n} = \dfrac{1}{b^n}$.

This definition also keeps Rule 2 valid.

Example 3 (*Compare Exercise 37*)
Simplify $(4^{-2})^3$

Solution $(4^{-2})^3 = 4^{-6} = \dfrac{1}{4^6}$

We have defined negative exponents so that the rules still work. Remembering how the rules work can help you remember what negative exponents mean.

Thus,

$$2^4/2^3 = (2 \cdot 2 \cdot 2 \cdot 2) \div (2 \cdot 2 \cdot 2) = 2 = 2^{4-3} = 2^4 \cdot 2^{-3}$$

may help you remember $2^{-3} = \frac{1}{2^3}$.

Fractional Exponents

Next we would like to see what fractional exponents should mean. We know that $4 = 4^1 = 4^{3/3}$. If we want Rule 2 to remain valid, we have $4 = 4^{3/3} = (4^{1/3})^3$. Thus, $4^{1/3} \cdot 4^{1/3} \cdot 4^{1/3} = 4$, and $4^{1/3}$ is called the cube root of 4.

Definition

> If n is a positive integer, $b^{1/n}$ is the nth root of b; $b^{1/n} = c$ where $c^n = b$

Special notation: $\sqrt{b}$ is used to mean the positive square root of b; $\sqrt{b} = b^{1/2}$. $\sqrt[n]{b}$ is used to mean the nth root of b for positive integers $n \geq 3$; $\sqrt[n]{b} = b^{1/n}$

Example 4 (*Compare Exercise 24*)
Compute $\sqrt{25}$, $\sqrt[3]{8}$, $81^{1/4}$

Solution

$$\sqrt{25} = 5 \qquad \textit{Note: } \sqrt{25} \text{ means 5, not } \pm 5.$$

$$\sqrt[3]{8} = 2$$

$$81^{1/4} = 3$$

With this definition for $b^{1/n}$, we extend the definition to all rational numbers $b^{k/n}$ by making sure Rule 2 still holds.

Definition

> $b^{k/n} = (b^{1/n})^k$

Example 5 (*Compare Exercise* 21)
Compute $32^{3/5}$; $64^{-2/3}$

Solution

$$32^{3/5} = (32^{1/5})^3 = 2^3 = 8.$$

$$64^{-2/3} = (64^{1/3})^{-2} = 4^{-2} = \frac{1}{4^2} = \frac{1}{16}$$

Special Remark on *b* when the Exponent is a Fraction

We defined $b^{1/2}$ to be the square root of b; a special remark must be made because $\sqrt{b}$ is not always a real number. We are dealing exclusively with real numbers in this text, and $\sqrt{-4}$ is not a real number. In fact, you can never have an even root of a negative number. This problem doesn't arise with odd roots. For example, $(-32)^{1/5} = -2$.

Example 6 (*Compare Exercise 27*)
Simplify $(-16)^{1/2}$; $(-8)^{2/3}$; $(27)^{-1/3}$

Solution

(a) $(-16)^{1/2} = \sqrt{-16}$ is not defined

(b) $(-8)^{2/3} = ((-8)^{1/3})^2 = (-2)^2 = 4$

(c) $27^{-1/3} = \dfrac{1}{27^{1/3}} = \dfrac{1}{3}$

0–6 Exercises

Simplify the following expressions:

1. 5^2
2. 4^3
3. 8^1
4. 6^0
5. 9^0
6. 7^{-1}
7. 3^{-1}
8. 5^{-2}
9. 4^{-2}
10. 8^{-3}
11. 9^{-2}
12. 2^{-3}
13. $9^{1/2}$
14. $16^{1/2}$
15. $32^{1/5}$
16. $4^{3/2}$
17. $8^{2/3}$
18. $16^{3/4}$
19. $(-27)^{1/3}$
20. $(-3)^3$
21. (*See Example 5*) $4^{5/2}$
22. $8^{-2/3}$
23. $(\sqrt{3})^4$
24. (*See Example 4*) $(\sqrt{6})^4$
25. $16^{-3/4}$
26. $(-16)^{1/4}$
27. (*See Example 6*) $(-27)^{-1/3}$
28. (*See Example 2*) $(x^2)^3$
29. $(y^5)^4$
30. $(x^6)(x^7)$
31. $(x^6)^7$
32. $-\sqrt{81}$
33. $\sqrt{-81}$
34. (*See Example 1*) $5^2 \cdot 5^{-1}$
35. $3^4 \cdot 3^{-5}$
36. $x^3 \cdot x^{-2}$
37. (*See Example 3*) $(8^2)^{-1}$

IMPORTANT TERMS

0–1	**Natural numbers**	**Rational numbers**	**Irrational numbers**
	Real numbers		
0–2	**Variable**	**Solution**	**Linear equation**
0–3	**Cartesian coordinate system**	**Rectangular coordinate system**	**Origin**
	***x*-axis**	***y*-axis**	**Absicssa**
	Ordinate	**Quadrants**	
0–4	**>**	**<**	**≥**
	≤	**Inequalities**	**Properties of inequalities**

	Interval notation		
0–5	**Polynomial arithmetic**	**Addition and multiplication of polynomials**	**Factoring polynomials**
	Solving quadratic equations	**Quadratic formula**	**Rational expressions**
0–6	**Exponents**	**Fractional exponents**	**Negative exponents**
	Rules for adding and multiplying exponents		

1

Topics in this chapter can be applied to:
Pricing and Profit • Revenue and Total Sales • Salary and Commissions • Production and Cost • Depreciation • Manufacturing and Operating Costs • Revenue • Break-Even Analysis • Cost-Volume Function • Simple Interest • Wealth Distribution Functions • Monopoly Functions • Discount Rates • Consumer Awareness and Advertising

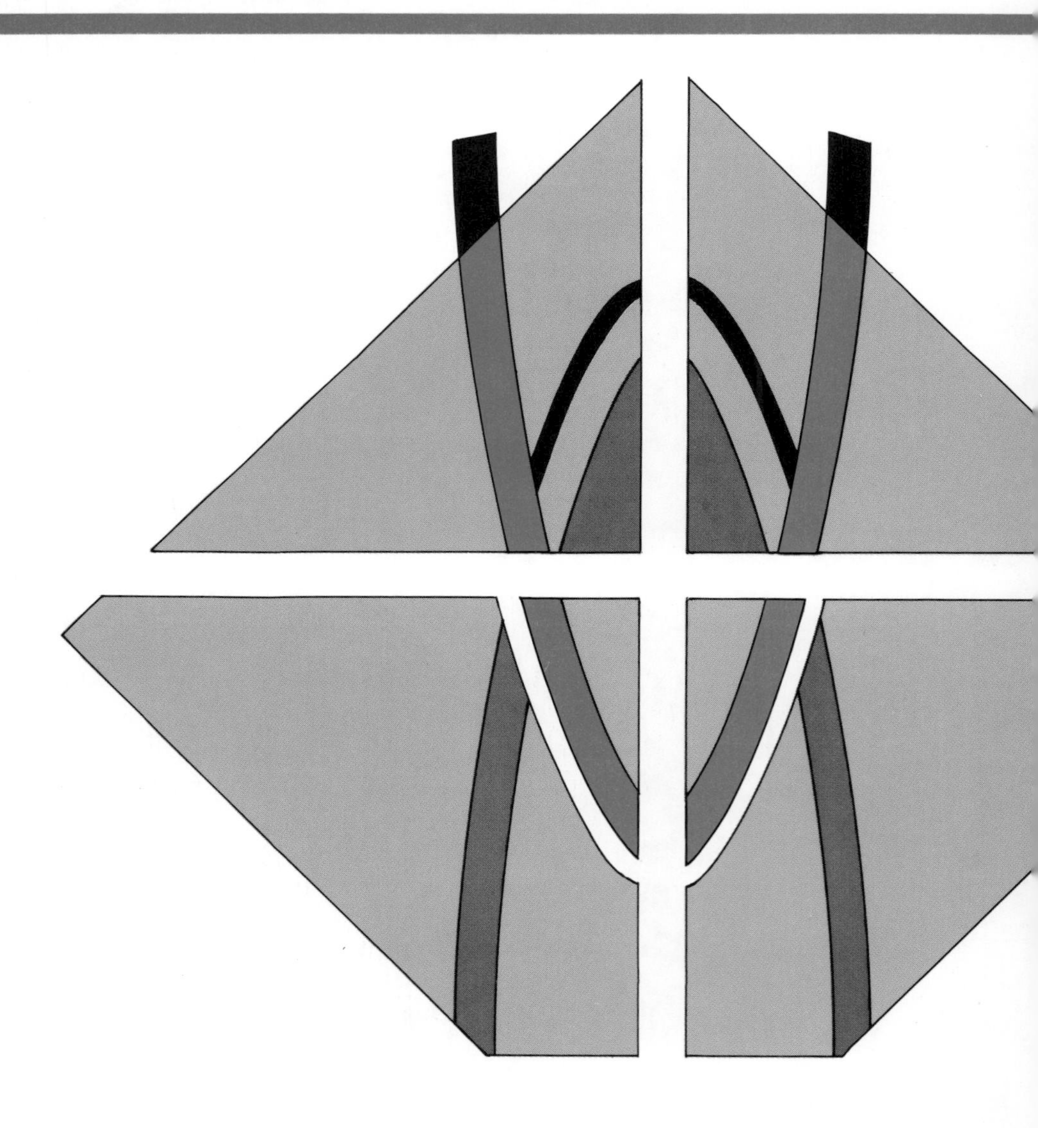

Pre-Calculus Topics

- **Functions**
- **Graphs and Lines**
- **Applications of Linear Functions**
- **Graphs of Nonlinear Functions**
- **Inequalities: The Cut-Point Method**

1–1 Functions

Introduction
Function Evaluation
Domain and Range

Introduction

Suppose that you are the manager in charge of the toy department, and your buyer has just sent you a memo saying that the hottest toy for the fall will be Burpy-baby. The buyer has ordered a large shipment, so you have to sell the Cuddle-me dolls you have on hand to make room for this new shipment. Naturally, you want to make as much money on your present stock as possible, but you have some decisions to make. If you keep your prices as high as they are now, you will make a good profit on each doll sold, but you won't sell very many. Furthermore, you will have to take a loss on all the ones you don't sell. On the other hand, if you lower the price too much, you will suffer a loss even if you sell all the dolls you have on hand. How do you decide what price to charge so that you make as much money as possible?

Calculus can help you answer this question. In fact, calculus will give you mathematical tools that will enable you to answer questions not only about the relationship between selling price and profit, but also about the relationship between any two quantities where one depends on the other. Mathematicians have developed a type of shorthand which uses letters for dealing with quantities like these, and with the relationship between them. To use this shorthand, let x stand for the selling price of the Cuddle-me's, and let y stand for the profit. Mathematicians use the word *function* to express the dependence of y on x. Thus y is a function of x; the profit is a function of the selling price. This statement is condensed even further by using the single letter f to stand for the phrase "a function of," and writing $y = f(x)$. (Read this "y equals f of x.")

Suppose, using our example, you determine that if you sell the Cuddle-me's for \$15 each, your total profit will be \$130. You can express that information in function notation by writing $130 = f(15)$. Similarly, the expression $125 = f(16)$ would mean a price of \$16 would result in a total profit of \$125. Notice that a higher price may mean fewer sales, and hence may mean a smaller total profit.

Function Evaluation

To clarify the concept of function it may be helpful to think of a function as a machine that takes a certain number x as input and gives the number $f(x)$ as the output (Figure 1–1).

Generally there is some rule or formula that tells you how to compute the output in terms of the input. For instance, instead of writing "the function that squares a number and subtracts 6 from that result," we write "the function defined by $f(x) = x^2 - 6$." The equation $f(x) = x^2 - 6$ is called *the rule for evaluating the function*, or more simply, *the rule for the function.* Here are some examples to help you become familiar with this notation.

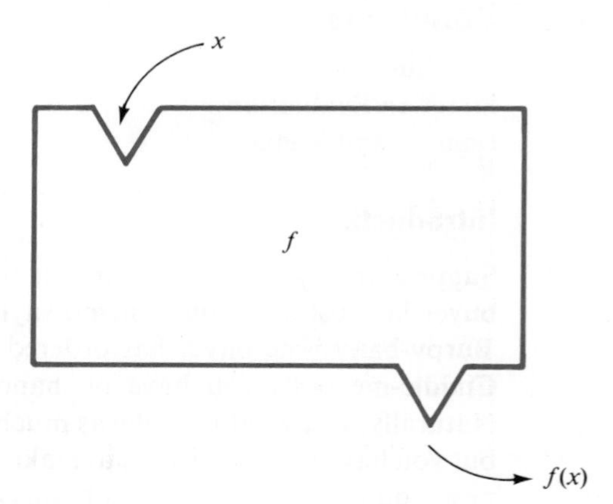

Figure 1–1

Example 1 (*Compare Exercise 17*)
What is the rule for the function—the machine—that multiplies the input by 3 and then adds 5 to this result?

Solution Let x stand for the input to the function; the rule would be

$$f(x) = 3x + 5.$$

We continue to use the function defined by $f(x) = 3x + 5$ in order to demonstrate some evaluations. Letting $x = 2$, we have $f(2) = 3 \cdot 2 + 5$; $f(2) = 11$. We say the value of f at $x = 2$ is 11. To evaluate f at $x = 6$, we have $f(6) = 3 \cdot 6 + 5 = 23$. Similarly $f(0) = 3 \cdot 0 + 5 = 5$, and $f(-4) = 3(-4) + 5 = -7$.

Example 2 (*Compare Exercise 1*)
If f is the function defined by $f(x) = 4 - 2x^2$, find each of the following values of f:

(a) $f(1)$ **(b)** $f(3)$ **(c)** $f(-5)$

Solution **(a)** Replacing the "x" by "1" in the rule, we have

$$f(1) = 4 - 2(1)^2 = 4 - 2 = 2$$
$$f(1) = 2$$

(b)
$$f(3) = 4 - 2(3)^2 = 4 - 2 \cdot 9 = 4 - 18 = -14$$
$$f(3) = -14$$

(c)
$$f(-5) = 4 - 2(-5)^2 = 4 - 2 \cdot 25 = 4 - 50 = -46$$
$$f(-5) = -46$$

Often the function rule is written without the $f(x)$ notation by just showing how the value of y depends on the value of x. We shall do Example 2 again using this alternate notation.

Example 3 (*Compare Exercise 3*)
If $y = 4 - 2x^2$, what is y when

(a) $x = 1$? **(b)** $x = 3$? **(c)** $x = -5$?

Solution Using the computations from Example 2,

(a) $y = 2$ if $x = 1$
(b) $y = -14$ if $x = 3$
(c) $y = -46$ if $x = -5$

The relationship between profit and price, for example, is usually expressed by a formula like $y = -x^2 + 26x - 35$, where y stands for the profit and x for the price. Later we will show how these formulas are obtained. Right now we concentrate on how to read them. (In fact, this is the formula that gives the profit in our doll example).

Example 4 (*Compare Exercise 21*)
Using $y = -x^2 + 26x - 35$, find the profit y for the various prices $x = 16$, $x = 10$, $x = -3$.

Solution When $x = 16$,

$$\begin{aligned} y &= -(16)^2 + 26(16) - 35 \\ &= -256 + 416 - 35 \\ &= 125 \end{aligned}$$

When $x = 10$,

$$\begin{aligned} y &= -(10)^2 + 26(10) - 35 \\ &= -100 + 260 - 35 \\ &= 125 \end{aligned}$$

Notice that different selling prices may yield the same total profit. When $x = 16$, the profit per doll is greater than when $x = 10$, but fewer dolls are sold at the higher price.

When $x = -3$,

$$\begin{aligned} y &= -(-3)^2 + 26(-3) - 35 \\ &= -9 - 78 - 35 \\ &= -122 \end{aligned}$$

(Be careful computing with negative numbers; it's easy to make a mistake with the signs.)

Domain and Range

Notice that in Example 4, even caring about the value of y when $x = -3$ is rather silly, as this would represent your profit if you paid people $3 for each Cuddle-Me they carried out of your store. The corresponding value $y = -122$ would represent a "profit" of $-\$122$; this is usually called a loss of $122. It would seem reasonable in this example to require that the selling price be greater than or equal to 0; that is, don't allow negative numbers as input. Letting $x = 0$ would correspond to giving the dolls away, but maybe you would do that for public relations or to attract people into your store rather than just throwing the dolls away.

In general, the set of values that are allowed as input into the function is called *the domain of the function.* The set of values that result as the output is called *the range of the function.* Don't worry about how to find the range of a function right now; we will come back to this problem. But you do have to make sure you understand the concept of domain. In the most usual case, the domain is the largest set of numbers x for which the expression $f(x)$ makes sense.

Example 5 (*Compare Exercise 7*)

If $f(x) = \dfrac{x+2}{x-3}$, what is the domain of the function?

Solution If $x = 3$ then $x - 3 = 0$; this means that you cannot compute $f(3)$; 5 divided by 0 doesn't make sense. **You can never divide by 0!!!!** Thus, 3 is not in the domain of this function. Every other number can be substituted for x, so the domain of this function is all real numbers except 3—all $x \neq 3$.

Example 6 What is the domain of f if $f(x) = \sqrt{x}$?

Solution The domain of the square root function is all $x \geq 0$; a negative number does not have a real square root. Using interval notation, we can write the domain as $[0, \infty)$.

Example 7 (*Compare Exercise 9*)
What is the domain of f if $f(x) = \sqrt{x - 5}$?

Solution If x is in the domain of f, then $x - 5$ cannot be negative; the domain of f is all x so that $x - 5 \geq 0$, that is, all $x \geq 5$. The domain of f is $[5, \infty)$.

As we saw in the doll example, sometimes there are values of x that you don't want in the domain because $f(x)$ doesn't make sense in the application. This restriction on the domain is indicated by explicitly giving the domain at the same time the formula for $f(x)$ is given. For example, we could indicate that the selling price of the dolls can't be negative by writing $f(x) = -x^2 + 26x - 35, x \geq 0$. If the domain is given explicitly, be careful that you don't try to evaluate the function outside its domain.

Example 8 (*Compare Exercise 31*)
If f is defined by $f(x) = 3x^2 + 5x - 2,\ 2 \leq x \leq 7$, what are $f(3)$, $f(1)$ and $f(2)$?

Solution $f(3) = 40$, as you expect, but $f(1)$ is not defined; 1 is not in the domain of this function. The domain is only those numbers from 2 to 7 inclusive. Because 2 is in the domain, we can use the rule to compute $f(2)$; $f(2) = 20$.

So far we have used only the letters f, x, and y, but there's no need to be so restrictive. Indeed, you will run into applications later where you will want to talk about three functions in the same problem—the cost function, the revenue function, and the profit function. It would certainly be confusing to use the same letter to represent all three! Likewise, you may want to use different letters for your variables. For example, if you are interested in representing the cost of a telephone call in terms of the call's length, you may want to name the function C (for cost), and you could let the input variable be t (for time) as in Example 9.

Example 9 Suppose the cost function is defined by $C(t) = .65t + .25$, where t is measured in minutes and $C(t)$ is measured in dollars.

(a) What does an 8-minute call cost?
(b) What is the cost of a one-hour call?
(c) How long can you talk for $10?

Solution
(a) $C(8) = (.65)8 + .25 = 5.45$, so the cost of an 8-minute call is $5.45.
(b) $C(60) = 39.25$, so the cost of a one-hour call is $39.25. Be careful with the units; time must be expressed in minutes.
(c) To answer this question, we need to solve the equation $C(t) = 10$. Using the formula for $C(t)$, we have

$$\begin{aligned} .65t + .25 &= 10 \\ .65t &= 9.75 \\ t &= 15 \end{aligned}$$

You can talk for 15 minutes.

The computations in the next chapter will demand a good understanding of this notation. So that you can better concentrate on the meaning of those computations, we give more examples that stress the mechanics of function evaluation.

Example 10 (*Compare Exercise 35*)
If $f(x) = 3x + 7$, what is

(a) $f(a)$?
(b) $f(t)$?
(c) $f(x + t)$?

Solution **(a)** Replacing x by a, we have $f(a) = 3a + 7$.
(b) Replacing x by t, we have $f(t) = 3t + 7$.
(c) Replacing x by $x + t$, we have $f(x + t) = 3(x + t) + 7 = 3x + 3t + 7$.

The evaluations $f(a)$ and $f(t)$ were meant to warm you up for $f(x + t)$. The rule states that $f(\text{input}) = 3 \cdot (\text{input}) + 7$, no matter how the input is expressed, by x, a, t, or even $x + t$.

Warning! There is no distributive law here; $f(x + t) \neq f(x) + f(t)$. We saw that $f(x + t) = 3x + 3t + 7$, but on the other hand, $f(x) + f(t) = (3x + 7) + (3t + 7) = 3x + 3t + 14$

Example 11 (*Compare Exercise 39*)
If $f(x) = 2x^2 - x$, what is $f(3 + h)$?

Solution Replacing x by $3 + h$,

$$\begin{aligned} f(3 + h) &= 2(3 + h)^2 - (3 + h) \\ &= 2(9 + 6h + h^2) - 3 - h \\ &= 18 + 12h + 2h^2 - 3 - h \\ &= 15 + 11h + 2h^2 \end{aligned}$$

Warning! You must compute $(3 + h)^2$, not $3^2 + h^2$. We encourage use of parentheses to help you remember this. Also, parentheses help make sure that you write $-(3 + h) = -3 - h$. Without parentheses, the attempt to write $f(3 + h)$ could look like $2 \cdot 3 + h^2 - 3 + h$, which could be interpreted to mean $6 + h^2 - 3 + h$, or $h^2 + h + 3$.

We conclude with examples of how the rule for a function can be determined in a given application. Notice how the application determines the domain.

Example 12 If a newspaper sells for 30¢, express the company's daily revenue in terms of the number of newspapers sold daily.

Solution We let x be the number of newspapers sold daily, and R be the daily revenue in dollars. We can write $R(x) = .30x, \quad x \geq 0$.

Example 13 (*Compare Exercise 42*)
A salesman's monthly salary is \$500 plus 4% of his total sales.

(a) Express his salary as a function of his total sales.
(b) What must his total sales be in order to earn a monthly salary of \$2,000?

Solution **(a)** Let S be his monthly salary in dollars and x be his total sales, also in dollars, for the month. Then

$$S(x) = 500 + .04x, \quad x \geq 0$$

(b) We must solve the equation $S(x) = 2000$.

$$\begin{aligned} 500 + .04x &= 2000 \\ .04x &= 1500 \\ 4x &= 150{,}000 \\ x &= 37{,}500 \end{aligned}$$

His total sales for the month must be \$37,500.

1–1 Exercises

I.

1. (*See Example 2*) Let $f(x) = 3x + 7$. Evaluate each of the following.

   (a) $f(2)$ **(b)** $f(-4)$ **(c)** $f\left(\frac{1}{3}\right)$

2. If $f(x) = 6x - 5$, evaluate

   (a) $f(-2)$ **(b)** $f(3)$ **(c)** $f\left(\frac{1}{2}\right)$

3. (*See Example 3*) If $y = 7 - 2x$, what is y when

   (a) $x = 4$ **(b)** $x = -3$ **(c)** $x = \frac{5}{2}$

4. If $f(x) = 9 - 3x$, what is y when

   (a) $x = 5$ **(b)** $x = 4$ **(c)** $x = \frac{10}{3}$

What is the domain of each function given by the following rules?

5. $f(x) = 7x + 5$

6. $f(x) = 5x - 6x^2$

7. (*See Example 5*)
$f(x) = \frac{3}{x-9}$

8. $f(x) = \frac{7}{x+5}$

9. (*See Example 7*)
$f(x) = \sqrt{x-8}$

10. $f(x) = \sqrt{x+5}$

11. $f(x) = \frac{6}{x^2+4}$

12. $f(x) = \sqrt{x^2+9}$

13. $f(x) = x^2, \quad -1 \le x \le 5$

14. $f(x) = \frac{1}{x+4}, \quad 2 \le x \le 8$

II.

Write a rule for a function that performs the following operations on the input.

15. The function squares the input.
16. The function subtracts 8 from the input.
17. (*See Example 1*) The function multiplies the input by 3 and then adds 2 to the result.
18. The function adds 2 to the input, and then multiplies the result by 3.
19. The function squares the input and then subtracts 9 from the result.
20. The function subtracts 4 from the input and then squares the result.

Perform each evaluation as indicated.

21. (*See Example 4*) If $f(x) = 9x^2 - 3x + 7$, evaluate

(a) $f(4)$ **(b)** $f(-2)$ **(c)** $f\left(\frac{1}{3}\right)$

22. If $f(x) = 8x^2 - 6x + 5$, evaluate

(a) $f(3)$ **(b)** $f(-3)$ **(c)** $f(\frac{1}{2})$

23. If $f(x) = \sqrt{x+9}$, evaluate

(a) $f(7)$ **(b)** $f(-5)$ **(c)** $f(-25)$

24. If $f(x) = \sqrt{2x+10}$, evaluate

(a) $f(13)$ **(b)** $f(-5)$ **(c)** $f(-7)$

What is the domain of each function given by the following rules?

25. $f(x) = \frac{3x-6}{x^2-9}$

26. $f(x) = \frac{5x+10}{x^2-4}$

27. $f(x) = \frac{x^2-4}{x^2-2x-3}$

28. $f(x) = \frac{x^2-16}{x^2-2x+1}$

29. $f(x) = \dfrac{3x + 5}{\sqrt{x - 4}}$

30. $f(x) = \dfrac{x - 7}{\sqrt{x + 3}}$

III.

Perform each evaluation as indicated.

31. (*See Example 8*) If $f(x) = 5x + 6, \quad -2 \le x \le 3$, evaluate
    (a) $f(2)$ **(b)** $f(-2)$ **(c)** $f(4)$

32. If $f(x) = 2x - 4, \quad 1 \le x \le 5$, evaluate
    (a) $f(0)$ **(b)** $f(5)$ **(c)** $f(6)$

33. If $f(x) = x^2 + 1, \quad -1 \le x < 4$, evaluate
    (a) $f(3)$ **(b)** $f(-3)$ **(c)** $f(4)$

34. If $f(x) = -x^2 - 2x + 6, \quad -3 < x \le 3$, evaluate
    (a) $f(2)$ **(b)** $f(-\frac{5}{2})$ **(c)** $f(-4)$

35. (*See Example 10*) If $f(x) = 3x + 7$, evaluate
    (a) $f(w)$ **(b)** $f(x + 2)$

36. If $f(x) = 5x - 8$, evaluate
    (a) $f(u)$ **(b)** $f(x - 3)$

37. If $f(x) = 4 - 7x$, evaluate
    (a) $f(x + 3)$ **(b)** $f(x - 3)$

38. If $f(x) = 5 - 4x$, evaluate
    (a) $f(x + 2)$ **(b)** $f(x - 2)$

39. (*See Example 11*) If $f(x) = x^2 - 4x$, evaluate
    (a) $f(2 + h)$ **(b)** $f(x + h)$

40. If $f(x) = 3x - 2x^2$, evaluate
    (a) $f(x + t)$ **(b)** $f(3 + t)$

41. If $f(x) = 2x^2 - 4x + 5$, evaluate
    (a) $f(3 + h)$ **(b)** $f(x + w)$

42. (*See Example 13*) A student has a summer job selling magazine subscriptions by telephone. She makes $4 per hour plus $2.50 for every customer she gets to subscribe to the magazine. She works 6 hours every day.
    (a) Express her daily salary in dollars as a function of the number of subscriptions she sells.
    (b) How many magazine subscriptions must she sell to make $44 in one day?

43. A person goes to an amusement park that costs $3 for admission and $.50 for each ride.

 (a) Express the person's expenses in dollars as a function of the number of rides he or she takes.

 (b) How many rides can he or she take with $10 to spend?

1–2 Graphs and Lines

Definition of a Graph

"A picture is worth a thousand words" may be an overworked phrase, but it does convey an important idea. You may even occasionally use the expression "Oh, I see!" when you really grasp a difficult concept. A graph shows a picture of a function, and can help you to understand the behavior of the function. Imagine how difficult it would be to convey all the information (and the drama!) in the following example without the aid of the graphs.

The 1975 Masters Golf Tournament in Augusta, Georgia, was one of the most dramatic events in the history of golf. In early rounds several players held or shared the lead, but, by the end, all eyes were focused on a battle among three of the finest players in the game: Jack Nicklaus, Tom Weiskopf, and Johnny Miller.

The following graphs (one for each player) tell the story. Follow Miller's plunge over the opening holes and his record-breaking six straight birdies later. Follow the slow, descending plateaus of Weiskopf's flawless game over much of the third day. Follow Nicklaus's early dominance and his seesaw battle with Weiskopf. Would Miller have won on a 73rd hole?

This example makes another important point. The graph seems to indicate that Miller was $2\frac{1}{2}$ over par after $17\frac{1}{2}$ holes; such an interpretation would be silly—golfers never talk about half of a hole. It is much harder to get information from a graph drawn with just dots. Remember those drawings you made when you were a child by "connecting-the-dots"? The picture made a lot more sense after you drew in the connecting lines. In Figure 1–2 we sacrificed some technical accuracy by "connecting-the-dots" but got a better picture of what happened by doing so. Professional users of mathematics do the same thing; an accountant may let $C(x)$ represent the cost of manufacturing x items, or a manager may let $E(x)$ represent an efficiency index when x people are involved in a large project. In reality, the domains of these functions involve only positive integers. But many methods of mathematics require the domain of the function to be an interval or intervals, rather than isolated points. These methods have proven so powerful in solving problems that people set up their functions using such domains. They

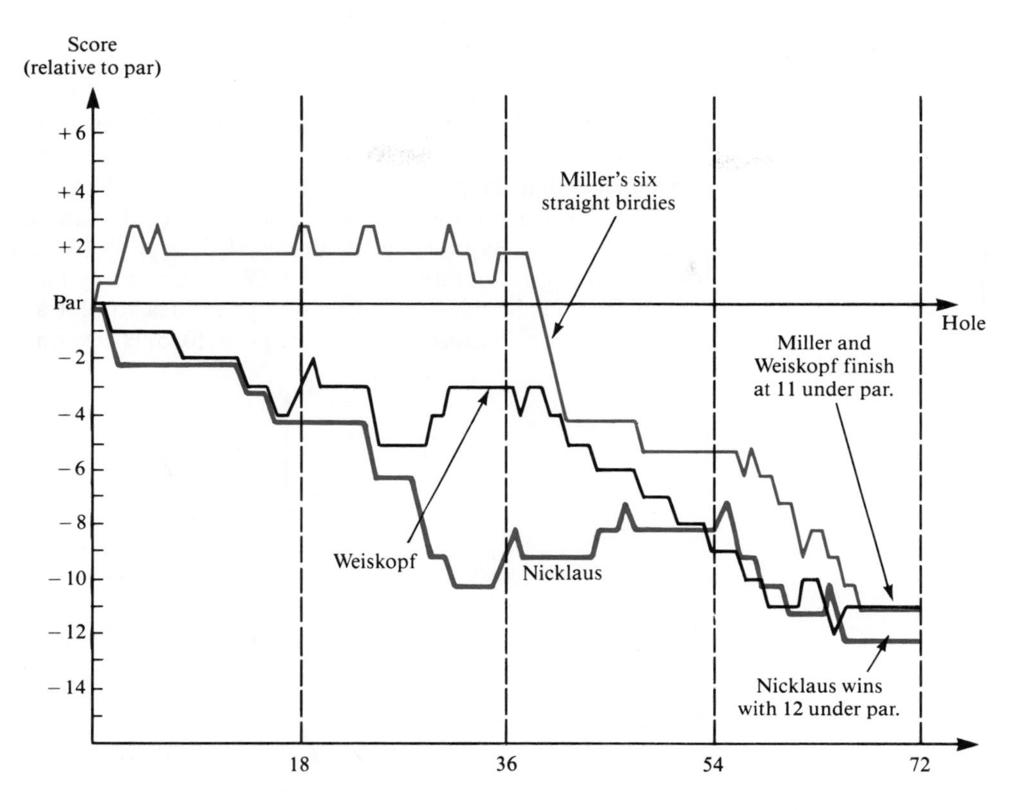

Figure 1–2

may then have to use some common sense in interpreting their answer. If the manager finds that the most efficient number of people to assign to a project is 54.87, she will probably end up using either 54 or 55 people.

Definition

> The **graph of a function f** is the set of points (x, y) in the plane that satisfy the equation $y = f(x)$. In other words, the graph of f is the collection of points of the form $(x, f(x))$.

As we develop different mathematical techniques throughout this text, we will use some concrete applications. This in turn will require some familiarity with the functions involved and some idea of the shape of their graphs. We start with the simplest functions and graphs.

Linear Functions and Straight Lines

Definition

> A function is called a **linear function** if its rule—its defining equation—can be written $f(x) = mx + b$. Such a function is called linear because its graph is a straight line.

Example 1 (*Compare Exercise 1*)
Draw the graph of $f(x) = 2x + 5$.

Solution The graph will be a straight line, and it takes just two points to determine a straight line. If we let $x = 1$, then we have $f(1) = 7$; if we let $x = 4$, then $f(4) = 13$. This means that the points (1, 7) and (4, 13) are on the graph of $f(x) = 2x + 5$. Since we also use y for $f(x)$, we could also say that these points are on the line $y = 2x + 5$. By plotting the points (1, 7) and (4, 13) and drawing the line through them, we obtain Figure 1–3. (We can use any pair of x values to get two points on the line.) It is usually a good idea to plot a third point to help catch any error. Because $f(0) = 5$, the point (0, 5) is also on the graph of f.

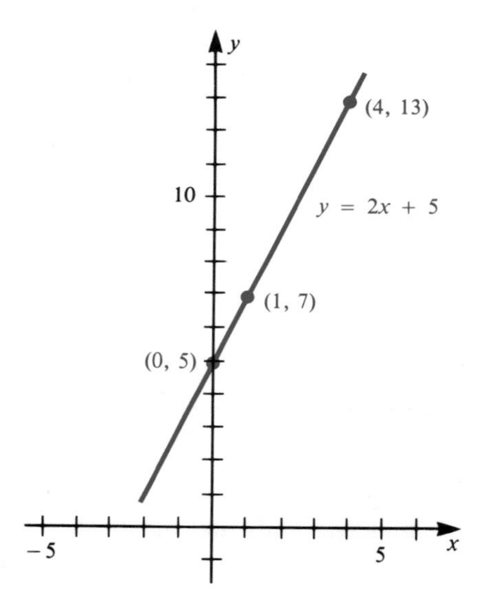

Figure 1–3

Slope and Intercept

When the equation of a line is written in the form $y = 2x + 5$ ($y = mx + b$ is the general form), the constants 2 and 5 (m and b in general) give key information about the line. The constant b (5 in this case) is the value of y when $x = 0$. Thus, $(0, b)$ (or (0, 5) in our example) is a point on the line. We refer to b as the **y-intercept** of the line; it gives the point where the line intercepts the y-axis. The constant m gives information about the direction, or slant, of a line. We call m the **slope** of the line. We will give more information about the slope later.

Example 2 (*Compare Exercise 7*)
What are the slope and y-intercept of each of the following lines?

(a) $y = 3x - 5$
(b) $y = -6x + 15$

Solution **(a)** For the line $y = 3x - 5$, the slope $m = 3$ and the y-intercept $b = -5$.
(b) For the line $y = -6x + 15$, the slope is -6 and the y-intercept is 15.

Warning! The coefficient of x in a general linear equation does not automatically give you the slope of the line. When the equation of the line is in the form $y = mx + b$, the coefficient of x is the slope of the line and the constant term is the y-intercept. If the equation is in another form, it is a good idea to change to this form to determine the slope and y-intercept.

Example 3 (*Compare Exercise 11*)
What are the slope and y-intercept of the line $3x + 2y - 4 = 0$?

Solution We must write the equation $3x + 2y - 4 = 0$ in the slope-intercept form, $y = mx + b$
Solve the equation

$$3x + 2y - 4 = 0$$

for y.

$$2y = -3x + 4$$
$$y = -\tfrac{3}{2}x + 2$$

Thus, the slope-intercept form is $y = -\frac{3}{2}x + 2$. Now we can say the slope is $-3/2$ and the y-intercept is 2.

The slope relates to the way the line slants in the following manner. Select two points on the line $y = 2x + 5$, such as (1, 7) and (4, 13).

Compute the difference in the y-coordinates of the two points: $13 - 7 = 6$. (See Figure 1–4.) Now compute the difference in x-coordinates: $4 - 1 = 3$. The quotient $\frac{6}{3} = 2$ is m, the slope of the line $y = 2x + 5$. Following this procedure with any other two points on the line $y = 2x + 5$ will also yield the answer 2.

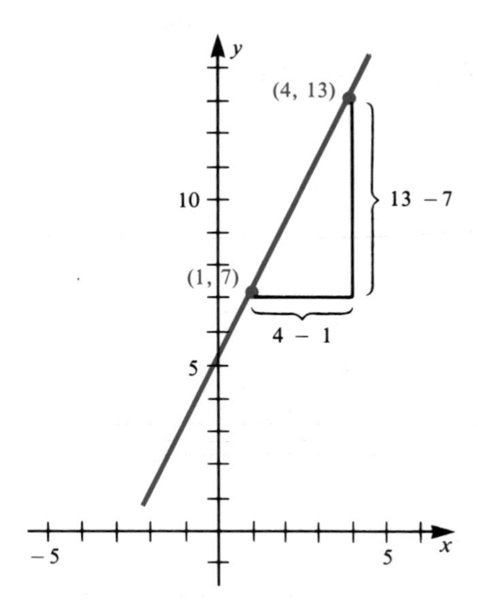

Figure 1–4

The following is the general formula showing how to compute the slope of a line.

Slope Formula

> Choose two points P and Q on the line. Let (x_1, y_1) be the coordinates of P and (x_2, y_2) be the coordinates of Q. The **slope** of the line, m, is given by the equation
>
> $$m = \frac{y_2 - y_1}{x_2 - x_1} \qquad \text{where} \qquad x_2 \neq x_1$$

The slope is the difference in the y-coordinates divided by the difference in the x-coordinates.

Be sure you subtract the x and y coordinates in the same order. Figure 1–5 shows the geometric meaning of this quotient.

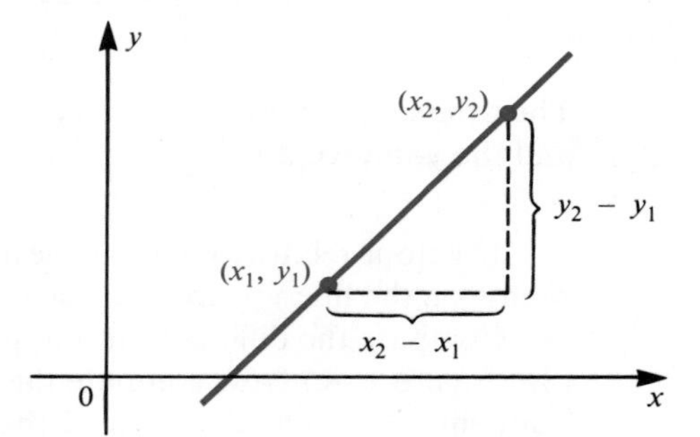

Figure 1–5

Example 4 (*Compare Exercise 15*)
Find the slope of the line through the points (1, 2) and (5, 6).

Solution We let $(x_1, y_1) = (1, 2)$ and $(x_2, y_2) = (5, 6)$. From the definition for m,

$$m = \frac{6 - 2}{5 - 1} = \frac{4}{4} = 1$$

Figure 1–6 shows the geometry.

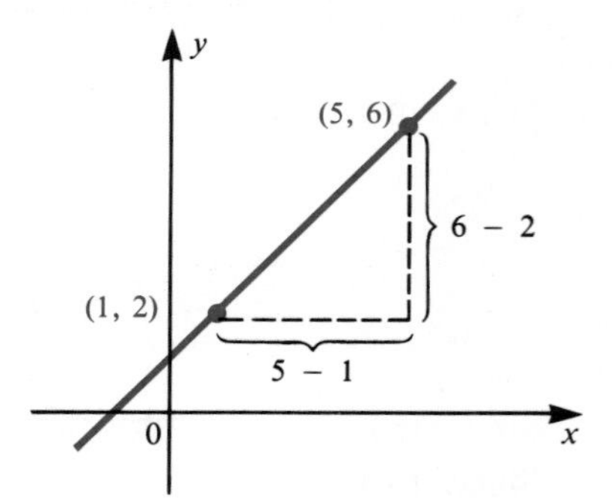

Figure 1–6

Notice that it doesn't matter which point we call (x_1, y_1) and which one we call (x_2, y_2); it doesn't affect the computation of m. If we label the points differently in Example 4, the computation becomes

$$m = \frac{2 - 6}{1 - 5} = \frac{-4}{-4} = 1$$

The answer is the same.

Be sure to subtract the x- and y-coordinates in the same order.

Example 5 Determine the slope of the line through the points (4, 5) and (6, 2).

Solution If we let (x_1, y_1) be (4, 5) and (x_2, y_2) be (6, 2), we get

$$m = \frac{y_2 - y_1}{x_2 - x_1} = \frac{2 - 5}{6 - 4} = \frac{-3}{2} = -\frac{3}{2}$$

The geometry is shown in Figure 1–7.

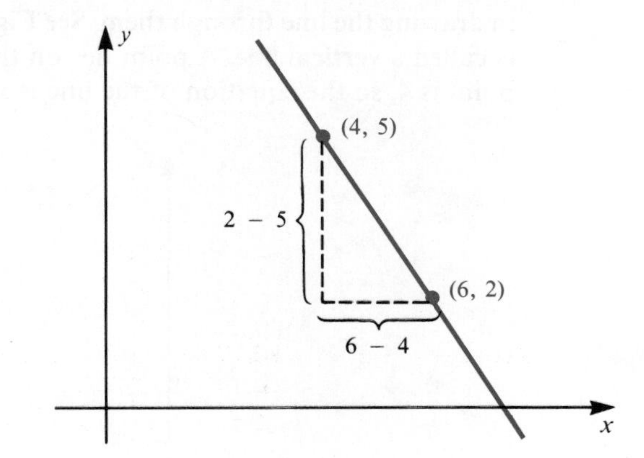

Figure 1–7

Example 6 (*Compare Exercise 19*)
Determine the slope of the line through the points (2, 5) and (6, 5).

Solution The slope of the line is

$$m = \frac{5 - 5}{6 - 2} = \frac{0}{4} = 0$$

Whenever $m = 0$, the equation $f(x) = 0x + b$ is written more simply as $f(x) = b$; f is called a constant function. The graph of a constant function is a line parallel to the x-axis; such a line has an equation of the form $y = b$ and is called a **horizontal line**. See Figure 1–8.

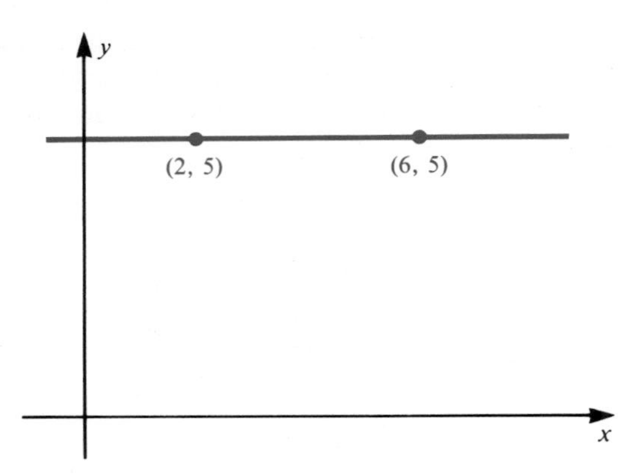

Figure 1–8

Example 7 (*Compare Exercise 23*)
Determine the equation of the line through the points (4, 1) and (4, 3).

Solution We can try to use the rule for computing the slope, but we obtain the quotient $\frac{3-1}{4-4} = \frac{2}{0}$, which doesn't make sense because division by zero is not defined. The slope is not defined. When we plot the two points, however, we have no difficulty in drawing the line through them. See Figure 1–9. The line, parallel to the y-axis, is called a **vertical line**. A point lies on this line when the first coordinate of the point is 4, so the equation of the line is $x = 4$.

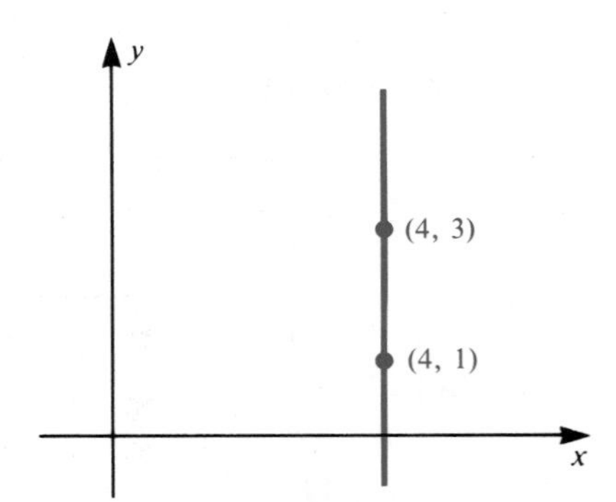

Figure 1–9

Whenever $x_2 = x_1$, you get a 0 in the denominator when computing the slope, so we say that the *slope does not exist* for such a line.

Be careful! A vertical line does *not* have a slope, but it does have an equation.

The slope of a line can be positive, negative, zero, or even not exist. These situations are depicted in Figure 1–10.

This figure shows the relationship between the slope and the slant of the line. If $m > 0$, the graph slants up as x moves to the right. If $m < 0$, the graph slants down as x moves to the right. If $m = 0$, the graph remains at the same height. If m does not exist, the line is vertical and the line is *not* the graph of a linear function.

The equation of a vertical line cannot be written in form $f(x) = mx + b$ for there is no m.

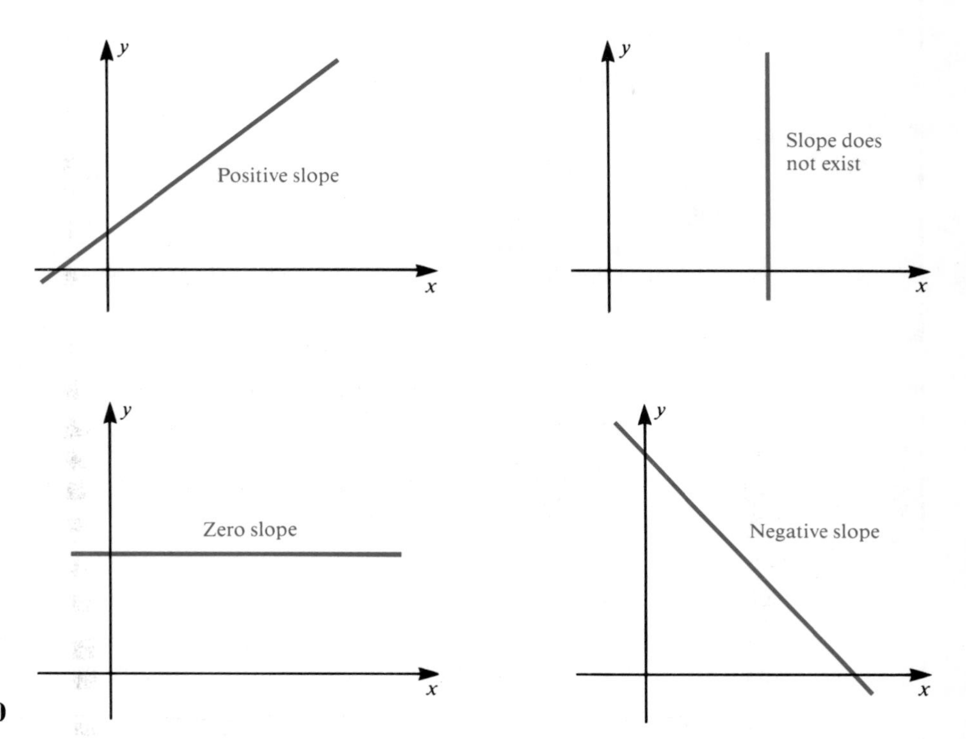

Figure 1–10

We conclude this section by showing how to find equations of lines. Linear functions arise in many applied settings. When an application provides the appropriate two pieces of information, you can find an equation of the corresponding line. Two pieces of information that determine a line are

1. the slope and a point on the line, or
2. two points on the line.

We will show each form in a particular application and then give the general method of solving the problem.

Example 8 **(point-slope given)**
(*Compare Exercise 58*)
A manufacturer of watches knows that it costs \$3.14 in materials to make each watch. Also, the manufacturer must pay \$7415 per week on other costs such as property taxes, payroll, and so on. Find the weekly costs of the company as a function of the number of watches produced.

Solution Let x be the number of watches produced each week, and C be the total weekly costs for the company in dollars. The amount \$7415 is called the **fixed cost** for the company; it does not depend on the number of watches produced. The other cost to the company is called the **variable cost**, and depends directly on the number of items produced. In this example the variable cost is $3.14x$. The **total cost** is the sum of the fixed cost and the variable cost. Thus, C and x are related by the equation $C = 3.14x + 7415$. C is a linear function of x with slope $m = 3.14$ and $(0, 7415)$ a point on the line.

Example 9 (*Compare Exercise 27*)
Determine an equation of the line with slope 3 and y-intercept 2.

Solution This example gives the same kind of information as Example 8. Note that the fixed cost corresponds to b in the equation $f(x) = mx + b$. The phrase "y-intercept 2" means that the point $(0, 2)$ is on the line. Since $m = 3$ and $b = 2$, the equation is

$$y = 3x + 2$$

Example 10 (*Compare Exercise 33*)
Determine an equation of the line that has slope -2 and passes through the point $(-3, 5)$.

Solution The value of $m = -2$, so the line has an equation of form

$$y = -2x + b$$

The solution will be complete when the value of b is found. Since the point $(-3, 5)$ lies on the line, $x = -3$ and $y = 5$ must be a solution to $y = -2x + b$. Just substitute those values into $y = -2x + b$ to obtain

$$5 = (-2)(-3) + b$$
$$5 = 6 + b$$
$$5 - 6 = b$$

Thus, $b = -1$ and the equation of the line is $y = -2x - 1$.

There is a formula, called the point-slope formula, that may be used to write an equation of the line in one step when the slope and point are given.

Point-Slope Formula

We will work with the same information in Example 9, $m = -2$ and the point $(-3, 5)$. We use the slope formula with $(-3, 5)$ as (x_1, y_1) and an arbitrary point (x, y) as (x_2, y_2). Because $m = -2$, we can write

$$-2 = \frac{y - 5}{x - (-3)} = \frac{y - 5}{x + 3}$$

Multiply both sides by $x + 3$ to obtain

$$-2(x + 3) = y - 5$$

The formula is usually written as

$$y - 5 = -2(x + 3)$$

Point-Slope Formula

> If a line has slope m and passes through (x_1, y_1), an equation of the line is
> $$y - y_1 = m(x - x_1)$$

Example 11 (*Compare Exercise 37*)
Find an equation of the line with slope 4 that passes through $(-1, 5)$.

Solution $x_1 = -1$ and $y_1 = 5$; $m = 4$, so the point-slope formula gives us

$$y - 5 = 4(x - (-1))$$

The arithmetic to write this equation in the form $y = mx + b$ is

$$y - 5 = 4x + 4$$
$$y = 4x + 4 + 5$$
$$y = 4x + 9$$

Example 12 **(2 points given)**
Businesses can deduct the cost of equipment in various ways; one method is called straight-line depreciation. For example, a piano teacher buys a new piano for \$5000 and can depreciate the value of the piano for the next 6 years, when it will be worth \$800. Find the value of the piano as a function of its age.

Solution Let t be the number of years after the piano was purchased and let v represent the value of the piano. When $t = 0$, $v = 5000$, so the graph contains the point (0, 5000). When $t = 6$, $v = 800$ so the graph also contains the point (6, 800). The graph is a straight line with slope

$$\frac{5000 - 800}{0 - 6} = \frac{4200}{-6} = -700$$

The negative slope means that the value of the piano is decreasing.

Now that we have the slope we can use either of the two points to generate an equation of the line. Using the point (0, 5000), we have $v - 5000 = -700(t - 0) = -700t$; $v = -700t + 5000$. This equation is valid for the application when $0 \le t \le 6$.

Just as we had a point-slope formula, we have a formula for finding an equation of a line using the coordinates of two points on the line.

Two-Point Formula

> If a line passes through the points (x_1, y_1) and (x_2, y_2), with $x_1 \neq x_2$, an equation of the line is
> $$y - y_1 = \frac{y_2 - y_1}{x_2 - x_1}(x - x_1)$$

Observe: This is basically the point-slope form of a line using

$$m = \frac{y_2 - y_1}{x_2 - x_1}$$

and (x_1, y_1) as the point on the line. You can think of this as a variation of the point-slope formula. First, use the two points to compute m, and then use one of the points and the point-slope formula.

Example 13 (*Compare Exercise 41*)
Determine an equation of the straight line through the points (1, 3) and (4, 7).

Solution Let (x_1, y_1) be the point (1, 3) and let (x_2, y_2) be the point (4, 7); then

$$m = \frac{7 - 3}{4 - 1} = \frac{4}{3},$$

so $y - 3 = \frac{4}{3}(x - 1)$ is an equation of this line. We can simplify by multiplying both sides by 3, obtaining

$$\begin{aligned} 3y - 9 &= 4x - 4 \\ 3y &= 4x + 5 \end{aligned}$$

We conclude this section with a discussion of parallel lines.

Parallel Lines

Definition

> Two lines are **parallel** if they have the same slope or if they are both vertical lines.

Example 14 (*Compare Exercise 47*)
Is the line through the points (1, 2) and (3, 3) parallel to the line through the points (−3, 2) and (5, 6)?

Solution Let L_1 be the line through (1, 2) and (3, 3), and let m_1 be its slope.

$$m_1 = \frac{3 - 2}{3 - 1} = \frac{1}{2}$$

Let L_2 be the line through (−3, 2) and (5, 6), and let m_2 be its slope.

$$m_2 = \frac{6 - 2}{5 - (-3)} = \frac{4}{8} = \frac{1}{2}$$

The slopes are identical; thus, the lines are parallel. (See Figure 1–11.)

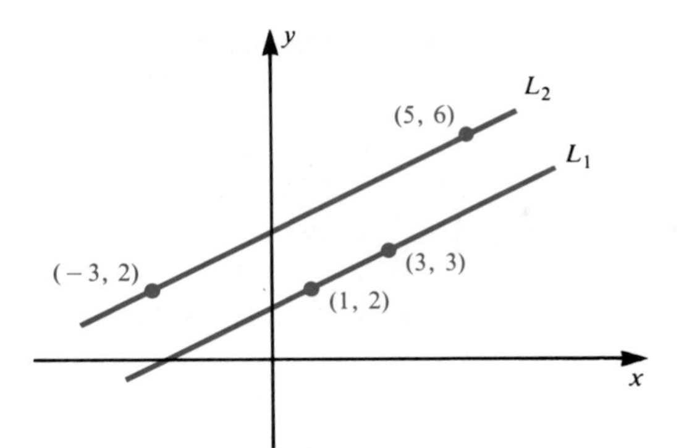

Figure 1–11

Example 15 Is the line through (5, 4) and (−1, 2) parallel to the line through (3, −2) and (4, 4)?

Solution The slope of the first line is

$$m_1 = \frac{4-2}{5-(-1)} = \frac{2}{6} = \frac{1}{3}$$

The slope of the second line is

$$m_2 = \frac{-2-4}{3-4} = \frac{-6}{-1} = 6$$

Since the slopes are not equal, the lines are not parallel.

1–2 Exercises

I.

Draw the graphs of the lines in Exercises 1 through 6.

1. (*See Example 1*) $f(x) = 3x + 8$
2. $f(x) = 4x - 2$
3. $f(x) = x + 7$
4. $f(x) = -2x + 5$
5. $f(x) = -3x - 1$
6. $f(x) = \frac{2}{3}x + 4$

Find the slope and y-intercept for the lines in Exercises 7 through 10.

7. (*See Example 2*) $y = 7x + 22$
8. $y = 13x - 4$
9. $y = -\frac{2}{5}x + 6$
10. $y = -\frac{1}{4}x - \frac{1}{3}$

Find the slope and y-intercept of the lines in Exercises 11 through 14.

11. (*See Example 3*) $2x + 5y - 3 = 0$
12. $4x + y - 3 = 0$
13. $x - 3y + 6 = 0$
14. $5x - 2y = 7$

Determine the slopes of the straight lines through the pairs of points in Exercises 15 through 22.

15. (*See Example 4*) $(1, 1), (2, 3)$

16. $(-1, 0), (1, 2)$

17. $(-1, -1), (-1, -3)$

18. $(4, -1), (6, -1)$

19. (*See Example 6*) $(5, -2), (-3, -2)$

20. $(8, 3), (1, 3)$

21. $(-1, 0), (-4, 0)$

22. $(0, 0), (17, 0)$

Find an equation of the line through the given points in Exercises 23 through 26.

23. (*See Example 7*) $(3, 2), (3, 5)$

24. $(-4, 6), (-4, 9)$

25. $(10, 0), (10, 7)$

26. $(-6, -1), (-6, 13)$

Find equations of the lines in Exercises 27 through 32 with the given slope and y-intercept.

27. (*See Example 9*) slope 4 and y-intercept 3

28. slope -2 and y-intercept 5

29. $m = -1, b = 6$

30. $m = -\frac{3}{4}, b = 7$

31. $m = \frac{1}{2}, b = 0$

32. $m = 3.5, b = -1.5$

Find equations of the lines in Exercises 33 through 36 with the given slope and passing through the given point.

33. (*See Example 10*) slope -4 and point $(2, 1)$

34. slope 6 and point $(-1, -1)$

35. slope $\frac{1}{2}$ and point $(5, 4)$

36. slope -1.5 and point $(2.6, 5.2)$

Use the point-slope formula to find the equations of the lines in Exercises 37 through 40 with given slope and point.

37. (*See Example 11*) slope 7 and point $(1, 5)$

38. slope -2 and point $(3, 1)$

39. slope $\frac{1}{5}$ and point $(9, 6)$

40. slope $-\frac{2}{3}$ and point $(-1, 4)$

In Exercises 41 through 46, determine the equations of the straight lines through the pairs of points, and sketch the lines.

41. (*See Example 13*) $(-1, 0), (2, 1)$

42. $(3, 0), (1, -1)$

43. $(0, 0), (1, 2)$

44. $(1, 3), (1, 5)$

45. $(2, 4), (5, 4)$

46. $(0, 3), (7, 0)$

47. (*See Example 14*) Is the line through the points $(8, 2)$ and $(3, -3)$ parallel to the line through $(6, -1)$ and $(16, 9)$?

48. Is the line through $(5, 4)$ and $(1, -2)$ parallel to the line through $(1, 2)$ and $(6, 8)$?

49. Is the line through $(9, -1)$ and $(2, 8)$ parallel to the line through $(3, 5)$ and $(10, -4)$?

Determine if the pairs of lines in Exercises 50 through 53 are parallel.

50. $y = 6x + 22$
$y = 6x - 17$

51. $3x + 2y = 5$
$6x + 4y = 15$

52. $x - 2y = 3$
$2x + y = 1$

53. $3x - 5y = 4$
$-6x + 10y = -8$

54. Write the equation of the line through $(-1, 5)$ that is parallel to $y = 3x + 4$.

55. Write the equation of the line through (2, 6) that is parallel to $3x + 2y = 17$.
56. Write the equation of the line with y-intercept 8 that is parallel to $5x + 7y = -2$.
57. Write the equation of the line through $(-4, 5)$ that is parallel to the line through (6, 1) and $(2, -3)$.

II.

58. (*See Example 8*) A manufacturer of lawnmowers estimates that the material for each lawnmower costs \$85.00. The manufacturer's fixed operating costs are \$4250 per week. Find the weekly costs of the company.
59. A hamburger place estimates that the materials for each hamburger cost \$0.67 and fixed daily operating expenses are \$480. Find the daily costs of the place.
60. (*See Example 12*) Don's Auto Parts bought a new delivery van for \$12,000. Don expects the van to be worth \$1500 in five years. Find the value of the van as a function of its age.
61. A wholesale company paid \$1340 for 500 items, and later bought 800 of the same item for \$1760. Assume the cost is a linear function of the number of items. Write the equation.
62. The calorie content of four large shrimp is 36 calories; that for eleven shrimp is 99 calories. Write the calorie content as a linear function of the number of shrimp.
63. A male college student requires 3000 calories per day to maintain his daily activities and to maintain a constant weight. He wants to gain weight. Each pound of body fat requires an additional 3500 calories. Write his daily calorie intake as a function of the number of pounds gained per day.

III.

64. (*See Example 8*) In a cost equation like $C(x) = 3x + 15$, 15 is the fixed cost and $3x$ is the variable cost. The coefficient of x, 3 in this case, is the **unit cost**.

    Find the cost equation for the following:

    (a) Fixed cost = \$745, unit cost = \$4.50
    (b) Fixed cost = \$9248, unit cost = \$22.75
    (c) Fixed cost = \$860, unit cost = \$2.47

65. The slope of a line is the amount of change in y when x increases by one unit. For $y = 3x + 7$, y increases by 3 when x increases by 1. For $y = -2x + 4$, y decreases by 2 when x increases by 1. (The negative value indicates a decrease in y.)

    Find the change in y when x increases by 1:

    (a) $y = 4x - 5$
    (b) $y = -3x + 4$

(c) $y = \frac{2}{3}x + 7$
(d) $y = -\frac{1}{2}x + \frac{2}{5}$
(e) $3y + 2x - 4 = 0$
(f) $y = 17$

66. A female college student needs 2100 calories per day to maintain her present weight and daily activities. She wants to lose x pounds per day. Each pound of body fat is equivalent to 3500 calories. Write her daily calorie intake as a linear function of the number of pounds of weight lost per day.

67. An auto rental firm charges a fixed daily rate and a mileage charge. One customer rents a car for one day and drives it 125 miles. His bill is \$35.75. Another customer rents a car for one day and drives 265 miles. Her bill is \$51.65. Write the linear equation relating miles driven and total cost.

1–3 Applications of Linear Functions

Cost-Volume Function
Revenue Function
Break-Even Analysis
Simple Interest
Straight-Line Depreciation

In practical problems the relationship between the variables can be quite complicated. For example, the variables and their relationship that affect the stock market still defy the best of analysts. However, many times a linear relationship can be used to provide reasonable and useful information for solving practical problems. This section contains several applications of the linear function.

Cost-Volume Function

A manufacturer of mopeds conducted a study of production costs and found that fixed costs averaged \$5600 per week and material costs averaged \$359 per moped. This information can be stated as

$$C = 359x + 5600$$

where x represents the number of mopeds produced, also called the **volume**, and C is the total cost of producing x mopeds. A linear function like this is used when

1. there are **fixed costs** such as rent, utilities, and salaries which are the same each week independent of the number of items produced;
2. there are **variable costs** which depend on the number of items produced, such as the cost of materials for the items, packaging, and shipping costs.

The moped example is a linear **cost-volume function** (often simply called the cost function). It is appropriate when the general form of the cost function C is given by

$$C(x) = ax + b$$

where

x is the number of items (volume),
b is the fixed cost in dollars,
ax is the variable cost in dollars,
a is the *unit* cost (the cost per item) in dollars,
$C(x)$ is the total cost in dollars of producing x items.

Notice the form of the cost function. It is essentially the slope-intercept form of a line where the slope is the unit cost and the intercept is the fixed cost.

Example 1 (*Compare Exercise 1*)
If the cost function of manufacturing mopeds is given by $C(x) = 359x + 5600$, then

(a) determine the cost of producing 700 mopeds per week.
(b) determine how many mopeds were produced if the production cost for one week was \$200,178.

Solution **(a)** Substitute $x = 700$ into the cost equation to obtain

$$\begin{aligned} C(700) &= 359(700) + 5600 \\ &= 251{,}300 + 5600 \\ &= 256{,}900 \end{aligned}$$

So, the total cost is \$256,900.

(b) This information gives us $C(x) = 200{,}178$, so we have

$$200{,}178 = 359x + 5600$$

We need to solve for x.

$$\begin{aligned} 200{,}178 - 5600 &= 359x \\ 194{,}578 &= 359x \\ x &= \frac{194{,}578}{359} = 542 \end{aligned}$$

so 542 mopeds were produced.

Revenue Function

If a sporting goods store sells mopeds for \$798 each, the total income (**revenue**) in dollars from mopeds is 798 times the number of mopeds sold. This illustrates the general concept of a **revenue function**; it gives the total revenue obtained from the sale of x items. In the moped example, the revenue function is given by

$$R(x) = 798x$$

where x represents the number of mopeds sold, 798 is the selling price in dollars for each item, and $R(x)$ is the total revenue in dollars from x items.

Example 2 (*Compare Exercise 5*)
The sporting goods store has a sale on mopeds at \$725 each.

(a) Give the revenue function.
(b) The store sold 23 mopeds. What was the total revenue?
(c) One salesperson sold \$5075 worth of mopeds. How many did she sell?

Solution **(a)** The revenue function is given by

$$R(x) = 725x$$

(b) The revenue for 23 mopeds is obtained from the revenue function when $x = 23$:

$$R(23) = 725(23) = 16{,}675$$

The revenue in this case is \$16,675.

(c) This gives $R(x) = 5075$ so

$$725x = 5075$$

$$x = \frac{5075}{725} = 7$$

So, she sold 7 mopeds.

Break-Even Analysis

Break-even analysis answers a common management question: At what sales volume will we break even? When do revenues equal costs? Greater sales will induce a profit while lesser sales will show a loss.

The break-even point occurs when the cost equals the revenue, so the cost and revenue functions can be used to determine the break-even point. In function notation this is written as $C(x) = R(x)$.

Example 3 (*Compare Exercise 8*)
A department store pays \$99 each for tape decks. Their monthly fixed costs are \$1250. They sell the tape decks for \$189.95 each.

(a) What is the cost-volume function?
(b) What is the revenue function?
(c) What is the break-even point?

Solution Let x represent the number of tape decks sold.

(a) The cost function is given by

$$C(x) = 99x + 1250$$

(b) The revenue function is defined by

$$R(x) = 189.95x$$

(c) The break-even point occurs when cost equals revenue,

$$C(x) = R(x)$$

Writing out the functions gives

$$99x + 1250 = 189.95x$$

The solution of this equation gives the break-even point.

$$\begin{aligned} 99x + 1250 &= 189.95x \\ 1250 &= 189.95x - 99x \\ 1250 &= 90.95x \\ x &= \frac{1250}{90.95} \\ &= 13.74 \end{aligned}$$

Because x represents the number of tape decks, we use the next integer, 14, as the number sold per month to break even. If more than 14 are sold, there will be a profit. If fewer than 14 are sold, there will be a loss.

Example 4 (*Compare Exercise 15*)
A temporary secretarial service has a fixed weekly cost of \$730. The wages and benefits of the secretaries amount to \$3.93 per hour. A firm who employs a secretary pays Temporary Service \$6.40 per hour. How many hours per week of secretarial service must Temporary Service place in order to break even?

Solution First, write the rule for the cost and revenue functions. The fixed cost is \$730 and the unit cost is \$3.93, so the cost function is given by

$$C(x) = 3.93x + 730$$

where x is the number of hours placed each week. The revenue function is given by

$$R(x) = 6.40x.$$

Equating cost and revenue we have

$$3.93x + 730 = 6.40x$$

This equation reduces to

$$\begin{aligned} 6.40x - 3.93x &= 730 \\ 2.47x &= 730 \\ x &= \frac{730}{2.47} \\ &= 295.5 \end{aligned}$$

Temporary Service must place secretaries for a total of 296 hours per week (rounded up) in order to break even.

Simple Interest

Our modern economy depends on borrowed money. Very few families would own a house or a car without credit. Business depends on borrowed money for day-to-day operations and for long term expansion. When we borrow money from a bank, we pay them for the use of it. Interest is the "rent" paid to use money for a period of time. The amount of interest is usually based on a specified percent of the money borrowed. Here, we discuss only simple interest; we will talk about compound interest later in the text. If you borrow \$800 at a simple interest rate of 10% per year, the interest paid is

$$I = 800(.10)t$$

where I is the interest in dollars and t is the length of the loan in years. If \$800 is borrowed at 10% per year for three years, the amount of interest is found by computing

$$I = 800(.10)(3) = 240$$

The amount of interest paid is \$240.

In general, the interest is

$$I = Prt$$

where P is the amount borrowed (principal), r is the interest rate written in decimal form, and t is the length of the loan.

Example 5 (*Compare Exercise 16*)
Find the interest paid on a \$1500 loan if the interest rate is 8.5% and the loan is for 2.5 years.

Solution The decimal form of 8.5% is .085 so the formula for interest becomes

$$\begin{aligned} I &= 1500(.085)(2.5) \\ &= 318.75 \end{aligned}$$

The interest paid is \$318.75.

Notice that when P and r are fixed, I is a linear function of t.

Example 6 (*Compare Exercise 23*)
If Brady borrows \$2000 at a 12% interest rate to pay college expenses, the total amount due at the end of the loan is the orginal amount borrowed plus the interest. This amount due, A, for t years is

$$A = 2000 + 2000(.12)t$$

If the loan is for 2 years, the amount due, in dollars, is

$$\begin{aligned} A &= 2000 + 2000(.12)(2) \\ &= 2000 + 480 = 2480 \end{aligned}$$

The general expression for the amount due at the end of t years is

$$A = P + Prt$$

Notice that when P and r are fixed, A is a linear function of t.

Straight-Line Depreciation

When a corporation buys a fleet of cars, they expect them to decline in value due to wear and tear. If they purchase new cars for \$11,500 each, they may expect them to be worth only \$2500 three years later. This decline in value is called **depreciation**. The value of an item after deducting depreciation is called its **book value**. In three years each car depreciated \$9000, and its book value at the end of three years was \$2500. For tax and accounting purposes a company will report depreciation and book value each year during the life of an item. The Internal Revenue Service allows several methods of depreciation. The simplest is **straight-line depreciation**. This method assumes that the book value is a linear function of time, that is,

$$BV = mx + b$$

where BV is the book value and x is the number of years. For example, each car had a book value of \$11,500 when $x = 0$ ("brand new" occurs at zero years). When $x = 3$, its book value declined to \$2500. This information is equivalent to giving two points (0, 11500) and (3, 2500) on the straight line representing book value. (See Figure 1–12.) We obtain the linear equation of the book value by finding the equation of a line through these two points. The slope of the line is

$$\begin{aligned} m = \frac{y_2 - y_1}{x_2 - x_1} &= \frac{2500 - 11{,}500}{3 - 0} \\ &= \frac{-9000}{3} = -3000 \end{aligned}$$

and the y-intercept is 11,500 so the equation is

$$BV = -3000x + 11{,}500$$

The book value at the end of two years is

$$BV = -3000(2) + 11{,}500 = -6000 + 11{,}500 = 5500$$

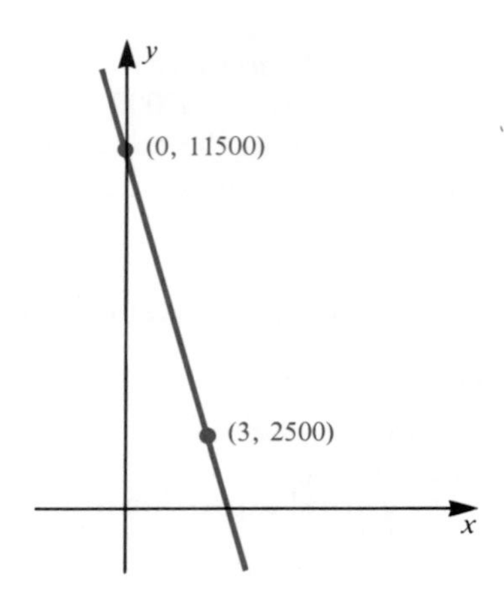

Figure 1–12

The negative value of the slope indicates that the book value is decreasing by \$3000 each year. This annual decrease is the **annual depreciation**.

Generally, a company will estimate the number of years of useful life of an item. The value of the item at the end of its useful life is called its **scrap value**. The values of x are restricted to $0 \le x \le n$, where n is the number of years of useful life.

Example 7 (*Compare Exercise 19*)
Acme Manufacturing Co. purchases a piece of equipment for \$28,300 and estimates its useful life as 8 years. At the end of its useful life, its scrap value is estimated at \$900.

(a) Find the linear equation expressing the relationship between book value and time.

(b) Find the annual depreciation.

(c) Find the book value for the first, fifth, and seventh year.

Solution **(a)** The line passes through the two points (0, 28300) and (8, 900). Thus, the slope is

$$m = \frac{900 - 28{,}300}{8 - 0} = -3425$$

and the y-intercept is 28,300 giving the equation

$$BV = -3425x + 28{,}300$$

(b) The annual depreciation is obtained from the slope and is \$3425.

(c) For year 1, $BV = -3425(1) + 28{,}300 = 24{,}875$.
For year 5, $BV = -3425(5) + 28{,}300 = 11{,}175$.
For year 7, $BV = -3425(7) + 28{,}300 = 4325$.
Thus, after 7 years the book value of the equipment is \$4325.

1–3 Exercises

I.

1. (*See Example 1*) The weekly cost function of manufacturing x bicycles is given by

$$C(x) = 43x + 2300$$

   (a) Determine the cost of producing 180 bicycles per week.
   (b) One week the total production cost was \$11,889. How many bicycles were produced that week?

2. A software company produces a home-accounting system. Their cost function for producing x systems per month is given by

$$C(x) = 16.25x + 28{,}300$$

   (a) Determine the cost of producing 2500 systems per month.
   (b) One month their production costs were \$63,010. How many systems did they produce?

3. A company has determined that the relationship between cost and volume (the cost-volume formula) for a certain product is $C(x) = 3x + 400$.

   (a) Determine the fixed cost and the unit cost.
   (b) Find the total costs when the production is 600 units and 1000 units, respectively.

4. The cost-volume formula for a certain product is $y = 2.5x + 750$. Determine the fixed cost and the unit cost. Find the total costs when the production is

   (a) 100 units
   (b) 300 units
   (c) 650 units

5. (*See Example 2*) A store sells jogging shoes for \$32 per pair.

   (a) Write the rule that gives the revenue function.
   (b) The store sold 78 pairs. What was the revenue?
   (c) One day the store sold \$672 worth of jogging shoes. How many pairs did they sell?

6. Tony's Cassette Warehouse sells cassettes for \$6.25 each.

   (a) Write the rule for the revenue function.
   (b) What is the revenue from selling 265 cassettes?

7. Cold Pizza sells frozen pizzas for \$3.39 each.

   (a) Write the rule that gives the revenue function.
   (b) What is the revenue from selling 834 pizzas?

8. (*See Example 3*) A clothing store pays \$57 each for sports coats and has a fixed monthly cost of \$780. They sell the coats for \$79 each.

   (a) What is the rule for the cost-volume function?

   (b) What is the rule for the revenue function?

   (c) What is the break-even point?

9. The monthly expenses of The Campus Copy Shop are given by the cost equation

$$C(x) = 3690 + 0.025x$$

   where x is the number of pages copied in a month. The revenue function is given by

$$R(x) = .055x$$

   Find the break-even point of The Campus Copy Shop.

10. The Academic T-Shirt Company did a cost study and found that it cost \$1400 to produce 600 "I Love Math" T-shirts. The total cost is \$1600 when the volume is 700 T-shirts. Determine the linear equation that describes the relationship of cost to volume.

    (a) What is the fixed cost?

    (b) What is the unit cost?

11. A Toy Co. estimates that total costs are \$1000 when their volume is 500 Fastback cars, and \$1200 when their volume is 900 cars.

    (a) Determine the cost-volume formula.

    (b) What are the fixed cost and the unit cost?

    (c) What are the estimated total costs when the volume is 1200 units?

12. Find the cost equation if the fixed cost is \$700 and the cost-per-unit volume is \$2.50. What are the total costs when the volume produced is 400 units?

13. Find the cost-volume formula if the fixed cost is \$500 and the cost-per-unit volume is \$4. What are the total costs when the volume produced is 800 units?

14. A company has a cost function given by

$$C(x) = 22x + 870$$

    and a revenue function given by

$$R(x) = 37.50x$$

    Find the break-even point.

15. (*See Example 4*) A Computer Shop sells computers. The shop has fixed costs of \$1500 per week. Their average cost per computer is \$649 and the average selling price is \$899.

    (a) Write the rule for the cost function.

(b) Write the rule for the revenue function.
(c) Find the cost of selling 37 computers per week.
(d) Find the revenue from selling 37 computers.
(e) Find the break-even point.

16. (*See Example 5*) Compute the simple interest on $4800 for 1.5 years at 11% interest rate.
17. Compute the simple interest on $500 for 2 years at 7% interest rate.
18. Compute the simple interest on $950 for 1.75 years at 8% interest rate.
19. (*See Example 7*) A TV costs $425, has a scrap value of $25, and a useful life of 8 years. Find

    (a) the linear equation relating book value and number of years.
    (b) the annual depreciation.
    (c) the book value for year 3.

20. A machine costs $1500 and has a useful life of 10 years. If it has a scrap value of $200, find

    (a) the linear equation relating book value and number of years.
    (b) the annual depreciation.
    (c) the book value after 7 years.

21. An automobile costs $9750, has a useful life of 6 years, and a scrap value of $300. Find

    (a) the linear equation relating book value and number of years.
    (b) the annual depreciation.
    (c) the book value for years 2 and 5.

II.

22. Jones borrows $3500 for 3 years at 8% simple interest. How much interest does Jones pay?
23. (*See Example 6*) Angie borrowed $750 for 2 years at 9% interest. What is the amount due at the end of two years?
24. Joe obtained a loan of $2500 for 4.5 years at 9% simple interest. What amount was due at the end of that time?
25. Alfred borrowed $1850 at 10.5% simple interest. The amount due at the end of the loan was $2529.88. What was the length of the loan?
26. A health club membership costs $35 per month.

    (a) What is the monthly revenue function for the club?
    (b) What is the monthly revenue if the club has 1238 members?
    (c) The revenue increased $595 in one month. What was the increase in membership?

III.

27. A company's records showed that the daily fixed costs for one of their production lines was $1850 and the total cost of one day's production of 320 items was $3178. What is the cost-volume function?

28. How much should be invested at 8% simple interest in order to have $2000 in 18 months?

29. A specialty store sells personalized telephones. Their weekly cost function is given by

$$C(x) = 28x + 650$$

and the break-even point is $x = 65$ phones per week. What is the revenue function?

30. The break-even point for a tanning salon is 260 memberships, which gives them $3120 monthly revenue. If they sell only 200 memberships, they will lose $330 per month.

(a) What is their revenue function?

(b) What is their cost function?

31. The profit function is revenue minus cost, that is,

$$P(x) = R(x) - C(x)$$

(a) The cost and revenue functions for Acme Manufacturing are given by

$$C(x) = 28x + 465$$
$$R(x) = 52x$$

(i) What is the profit function?

(ii) What is the profit from selling 25 items?

(b) The weekly expenses of selling x bicycles in The Bike Shop are given by

$$C(x) = 1200 + 130x$$

and revenue is given by

$$R(x) = 210x$$

(i) What is the profit function?

(ii) Find the profit from selling 18 bicycles in a week.

(c) Another bicycle shop has monthly fixed costs of $5200 and unit costs of $145. They sell their bicycles for $225 each.

(i) Write the profit function.

(ii) What is the profit from selling 75 bicycles per month?

1–4 Graphs of Nonlinear Functions

Quadratic Functions
Lorenz Curves
Higher Degree Polynomials
Functions Defined Piecewise

One of the goals of a calculus course is to give the student the ability to sketch a graph of a function accurately and fairly quickly. In Chapter 4, we will see how to identify a few crucial points to plot, and then how to use these points to fill in the rest of the curve without having to plot additional points. There are some functions, however, whose graphs can be drawn without using calculus. In the last two sections we dealt with linear functions and straight lines. In this section we introduce some more types of functions and their graphs.

Quadratic Functions

A function is called a **quadratic function** if its rule can be written $f(x) = ax^2 + bx + c$ with $a \neq 0$; its graph is called a **parabola**.

Quadratic functions are often used to describe revenue, or income, from sales because the revenue is obtained as the product of two variables; revenue equals the sales price per item times the number of items sold. A Girl Scout who sells 40 boxes of cookies at $3 per box has a revenue of $120. More generally, if a seller sells x items at a price of p per item, the seller's revenue is $x \cdot p$. In most cases, there is a relationship between x and p. If the price per item is low, the seller may be able to sell a large number of items. But as the seller raises the price, the demand for the item goes down. In applications it is often assumed that demand is linearly related to price. This relation may be expressed by an equation like $x = 180 - 3p$. If we let R stand for revenue, then $R = x \cdot p = (180 - 3p)p$, or $R = 180p - 3p^2$. R is a quadratic function of p, with $a = -3$, $b = 180$, and $c = 0$. Assuming that demand is a linear function of price means that revenue is a quadratic function of the price.

Example 1 (*Compare Exercise 1*)
The number of television sets a store can sell in a month, x, is related to the price, p, by the linear equation $x = 200 - \frac{1}{5}p$.

(a) How many sets will the store sell if the price is $250 per set?
(b) What will be the revenue from TV sales if the price is $250 per set?
(c) Express the revenue R as a function of p.

Solution **(a)** If $p = 250$, then

$$\begin{aligned} x &= 200 - (\tfrac{1}{5})(250) \\ &= 200 - 50 = 150 \end{aligned}$$

The store will sell 150 sets.

(b) 150 sets at \$250 per set produces a revenue R of $(150)(250) = 37{,}500$ dollars per month.

(c) In general, $R = x \cdot p$. Substituting $x = 200 - \frac{1}{5}p$ gives

$$R = [200 - \tfrac{1}{5}p] \cdot p = 200p - \tfrac{1}{5}p^2$$

The manager in charge of television sales may be interested in knowing what price will produce the largest revenue. In order to find this price, we need to look at the general shape of parabolas. The graph of $f(x) = ax^2 + bx + c$, a parabola, has the general shape of a bowl that either opens upward or downward. If $a > 0$, the parabola opens upward; if $a < 0$ it opens downward (Figure 1–13).

The vertex is the point where the parabola changes direction. If $a > 0$, the vertex is the lowest point on the graph; if $a < 0$, the vertex is the highest point on the graph. The first coordinate of the vertex is given by $x = -b/(2a)$. You may have seen this result derived in an algebra course by a method called "completing-the-square." Later, you will be able to derive this result easily using calculus.

We could also give a formula for the y-coordinate of the vertex, but the fewer formulas to be memorized the better. Once you have the x-coordinate, $-b/(2a)$, just evaluate $f(-b/(2a))$ to find the y-coordinate.

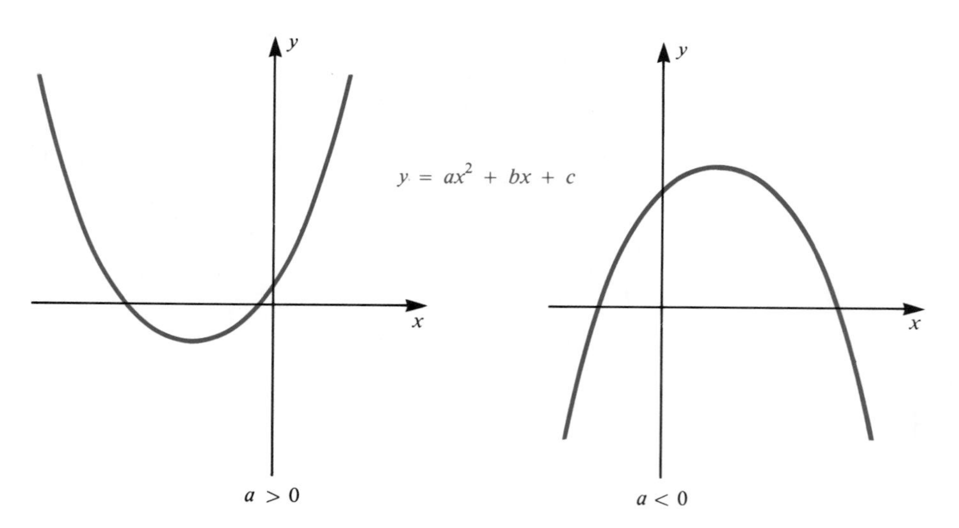

Figure 1–13

Example 2 (*Compare Exercise 7*)
Sketch a rough graph of $f(x) = 2x^2 + 4x + 5$.

Solution Here $a = 2$, $b = 4$, $c = 5$. Because $a > 0$ the parabola opens up. (*Compare Exercise 3*). The x-coordinate of the vertex is $\dfrac{-b}{2a} = \dfrac{-4}{2 \cdot 2} = -1$. The y-coordinate is $f(-1) = 2 - 4 + 5 = 3$. Thus the vertex is the point $(-1, 3)$. We can now sketch a rough graph of this parabola, as shown in Figure 1–14.

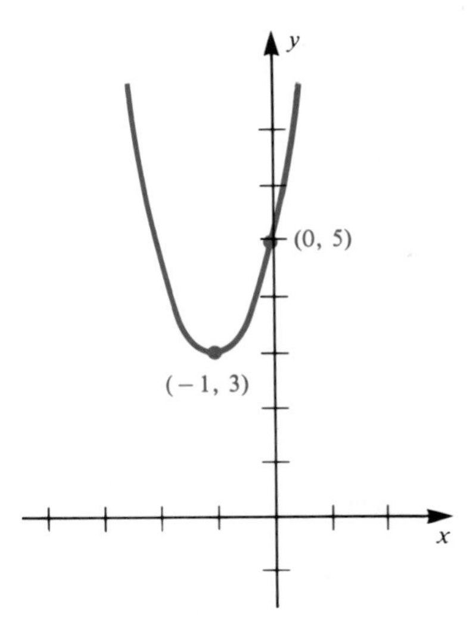

Figure 1–14

If we wanted a more accurate picture we could accurately plot a few more points. Usually when sketching a parabola, you should plot the vertex and the points where the parabola meets the axes. We indicate this requirement for the graph by leaving out the word "rough."

Example 3 (*Compare Exercise 11*)
Sketch the graph of $f(x) = -2x^2 + 4x + 6$.

Solution Here $a = -2$, $b = 4$ and $c = 6$. Since $a < 0$, the parabola opens down. The x-coordinate of the vertex is $-b/(2a) = -4/-4 = 1$; the y-coordinate is $f(1) = 8$. The vertex is the point (1, 8). To find where the parabola crosses the y-axis, we set $x = 0$ and evaluate $f(0)$; $f(0) = 6$. The point (0, 6) is on the graph. In general, the y-intercept is simply the constant term c. Finding the x-intercepts is not as easy. This time we know the value of y and are looking for the value of x that satisfies $f(x) = y$. In this instance, we want $y = 0$ and need to solve

$$
\begin{aligned}
-2x^2 + 4x + 6 &= 0 && \text{Divide by } -2 \\
x^2 - 2x - 3 &= 0 && \text{Now factor} \\
(x - 3)(x + 1) &= 0 \\
x = 3 \text{ and } x &= -1
\end{aligned}
$$

The parabola meets the x-axis at the points $(-1, 0)$ and $(3, 0)$. We can now draw Figure 1–15.

Warning! When solving the equation $f(x) = 0$, we were able to employ the technique of dividing both sides of the equation by -2. You are **not** allowed to divide the function by -2. The function $y = -2x^2 + 4x + 6$ is **not** the same as the function $y = x^2 - 2x - 3$. There is a difference between **solving** an equation and writing the rule for **evaluating** the function.

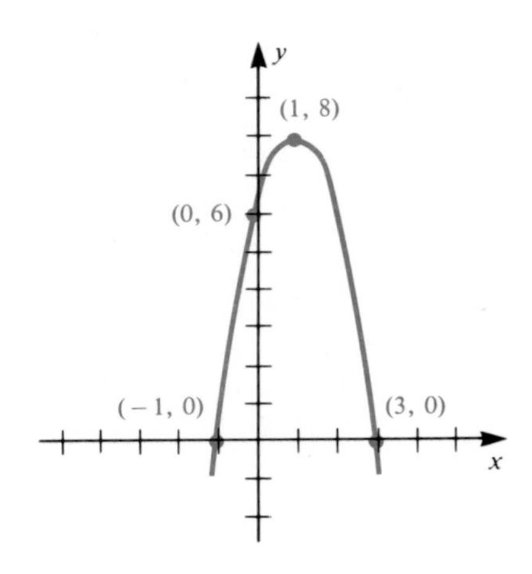

Figure 1–15

In general, the solutions to the equation $f(x) = 0$ are called **the zeros of the function**. So, using the last example, the zeros of the function $f(x) = -2x^2 + 4x + 6$ are the numbers $x = -1$ and $x = 3$. If you cannot factor $ax^2 + bx + c$, you can find the solutions to $ax^2 + bx + c = 0$ by using the quadratic formula, which gives the solution in the form

$$x = \frac{-b \pm \sqrt{b^2 - 4ac}}{2a}$$

Example 4 (*Compare Exercise 17*)
Let $f(x) = 2x^2 - 8x - 5$.

(a) What are the zeros of f?
(b) Draw the graph of f.

Solution **(a)** We use the quadratic formula to find the zeros with $a = 2$, $b = -8$ and $c = -5$.

$$x = \frac{-(-8) \pm \sqrt{(-8)^2 - 4(2)(-5)}}{2(2)}$$

$$= \frac{8 \pm \sqrt{64 + 40}}{4}$$

$$= \frac{8 \pm \sqrt{104}}{4} \qquad \sqrt{104} = \sqrt{4 \cdot 26} = 2\sqrt{26}$$

$$= \frac{8 \pm 2\sqrt{26}}{4} \qquad \text{Divide numerator and denominator by 2}$$

$$= \frac{4 \pm \sqrt{26}}{2}, \qquad \text{or}$$

$$= 2 \pm \frac{\sqrt{26}}{2}$$

The zeros of f are $2 + \dfrac{\sqrt{26}}{2}$ and $2 - \dfrac{\sqrt{26}}{2}$.

(b) The x-coordinate of the vertex is $-b/(2a) = -(-8)/(2 \cdot 2) = 2$. The y-coordinate is $f(2) = 2(2)^2 - 8(2) - 5 = -13$. The vertex is the point $(2, -13)$. $f(0) = -5$, so the y-intercept is -5, and the point $(0, -5)$ is on the graph.

The x-intercepts are $2 + \sqrt{26}/2$ and $2 - \sqrt{26}/2$. The parabola crosses the x-axis at the points $(2 + \sqrt{26}/2, 0)$ and $(2 - \sqrt{26}/2, 0)$.

The graph is shown in Figure 1–16.

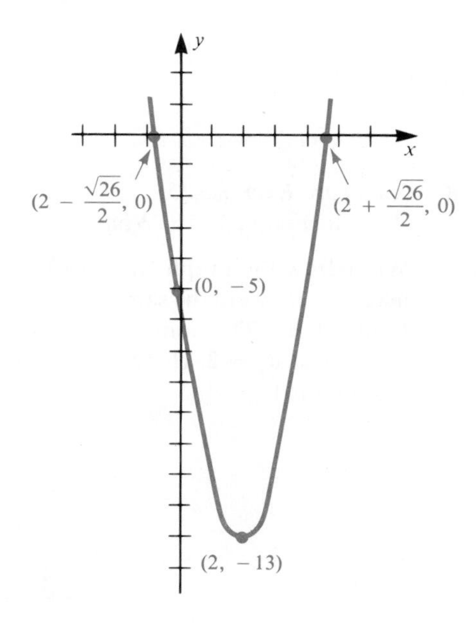

Figure 1–16

Notice that the x-intercepts are

$$\frac{-b}{2a} + \frac{\sqrt{b^2 - 4ac}}{2a}$$

and

$$\frac{-b}{2a} - \frac{\sqrt{b^2 - 4ac}}{2a}$$

The x-intercepts are the same distance from $-b/(2a)$, the first coordinate of the vertex. In Example 3, the intercepts were 2 units away from 1 ($1 + 2 = 3$ and $1 - 2 = -1$). In Example 4, the intercepts were $\sqrt{26}/2$ units away from 2. This is an example of symmetry. Parabolas are symmetric about the vertical line through the vertex. This means that at any height, the distances d_1 and d_2 in Figure 1–17 are the same.

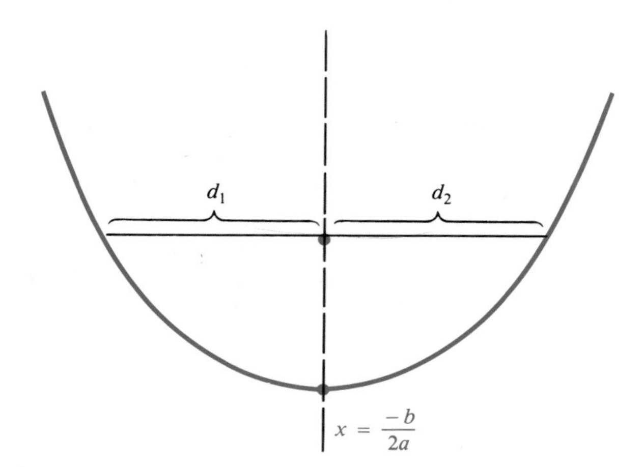

Figure 1–17

Example 5 (*Compare Exercise 15*)
Refer to Figure 1–16. What is the other point whose second coordinate is -5?

Solution We redraw the graph, and label the other point P. P and $(0, -5)$ are the same height so they are the same distance from the vertical line through the vertex. See Figure 1–18. The point $(0, -5)$ is 2 units to the left of the vertical line through the vertex; $d_1 = 2$. Hence $d_2 = 2$ also. P is 2 units to right of the vertical line; $P = (4, -5)$.

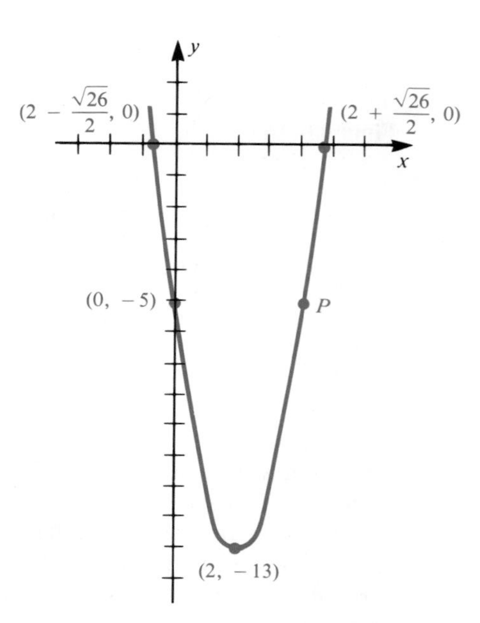

Figure 1–18

Example 6 (*Compare Exercise 18*)
Let $f(x) = 3x^2 - x + 6$. What are the zeros of f?

Solution Again use the quadratic formula, this time with $a = 3$, $b = -1$ and $c = 6$.

$$x = \frac{-(-1) \pm \sqrt{(-1)^2 - 4(3)(6)}}{2 \cdot 3}$$
$$= \frac{1 \pm \sqrt{1 - 72}}{6}$$
$$= \frac{1 \pm \sqrt{-71}}{6}$$

Since $\sqrt{-71}$ is not a real number, the graph of $y = 3x^2 - x + 6$ never crosses the x-axis. The function has no real zeros.

We now look at a particular application of graphing.

Lorenz Curves

Economists, sociologists, and political scientists are especially interested in the distribution of wealth within a given country. The curves used to study this distribution are called **Lorenz curves**; they are the graphs of **wealth distribution functions**. This general class of curves can also be used to measure other concentrations of wealth or power, but let us deal with a specific example before explaining their general use.

We start by explaining what a wealth distribution function is. There are various measures of wealth; our example will deal with gross family income. For every number x in $[0, 1]$, define $W(x)$ to be the percentage of a country's total income earned by the lower $100x\%$ of the families when families are ranked by income. Thus, $W(.3) = .2$ would mean that the lower 30% of families (in terms of income) earn 20% of the nation's total income. Every such function must satisfy two conditions:

$W(0) = 0$ (the lower 0% of population earn 0% of the income) and
$W(1) = 1$ (100% of the population accounts for 100% of the income).

Thus, every Lorenz curve goes through the origin, (0, 0), and the point (1, 1). The graph of a typical Lorenz curve looks like that shown in Figure 1–19.

If income were distributed perfectly equally, then the lower $100x\%$ of the population would account for $100x\%$ of the income. The wealth distribution function would be $W(x) = x$. The graph of this function is called the line of perfect equality, and its graph is shown in Figure 1–20.

For purposes of comparison, economists often draw the line of perfect equality on the same graph as the Lorenz curve for a specific country, as in Figure 1–21. The area of the shaded region has particular economic significance. We will show how to compute this area and talk more about its significance in Chapter 6.

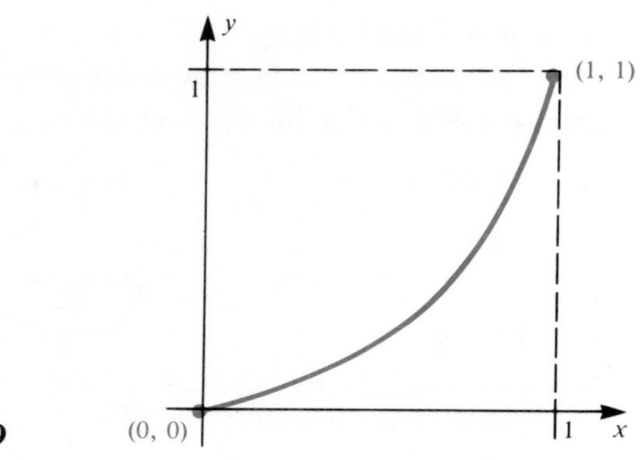

Figure 1–19

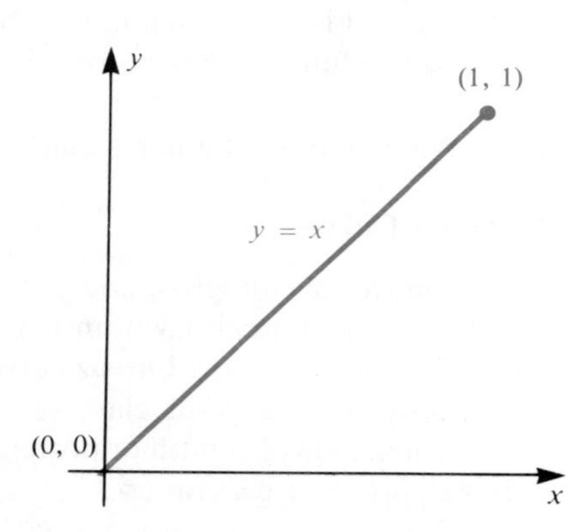

Figure 1–20

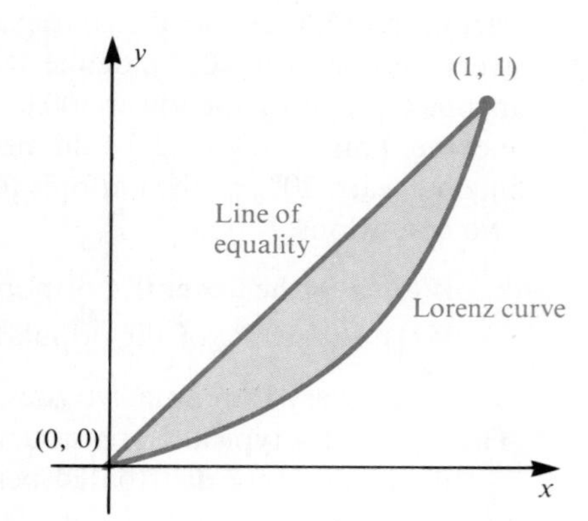

Figure 1–21

We look now at a Lorenz curve and show you how to read its graph in two ways.

Example 7 (*Compare Exercise 35*)
The Lorenz curve for a certain country can be approximated by

$$W(x) = .75x^2 + .25x$$

(a) What percent of total income is earned by the lower 20% of the families?
(b) What percent of total income is earned by the lower 50% of the families?
(c) Solve $W(x) = .5$ and interpret.

Solution **(a)** Compute $W(.2)$.

$$W(.2) = .75(.04) + (.25)(.2)$$
$$= .03 + .05 = .08$$

The lowest 20% of the families earn 8% of total income.

(b) $W(.5) = (.75)(.25) + (.25)(.5) = .3125$. The lower 50% of the families earn 31.25% of the total income.

(c) To solve $W(x) = .5$, set

$$.75x^2 + .25x = .5 \qquad \text{Multiply by 100}$$
$$75x^2 + 25x = 50 \qquad \text{Divide by 25}$$
$$3x^2 + x = 2$$
$$3x^2 + x - 2 = 0$$
$$(3x - 2)(x + 1) = 0$$
$$x = \frac{2}{3}, x = -1$$

Reject $x = -1$ because $x = -1$ is not in the domain of W.

Thus, $W(\frac{2}{3}) = .5$. Since $\frac{2}{3} = 66\frac{2}{3}\%$, we have the following interpretation: The lowest $66\frac{2}{3}\%$ of families account for 50% of the total income.

We look at the results of Example 7 geometrically. The graph of $y = .75x^2 + .25x$ is shown in Figure 1–22. The arrows indicate the two different ways the graph is used. To answer questions (a) and (b), you start with a value of x and find the corresponding value of y. To answer question (c), you start with the value of $y = .5$ and follow the arrow to the correct value of x.

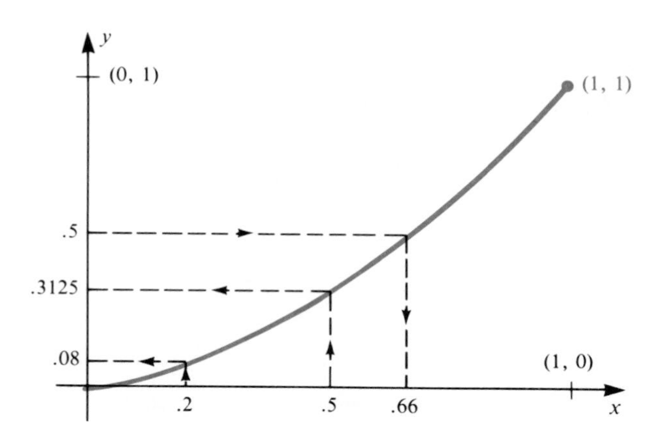

Figure 1–22

Lorenz curves are also used to measure other concentrations of wealth or power. For example, you can determine how monopolistic a given industry is by defining a monopoly function:

$M(x) = y$ if the lower $100x\%$ of the companies in the industry produce $100y\%$ of the total output of that industry.

Higher Degree Polynomials

Generally speaking, functions that arise in applications are constructed to approximate some data. The person choosing the function to describe the real situation should know the general behavior of different types of functions. The function chosen shouldn't have properties not found in the application. For example, here is a table giving data for family incomes in Brazil in 1972.*

Percentile of Population	Percentile of Income
0	0
20	2
40	7
60	16.4
80	33.4
100	100

Figure 1–23 shows these data points plotted, and also shows the parabola that best approximates these points. ("Best" here is meant in a technical sense

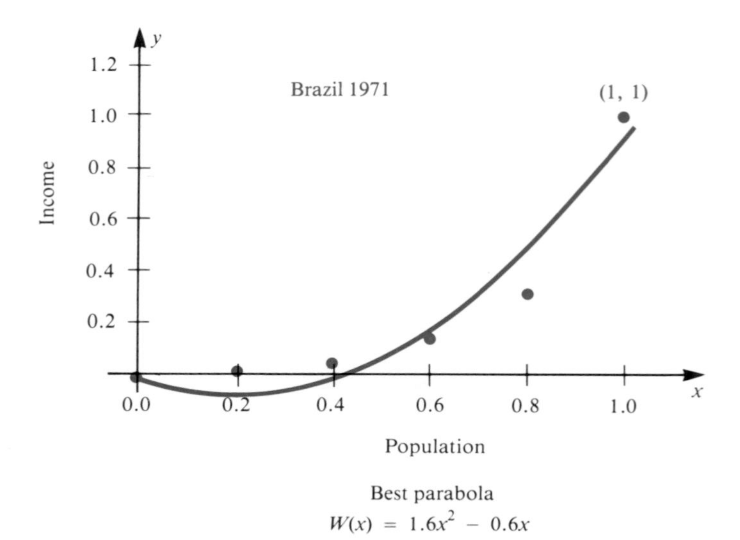

Figure 1–23

* These percentages were taken from W. van Ginneken and J. Park, eds., *Generating Internationally Comparable Income Distribution Estimates* (Geneva: International Labour Office, 1984).

from advanced statistics.) You don't need to worry about how this parabola was constructed, but it obviously has at least one undesirable characteristic. The graph goes below the x-axis. It looks as if the lower 20% of the families accounted for a negative percentage of the total income! Compare with Figure 1–24, which shows the best fit to the data by a third-degree polynomial, and Figure 1–25 which shows the best fit by a fourth-degree polynomial. These curves seem to describe reality better. Different situations may call for different types of functions.

We will deal with the general problem of graphing polynomials of degree greater than two in Chapter 4. Here we simply ask you to plot some points, and then draw a "smooth" curve through them.

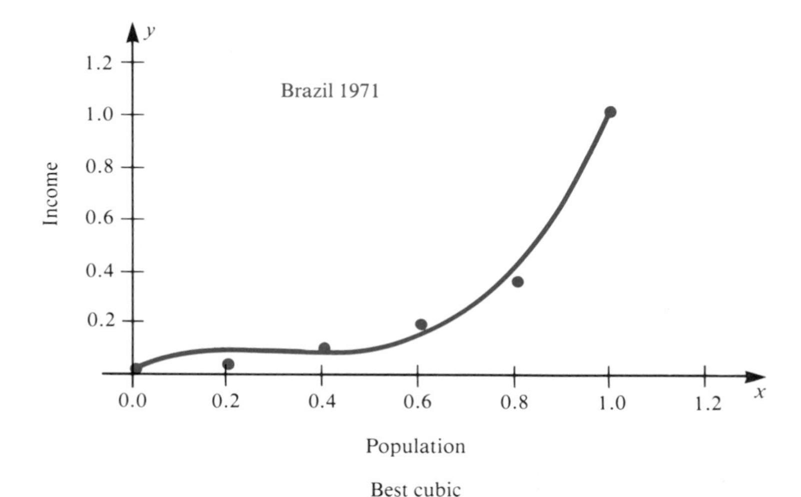

Figure 1–24 $W(x) = 2.7x^3 - 2.4x^2 + .7x$

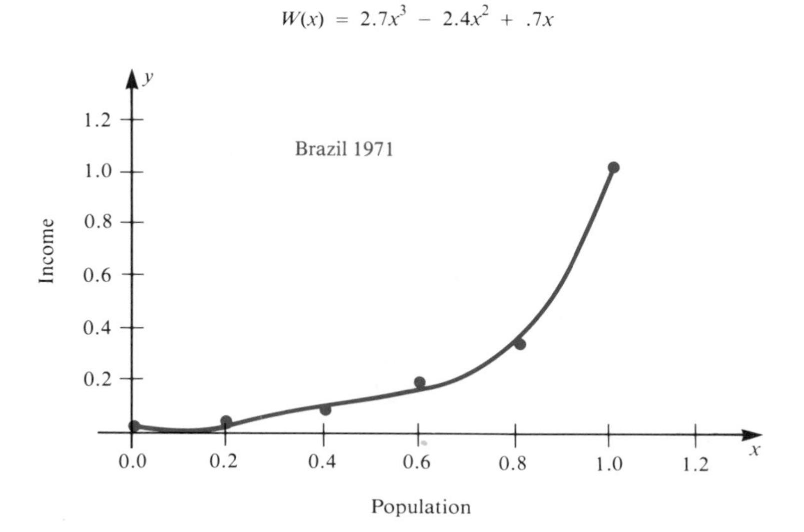

Figure 1–25 $W(x) = 5.3x^4 - 7.7x^3 + 3.9x^2 - .5x$

Example 8 (*Compare Exercise 25*)
Sketch the graph of $y = x^3 - 6x^2 + 9x + 2$

Solution We make a table of values

x	-1	0	1	2	3	4
$f(x) = x^3 - 6x^2 + 9x + 2$	-14	2	6	4	2	6

A sketch of the curve using these points is found in Figure 1–26.

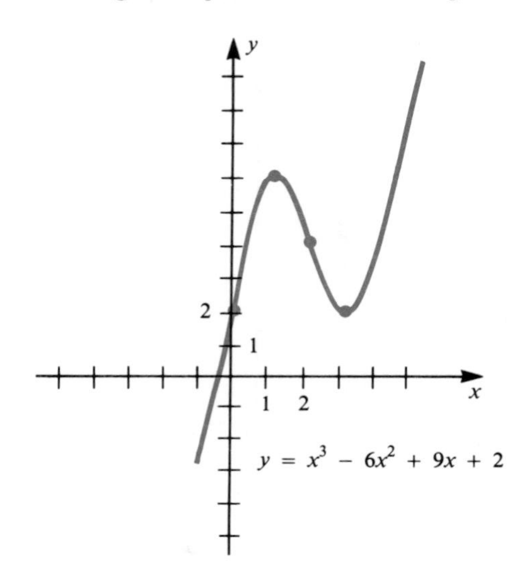

Figure 1–26

Functions Defined Piecewise

Sometimes the rule for evaluating a function depends on the value of x. A given function may have different rules for computing with different numbers. Such functions often arise in business applications because, for example, so many businesses give discount rates for large bulk orders.

Example 9 (*Compare Exercise 27*)
A print shop makes copies at a rate of 5¢ per page if twelve hundred or fewer copies are ordered. If more than twelve hundred copies are ordered, the charges drop to 4¢ per page. Describe the cost function and draw its graph.

Solution Let x be the number of hundreds of copies ordered, and let $C(x)$ be the cost in dollars of x pages. We write the rule for $C(x)$ as

$$C(x) = \begin{cases} 5x \text{ if } 0 \leq x \leq 12 \\ 4x \text{ if } x > 12 \end{cases}$$

Thus, if $x = 8$, then $C(8) = 5 \cdot 8 = 40$; the charge for 800 copies is \$40. If $x = 20$, we use the rule $C(20) = 4 \cdot 20 = 80$. The charge for 2000 copies is \$80.

The graph of C is given in Figure 1–27. Notice that the scale used on the y-axis is different from the scale used on the x-axis. This is often necessary in applications.

This graph is another example where in reality the application would require that x can only have a finite number of values, but we draw the graph through other points to get a better picture. Note that the second piece of the graph has an empty dot for its left-hand point; this signifies that the point, which is (12, 48), is not on the graph. The solid dot above it at (12, 60) indicates that (12, 60) is on the graph.

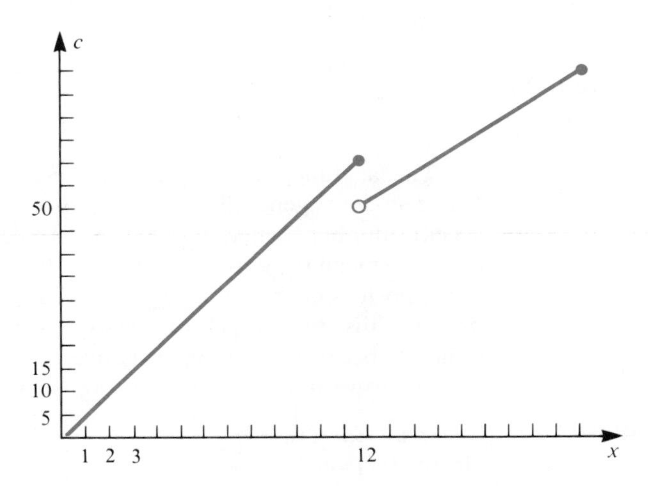

Figure 1–27

This notation is also used in the next example which shows that one function may have several "pieces" to its graph.

Example 10 A cosmetics company has studied public awareness of its product. The awareness was measured on a scale of 0 to 5, with 0 indicating that consumers were not aware of the product and 5 meaning that consumers were very aware of the product. The company found that after running a TV commercial, it could put general consumer awareness at 5, but that awareness dropped off linearly. Four days after the commercial, consumer awareness was at the 3 level. The company decided to run the commercial every four days. If t measures time in days after the original commercial, and $A(t)$ is consumer awareness, the company found that the awareness function was given by

$$A(t) = \begin{cases} -\frac{1}{2}t + 5, 0 \leq t < 4 \\ -\frac{1}{2}t + 7, 4 \leq t < 8 \\ -\frac{1}{2}t + 9, 8 \leq t < 12 \end{cases}$$

Graph the awareness function.

Solution The graph is assembled in pieces. First we graph the line segment $A(t) = -\frac{1}{2}t + 5$ that is above the interval $0 \leq t \leq 4$. Then, because this rule is only used for $0 \leq t < 4$ in the actual graph, we "knock off" the endpoint (4, 3) of this segment by drawing an empty dot for this point. Next we draw the line segment $A(t) = -\frac{1}{2}t + 7$ above $4 \le t \le 8$, and repeat the empty dot drawing process for the point (8, 3). Repeating this procedure we obtain Figure 1–28.

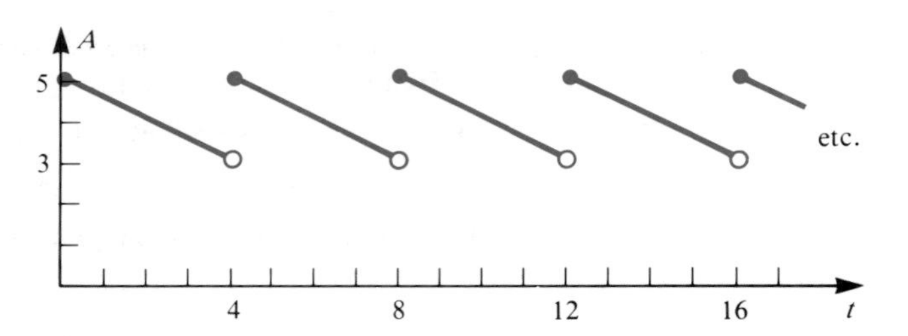

Figure 1–28

The last function we look at in these examples is the **absolute value function**. While students generally have no problem evaluating the absolute value of any specific number, sometimes the general rule for evaluation does give trouble. We use $|x|$ to mean the **absolute value of** $\boldsymbol{x}$. If $x \ge 0$, then $|x|$ is x. If $x < 0$, then $|x|$ has the opposite sign from x. The way to change the sign of a number is to multiply it by -1. Thus, if $x < 0, |x| = (-1) \cdot x = -x$. This is the part of the rule that causes difficulty because $-x$ looks negative—it has that minus sign in front of it. But remember we started with "if $x < 0$," so x itself is negative, and $-x$ is positive!

Example 11 (*Compare Exercise 30*)
Graph $f(x) = |x|$

Solution We have just seen that we can write the rule for f by

$$f(x) = \begin{cases} x & \text{if} \quad x \ge 0 \\ -x & \text{if} \quad x < 0 \end{cases}$$

The graph is given in Figure 1–29. Note that the "two pieces" meet to form one piece; the point (0, 0) does not deserve an empty dot because it is on the graph.

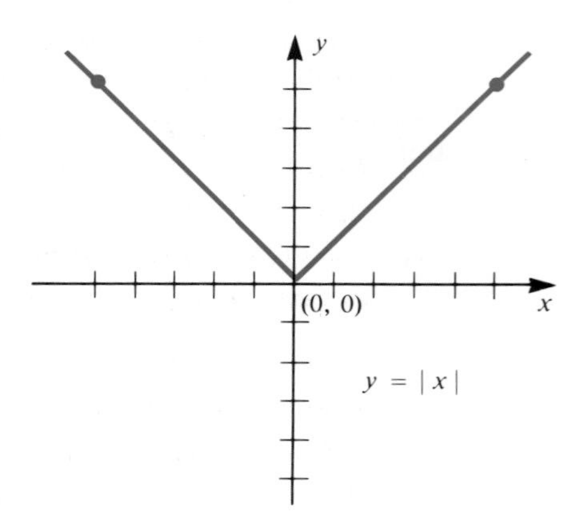

Figure 1–29

1–4 Exercises

I.

1. (*See Example 1*) If $R = x \cdot p$ and $x = 300 - (\frac{1}{6})p$, express R as a function of p.
2. If $R = x \cdot p$ and $x = 400 - p^2$, express R as a function of p.
3. (*See Example 2*) If $y = 5x^2 - 9x + 2$, does the parabola open down or up?
4. If $y = -2x^2 + 10x + 3$, does the parabola open down or up?
5. If $y = 5 + 4x - 6x^2$, does the parabola open down or up?
6. If $y = -8x - 7 + 2x^2$, does the parabola open down or up?
7. (*See Example 2*) What is vertex of the parabola $y = x^2 - 6x + 7$?
8. What is the vertex of the parabola $y = -2x^2 + 8x + 9$?
9. What is the vertex of the parabola $y = x^2 + 4x - 7$?
10. What is the vertex of the parabola $y = -3x^2 - 12x + 10$?

II.

11. (*See Example 3*) Sketch the graph of $f(x) = x^2 - 4x + 5$.
12. Sketch the graph of $y = 9 - x^2$.
13. Sketch the graph of $y = 3 - 2x - x^2$.
14. Sketch the graph of $y = 2x^2 - 3x - 2$.
15. (*See Example 5*) A certain parabola has vertex (3, 7). The point (5, 13) is on the parabola. What other point on the parabola has 13 for its y-coordinate?
16. A certain parabola has vertex $(-3, -5)$. The point (1, 11) is on the parabola. What other point on the parabola has 11 for its y-coordinate?
17. (*See Example 4*) What are the zeros of $f(x) = 6x^2 + 7x - 20$?
18. (*See Example 6*) What are the zeros of $f(x) = 3x^2 + 5x + 7$?
19. What are the zeros of $f(x) = 4 + 8x - 3x^2$?
20. What are the zeros of $f(x) = -5x^2 + x + 2$?
21. What are the zeros of $f(x) = x^2 + 2x + 5$?
22. What are the zeros of $f(x) = x^2 + x - 5$?

III.

Graph each of the functions defined in Exercises 23 through 34.

23. $f(x) = x^2 - x - 7$
24. $f(x) = x^2 + 2x - 5$
25. (*See Example 8*) $f(x) = x^3 - 3x^2 + 2x$
26. $f(x) = x^3 + 3x^2 - 9x + 1$

27. (*See Example 9*) $f(x) = \begin{cases} 2x + 1, & x \le 1 \\ 3 - x, & x > 1 \end{cases}$

28. $f(x) = \begin{cases} x^2, & x \le 2 \\ 4, & x > 2 \end{cases}$

29. $f(x) = \begin{cases} 3x + 1, & x < 2 \\ 9 - x, & x \ge 2 \end{cases}$

30. (*See Example 11*) $f(x) = |x - 3| = \begin{cases} x - 3, & x - 3 \ge 0 \\ -(x - 3), & x - 3 < 0 \end{cases}$

31. $f(x) = |x + 5|$

32. $f(x) = |2x + 1|$

33. $f(x) = \sqrt{x^2}$

34. $f(x) = \sqrt{(x - 1)^2}$

35. (*See Example 7*) A certain country's Lorenz curve is given by

$$W(x) = (\tfrac{2}{3})x^2 + (\tfrac{1}{3})x$$

(a) What percent of the total income does the lower 30% of the families account for?

(b) What percent of the total income does the upper 10% of the families account for?

(c) Solve $W(x) = \frac{1}{3}$ and interpret.

36. A certain country's Lorenz curve is given by

$$W(x) = .8x^2 + .2x$$

(a) What percent of the total income does the lower 50% of the families account for?

(b) What percent of total income does the upper 75% of the families account for?

(c) The lower x percent of the families account for 60% of the income. Solve for x.

37. If $R = x \cdot p$ and $x = 300 - \frac{1}{6}p$, express R as a function of x.

38. If $R = x \cdot p$ and $x = 400 - p^2$, express R as a function of x.

1–5 Inequalities: The Cut-Point Method

We are able to solve linear inequalities in a fairly straightforward manner. The steps involved are the same as those for solving an equation, with one important exception. When we have to multiply or divide both sides of the inequality by some number, what we do with the inequality sign depends on whether the number we're using is positive or negative. In solving more general inequalities, we may have to multiply by a variable, and that can cause trouble because we don't know if the variable represents a positive or a negative number. Let's look at the potential trouble by trying to solve the inequality:

$$\frac{1}{x} + 6 \leq 9$$

A first step would be to subtract 6 from both sides:

$$\frac{1}{x} \leq 3$$

A second step would be to multiply both sides by x:

$$1 \leq 3x$$

Lastly, divide both sides by 3

$$\frac{1}{3} \leq x$$

Warning! These steps are valid for solving the **equation** $\frac{1}{x} + 6 = 9$ but are **NOT** valid for solving the **inequality**. Before giving the technical reason why these steps do not give the right solution to the inequality, we point out how **the answer is wrong**.

If x is some number larger than $\frac{1}{3}$ (5 for example) then $\frac{1}{x} + 6$ is in fact less than 9; $\frac{1}{5} + 6 = 6\frac{1}{5}$, and $6\frac{1}{5} < 9$ is true. What's wrong is that these steps overlooked some solutions, -2 for example. Note that -2 should be in the answer set because $\frac{1}{-2} + 6 = 5\frac{1}{2}$ is less than 9.

Now that we know the answer is wrong, let's see where the trouble is. There is no difficulty with the first step—if you add or subtract the same number on both sides of an inequality, the inequality remains valid. It's the second step that contains the trouble spot. Multiplication on both sides of an inequality has to be handled carefully. The rule states that starting with $a < b$ we get $ac < bc$ **if c is positive**, but we get $ac > bc$ **if c is negative**. For example, if we start with $2 < 4$ and multiply by 3, we get $6 < 12$. But if we start with $2 < 4$ and multiply by -3, we get $-6 > -12$. The inequality does not stay the same; it must reverse direction.

Trying to solve the inequality $\frac{1}{x} + 6 \leq 9$ by worrying about the two cases when $x > 0$ and when $x < 0$ can be cumbersome. Analyzing each case and putting them all back together correctly can be *very* difficult with more complicated inequalities. Yet solving inequalities is very important, both as a step in solving a larger problem (Chapter 4 is full of such problems) and as an end in themselves (for example, knowing what prices keep revenue > expenses is important). The method we will use to solve inequalities requires calculus for *its justification* but requires no calculus for its use. We introduce the method now rather than later because the emphasis then will be on techniques of calculus. We choose this process because we recognize that it is easier for people to solve equations than to solve inequalities by cases.

The procedure consists of 6 steps:

1. Change the inequality to equality, and solve this equation. Call each solution **a solution cut point**.
2. Determine all numbers c such that when c is substituted into the inequality, there is an expression that can't be computed (c is not in the "domain" of the inequality). Call each such c **a domain cut point**.
3. A number obtained from Step 1 or Step 2 is called a cut point. Plot all the cut points on the number line. These points are called cut points because they cut what remains of the line into intervals.
4. Each interval between successive cut points is either completely in the solution set or completely outside the solution. To see which is the case, pick a test number from each such interval, and see if this particular number satisfies the original inequality.
5. Test each solution cut point in the original inequality. Notice that by the very way that they are defined, the domain cut points cannot possibly be solutions of the inequality.
6. Write down the answer using the results from Steps 4 and 5.

General procedures like these are always clearer after seeing them in action.

Example 1 (*Compare Exercise 17*)

Solve $\dfrac{1}{x} + 6 \leq 9$.

Solution

STEP 1 Change to $\dfrac{1}{x} + 6 = 9$, and solve. Copying the steps used above, we find that $x = 1/3$ is a solution cut point.

STEP 2 We cannot substitute $x = 0$ into the inequality because $1/0$ cannot be computed; $1/0$ is not a number. Thus, $x = 0$ is a domain cut point.

STEP 3 There are two cut points, 0 and $1/3$.

STEP 4 There are three intervals determined by cutting the line at 0 and $\frac{1}{3}$. The intervals are $(-\infty, 0)$, $(0, \frac{1}{3})$, and $(\frac{1}{3}, \infty)$. We must pick a number from each interval and then test that number in the inequality $\dfrac{1}{x} + 6 \leq 9$.

(a) Pick a number in $(-\infty, 0)$, say -1, and test this number in the inequality. With $x = -1$, $\dfrac{1}{x} + 6 \leq 9$ becomes the statement $\frac{1}{-1} + 6 \leq 9$, which is true. Therefore, all the numbers in $(-\infty, 0)$ satisfy the inequality.

(b) Pick a number in $(0, \frac{1}{3})$, say $\frac{1}{5}$, and test this number in the inequality. With $x = \frac{1}{5}$, $\dfrac{1}{x} + 6 = 5 + 6 = 11$, and $11 \leq 9$ is false. Therefore, no number in $(0, \frac{1}{3})$ satisfies the inequality.

(c) Pick a number in $(\frac{1}{3}, \infty)$, say 8, and test this number in the inequality. With $x = 8$, $\dfrac{1}{x} + 6 = \frac{1}{8} + 6$ and $6\frac{1}{8} \leq 9$ is true. Therefore, all the numbers in $(\frac{1}{3}, \infty)$ satisfy the inequality.

STEP 5 Test the cut points in the inequality.

(a) $\frac{1}{3}$ is in the solution because with $x = \frac{1}{3}$, $\dfrac{1}{x} + 6 = 3 + 6 = 9 \leq 9$ is true.

(b) 0 is not in the solution (a cut point of this type can never be in the solution).

STEP 6 x is a solution if x is in $(-\infty, 0)$ or in $[\frac{1}{3}, \infty)$.

Example 2 (*Compare Exercise 9*)
Solve the inequality $x^2 + 5x - 6 \leq 0$.

Solution

STEP 1 Change to $x^2 + 5x - 6 = 0$, and solve.

$$x^2 + 5x - 6 = 0$$
$$(x + 6)(x - 1) = 0$$

There are two solutions: $x = -6$ and $x = 1$.

STEP 2 This inequality can be evaluated no matter what number is substituted for x. There are no cut points from this step.

STEP 3 There are two cut points, $x = -6$ and $x = 1$.

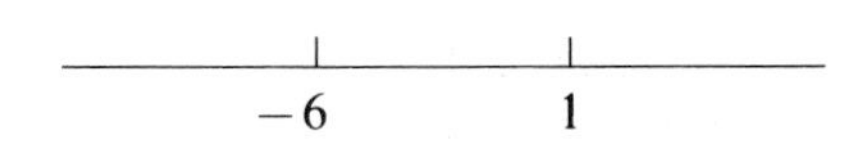

STEP 4 **(a)** Test a point from $(-\infty, -6)$, say -10, in the inequality:

$$(-10)^2 + 5(-10) - 6 = 44 \leq 0$$

is false. This interval is not in the solution.

(b) Test a number in $(-6, 1)$, say 0, in the inequality: $0^2 + 5 \cdot 0 - 6 = -6 \leq 0$ is true, so the interval $(-6, 1)$ is in the solution.

(c) Test a number bigger than 1, say 2, in the inequality: $4 + 10 - 6 = 8 \leq 0$ is false. No number bigger than 1 is in the solution.

STEP 5 Test -6 and 1. From Step 2 we know that they make the left-hand side equal to 0, and $0 \leq 0$ is true. Both -6 and 1 are solutions.

STEP 6 The answer is all x in $[-6, 1]$.

Example 3 (*Compare Exercise 19*)

Solve $\frac{1}{2x-6} \le \frac{1}{x+2}$.

Solution

STEP 1 Solve $\frac{1}{2x-6} = \frac{1}{x+2}$. Cross multiply to get

$$x + 2 = 2x - 6 \qquad \text{so}$$
$$8 = x$$

STEP 2 Determine all x not in the domain of the inequality:
We cannot have $2x - 6 = 0$, so $x = 3$ is a cut point.
We cannot allow $x + 2 = 0$, so $x = -2$ is a cut point.

STEP 3

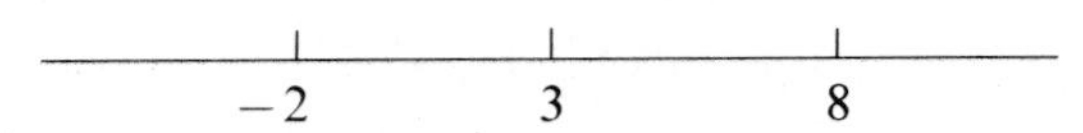

STEP 4 **(a)** Test -5 (or any number less than -2)

$$\frac{1}{-10-6} < \frac{1}{-5+2}$$
$$-\frac{1}{16} < -\frac{1}{3} \quad \text{False.}$$

(b) Test 0 (or any number between -2 and 3)

$$-\frac{1}{6} \le \frac{1}{2} \quad \text{True: } \mathbf{(-2, 3)} \textbf{ is in the solution}$$

(c) Test 4 (or any number between 3 and 8)

$$\frac{1}{8-6} \le \frac{1}{4+2}$$
$$\frac{1}{2} \le \frac{1}{6} \quad \text{False}$$

(d) Test 10 (or any number bigger than 8)

$$\frac{1}{14} \le \frac{1}{12} \quad \text{True: } \mathbf{(8, \infty)} \textbf{ is in the solution}$$

STEP 5 The solution cut points from Step 1 give equality, and this inequality includes the equality, so $x = 8$ is a solution. Again, the domain cut points from Step 2 cannot possibly be solutions.

STEP 6 x is a solution if x is in $(-2, 3)$ or in $[8, \infty)$.

The next example gives an inequality of a particular type you may encounter in Chapter 4.

Example 4 (*Compare Exercise 25*)
Solve

$$\sqrt{2x+12} + \frac{x}{\sqrt{2x+12}} \le 0$$

Solution

STEP 1 Solve

$$\sqrt{2x+12} + \frac{x}{\sqrt{2x+12}} = 0 \qquad \text{Multiply by } \sqrt{2x+12}\text{, which gives}$$

$$2x + 12 + x = 0$$
$$3x + 12 = 0$$
$$x = -4$$

STEP 2 Find all x for which

$$\sqrt{2x+12} + \frac{x}{\sqrt{2x+12}}$$

is not defined. We cannot allow $2x + 12$ to be negative because we cannot take the square root of a negative number. Also we cannot let $2x + 12 = 0$ because $\sqrt{0} = 0$, and we cannot divide by 0. Thus, we must have $2x + 12 > 0$, or $x > -6$. Therefore, all numbers less than or equal to -6 are cut points.

STEP 3

Cut points -6 -4

STEP 4
(a) Do not test any number less than -6; they are excluded by Step 2.
(b) Test -5: To see if $\sqrt{2} + -(5/\sqrt{2}) \le 0$ is true or false, multiply by $\sqrt{2}$ ($\sqrt{2}$ is positive; the sign of the inequality remains the same).

$$\sqrt{2}\cdot\sqrt{2} + -5 = 2 - 5 = -3 \le 0$$

is true. **Thus the interval $(-6, -4)$ is in the solution.**
(c) Test 0:

$$\sqrt{12} + \frac{0}{\sqrt{12}} = \sqrt{12} \le 0 \text{ is false.}$$

STEP 5 We know from Step 1 that $x = -4$ gives a solution; $0 \le 0$ is true.

STEP 6 The answer is all x in the interval $(-6, -4]$.

Notice that in Step 2 of Example 4, we had to solve the inequality $2x + 12 > 0$. We did so in the same manner as presented for solving linear inequalities in Chapter 0. You need not make the solution to linear inequalities more lengthy by going through this cut point and testing process. We recommend that you do use this process however for all inequalities except linear inequalities.

1–5 Exercises

I.

Solve each of these linear inequalities without using cut points. Then use the cut point method of this section. Compare answers.

1. $2x - 8 \le 0$
2. $3x + 12 > 0$
3. $5x + 10 < 0$
4. $4x - 24 > 0$
5. $6 - 2x < 0$
6. $15 - 3x > 0$
7. $8x + 3 \le 2x - 9$
8. $3x - 18 \le 5x + 8$

II.

Use the cut point method to solve the following inequalities:

9. (*See Example 2*) $x^2 - x - 6 \le 0$
10. $x^2 + 2x - 3 \ge 0$
11. $2x^2 - 7x - 4 > 0$
12. $2x^2 - 3x - 5 < 0$
13. $4 - x^2 < 0$
14. $9 - x^2 \ge 0$
15. $x^2 + 2x \le 8$
16. $x^2 + x > 12$
17. (*See Example 1*) $\dfrac{2}{x+5} \le 6$
18. $\dfrac{1}{x+7} \le 9$
19. (*See Example 3*) $\dfrac{1}{3x-6} \le \dfrac{1}{x+2}$
20. $\dfrac{1}{5x-15} \le \dfrac{1}{2x-12}$

III.

Solve the following inequalities using the cut point method:

21. $\dfrac{4}{x} + 1 \le 0$
22. $\dfrac{5}{x-3} \le x + 1$
23. $x^2 - 4x \le 8$
24. $2x^2 + x \le 10$
25. $\sqrt{x+9} + \dfrac{x}{2\sqrt{x+9}} \le 0$
26. $\sqrt{2x-12} + \dfrac{x}{\sqrt{2x-12}} > 0$
27. $(x-8)^{1/3} + \dfrac{x}{3(x-8)^{2/3}} < 0$
28. $(x-1)^{1/3} + \dfrac{x}{3(x-1)^{2/3}} \le 0$
29. A company describes its costs by $C(x) = 40x + 200$ and its revenue by $R(x) = 10x^2 - 150x$, where x is daily production in thousands of units, and $C(x)$ and $R(x)$ are measured in hundreds of dollars. For what x does the company make a profit? That is, for what x is $R(x) > C(x)$?
30. If a company's revenue is given by $R(x) = 20x^2 - 70x$ and its cost by $C(x) = 190x + 280$, where x is the number of hundreds of units manufactured per month, what monthly production output will allow the company to make a profit?

IMPORTANT TERMS

1–1	**Function**	**Domain**	**Range**
	Rule of a function	**Functional value**	
1–2	**Graph**	**Linear function**	**Slope**
	***y*-intercept**	**Point-slope formula**	**Two-point formula**
	Fixed cost	**Variable cost**	**Parallel lines**
1–3	**Cost-volume function**	**Fixed costs**	**Variable costs**
	Unit costs	**Revenue function**	**Break-even analysis**
	Break-even point	**Simple interest**	**Depreciation**
	Book value	**Straight-line depreciation**	**Annual depreciation**
1–4	**Quadratic function**	**Parabola**	**Quadratic formula**
	Zeros of a function	**Lorenz curve**	**Absolute value function**
	Piecewise defined functions		
1–5	**Inequalities**	**Cut points**	**Test points**

REVIEW EXERCISES

1. If $f(x) = (x + 2)/(x - 1)$ and $g(x) = 5x + 3$, find $f(2) + g(3)$.
2. If $f(x) = (x + 5)(2x - 1)$, find the domain of f.
3. Apples cost \$1.20 per pound so the price of a bag of apples is

$$f(x) = 1.2x$$

where x is the weight in pounds and $f(x)$ the purchase price in dollars.

(a) What is $f(3.5)$?

(b) A bag of apples cost \$3.30. How much did it weigh?

4. Tuition and fees charges at a university are given by

$$f(x) = 135x + 450$$

where x is the number of semester hours enrolled and $f(x)$ is the total cost of tuition and fees.

(a) Find $f(15)$.

(b) If a student's bill for tuition and fees was \$2205, in how many semester hours was he enrolled?

5. Write an equation of the function described by the following statements.

(a) The function adds 5 to the input and squares the result.

(b) The function subtracts 4 from the input and takes the square root of the result.

6. If $f(x) = 3x + 5$, what are $f(1 + t)$? $f(a + h)$?
7. If $f(x) = 2x^2 - 4x + 3$, what are $f(4 + D)$? $f(a + q)$?
8. Sketch the graph of
   - **(a)** $f(x) = 2x - 5$.
   - **(b)** $6x + 10y = 30$.
9. Find the slope and y-intercept for the following lines.
   - **(a)** $y = -2x + 3$
   - **(b)** $4y = 5x + 6$
   - **(c)** $6x + 7y + 5 = 0$
10. Find the slope of the line through the following pairs of points.
    - **(a)** $(2, 7)$ and $(-3, 4)$
    - **(b)** $(6, 8)$ and $(-11, 8)$
    - **(c)** $(4, 2)$ and $(4, 6)$
11. Find the equation of the following lines.
    - **(a)** with slope $-\frac{3}{4}$ and y-intercept 5
    - **(b)** with slope 8 and y-intercept -3
    - **(c)** with slope -2 and passing through $(5, -1)$
    - **(d)** with slope 0 and passing through $(11, 6)$
    - **(e)** passing through $(5, 3)$ and $(-1, 4)$
    - **(f)** passing through $(-2, 5)$ and $(-2, -2)$
    - **(g)** passing through $(2, 7)$ and parallel to $4x - 3y = 22$
12. Determine if the following pairs of lines are parallel.
    - **(a)** $7x - 4y = 12$ and $-21x + 12y = 17$
    - **(b)** $3x + 2y = 13$ and $2x - 3y = 28$
13. Is the line through $(5, 19)$ and $(-2, 7)$ parallel to the line through $(11, 3)$ and $(-1, -5)$?
14. A manufacturer has fixed costs of \$12,800 per month and a unit cost of \$36 per item produced. What is the cost function?
15. The cost of producing x bags of Hi-Gro fertilizer per week is given by

    $$C(x) = 3.60x + 2850$$

    - **(a)** What is the cost of producing 580 bags per week?
    - **(b)** If the production costs for one week amounted to \$2648, how many bags were produced?
16. A T-shirt shop pays \$6.50 each for T-shirts. Their weekly fixed expenses are \$675. They sell the T-shirts for \$11.00 each.
    - **(a)** What is their revenue function?
    - **(b)** What is their cost function?
    - **(c)** What is their break-even point?

17. What is the simple interest on a \$4500 loan at 9% interest rate for 18 months?

18. What is the amount due at the end of two years for a \$1500 loan at 7% simple interest?

19. A company purchases a piece of equipment for \$17,500. The useful life is eight years and the scrap value at the end of eight years is \$900.

(a) Find the equation relating book value and the age of the equipment using straight-line depreciation.

(b) What is the annual depreciation?

(c) What is the book value after the fifth year?

20. Graph the following:

(a) $y = x^3 + 2$

(b) $y = 5 + 6x - x^2$

(c) $f(x) = x^2 - 3x - 10$

(d) $f(x) = x^2 + x + 1$

21. Solve the following inequalities:

(a) $\frac{1}{x} - 3 \leq 2$

(b) $x^2 - 4x - 4 \geq 0$

(c) $\frac{9}{\sqrt{x}} - \sqrt{x} > 0$

(d) $\frac{3}{x+1} \leq \frac{5}{x-2}$

2

Topics in this chapter can be applied to:
Production Strategies • Maximizing Profit • Pricing and Sales Volume • Rates of Inflation • Marginal Costs • Consumer Awareness

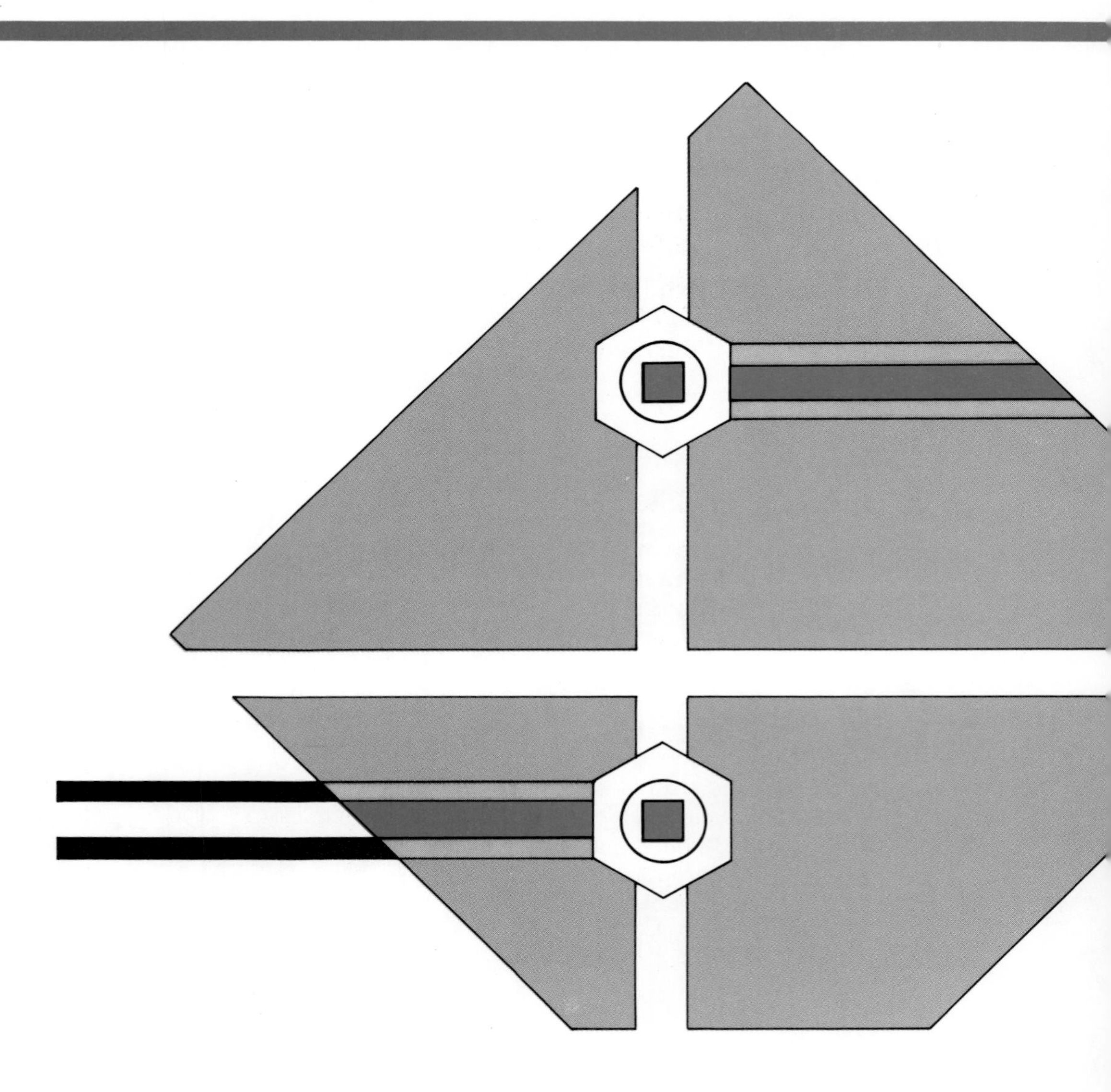

The Derivative

- **Introduction**
- **Rates of Change and the Derivative**
- **Limits**
- **One-Sided Limits and Continuity**
- **Derivatives and Tangent Lines**

2–1 Introduction

The two main concepts in calculus are the derivative and the integral. We will start our study of calculus with the derivative and will introduce the integral later, in Chapter 6.

Western civilization's view of the universe changed dramatically during the 16th and 17th centuries. People began to make careful observations of the physical world and to describe what they observed. They developed new mathematics to help with these descriptions and then further used this new mathematics to make accurate predictions of physical phenomena. The scientific and technological revolution had begun. The last half of the 20th century has seen rapid growth in the use of mathematics by what are called the social and behavioral sciences and by business. As in the physical sciences, mathematics is used both to describe and to predict. Mathematics has become especially important in economics. In fact, since the Nobel Prize in Economics was established in 1968, many of the awards have been given to individuals for their application of mathematics or for their invention of new mathematical methods.

2–2 Rates of Change and the Derivative

Average Rate of Change
Instantaneous Rate of Change
The Derivative

The idea we are about to study, the derivative, is the mathematical tool for talking about rate of change. "Rate of change" may be described by a special word within a particular application. For example, if physicists are describing motion, then they use the word "velocity" for the rate of change of position and "acceleration" for the rate of change of the velocity. In financial applications, people use the words "interest added" to describe the rate of change in their charge accounts (other than changes caused by payments or new charges); economists use "marginal profit" for the rate of change of profit. Not only does the derivative provide the mathematical language for describing rate of change, but we will also see later how the derivative can be used to decide what production strategy will maximize profit, to estimate how a small change in price will affect sales, and to describe how various rates of change are related.

Average Rate of Change

One way of describing change is by talking about the average rate of change over an interval. For example, if a particular stock sold for \$40 a share on Monday morning and sold for \$43.50 the following Friday, then the average daily rate of change of the price of the stock can be computed as follows:

$$\frac{\text{difference in price}}{\text{number of days}} = \frac{3.50}{5} = 0.70.$$

Thus the average rate of change of the stock is \$.70, or 70¢, per day. If the stock sold for \$40 on Monday and \$38 on Friday, the average rate of change would be

$$\frac{38 - 40}{5} = -\frac{2}{5} = -0.40$$

for an average rate of change of $-$\$.40 per day.

When the price of the stock increases in value, the average rate of change is positive; when the price of the stock decreases, the average rate of change is negative.

For describing the average rate of change of an arbitrary function, we make the following definition.

Definition

> The **average rate of change of the function f on the interval $[a, b]$** is
>
> $$\frac{f(b) - f(a)}{b - a}$$
>
> the "change in y" divided by the corresponding "change in x."

Example 1 (*Compare Exercise 3*)
If $f(x) = 3x + 5$, what is the average rate of change of f on

(a) the interval [2, 6]?
(b) the interval [1, 8]?

Solution **(a)** Here $a = 2$ and $b = 6$;

$$\frac{f(6) - f(2)}{6 - 2} = \frac{23 - 11}{4} = \frac{12}{4} = 3$$

(b) Apply the same formula with $a = 1$ and $b = 8$:

$$\frac{f(8) - f(1)}{8 - 1} = \frac{29 - 8}{7} = \frac{21}{7} = 3$$

Both answers are the slope of the graph of the function! This example shows an important principle. For a linear function, the change in y divided by the change in x is independent of the interval.

Theorem

> If the graph of f is a straight line, the average rate of change of f on any interval is the same number—the slope of the straight line.

Example 2 (*Compare Exercise 5*)
If the revenue R is expressed as a function of the price p by $R = 50p - 3p^2$, what is the average change in R if p increases from 4 to 6?

Solution The wording is slightly different from the previous example, but the question is asking for the average rate of change of R on the interval [4, 6].

$$\frac{R(b) - R(a)}{b - a} = \frac{R(6) - R(4)}{6 - 4}$$

$$= \frac{192 - 152}{2}$$

$$= \frac{40}{2} = 20$$

Example 3 (*Compare Exercise 5*)
Using the same revenue function as in Example 2, compute the average change in revenue as the price increases from 10 to 12.

Solution

$$\frac{R(12) - R(10)}{12 - 10} = \frac{168 - 200}{2} = -\frac{32}{2} = -16$$

Note: The graph of this revenue function is not a straight line, and the function will generally have different average rates of change on different intervals.

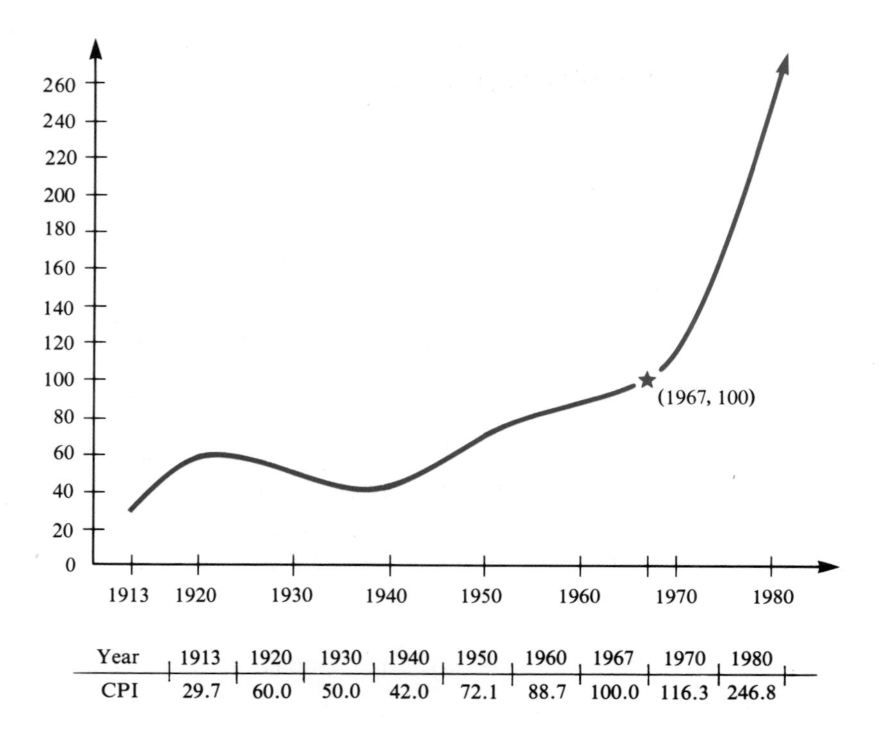

Year	1913	1920	1930	1940	1950	1960	1967	1970	1980
CPI	29.7	60.0	50.0	42.0	72.1	88.7	100.0	116.3	246.8

Figure 2–1

Sometimes important information is lost by looking at the average rate of change, as the following examples show.

One measure of inflation (the change in purchasing power of money) in the United States is provided by the Consumer Price Index (CPI) which is issued each month by the Bureau of Labor Statistics. The CPI uses 1967 as its base. For example, the CPI for 1983 was 298.4. This means that in 1983, it cost $298.40 to purchase the same goods that cost $100 in 1967. Similarly the CPI in 1920 was 60.0, so that the goods that cost $100 in 1967 would have only cost $60 in 1920. Figure 2–1 is the graph of the CPI from 1913 (the year it was first issued) to 1980, along with a table of actual values of the CPI for selected years.

Example 4 (*Compare Exercise 11*)
What was the average change in the CPI from 1920 to 1960?

Solution The average change from 1920 to 1960 was

$$\frac{\text{CPI}(1960) - \text{CPI}(1920)}{60 - 20} = \frac{88.7 - 60}{40}$$

$$= \frac{28.7}{40}$$

$$\approx 0.72 \text{ (dollars per year)}$$

Example 5 (*Compare Exercise 11*)
What was the average change in the CPI from 1920 to 1940?

Solution

$$\frac{\text{CPI}(1940) - \text{CPI}(1920)}{40 - 20} = \frac{42 - 60}{20} = -.9 \text{ (dollars per year)}$$

The CPI declined an average of $.90 per year for these 20 years.

Example 6 (*Compare Exercise 11*)
What was the average change in the CPI from 1940 to 1960?

Solution

$$\frac{\text{CPI}(1960) - \text{CPI}(1940)}{60 - 40} = \frac{88.7 - 42.0}{20} \approx 2.34 \text{ (dollars per year)}$$

The CPI rose an average of approximately $2.34 per year for these 20 years.

The answers to Examples 5 and 6 give us more information than did just the answer to Example 4. In fact, we would get an even better picture of inflation over these 40 years if we were given several averages, each taken over small time intervals.

Let's look at another example that is probably more familiar to you—the speed of an automobile. One of the authors (Cannon) lives about seven miles from work, and his trip to the office in the morning usually takes about 15 minutes, or 1/4 hour. Cannon's average speed on this trip is

$$\frac{7 \text{ miles}}{1/4 \text{ hr.}} = 28 \text{ mph}$$

This average again hides a lot of information. The last two miles are on an interstate highway, where Cannon's speed is 55 mph. The middle of the trip is on a heavily used road with a speed limit of 40 mph, yet which has nine traffic lights in a 3 mile stretch. Cannon spends a good portion of the 15 minutes at red lights. You would get a better idea of this trip if you were given his average speed in 3 intervals of 5 minutes each. You would get an even better idea of this trip if you were given his average speed for 15 intervals of 1 minute each. But even with these fifteen averages, it would be hard for you to distinguish a minute during which his speed was low the whole time, say due to heavy traffic, from a minute which had a short period at a light followed by rapid return to 40 mph.

But we've been asking you to talk about speed by only letting you look at the odometer (to measure distance travelled) and the clock (to measure time intervals). This is a good way to talk about average velocity, but if you want to know how Cannon's speed varied on this trip you would have to read the speedometer during the trip. The speedometer gives what the speed, the **rate of change** in the position, of the car is **at every instant**! The derivative of a function is like the speedometer of a car—the derivative gives us the rate of change of the function at every point.

Later sections in the text will give you rules to follow to compute the derivative of a function, but to understand how to apply derivatives you should know what it is you are computing.

Example 7 (*Compare Exercise 13*)
Measurements have shown that t seconds after a rock has been dropped, it will have fallen $16t^2$ feet. This means that 3 seconds after the rock is dropped it has travelled $16 \cdot 9 = 144$ feet. How fast is it travelling 3 seconds after it is dropped?

Solution The information given is about the distance travelled and the time, so we proceed as with the odometer information—we compute some average speeds. After 3 seconds the rock has dropped 144 feet and after 4 seconds it has dropped 256 feet. The distance the rock travels in the one second from time $t = 3$ to time $t = 4$ is $256 - 144$. Therefore, the average speed as t increases from 3 to 4 is

$$\frac{(16 \cdot 4^2) - (16 \cdot 3^2)}{4 - 3} = \frac{256 - 144}{1} = 112 \text{ (ft/sec)}$$

From 3 to 3.5 secs, the average speed is

$$\frac{16(3.5)^2 - 16(3)^2}{.5} = 104 \text{ (ft/sec)}$$

For $3 \leq t \leq 3.1$, the average speed is 97.6 ft/sec. For $3 \leq t \leq 3.01$, the average speed is 96.16 ft/sec. On the other side of 3, for $2.9 \leq t \leq 3$, the average speed is 94.4 ft/sec. For $2.99 \leq t \leq 3$, the average speed is 95.84 ft/sec. You might now guess that the speed of the rock when $t = 3$ is about 96 ft/sec.

Some algebra will remove the guess work.

First some notation that is useful when discussing rates of change. The symbol Δ means "the change in" and is immediately followed by a variable. Thus Δt means "the change in t," Δy means "the change in y," Δx means "the change in x," etc. When dealing with linear functions, we have seen in Example 1 that the change in y divided by the change in x is the slope of the line; $\Delta y/\Delta x = m$.

Now back to the falling rock. The length of the interval between the two moments $t = 3$ and $t = 3 + \Delta t$ is Δt. The distance fallen $3 + \Delta t$ seconds after the rock is dropped is

$$16(3 + \Delta t)^2 = 16(9 + 6\Delta t + (\Delta t)^2) = 144 + 96\Delta t + 16(\Delta t)^2 \text{ (feet).}$$

The rock has fallen 144 feet after 3 seconds. So the distance travelled in the time interval from 3 to $3 + \Delta t$ is

$$144 + 96\Delta t + 16(\Delta t)^2 - 144 = 96\Delta t + 16(\Delta t)^2 \text{ (feet).}$$

To find the average speed over this interval we divide this distance by the change in time, Δt.

$$\text{Average velocity} = \frac{96\Delta t + 16(\Delta t)^2}{\Delta t}$$

$$= 96 + 16\Delta t \text{ (ft/sec)}$$

Remember that Δt is one term—it does not mean Δ multiplied by t. For the interval $3 \leq t \leq 3.1$, $\Delta t = .1$; using $\Delta t = .1$ in the expression $96 + 16\Delta t$ gives us the average velocity of

$$96 + 16(.1) = 96 + 1.6 = 97.6 \text{ (ft/sec)}$$

as we computed just above!

For the interval $3 \le t \le 3.001$, Δt is $\frac{1}{1000}$th of a second, and the average velocity on this interval is

$$96 + 16(.001) = 96.016 \text{ ft/sec}$$

If Δt is close to 0, then $16 \cdot \Delta t$ is also close to 0, so the average velocity, $(96 + 16 \cdot \Delta t)$ ft/sec, is close to 96 ft/sec. This leads us to call 96 ft/sec **the instantaneous velocity** of the rock when $t = 3$.

We can condense these computations now that we have gone through one detailed example. First, just one more bit of notation. In the following, to avoid piling on parentheses, interpret the symbol Δt^2 to mean $(\Delta t)^2$, not $\Delta(t^2)$.

Example 8 (*Compare Exercise 13*)

(a) What is the rock's average velocity on the interval $[2, 2 + \Delta t]$?

(b) How fast is the rock going when $t = 2$?

Solution After 2 seconds, the rock has dropped $16 \cdot 2^2 = 64$ feet. After $2 + \Delta t$ seconds, the rock has dropped

$$16(2 + \Delta t)^2 = 16(4 + 4\Delta t + \Delta t^2) = 64 + 64\Delta t + 16\Delta t^2 \text{ feet}$$

In the interval $[2, 2 + \Delta t]$, the rock has travelled

$$(64 + 64\Delta t + 16\Delta t^2) \text{ feet} - 64 \text{ feet} = (64\Delta t + 16\Delta t^2) \text{ feet}$$

(a) The average velocity is

$$\frac{64\Delta t + 16\Delta t^2}{\Delta t} = 64 + 16\Delta t \text{ ft/sec}$$

(b) When Δt is close to 0, $64 + 16\Delta t$ is close to 64. The rock is travelling at 64 ft/sec when $t = 2$.

We have been led to a definition of instantaneous velocity by computing the average velocity over very small intervals. Velocity is the special case of the rate of change of a distance-travelled function. We define the instantaneous rate of change of an arbitrary function in an analogous manner.

Instantaneous Rate of Change

To define the rate of change of a function f at the "instant" when $x = a$, we compute the average rate of change of f on the interval $[a, a + \Delta x]$ if $\Delta x > 0$, or $[a + \Delta x, a]$ if $\Delta x < 0$. If, when Δx is close to 0, the numbers we get as average rates of change are all close to one fixed number, we call that fixed number the instantaneous rate of change of f when $x = a$. In Example 7, for instance, when Δt was close to 0, the numbers we obtained as average rates of change were close to the one fixed number 96.

The average rate of change of the function f on the interval with endpoints a and $a + \Delta x$ is given by the expression

$$\frac{f(a + \Delta x) - f(a)}{\Delta x}$$

In summation, we have the following definition.

Definition

> The **instantaneous rate of change of the function f when $x = a$** is the number L if $\dfrac{f(a + \Delta x) - f(a)}{\Delta x}$ is close to L whenever Δx is close to 0.

Example 9 If $f(x) = 10x + 6$, what is the instantaneous rate of change of f when $x = 7$?

Solution The solution can be broken into 3 steps:

STEP 1 Compute and simplify $f(a + \Delta x) - f(a)$.

The value of a is 7. $f(7) = 76$.

$$f(7 + \Delta x) = 10(7 + \Delta x) + 6 = 76 + 10\Delta x$$
$$f(7 + \Delta x) - f(7) = 76 + 10\Delta x - 76 = 10\Delta x$$

STEP 2 Divide the result of Step 1 by Δx.

The result of Step 1 is $10\Delta x$. Here $\dfrac{10\Delta x}{\Delta x} = 10$.

STEP 3 Find L by letting Δx be close to 0.

Here, no matter what the value of Δx is, the expression $\dfrac{f(7 + \Delta x) - f(7)}{\Delta x} = 10$.

Surely we can say "10 is close to 10." When Δx is close to 0,

$$\frac{f(7 + \Delta x) - f(7)}{\Delta x}$$

is close to 10.

ANSWER The instantaneous rate of change of f when $x = 7$ is 10.

In this example, f is a linear function. We saw earlier that for linear functions the average rate of change is a constant, the slope. There is nothing special about the numbers in Example 9; we can make the following general statement.

Theorem

> If $f(x) = mx + b$, then the instantaneous rate of change of f when $x = a$ is the slope m, regardless of the value of a.

Now we look at an example, like the falling rock example, where the instantaneous rate of change depends on the number a, which is the usual situation.

Example 10 (*Compare Exercise 17*)
Find the instantaneous rate of change of $f(x) = 5x^2$

(a) when $x = 3$, and
(b) when $x = 2$.

Solution **(a)** First, let $a = 3$.

STEP 1 Compute and simplify $f(a + \Delta x) - f(a)$.

$$\begin{aligned} f(a) &= f(3) = 45 \\ f(a + \Delta x) &= f(3 + \Delta x) = 5(3 + \Delta x)^2 \\ &= 5(9 + 6\Delta x + \Delta x^2) \qquad [\text{Remember } \Delta x^2 = (\Delta x)^2] \\ &= 45 + 30\Delta x + 5\Delta x^2 \qquad [\text{Remember } \Delta x \text{ is one number}] \\ f(a + \Delta x) - f(a) &= f(3 + \Delta x) - f(3) = 45 + 30\Delta x + 5\Delta x^2 - 45 \\ &= 30\Delta x + 5\Delta x^2 \end{aligned}$$

STEP 2 Divide the result of Step 1 by Δx.

$$\frac{30\Delta x + 5(\Delta x)^2}{\Delta x} = \frac{(30 + 5\Delta x)\Delta x}{\Delta x} = 30 + 5\Delta x$$

STEP 3 If Δx is close to 0, then $30 + 5\Delta x$ is close to 30.

ANSWER The instantaneous rate of change of $f(x) = 5x^2$ when $x = 3$ is 30.

(b) Second, to find the instantaneous rate of change when $x = 2$, we go through the same steps using the number 2 in place of the number 3.

STEP 1

$$\begin{aligned} f(2) &= 20 \\ f(2 + \Delta x) &= 5(2 + \Delta x)^2 \\ &= 5(4 + 4\Delta x + \Delta x^2) = 20 + 20\Delta x + 5\Delta x^2 \\ f(2 + \Delta x) - f(2) &= 20\Delta x + 5\Delta x^2 \end{aligned}$$

STEP 2

$$\frac{20\Delta x + 5\Delta x^2}{\Delta x} = 20 + 5\Delta x$$

STEP 3 If Δx is close to 0, then $20 + 5\Delta x$ is close to 20.

ANSWER The instantaneous rate of change of $f(x) = 5x^2$ when $x = 2$ is 20.

The Derivative

The language, like "instantaneous rate of change of f," in the last two examples is drawn out and somewhat awkward. For convenience then, we introduce some new notation and a new word.

Definition

> "The instantaneous rate of change of the function f at $x = a$" is called **the derivative of f at a** and is denoted simply by $f'(a)$.
> ($f'(a)$ is read "f prime of a.")

In Example 9 we found that if $f(x) = 10x + 6$, then $f'(7) = 10$ [in fact we found that $f'(a) = 10$ for all values of a]. In Example 10 we found that if $f(x) = 5x^2$, then $f'(3) = 30$ and $f'(2) = 20$.

Example 11 (*Compare Exercise 21*)
If $f(x) = 5x^2$, find $f'(-2)$.

Solution The function is the same as in Example 10, but the value of a is different. Here $a = -2$.

STEP 1

$$\begin{aligned} f(-2) &= 20; \\ f(-2 + \Delta x) &= 5(-2 + \Delta x)^2 \\ &= 5(4 - 4\Delta x + (\Delta x)^2) \\ &= 20 - 20\Delta x + 5(\Delta x)^2 \\ f(-2 + \Delta x) - f(-2) &= 20 - 20\Delta x + 5(\Delta x)^2 - 20 \\ &= -20\Delta x + 5(\Delta x)^2 \end{aligned}$$

STEP 2

$$\begin{aligned} \frac{-20\Delta x + 5(\Delta x)^2}{\Delta x} &= \frac{(-20 + 5\Delta x)\Delta x}{\Delta x} \\ &= -20 + 5\Delta x \end{aligned}$$

STEP 3 If Δx is close to 0, then $-20 + 5\Delta x$ is close to -20.

ANSWER $f'(-2) = -20$

Example 12 (*Compare Exercise 23*)
If $f(x) = 3x^2 - 4x$, find $f'(2)$.

Solution The function is a bit more complicated, but the procedure is the same as in the previous three examples.

STEP 1 Compute $f(a)$, $f(a + \Delta x)$ and then simplify $f(a + \Delta x) - f(a)$.

Here $a = 2$, so $f(2) = 4$;

$$\begin{aligned} f(2 + \Delta x) &= 3(2 + \Delta x)^2 - 4(2 + \Delta x) \\ &= 3(4 + 4\Delta x + \Delta x^2) - 8 - 4\Delta x \\ &= 12 + 12\Delta x + 3\Delta x^2 - 8 - 4\Delta x \\ &= 4 + 8\Delta x + 3\Delta x^2 \\ f(2 + \Delta x) - f(2) &= 4 + 8\Delta x + 3\Delta x^2 - 4 \\ &= 8\Delta x + 3\Delta x^2 \end{aligned}$$

STEP 2 Divide result of Step 1 by Δx.

$$\begin{aligned} \frac{8\Delta x + 3\Delta x^2}{\Delta x} &= \frac{(8 + 3\Delta x)\Delta x}{\Delta x} \\ &= 8 + 3\Delta x \end{aligned}$$

STEP 3 If Δx is close to 0, then $8 + 3\Delta x$ is close to 8.

ANSWER $f'(2) = 8$

Example 13 (*Compare Exercise 25*)
If $f(x) = 36/x$, find $f'(3)$.

Solution

STEP 1

$$f(3) = 12$$

$$f(3 + \Delta x) = \frac{36}{3 + \Delta x}$$

$$f(3 + \Delta x) - f(3) = \frac{36}{3 + \Delta x} - 12$$

In this case, "simplify" means to combine terms into one fraction.

$$\frac{36}{3 + \Delta x} - 12 = \frac{36}{3 + \Delta x} - \frac{12(3 + \Delta x)}{3 + \Delta x}$$

$$= \frac{36 - 12(3 + \Delta x)}{3 + \Delta x}$$

$$= \frac{-12\Delta x}{3 + \Delta x}$$

STEP 2 Dividing by Δx is the same as multiplying $1/\Delta x$, which is often an easier way to do this step when Step 1 leaves you with a fraction.

$$\frac{-12\Delta x}{3 + \Delta x} \cdot \frac{1}{\Delta x} = \frac{-12}{3 + \Delta x}$$

STEP 3 If Δx is near 0, $\dfrac{-12}{3 + \Delta x}$ is near $\dfrac{-12}{3}$, or -4.

ANSWER $f'(3) = -4$

2–2 Exercises

I.

For each of the following functions, find the average rate of change of f on the indicated interval or intervals.

Function	Interval or Intervals	
1. $f(x) = -7x + 5$	**(a)** $[2, 6]$	**(b)** $[-1, 3]$
2. $f(x) = 3x - 8$	**(a)** $[0, 7]$	**(b)** $[-2, 3]$
3. (*See Example 1*) $f(x) = \frac{1}{2}x + 9$	**(a)** $[2, 6]$	**(b)** $[-4, 4]$
4. $f(x) = -\frac{2}{3}x - 5$	**(a)** $[-3, 9]$	**(b)** $[-6, 6]$

5. (*See Examples 2 & 3*) $f(x) = x^2 + 2x$ (a) $[1, 3]$ (b) $[0, 2]$
6. $f(x) = x^2 - 2x + 5$ (a) $[0, 4]$ (b) $[1, 5]$
7. $f(x) = x^2 - 4x + 6$ (a) $[0, 4]$ (b) $[-1, 5]$
8. $f(x) = x^2 - x - 2$ (a) $[-1, 2]$ (b) $[1, 4]$
9. $f(x) = \sqrt{x + 16}$ (a) $[0, 9]$ (b) $[-12, -7]$
10. $f(x) = \sqrt{x^2 + 9}$ (a) $[0, 4]$ (b) $[-4, 4]$

II.

Use the table from Figure 2–1 to get the values of the CPI for the next two exercises.

11. (*See Examples 4, 5, and 6*) What was the average rate of change in the CPI from 1920 to 1940? From 1920 to 1930? From 1930 to 1940?
12. What was the average rate of change in the CPI from 1950 to 1960? From 1960 to 1970? From 1950 to 1970?
13. (*See Examples 7 and 8*)
    (a) What is the average velocity of a dropped rock on the interval $[4, 4 + \Delta t]$?
    (b) How fast is the rock moving 4 seconds after it is dropped?
14. (a) What is the average velocity of a dropped rock for the interval $1 \leq t \leq 1 + \Delta t$?
    (b) How fast is the rock moving 1 second after it is dropped?
15. An automobile has travelled $40t^2 + 10t$ feet after t seconds.
    (a) How far has the auto travelled after 3 seconds?
    (b) How far has the auto travelled after $3 + \Delta t$ seconds?
    (c) What is its average velocity on the interval $[3, 3 + \Delta t]$?
    (d) How fast is it going when $t = 3$?
16. Use the same automobile as in Exercise 15.
    (a) How far has the auto travelled after 2 seconds?
    (b) How far has the auto travelled after $2 + \Delta t$ seconds?
    (c) What is its average velocity on the interval $[2, 2 + \Delta t]$?
    (d) How fast is it going when $t = 2$?

III.

In Exercises 17 through 20, find the instantaneous rate of change of each function at the indicated value of a.

17. (*See Example 10*) $f(x) = 4x^2, \quad a = 3$
18. $f(x) = -x^2, \quad a = 5$
19. $f(x) = x^2 - 3, \quad a = 1$
20. $f(x) = 4 - x^2, \quad a = 3$

21. (*See Example 11*)
If $f(x) = -x^2$, compute $f'(-1)$.

22. If $f(x) = 6x^2$, compute $f'(-2)$.

23. (*See Example 12*)
If $f(x) = x^2 - 7x$, compute $f'(3)$.

24. If $f(x) = 2x^2 + 5x + 1$, compute $f'(0)$.

25. (*See Example 13*)
If $f(x) = \frac{3}{x}$, compute $f'(2)$.

26. If $f(x) = \frac{4}{x}$, compute $f'(5)$.

27. If $f(x) = mx + b$, compute $f'(a)$.

28. If a company's revenue function is given by

$$R(x) = 100x - x^2$$

what is the instantaneous rate of change of revenue when $x = 20$?

29. Use the revenue function from Exercise 28 to compute the instantaneous rate of change of revenue when $x = 50$.

30. Use the revenue function from Exercise 28 to compute the instantaneous rate of change of revenue when $x = 80$.

31. If a company's cost function is given by

$$C(x) = 17x + 2800$$

what is the instantaneous rate of change of cost when

(a) $x = 10$? **(b)** $x = 30$? **(c)** $x = 60$?

2–3 Limits

Introduction
Definition of Limit
Graphical Interpretation of Limits
Rules for Computing Limits

Introduction

In the previous section, we wrote the derivative of f at a as $f'(a)$ and said $f'(a)$ has the value L if

$$\frac{f(a + \Delta x) - f(a)}{\Delta x}$$

is close to L whenever Δx is close to 0.

In the specific examples and exercises we have seen so far, the computation of $f'(a)$ has been fairly straightforward. The phrase "close to" however, is somewhat vague, and this ambiguity can cause problems. Historically, the derivative has been very useful in solving important applications, and people realized that a more precise explanation of derivative was needed. Mathematicians struggled with this problem for roughly two hundred years until a precise formulation was developed in the 19th century. Because the difficulties centered around use of the phrase "close to" and were not special to the notion of derivative, language was developed to deal specifically with the notion of "close to."

The quotient used in computing $f'(a)$ is

$$\frac{f(a + \Delta x) - f(a)}{\Delta x}$$

This quotient doesn't make sense if $\Delta x = 0$; for then the numerator would be $f(a) - f(a)$, which is 0, but the denominator would also be 0, and 0/0 doesn't stand for a number. A special word was made up to talk about what happens to the values of

$$\frac{f(a + \Delta x) - f(a)}{\Delta x}$$

when Δx is close to 0 but not equal to 0. That word is **limit**.

One of the reasons mathematics is such a powerful tool in so many different types of applications—psychology, economics, chemistry—is its very abstractness. For example, chemists are concerned with rates of change in chemical reactions; economists are concerned with rates of change in supply and demand; psychologists are concerned with rates of change of learning. Whatever mathematicians can prove about an "abstract" concept—like the derivative, which is a special type of limit—can then be used in different applications by different users of mathematics. Of course that same abstractness is a frustration for many people as they struggle to learn mathematics.

Definition of Limit

It may seem to you like a giant leap from a discussion about average velocity of an automobile to a discussion about limits, so let's recap where we are and where we're headed.

We wanted to talk about rates of change in various applications, so we were led to the mathematical notion of derivative. The notion of derivative involves a certain quotient whose denominator, Δx, is very close to 0. This in turn leads us to a general discussion of what happens to an arbitrary function f when x is close to an arbitrary number c.

Definition

> We say that the **limit of a function f at $x = c$** is the number L if the values of $f(x)$ are close to L whenever x is close to, but not equal to, c.

Another way to express the idea of limit is to say the limit of f at c is L if $f(x)$ approaches L as x approaches c. We write "the limit of f at $x = c$ is L" as "$\lim_{x \to c} f(x) = L$."

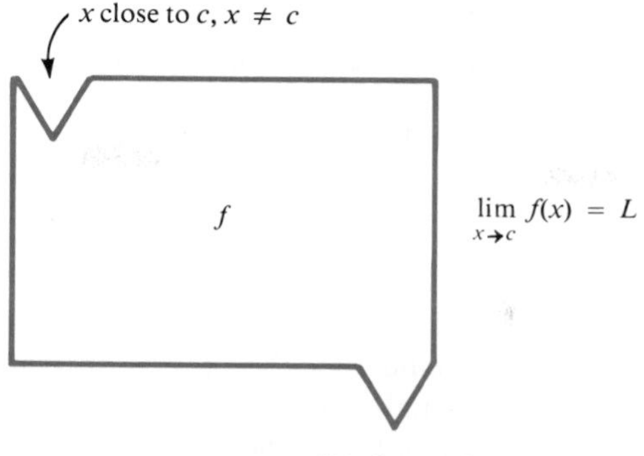

Figure 2–2

The description of a function as a machine may help. The picture is shown in Figure 2–2.

Important Notice that the definition of L has nothing to do with the value of the function at $x = c$, $f(c)$. The condition that $x \neq c$ specifically excludes considering $f(c)$ in deciding what L is. The reason to exclude the value $x = c$ comes from the definition of derivative;

$$\frac{f(a + \Delta x) - f(a)}{\Delta x}$$

isn't defined if $\Delta x = 0$.

We can now state the definition of derivative using limit terminology.

Definition

> The **derivative of f at a**, written $f'(a)$, is defined by
>
> $$f'(a) = \lim_{\Delta x \to 0} \frac{f(a + \Delta x) - f(a)}{\Delta x}$$

Limits have proven useful in describing other properties besides instantaneous rate of change, and so we devote the remainder of this section to the further study of limits.

Graphical Interpretation of Limits

As a first step, we would like to help you develop some geometric feeling for what the statement "$\lim_{x \to c} f(x) = L$" means. (There is a lot of symbolism used in that statement. Practice saying the statement in words whenever you come across it: $\lim_{x \to c} f(x) = L$ is read "the limit of $f(x)$ at c equals L." You are learning new words in a "foreign language"; not many people can master this language completely on one reading. And don't be afraid to stumble with these terms; almost everybody feels a bit awkward when first using unfamiliar words, but nobody becomes good at anything without practicing.)

Example 1 (*Compare Exercise 1*)
What is $\lim_{x \to 3} x^2$? That is, what is the limit of $f(x) = x^2$ as x approaches 3?

Solution We apply the definition, with $f(x) = x^2$ and $c = 3$.
If x is near 3, x^2 is near 9, so $L = 9$:

$$\lim_{x \to 3} x^2 = 9$$

This is shown graphically in Figure 2–3. (*Compare Exercise 19*).
Remember that points on the graph are of the form $(x, f(x))$—the second coordinate represents $f(x)$. Graphically, if x is close to 3, then (x, x^2) is close to the point (3, 9), and x^2 is close to 9.

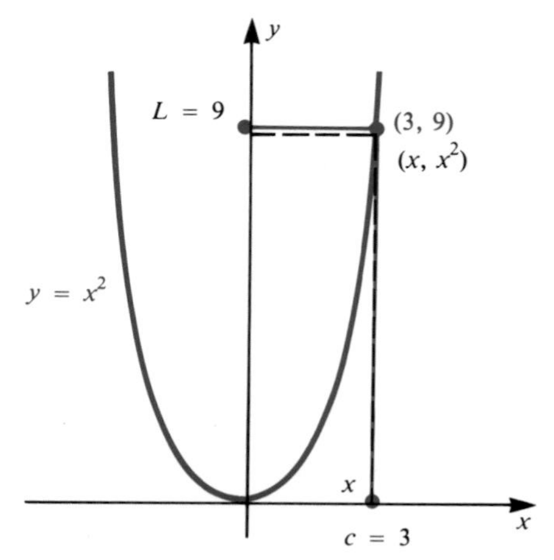

Figure 2–3

Example 2 (*Compare Exercise 5*)
What is $\lim_{x \to 2} (3x - 1)$?

Solution Applying the definition of limit, we know that if x is close to 2, then $3x$ is close to 6, so $3x - 1$ is close to 5.

$$\lim_{x \to 2} (3x - 1) = 5$$

See Figure 2–4.

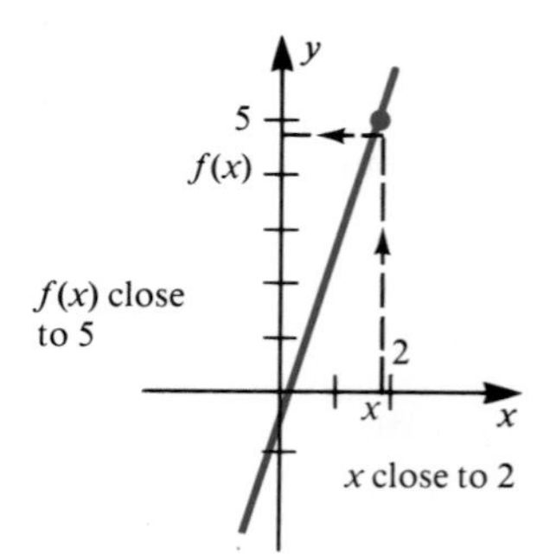

Figure 2–4

In these two examples we were able to find the limit by using the definition and "common sense." That is, if x is near 2, "common sense" leads us to believe that $3x - 1$ is near 5.

Now let's look at an example where the answer isn't as clear when we let x be a value near c.

Example 3 (*Compare Exercise 31*)

Find $\lim_{x \to 3} \frac{x^2 - 9}{x - 3}$.

Solution Now we are looking at a quotient, and when x is near 3 both the numerator and the denominator are close to 0. It is not immediately obvious what number

$$\frac{x^2 - 9}{x - 3}$$

is near if x is near 3. In fact, the expression

$$\frac{x^2 - 9}{x - 3}$$

makes no sense at all when $x = 3$. Since we can't tell what's going on just by looking, we rewrite this expression by factoring the numerator and trying to simplify.

$$\frac{x^2 - 9}{x - 3} = \frac{(x - 3)(x + 3)}{x - 3} = x + 3, \quad \text{if } x \neq 3$$

Thus, if $x \neq 3$, but x is close to 3, then

$$\frac{x^2 - 9}{x - 3} = x + 3$$

and $x + 3$ is close to 6. We can conclude that

$$\lim_{x \to 3} \frac{x^2 - 9}{x - 3} = \lim_{x \to 3} (x + 3) = 6$$

Warning! When trying to determine the number

$$\lim_{x \to 3} \frac{x^2 - 9}{x - 3}$$

you do **not** substitute $x = 3$. Again, examples like this are the reason that $x \neq c$ is part of the definition of limit.

The graph of

$$f(x) = \frac{x^2 - 9}{x - 3}$$

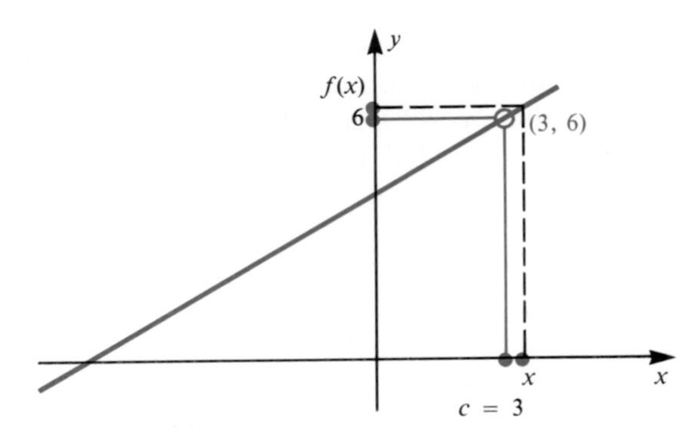

Figure 2–5

gives a picture of what is going on here. We saw above that $f(x) = x + 3$ whenever $x \neq 3$. The graph of $y = x + 3$ is a straight line, but since x can't be 3, then we must remove the point (3, 6) from the line. Thus the graph of

$$f(x) = \frac{x^2 - 9}{x - 3}$$

is a line with a "hole" in it—the point (3, 6) is missing from the line. See Figure 2–5.

The graph shows that if x is near 3, $f(x)$ is near 6, and that's what $\lim_{x \to 3} f(x) = 6$ means, even if $f(3)$ itself doesn't make sense. (*Compare Exercise 15*)

Rules for Computing Limits

Notice that what we did in Example 3 was to replace a complicated expression,

$$\frac{x^2 - 9}{x - 3}$$

by a simpler expression, $x + 3$, which still gave the values of $f(x)$ whenever $x \neq 3$. This is an important principle that is often used in computing limits. We state this principle as the following theorem.

Replacement Principle Theorem

> If f and g are two functions with $f(x) = g(x)$ for all x near c, $x \neq c$, and if $\lim_{x \to c} g(x) = L$, then $\lim_{x \to c} f(x) = L$ also.

You will use this theorem whenever you replace one expression for $f(x)$ by a simpler one.

In Example 3, we had

$$f(x) = \frac{x^2 - 9}{x - 3},$$

$g(x) = x + 3$, $c = 3$, and $L = 6$.

The replacement principle corresponds to the simplifying step in computing derivatives in the previous section.

Example 4 (*Compare Exercise 39*)
If $f(x) = 3x^2$, compute $f'(2)$.

Solution

$$f'(2) = \lim_{\Delta x \to 0} \frac{f(2 + \Delta x) - f(2)}{\Delta x}$$

$$= \lim_{\Delta x \to 0} \frac{3(2 + \Delta x)^2 - 3(2)^2}{\Delta x}$$

$$= \lim_{\Delta x \to 0} \frac{3(4 + 4\Delta x + \Delta x^2) - 12}{\Delta x}$$

$$= \lim_{\Delta x \to 0} \frac{12\Delta x + 3\Delta x^2}{\Delta x}$$

Next, replace $\frac{12\Delta x + 3\Delta x^2}{\Delta x}$ by $12 + 3\Delta x$; they are equal if $\Delta x \neq 0$

$$= \lim_{\Delta x \to 0} (12 + 3\Delta x)$$

$$= 12$$

Thus, $$f'(2) = 12$$

There are arithmetic steps that we used in these computations of limits, so we look more closely now at some arithmetic principles we can use when computing limits.

Example 5 (*Compare Exercise 8*)
Find $\lim_{x \to 2} \frac{x^2 - 3x + 7}{x + 8}$

Solution If x is near 2, x^2 is near 4 and $3x$ is near 6. Thus, $x^2 - 3x + 7$ is near $4 - 6 + 7 = 5$. Also when x is near 2, $x + 8$ is near 10. The numerator is near 5 and the denominator is near 10, so their quotient is near $\frac{1}{2}$. We have

$$\lim_{x \to 2} \frac{x^2 - 3x + 7}{x + 8} = \frac{1}{2}$$

We can generalize the procedures we went through in this example and state these procedures in the form of a theorem. This theorem shows that limits obey the usual rules of arithmetic. For example, if f and g are two functions, and x near c means both that $f(x)$ is close to some number L and that $g(x)$ is close to some number K, then we can conclude that $f(x) + g(x)$ is near $L + K$.

That sentence is awkward; using the limit notation, we have

$$\lim_{x \to c} (f(x) + g(x)) = \lim_{x \to c} f(x) + \lim_{x \to c} g(x).$$

A more informal way to remember this is "**the limit of the sum is the sum of the limits.**"

The other arithmetic operations of subtraction, multiplication, and division also carry over to limits in a natural way. For example, "**the limit of the product is the product of the limits.**" We summarize this discussion in the following theorem.

Arithmetic Principles Theorem

If $\lim_{x \to c} f(x) = L$ and $\lim_{x \to c} g(x) = K$, then

1. (sum)

$$\lim_{x \to c} [f(x) + g(x)] = \lim_{x \to c} f(x) + \lim_{x \to c} g(x) = L + K$$

2. (difference)

$$\lim_{x \to c} [f(x) - g(x)] = \lim_{x \to c} f(x) - \lim_{x \to c} g(x) = L - K$$

3. (product)

$$\lim_{x \to c} [f(x) \cdot g(x)] = \lim_{x \to c} f(x) \cdot \lim_{x \to c} g(x) = L \cdot K$$

4. (quotient)
If $K \neq 0$,

$$\lim_{x \to c} \frac{f(x)}{g(x)} = \frac{\lim_{x \to c} f(x)}{\lim_{x \to c} g(x)} = \frac{L}{K}$$

Two more observations about limits provide what we need for most computations.

Two Special Limits Theorem

1. If K is a constant, then $\lim_{x \to c} K = K$.

2. $\lim_{x \to c} x = c$.

This theorem says

1. if x is near c, then K is near K; and
2. if x is near c, then x is near c.

See Figure 2–6.

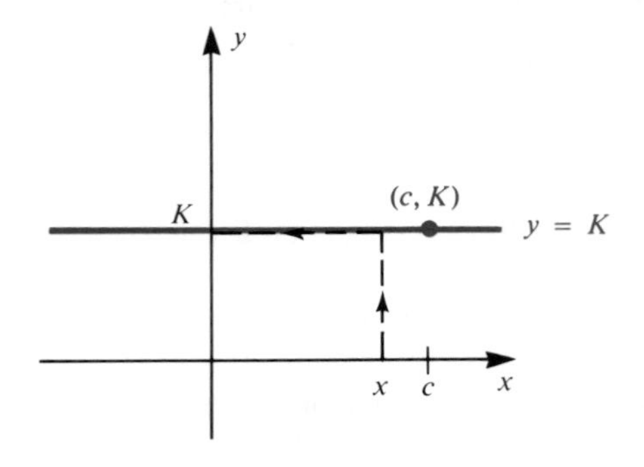

(1)

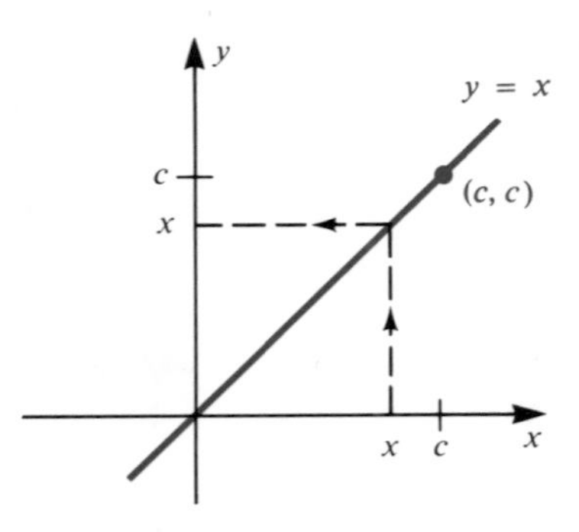

(2)

Figure 2–6

We also note that if $g(x)$ is a constant K, we have $\lim_{x \to c} [K \cdot f(x)] = K \cdot \lim_{x \to c} f(x)$ for every constant K.

Example 6 (*Compare Exercises 9 and 11*)
Compute $\lim_{x \to 2} (x^2 + 3x)$.

Solution We make repeated applications of the theorem. We first recognize the function $x^2 + 3x$ as the sum of the functions x^2 and $3x$ and use "the limit of the sum is the sum of the limits."

$$\lim_{x \to 2} (x^2 + 3x) = \lim_{x \to 2} x^2 + \lim_{x \to 2} 3x$$

Now $x^2 = (x \text{ times } x)$ and $3x = (3 \text{ times } x)$, so we use "the limit of the product is the product of the limits" rule twice

$$= \left(\lim_{x \to 2} x\right)\left(\lim_{x \to 2} x\right) + \left(\lim_{x \to 2} 3\right)\left(\lim_{x \to 2} x\right)$$

We have expressed the limit of the function $x^2 + 3x$ in terms of limits of simpler functions, namely x and 3. We now make use of the known limits, $\lim_{x \to 2} x = 2$ and $\lim_{x \to 2} 3 = 3$, to get

$$\lim_{x \to 2} (x^2 + 3x) = (2)(2) + (3)(2) = 10$$

Example 7 Compute $\lim_{x \to -1} (3x^3 + 4)$

Solution We write

$$\begin{aligned} \lim_{x \to -1} (3x^3 + 4) &= \lim_{x \to -1} 3x^3 + \lim_{x \to -1} 4 \\ &= \left(\lim_{x \to -1} 3\right)\left(\lim_{x \to -1} x\right)\left(\lim_{x \to -1} x\right)\left(\lim_{x \to -1} x\right) + \lim_{x \to -1} 4 \\ &= 3(-1)(-1)(-1) + 4 = -3 + 4 \\ &= 1 \end{aligned}$$

Example 8 (*Compare Exercise 21*)
If $\lim_{x \to 8} f(x) = 2$, what is $\lim_{x \to 8} (xf(x))$?

Solution $\lim_{x \to 8} (xf(x)) = \left(\lim_{x \to 8} x\right)\left(\lim_{x \to 8} f(x)\right) = 8 \cdot 2 = 16$

Example 9 (*Compare Exercise 23*)
If $\lim_{x \to 3} f(x) = 5$ and $\lim_{x \to 3} g(x) = 4$, what are

(a) $\lim_{x \to 3} (f(x) \cdot g(x))$?

(b) $\lim_{x \to 3} \dfrac{f(x)}{g(x)}$?

(c) $\lim_{x \to 3} (2f(x) - 6g(x))$?

Solution **(a)**
$$\lim_{x \to 3} (f(x) \cdot g(x)) = \lim_{x \to 3} f(x) \cdot \lim_{x \to 3} g(x) = 5 \cdot 4 = 20$$

(b)
$$\lim_{x \to 3} \frac{f(x)}{g(x)} = \frac{\lim_{x \to 3} f(x)}{\lim_{x \to 3} g(x)} = \frac{5}{4}$$

(c)
$$\begin{aligned}\lim_{x \to 3} (2f(x) - 6g(x)) &= \lim_{x \to 3} (2f(x)) - \lim_{x \to 3} (6g(x)) \\ &= 2 \lim_{x \to 3} f(x) - 6 \lim_{x \to 3} g(x) \\ &= 2 \cdot 5 - 6 \cdot 4 = -14\end{aligned}$$

These examples are special cases of a more general theorem that simplifies the computation of limits when the function is a polynomial.

By applying the arithmetic theorem many times, we could derive the following result, which lets us quickly compute limits of polynomial functions.

Theorem

If f is a polynomial function, that is, if
$$f(x) = a_n x^n + a_{n-1} x^{n-1} + \cdots + a_1 x + a_0$$
then, for any number c,
$$\lim_{x \to c} f(x) = f(c)$$

In short, if $f(x)$ is a polynomial, then $\lim_{x \to c} f(x) = f(c)$.

Example 10 Compute $\lim_{x \to -1} (3x^4 - 7x^2 + 2x + 3)$.

Solution The function is the polynomial
$$f(x) = 3x^4 - 7x^2 + 2x + 3$$
The value of the function at $x = -1$ is
$$\begin{aligned}f(-1) &= 3(-1)^4 - 7(-1)^2 + 2(-1) + 3 \\ &= -3\end{aligned}$$
Thus, $\lim_{x \to -1} (3x^4 - 7x^2 + 2x + 3) = -3$

We have a similar theorem if $f(x)$ is the quotient of polynomials.

Theorem

If $f(x) = \dfrac{P(x)}{Q(x)}$ and P and Q are polynomials with $Q(c) \neq 0$, then
$$\lim_{x \to c} f(x)) = f(c).$$

Example 11 Compute $\lim_{x\to 3} \dfrac{2x^2+3x-1}{x^3-7}$.

Solution Interpret

$$\frac{2x^2+3x-1}{x^3-7}$$

as the quotient of the polynomial $2x^2+3x-1$ by the polynomial x^3-7. Since $\lim_{x\to 3}(x^3-7) = 3^3-7 = 20$, the limit of the denominator is nonzero. Thus, we can apply the result

$$\lim_{x\to c}\frac{f(x)}{g(x)} = \frac{\lim_{x\to c} f(x)}{\lim_{x\to c} g(x)}$$

We get

$$\lim_{x\to 3}\frac{2x^2+3x-1}{x^3-7} = \frac{\lim_{x\to 3}(2x^2+3x-1)}{\lim_{x\to 3}(x^3-7)} = \frac{2\cdot 9+3\cdot 3-1}{27-7}$$

$$= \frac{26}{20}$$

Warning! It is important to realize the implication of the condition $\lim_{x\to c} g(x) \neq 0$ when applying the rule

$$\lim_{x\to c}\frac{f(x)}{g(x)} = \frac{\lim_{x\to c} f(x)}{\lim_{x\to c} g(x)}$$

If the limit of the denominator is zero, this method cannot be used; however, the limit of the quotient may or may not exist. We have to resort to alternative approaches to discuss the limit. Such quotient functions do play a fundamental role in calculus, so we now look at some examples to illustrate some of the concepts involved. **You must check that $Q(c) \neq 0$.**

Example 12 (*Compare Exercise 33*)
Compute $\lim_{x\to 2}\dfrac{x^2-4}{x-2}$.

Solution We cannot just let $x = 2$ and get an answer. Nor is it legitimate to say that the limit in the denominator is 0 and, since division by 0 is impossible, the limit of the quotient doesn't exist. Remember the definition of derivative, which is one of the most important applications of limits. This theorem tells you what happens if $Q(c) \neq 0$; it doesn't tell you what happens if $Q(c) = 0$. In fact here,

$$\lim_{x\to 2}\frac{x^2-4}{x-2} = \lim_{x\to 2}\frac{(x-2)(x+2)}{x-2} = \lim_{x\to 2}(x+2) = 4$$

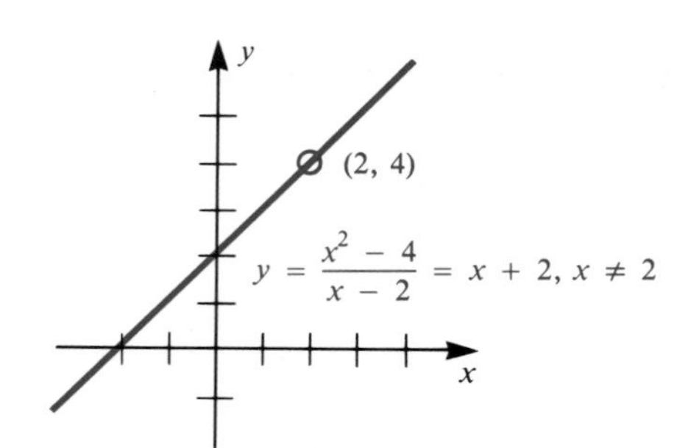

Figure 2–7

We had to rewrite the problem and simplify $\frac{x^2 - 4}{x - 2}$ to see what was really going on.

The graph of $f(x) = \frac{x^2 - 4}{x - 2}$ is given in Figure 2–7.

It is also true that limits behave the way you'd like them to when exponents are involved.

Example 13 Find $\lim_{x \to 3} (x + 5)^2$.

Solution If x is near 3, $x + 5$ is near 8, and so $(x + 5)^2$ is near 8^2, or 64. We conclude that $\lim_{x \to 3} (x + 5)^2 = 64$.

Example 14 Find $\lim_{x \to 11} \sqrt{x + 14}$.

Solution If x is near 11, $x + 14$ is near 25, so $\sqrt{x + 14}$ is near $\sqrt{25}$ or 5. We have $\lim_{x \to 11} \sqrt{x + 14} = 5$.

The general property is stated as follows.

Theorem

$$\lim_{x \to c} (f(x))^R = \left(\lim_{x \to c} f(x)\right)^R$$

whenever the expressions make sense. (That qualifying remark is included because, for instance, we can't take square roots of negative numbers.)

Example 15 (*Compare Exercise 25*)
If $\lim_{x \to 5} f(x) = 4$, what are:

(a) $\lim_{x \to 5} (f(x))^2$?

(b) $\lim_{x \to 5} \sqrt{f(x)}$?

Solution **(a)** $\lim_{x\to 5} (f(x))^2 = \left(\lim_{x\to 5} f(x)\right)^2 = 4^2 = 16$

(b) $\lim_{x\to 5} \sqrt{f(x)} = \sqrt{\lim_{x\to 5} f(x)} = \sqrt{4} = 2$

Sometimes we can have limits involving square roots that require some algebra to help us compute the answer.

Example 16 (*Compare Exercise 37*)

Find $\lim_{x\to 4} \dfrac{\sqrt{x} - 2}{x - 4}$.

Solution Here again the individual limits in the numerator and denominator are both 0. One trick that works is to rationalize the numerator. In previous courses you may have had to rationalize the denominator to simplify an expression. Here the square root is in the numerator, but the procedure is the same. Multiply numerator and denominator by $\sqrt{x} + 2$.

$$\lim_{x\to 4} \frac{\sqrt{x} - 2}{x - 4} = \lim_{x\to 4} \frac{(\sqrt{x} - 2)(\sqrt{x} + 2)}{(x - 4)(\sqrt{x} + 2)}$$

$$= \lim_{x\to 4} \frac{x - 4}{(x - 4)(\sqrt{x} + 2)} \qquad \text{(divide numerator and denominator by } x - 4)$$

$$= \lim_{x\to 4} \frac{1}{\sqrt{x} + 2} = \frac{1}{2 + 2} = \frac{1}{4}$$

Thus, $\lim_{x\to 4} \dfrac{\sqrt{x} - 2}{x - 4} = \dfrac{1}{4}$

Two Special Cases Involving Zero. Sometimes we are forced to conclude that $\lim_{x\to c} f(x)$ doesn't exist. That is, there isn't a number L that $f(x)$ is near whenever x is near c.

Example 17 (*Compare Exercise 41*)

Find $\lim_{x\to 0} \dfrac{1}{x}$.

Solution If x is slightly bigger than 0, then $1/x$ is a very large positive number. The closer x is to 0, the larger $1/x$ is. There is no single number we can call the limit of $1/x$ as x approaches 0. We say that

$$\lim_{x\to 0} \frac{1}{x}$$

doesn't exist.

Example 18 (*Compare Exercise 43*)

Find $\lim_{x \to 1} \frac{x^2 - 1}{x + 3}$.

Solution We can use the quotient theorem, and say that the limit is $0/4 = 0$. It's only when the **denominator** is 0 that we can't apply the theorem.

2–3 Exercises

I.

Use the definition of limit to find the following limits:

1. (*See Example 1*) $\lim_{x \to 2} x^2$
2. $\lim_{x \to 2} 3x$
3. $\lim_{x \to 7} (2x - 8)$
4. $\lim_{x \to -2} (5x - 3)$
5. (*See Example 2*) $\lim_{x \to 2} (x^2 - 9)$
6. $\lim_{x \to 3} (2x + 5)$
7. $\lim_{x \to -1} (x^2 + 3)$
8. (*See Example 5*) $\lim_{x \to -2} \frac{x^2 + 4}{x + 7}$
9. (*See Example 6*) $\lim_{x \to 3} 5$
10. $\lim_{x \to 1} -2$
11. (*See Example 6*) $\lim_{x \to 4} x$
12. $\lim_{x \to 0} x$

Use Figure 2–8 to find the following limits in Exercises 13 through 16.

13. $\lim_{x \to -1} f(x)$
14. $\lim_{x \to 0} f(x)$
15. (*See Example 3*) $\lim_{x \to 2} f(x)$
16. $\lim_{x \to 3} f(x)$

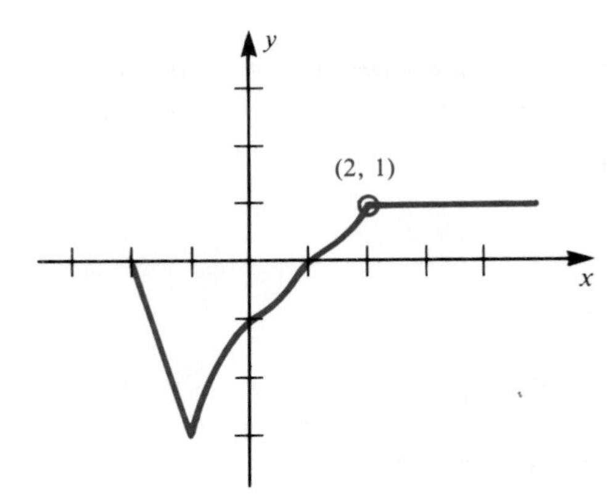

Figure 2–8

Use Figure 2–9 to find the following limits.

17. $\lim_{x \to 0} f(x)$
18. $\lim_{x \to 1} f(x)$
19. (*See Example 1*) $\lim_{x \to 3} f(x)$
20. $\lim_{x \to -3} f(x)$

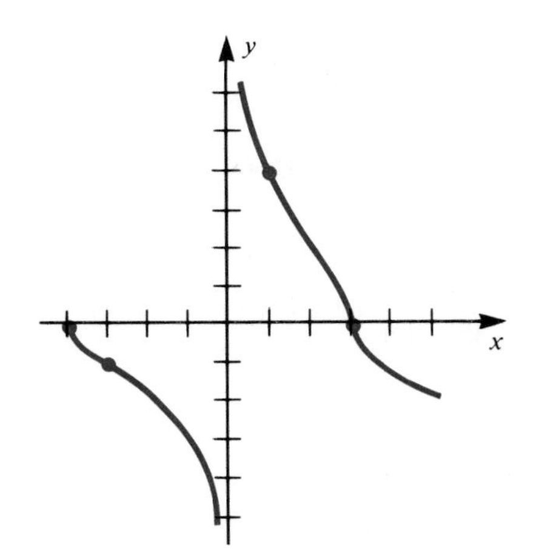

Figure 2–9

II.

21. (*See Example 8*) If $\lim_{x\to 3} f(x) = 5$, what is $\lim_{x\to 3} (6f(x))$?

22. If $\lim_{x\to 3} f(x) = 6$ and $\lim_{x\to 3} g(x) = 9$, what are

(a) $\lim_{x\to 3} (f(x) + g(x))$?

(b) $\lim_{x\to 3} (f(x) \cdot g(x))$?

23. If $\lim_{x\to 2} f(x) = -1$ and $\lim_{x\to 2} g(x) = 7$, what are

(a) $\lim_{x\to 2} (g(x) - f(x))$?

(b) (*See Example 9*) $\lim_{x\to 2} \dfrac{g(x)}{f(x)}$?

24. If $\lim_{x\to 2} f(x) = -3$ and $\lim_{x\to 2} g(x) = 8$, what are

(a) $\lim_{x\to 2} (5\,f(x) + 2g(x))$?

(b) $\lim_{x\to 2} (f(x) + x)$?

25. If $\lim_{x\to 4} f(x) = 9$ what are

(a) $\lim_{x\to 4} \sqrt{f(x)}$?

(b) (*See Example 15*) $\lim_{x\to 4} (f(x))^2$?

26. If $\lim_{x\to 3} f(x) = 2$ what are

(a) $\lim_{x\to 3} (4x + (f(x))^2)$?

(b) $\lim_{x\to 3} (f(x))^{-5}$?

27. If $\lim_{x\to 2} f(x) = 5$, what is $\lim_{x\to 2} (xf(x))$?

28. If $\lim_{x\to 3} f(x) = 12$, what is $\lim_{x\to 3} \dfrac{f(x)}{x}$?

29. (*See Example 13*) What is $\lim_{\Delta x\to 0} (6 + \Delta x)^2$?

30. What is $\lim_{\Delta x\to 0} (a + \Delta x)^2$?

III.

Compute each of the following limits.

31. (*See Example 3*) $\lim_{x \to 1} \frac{x^2 - 1}{x - 1}$

32. $\lim_{x \to 2} \frac{x^2 + x - 6}{x - 2}$

33. (*See Example 12*) $\lim_{x \to 0} \frac{x^2 + 5x}{x}$

34. $\lim_{\Delta x \to 0} \frac{3\Delta x - 7(\Delta x)^2}{\Delta x}$

35. $\lim_{\Delta x \to 0} \frac{6\Delta x + (\Delta x)^2 - 4(\Delta x)^3}{\Delta x}$

36. $\lim_{\Delta x \to 0} \frac{\sqrt{4 + \Delta x} - 2}{\Delta x}$

37. (*See Example 16*) $\lim_{x \to 9} \frac{\sqrt{x} - 3}{x - 9}$

38. $\lim_{x \to 25} \frac{\sqrt{x} - 5}{x - 25}$

Use the limit definition of derivative to find $f'(a)$ for the given function and the given value of a.

39. (*See Example 4*) $f(x) = 5x^2$, $a = 1$

40. $f(x) = x^2$, $a = -1$

Find the following limits or state that they do not exist.

41. (*See Example 17*) $\lim_{x \to 0} \frac{1}{x^2}$

42. $\lim_{x \to 3} \frac{1}{x - 3}$

43. (*See Example 18*) $\lim_{x \to 1} \frac{x - 1}{x}$

44. $\lim_{x \to 2} \frac{x - 1}{x^2}$

45. $\lim_{x \to 0} \frac{x^2 + 5x}{x + 3}$

46. $\lim_{x \to 0} \frac{x^2 + 5x}{x(x + 3)}$

47. $\lim_{x \to 0} \frac{x^2 + 5x}{x^2(x + 3)}$

48. $\lim_{x \to 4} \frac{(x - 4)^2}{x - 4}$

49. $\lim_{x \to 4} \frac{(x - 4)^2}{(x - 4)^2}$

50. $\lim_{x \to 4} \frac{(x - 4)^2}{(x - 4)^3}$

2–4 One-Sided Limits and Continuity

One-Sided Limits
Continuity

One-Sided Limits

We have seen that, although the formula for $f(x)$ looks complicated, the graph of

$$f(x) = \frac{x^2 - 9}{x - 3}$$

is simply a straight line with a hole in it. How does the idea of limit apply to a function whose graph jumps around? In Chapter 1, we looked at a consumer

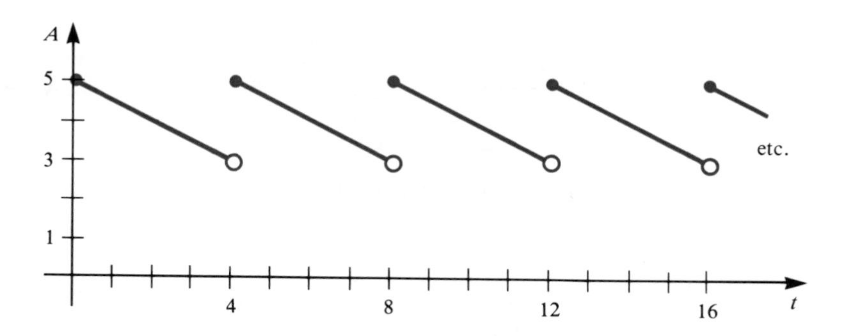

Figure 2–10

awareness function. The rule for evaluating this function was as follows:

$$A(t) = \begin{cases} -\frac{1}{2}t + 5, & 0 \le t < 4 \\ -\frac{1}{2}t + 7, & 4 \le t < 8 \\ -\frac{1}{2}t + 9, & 8 \le t < 11 \\ \text{etc.} \end{cases}$$

Its graph is shown in Figure 2–10.

What is happening to the function near $t = 4$? What is $\lim_{t \to 4} A(t)$?

The function now is called A, not f, and we are using t (for time) instead of x, but the idea is the same—if t is near 4, what is $A(t)$ near? In this example, knowing that t is near 4 doesn't tell us what $A(t)$ is near. If t is near 4 and a little bit *less than 4*, then $A(t) = -\frac{1}{2}t + 5$, so $A(t)$ is near 3. Look at the graph a little bit to the left of $t = 4$.

On the other hand, if t is near 4 and a little bit *greater than 4*, then $A(t) = -\frac{1}{2}t + 7$ is close to 5. Now look at the section of the graph just to the right of $t = 4$.

In an attempt to preserve the geometric flavor, mathematicians have coined the words "left-hand limit" and "right-hand limit." The notation for left-hand limit uses the symbol "$x \to c^-$" in place of "$x \to c$." The minus sign to the right and above c means that x is just a little bit **less** than c.

Definition

> The number L is called **the left-hand limit of f at c** if $f(x)$ is near L whenever x is near c and less than c; we write $\lim_{x \to c^-} f(x) = L$.

To return to the consumer-awareness example, we can now write

$$\lim_{t \to 4^-} A(t) = \lim_{t \to 4^-} (-\tfrac{1}{2}t + 5) = 3$$

The right-hand limit is defined in an analogous manner, using $x \to c^+$ to mean that x is just a little bit more than c.

Definition

The number L is called **the right-hand limit of f at c** if $f(x)$ is near L whenever x is near c and greater than c; we write $\lim_{x \to c^+} f(x) = L$.

In our consumer-awareness example we can now say

$$\lim_{t \to 4^+} A(t) = \lim_{t \to 4^+} (-\tfrac{1}{2}t + 7) = 5$$

All the rules from the previous section that we used to compute limits are valid for one-sided limits, and we will use them without repeating them here.

Because $A(t)$ is close to 5 if t is close to and greater than 4, but $A(t)$ is close to 3 if t is close to and less than 4, there is not a single number we can say $A(t)$ is close to if t is close to 4. Because there is no such single number, we say that $\lim_{t \to 4} A(t)$ doesn't exist. This example is one instance of a general statement that can help us in computing the limit or in deciding that the limit doesn't exist.

Theorem

$\lim_{x \to c} f(x) = L$ if and only if both $\lim_{x \to c^-} f(x) = L$ *and* $\lim_{x \to c^+} f(x) = L$.

Example 1 (*Compare Exercise 7*)
Using the consumer-awareness function above, what are

(a) $\lim_{t \to 8^-} A(t)$? **(b)** $\lim_{t \to 8^+} A(t)$? **(c)** $\lim_{t \to 8} A(t)$?

Solution

(a) $\lim_{t \to 8^-} A(t) = \lim_{t \to 8^-} (-\tfrac{1}{2}t + 7) = -4 + 7 = 3$

(b) $\lim_{t \to 8^+} A(t) = \lim_{t \to 8^+} (-\tfrac{1}{2}t + 9) = -4 + 9 = 5$

(c) Because the answers to (a) and (b) are not equal, $\lim_{t \to 8} A(t)$ does not exist.

Example 2 (*Compare Exercises 1 and 12*)
What is $\lim_{x \to 0} \dfrac{|x|}{x}$?

Solution

(This is a particularly important example, and we will look at it again in a slightly different setting in Example 10 of the next section.) Remember that $|x|$ means the absolute value of x, and the rule for evaluating $|x|$ depends on the sign of x. If x is positive, $|x| = x$; but if x is negative, $|x|$ has the opposite sign from x, so $|x| = -x$. Let $f(x) = |x|/x$.

$$\text{If } x > 0,\ f(x) = \frac{|x|}{x} = \frac{x}{x} = 1;\ \text{on the other hand}$$

$$\text{if } x < 0,\ f(x) = \frac{|x|}{x} = \frac{-x}{x} = -1$$

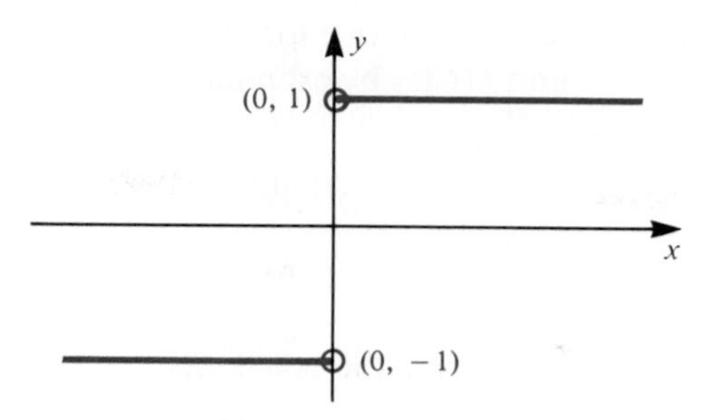

Figure 2–11

We can rewrite the rule for $f(x)$ as follows:

$$f(x) = \frac{|x|}{x} = \begin{cases} 1 & \text{if} \quad x > 0 \\ -1 & \text{if} \quad x < 0 \end{cases}$$

The graph of $f(x) = |x|/x$ is given in Figure 2–11.

Note: The empty dots at the points (0, 1) and (0, −1) indicate that these points are not in the graph of f. Furthermore, the number 0 is not in the domain of f, but we can still ask about $\lim_{x \to 0} f(x)$.

The value of $f(x)$ depends on which side of 0 that x comes from; this is a hint to us that we should consider the one-sided limits. Moreover, the geometry of the graph, the jump when $x = 0$, suggests that we should examine limts from each side of 0. For the right-hand limit,

$$\lim_{x \to 0^+} f(x) = \lim_{x \to 0^+} \frac{|x|}{x} = \lim_{x \to 0^+} \frac{x}{x} = \lim_{x \to 0^+} 1 = 1$$

For the left-hand limit,

$$\lim_{x \to 0^-} f(x) = \lim_{x \to 0^-} \frac{|x|}{x} = \lim_{x \to 0^-} \frac{-x}{x} = \lim_{x \to 0^-} -1 = -1$$

Because we get two different numbers, we say that $\lim_{x \to 0} |x|/x$ doesn't exist.

Warning! It would be a mistake to conclude that a change in the rule for evaluating a function means that the function doesn't have a limit where the rule changes. A change in the evaluation formula suggests the use of one-sided limits, but **use** them before making any conclusions.

Example 3 (*Compare Exercise 16*)

A certain manufacturing company has noticed that for the first 50 items it produces each day, the cost per item is a constant \$20, and then the cost per item starts decreasing. A term sometimes used for the cost of the xth item is **marginal cost**, written $MC(x)$. This particular company has estimated that its marginal cost (in dollars) is given by

$$MC(x) = \begin{cases} 20, & 0 \le x \le 50 \\ 1000/x, & x > 50 \end{cases}$$

What is $\lim_{x \to 50} MC(x)$?

Solution Because the rule for evaluating $MC(x)$ changes at $x = 50$, we will decide about $\lim_{x \to 50} MC(x)$ by examining left-hand and right-hand limits.

$$\text{(Right)} \quad \lim_{x \to 50^+} MC(x) = \lim_{x \to 50^+} \frac{1000}{x} = \frac{1000}{50} = 20$$

$$\text{(Left)} \quad \lim_{x \to 50^-} MC(x) = \lim_{x \to 50^-} 20 = 20$$

Comparing the two limits, we find

$$\lim_{x \to 50^+} MC(x) = \lim_{x \to 50^-} MC(x) = 20, \text{ so we conclude}$$

$$\lim_{x \to 50} MC(x) = 20$$

If the right-hand limit equals the left-hand limit, then their common value is the value of the limit. The graph of this function is shown in Figure 2–12.

A change in the computational rule does not necessarily mean a jump in the graph!

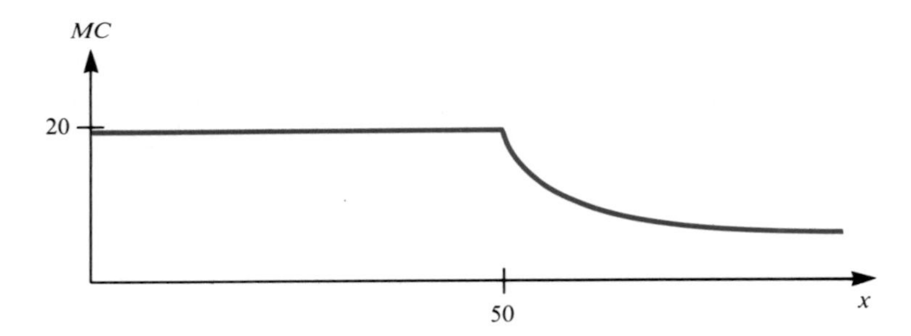

Figure 2–12

Example 4 (*Compare Exercise 1*)
Use the graph in Figure 2–13 to find the following limits.

(a) $\lim_{x \to 1^+} f(x)$ **(b)** $\lim_{x \to 1^-} f(x)$ **(c)** $\lim_{x \to 1} f(x)$

(d) $\lim_{x \to 2^+} f(x)$ **(e)** $\lim_{x \to 2^-} f(x)$ **(f)** $\lim_{x \to 2} f(x)$

Solution The rule for evaluating $f(x)$ is not given, so all we can do is look at the graph. Remember the distinction between the number $\lim_{x \to c} f(x)$ and the number $f(c)$. Here the heavy dots on the graph indicate that $f(1) = 5$ and $f(2) = 3$, but this doesn't tell us anything about the limits.

(a) $\lim_{x \to 1^+} f(x) = 5$ **(b)** $\lim_{x \to 1^-} f(x) = 3$

(c) $\lim_{x \to 1} f(x)$ doesn't exist

The answer to (c) follows because the answers to (a) and (b) are different. Remember not to look at $f(1)$ in deciding about $\lim_{x \to 1} f(x)$.

(d) $\lim_{x \to 2^+} f(x) = 4$ **(e)** $\lim_{x \to 2^-} f(x) = 4$ **(f)** $\lim_{x \to 2} f(x) = 4$

The answer to (f) follows because the answers to (d) and (e) are the same. The fact that $f(2) = 3$ has nothing to do with deciding that $\lim_{x \to 2} f(x) = 4$.

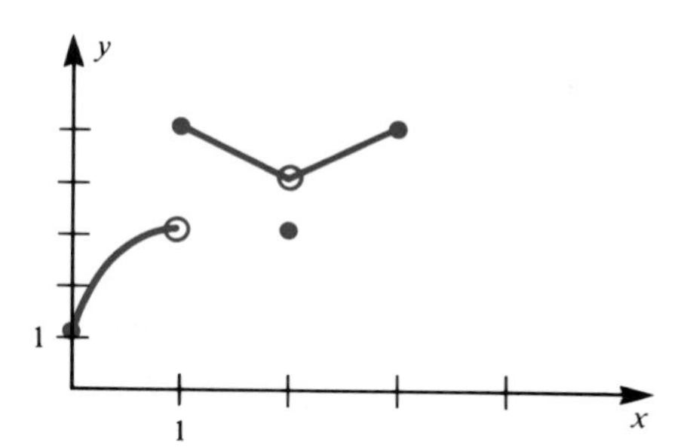

Figure 2–13

Continuity

If someone said to you, "The water flowed out of the faucet continuously for an hour," you would know that the flow of water was uninterrupted—the stream of water was unbroken. Mathematicians use the word **continuous** in a similar way; they say that a function is continuous if its graph is not broken, that is, if the graph just has one piece. Let's look at two examples of functions that are not continuous and see if we can discover what causes the graph to have a break in it.

Example 5 (*Compare Exercise 4*)
Give examples of graphs that are "broken"; the function is **not continuous.**

Solution

1. The graph can have a jump (see Figure 2–14).
   In both these cases $\lim_{x \to 0} f(x)$ does not exist.
2. The graph can have a hole in it (see Figure 2–15).
   When the graph has a hole in it, there is a point "missing" from the graph. Here $\lim_{x \to 3} f(x) = 6$, but $f(3)$ is not defined.

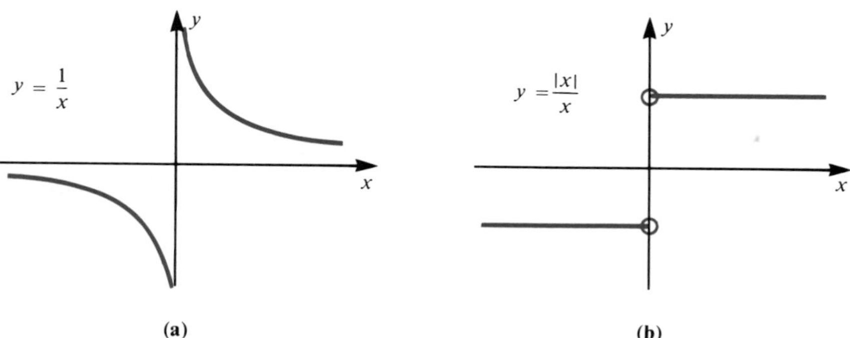

Figure 2–14 (a) (b)

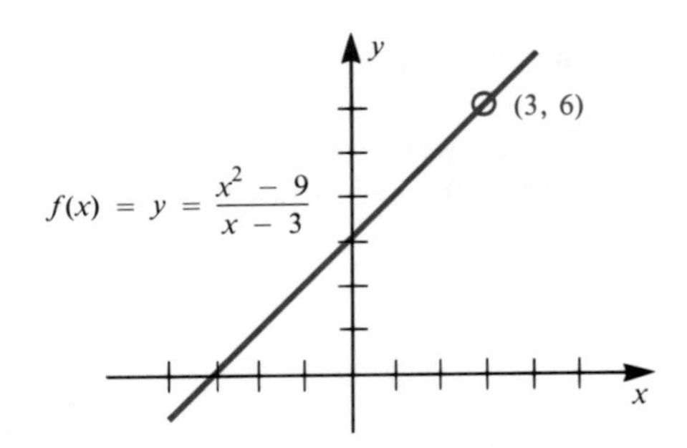

Figure 2–15

These examples are important because they show everything that can produce a break: Either the two pieces of the graph don't match up—the limit of $f(x)$ as $x \to c$ doesn't exist—or they do match up, but the graph has a hole in it—$f(c)$ is not what it should be.

So if we require that (1) the two pieces match up and that (2) the graph doesn't have a hole, then the graph can't have a break and the function will be continuous.

Mathematicians have discovered that these geometric properties (1) and (2) will hold if $f(c)$ and $\lim\limits_{x \to c} f(x)$ are the same number.

Definition

> f is **continuous at $x = c$** if $\lim\limits_{x \to c} f(x) = f(c)$.

Continuous functions are those that behave nicely; their limits can be found by "plugging in."

You must go through 3 steps to see if a function is continuous at the number c.

1. Check that $f(c)$ is defined.
2. Check that $\lim\limits_{x \to c} f(x)$ exists.
3. Finally, check that the numbers you found in (1) and (2) are the same; $f(c) = \lim\limits_{x \to c} f(x)$.

Example 6 (*Compare Exercise 9*)

If $f(x) = \dfrac{x^2 - 9}{x - 3}$,

(a) is f continuous at $x = 3$?

(b) is f continuous at $x = 5$?

Solution **(a)** We get this answer quickly, for in the first step, we see that $f(3)$ is not defined. Thus, f is not continuous at $x = 3$.

(b) First, note that $f(5) = \dfrac{25 - 9}{5 - 3} = 8$.

(More simply, $f(x) = x + 3$ whenever $x \neq 3$, so $f(5) = 5 + 3 = 8$.) Second,

$$\lim_{x \to 5} f(x) = \lim_{x \to 5} \frac{x^2 - 9}{x - 3} = \frac{16}{2} = 8$$

(the limit of a quotient = the quotient of the limits if the limit in the denominator is not 0). The first and second steps both had 8 as their answer. Thus, $\lim_{x \to 5} f(x) = f(5)$, and so f is continuous at $x = 5$.

The statements we had in the previous section about computing limits of quotients of polynomials show us that the above example is a special case of a general fact:

Theorem

> If $f(x) = P(x)/Q(x)$ where P and Q are polynomials, then f is continuous at c if and only if $Q(c) \neq 0$.

We can now give fast answers to the previous example:

$$\text{If } f(x) = \frac{x^2 - 9}{x - 3}, \qquad \text{then } f$$

is not continuous at $x = 3$ but is continuous at $x = 5$. Here $P(x) = x^2 - 9$, $Q(x) = x - 3$, and $Q(3) = 0$, but $Q(5) = 2 \neq 0$. This example shows that a function can be continuous at one number and not continuous at another number.

We will say that a function is **continuous on an interval I** if the function is continuous at all numbers in the interval I. Actually, for a function to be continuous at the endpoints of $[a, b]$ all that is required is that $\lim_{x \to a^+} f(x) = f(a)$ and $\lim_{x \to b^-} f(x) = f(b)$. You do not have to look at $\lim_{x \to a^-} f(x)$ or $\lim_{x \to b^+} f(x)$. Remember that a function is continuous on an interval I if the graph of the function over I has just one piece.

In this text, we will not get involved in many subtle distinctions. For the sake of completeness, however, we do point out that the consumer-awareness function with which we started this section is continuous on $[0, 4)$ and on $[4, 8)$ but is not continuous on $[0, 8)$.

As another example, if

$$f(x) = \frac{x^2 - 9}{x - 3}$$

then f is continuous on $(3, 9)$ and on $[1, 2]$ but is not continuous on $(0, 10)$. In fact, this function is continuous on any interval that doesn't contain the number 3.

Example 7 (*Compare Exercise 10*)

$$\text{Let } f(x) = \begin{cases} \dfrac{x^2 + x - 2}{x - 1}, & x \neq 1 \\ 3, & x = 1 \end{cases}$$

Discuss the continuity of f.

Solution If $x \neq 1$,

$$f(x) = \frac{x^2 + x - 2}{x - 1}$$

is the quotient of polynomials and the denominator is not 0, so f is continuous at all numbers $x \neq 1$.

What about $x = 1$?

First, we check that $f(1)$ is defined. Here $f(1) = 3$. Careful—don't use the same rule for $x = 1$ that you use for $x \neq 1$.

Second, we must check $\lim_{x \to 1} f(x)$. (If this limit is 3, f is continuous at $x = 1$; if this limit is not 3, f is not continuous at $x = 1$.)

$$\begin{aligned} \lim_{x \to 1} f(x) &= \lim_{x \to 1} \frac{x^2 + x - 2}{x - 1} \\ &= \lim_{x \to 1} \frac{(x + 2)(x - 1)}{x - 1} \\ &= \lim_{x \to 1} (x + 2) = 3 \end{aligned}$$

Third, $f(1) = \lim_{x \to 1} f(x)$, so f is continuous at $x = 1$. We summarize by saying that f is continuous for all x.

Warning! The rules for computing $f(x)$ may be handed to you in pieces, but the graph of f may still be connected. Odd-looking functions can still be continuous.

In fact, we draw the graph of f in Figure 2–16. If $x \neq 1$,

$$\frac{x^2 + x - 2}{x - 1} = x + 2$$

so the graph is the straight line $y = x + 2$ except for point (1, 3). But now, $f(1) = 3$ means that the point (1, 3) *is* on the graph. In fact, the function is $f(x) = x + 2$ but was just in disguise.

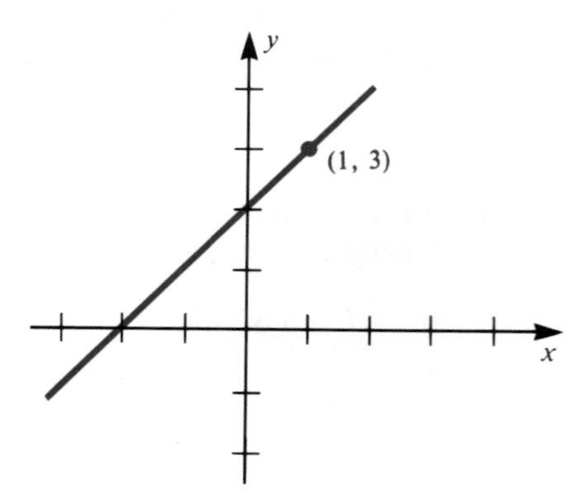

Figure 2–16

2–4 Exercises

I.

Use Figure 2–17 to complete the equations in Exercises 1 through 3.

1. (*See Examples 2 and 4*)

   (a) $\lim_{x\to 0^+} f(x) =$ **(b)** $\lim_{x\to 0^-} f(x) =$ **(c)** $\lim_{x\to 0} f(x) =$ **(d)** $f(0) =$

   (e) Is the function continuous at $x = 0$?

2. **(a)** $\lim_{x\to 2^-} f(x) =$ **(b)** $\lim_{x\to 2^+} f(x) =$ **(c)** $\lim_{x\to 2} f(x) =$ **(d)** $f(2) =$

   (e) Is the function continuous at $x = 2$?

3. **(a)** $\lim_{x\to 3^-} f(x) =$ **(b)** $\lim_{x\to 3^+} f(x) =$ **(c)** $\lim_{x\to 3} f(x) =$ **(d)** $f(3) =$

   (e) Is the function continuous at $x = 3$?

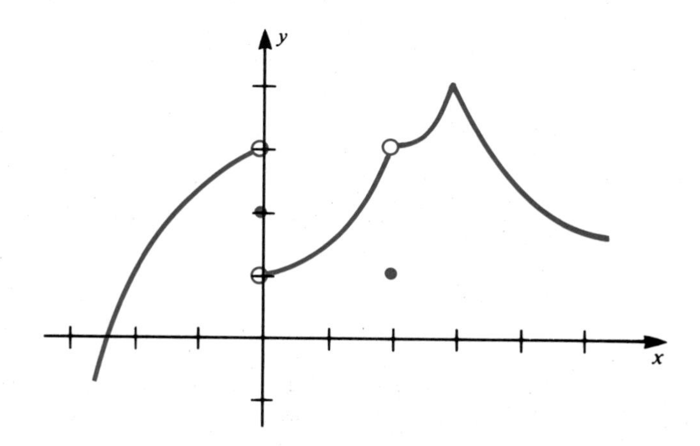

Figure 2–17

Use Figure 2–18 to complete the equations in Exercises 4 through 6.

4. (*See Example 5*)

   (a) $\lim_{x\to 0^-} g(x) =$ **(b)** $\lim_{x\to 0^+} g(x) =$ **(c)** $\lim_{x\to 0} g(x) =$ **(d)** $g(0) =$

   (e) Is the function continuous at $x = 0$?

5. **(a)** $\lim_{x\to 3^-} g(x) =$ **(b)** $\lim_{x\to 3^+} g(x) =$ **(c)** $\lim_{x\to 3} g(x) =$ **(d)** $g(3) =$

   (e) Is the function continuous at $x = 3$?

6. **(a)** $\lim_{x\to 5^-} g(x) =$ **(b)** $\lim_{x\to 5^+} g(x) =$ **(c)** $\lim_{x\to 5} g(x) =$ **(d)** $g(5) =$

   (e) Is the function continuous at $x = 5$?

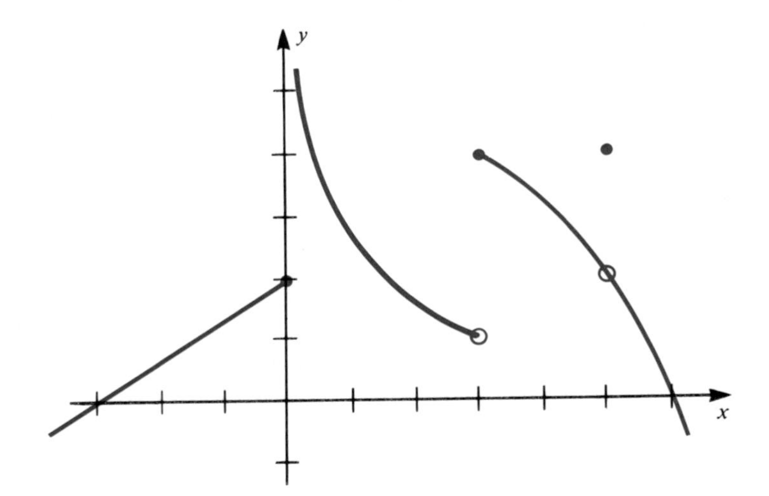

Figure 2–18

II.

7. (*See Example 1*)
Let f be defined by

$$f(x) = \begin{cases} 3x + 2, & x \le 1 \\ x^2 + 5, & x > 1 \end{cases}$$

What is

(a) $\lim_{x \to 1^-} f(x)$? **(b)** $\lim_{x \to 1^+} f(x)$? **(c)** $\lim_{x \to 1} f(x)$? **(d)** $f(1)$?
(e) Is f continuous at $x = 1$?

8. Let g be defined by

$$g(x) = \begin{cases} 16 - x^2, & x < 2 \\ 8, & x = 2 \\ 3x + 6, & x > 2 \end{cases}$$

What is

(a) $\lim_{x \to 2^-} g(x)$? **(b)** $\lim_{x \to 2^+} g(x)$? **(c)** $\lim_{x \to 2} g(x)$? **(d)** $g(2)$?
(e) Is g continuous at $x = 2$?

9. (*See Example 6*)
Let f be defined by

$$f(x) = \begin{cases} \dfrac{x^2 + x - 6}{x - 2}, & x \ne 2 \\ 7, & x = 2 \end{cases}$$

What is

(a) $\lim_{x \to 2} f(x)$? **(b)** $f(2)$? **(c)** Is f continuous at $x = 2$?

III.

10. (*See Example 7*)
Let f be defined by

$$f(x) = \begin{cases} \dfrac{x^2 - 3x}{x}, & x \neq 0 \\ B, & x = 0 \end{cases}$$

Is there a value of B so that f is continuous at $x = 0$? If so, what is it?

11. Let f be defined by

$$f(x) = \begin{cases} 2x + 3, & x \leq 2 \\ Ax - 9, & x > 2 \end{cases}$$

Is there a value of A so that f is continuous at $x = 2$? If so, what is it?

12. (*See Example 2*)
Let $f(x) = \dfrac{|x - 3|}{x - 3}$. Remember $|x - 3| = \begin{cases} x - 3 & \text{if} \quad x \geq 3 \\ -(x - 3) & \text{if} \quad x < 3 \end{cases}$
What is

(a) $\lim_{x \to 3^-} f(x)$? **(b)** $\lim_{x \to 3^+} f(x)$? **(c)** $\lim_{x \to 3} f(x)$?
(d) $\lim_{x \to 4^-} f(x)$? **(e)** $\lim_{x \to 4^+} f(x)$? **(f)** $\lim_{x \to 4} f(x)$?

13. Let $g(x) = \dfrac{|4 - x|}{4 - x}$
What is

(a) $\lim_{x \to 4^+} g(x)$? **(b)** $\lim_{x \to 4^-} g(x)$? **(c)** $\lim_{x \to 4} g(x)$?
(d) $\lim_{x \to 5^+} g(x)$? **(e)** $\lim_{x \to 5^-} g(x)$? **(f)** $\lim_{x \to 5} g(x)$?

14. Graph the function

$$f(x) = \begin{cases} 2x + 1, & x \leq 1 \\ 4 - x, & x > 1 \end{cases}$$

15. Graph the function

$$f(x) = \begin{cases} x^2, & x < 3 \\ 12 - x, & x \geq 3 \end{cases}$$

16. (*See Example 3*)
A company has found that its marginal cost function follows the rule

$$MC(x) = \begin{cases} 50 - .5x, & 0 \leq x \leq 60 \\ 20 + .01x^2, & x > 60 \end{cases}$$

Is the company's marginal cost function continuous at $x = 60$?

17. Management has found that the employees' general alertness (alertness is measured on a scale of 0–10) after being on the job for t hours follows the pattern given by

$$A(t) = \begin{cases} 5 + 2t - \frac{1}{3}t^2, & 0 \leq t \leq 6 \\ 8 - \dfrac{t}{2}, & 6 < t \leq 8 \end{cases}$$

Is this alertness function continuous on the interval [0, 8]?

2–5 Derivatives and Tangent Lines

Introduction

We do not live in a static world, but one that is always changing; the derivative is the mathematical tool used to quantify and describe rates of change. The derivative of f at a, written $f'(a)$, is defined by the equation

$$f'(a) = \lim_{\Delta x \to 0} \frac{f(a + \Delta x) - f(a)}{\Delta x}$$

and can be interpreted as the instantaneous rate of change of f when $x = a$. This same limit is also important geometrically, because there is a connection between $f'(a)$ and the graph of f. Understanding this relation will enable us to graph complicated functions by only plotting a few points. These graphs can then be used to give us a picture of the behavior of a function without burying us in a mass of data. The geometric interpretation of derivative paves the way to seeing how the derivative can be used to find optimal solutions to problems and to shed light on certain principles of economics. For example, one such principle we shall investigate later states: The biggest profit occurs when the marginal cost and marginal revenue are the same.

Slope of Tangent Lines

We turn now to a geometric interpretation of $[f(a + \Delta x) - f(a)]/\Delta x$. This quotient represents the change in functional values divided by the change in x. If we let $y = f(x)$ and interpret the symbol Δ (Δ is the Greek letter "delta" and is traditionally used in this setting) to mean "the change in," then

$$f(a + \Delta x) - f(a) = \text{the change in } y = \Delta y$$

Now we can rewrite the quotient

$$\frac{f(a + \Delta x) - f(a)}{\Delta x} \text{ as } \frac{\Delta y}{\Delta x}$$

The symbols $\Delta y/\Delta x$ can be read literally as "delta y over delta x" or by translating to "the change in y divided by the change in x." For a straight line, $\Delta y/\Delta x$ gives the slope of the line. Let's look at a picture to see what this all means; please refer to Figure 2–19.

We have labelled the point $(a, f(a))$ with the letter P, and used Q to label the point $(a + \Delta x, f(a + \Delta x))$. Then we have drawn the straight line L_Q through P and Q. The slope of L_Q is $\Delta y/\Delta x$. Now as Δx gets smaller the point Q slides along the graph $y = f(x)$ and approaches the point P, as is shown in Figure 2–20. The

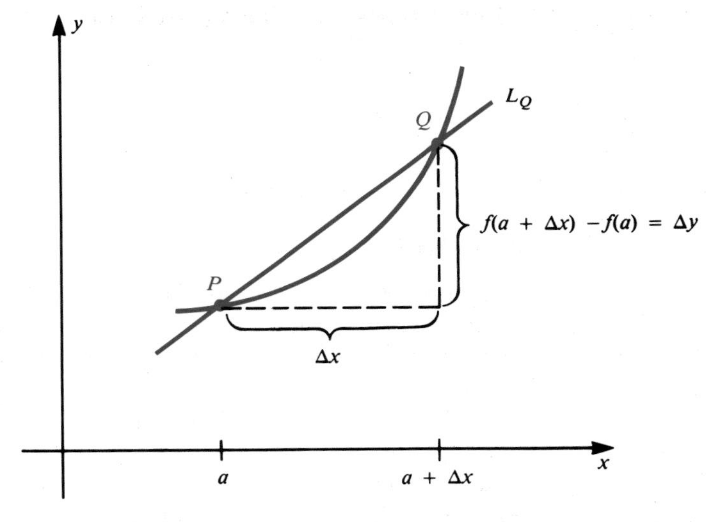

Figure 2–19

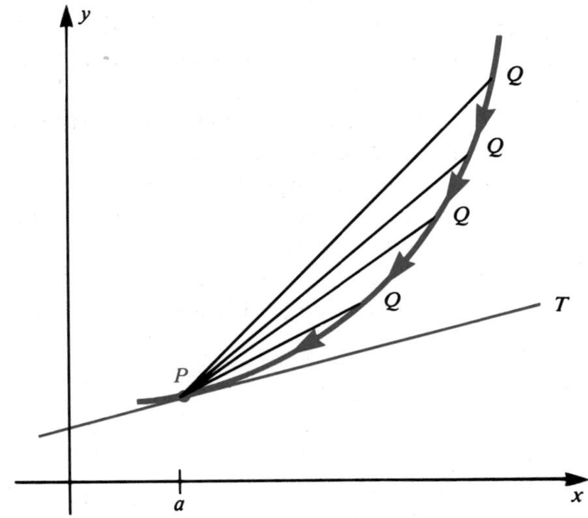

Figure 2–20

figure also shows the line tangent to the curve at the point $(a, f(a))$. Let's call this tangent line T and let m denote the slope of T. If Δx is close to 0, then the line through P and Q has a slope that is close to the slope of T. That is, if Δx is close to 0, then $\Delta y/\Delta x$ is close to m. We have already defined $f'(a)$ to represent the number that $\Delta y/\Delta x$ approaches as Δx approaches 0, so we must have $f'(a) = m$. The derivative of f at a is the slope of the line tangent to the graph of f at the point $(a, f(a))$. We use this to give a formal definition of tangent line.

Definition

The **line tangent to the graph of f at the point** $(a, f(a))$ is the line through $(a, f(a))$ with slope $f'(a)$.

Example 1 (*Compare Exercise 9*)
Given the graph $y = x^2 + 3$,

(a) find the slope of the line tangent to the graph at the point (2, 7);
(b) find an equation of this tangent line.

Solution The graph of f is shown in Figure 2–21, and the tangent line T has also been drawn. We want the slope of T.

(a) The slope of T is

$$f'(2) = \lim_{\Delta x \to 0} \frac{f(2 + \Delta x) - f(2)}{\Delta x}.$$

We now compute this limit using our three-step process for computing derivatives.

STEP 1 Compute (and simplify) $f(2 + \Delta x) - f(2)$.

$$f(2) = 2^2 + 3 = 7$$
$$f(2 + \Delta x) = (2 + \Delta x)^2 + 3 = 4 + 4\Delta x + (\Delta x)^2 + 3$$
$$= 7 + 4\Delta x + (\Delta x)^2$$

So, $f(2 + \Delta x) - f(2) = 4\Delta x + (\Delta x)^2$

STEP 2 Divide the result in Step 1 by Δx.

$$\frac{4\Delta x + (\Delta x)^2}{\Delta x} = \frac{(4 + \Delta x)(\Delta x)}{\Delta x} = 4 + \Delta x$$

STEP 3 Compute the limit as $\Delta x \to 0$.

$$\lim_{\Delta x \to 0} (4 + \Delta x) = 4$$

Thus, $f'(2) = 4$, and the slope of T is 4.

(b) The line T goes through the point (2, 7), and now we also know the slope of T is 4. We can use the point-slope form to find an equation for T; we have $y - 7 = 4(x - 2)$.

Warning! Remember parentheses; do not write $y - 7 = 4x - 2$.

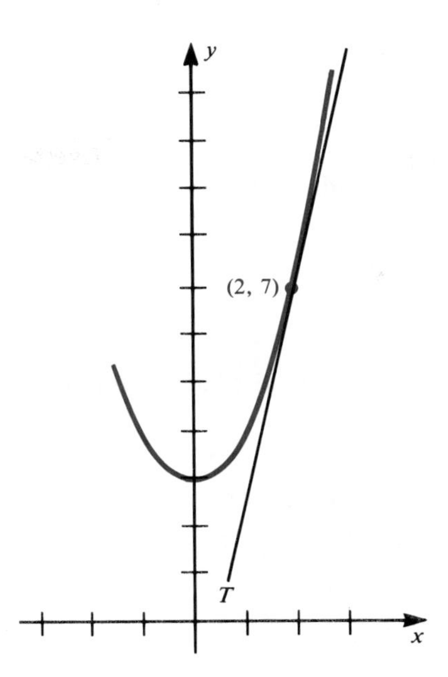

Figure 2–21

You will want to compute the derivative to understand the geometry; sometimes the geometry helps to understand the derivative.

Example 2 (*Compare Exercise 15*)
Given $f(x) = x^2 + 3$, find $f'(0)$.

Solution Refer again to the graph in Figure 2–21. $f'(0)$ is the slope of the line tangent to the graph at (0, 3). To the left of (0, 3), the tangent lines will have negative slope; to the right of (0, 3), the tangent lines will have positive slope. The line tangent to the curve at (0, 3) is a horizontal line; the slope of a horizontal line is 0. Thus, $f'(0) = 0$.

Formulas for Some Derivatives

Example 3 (*Compare Exercise 21*)
If $f(x) = x^2 - 2x - 3$, find the slope of the tangent lines when $x = 0$, when $x = 1$, and when $x = 4$.

Solution Rather than going through three computations to find $f'(0)$, $f'(1)$, and $f'(4)$ separately, let's discover the formula for $f'(a)$ in terms of a. We go through the three-step process for computing

$$\lim_{\Delta x \to 0} \frac{f(a + \Delta x) - f(a)}{\Delta x}$$

STEP 1 Compute $f(a + \Delta x) - f(a)$.

$$\begin{aligned} f(a) &= a^2 - 2a - 3 \\ f(a + \Delta x) &= (a + \Delta x)^2 - 2(a + \Delta x) - 3 \\ &= a^2 + 2a\Delta x + (\Delta x)^2 - 2a - 2\Delta x - 3 \end{aligned}$$

So, $f(a + \Delta x) - f(a) = a^2 + 2a\Delta x + (\Delta x)^2 - 2a - 2\Delta x - 3 - (a^2 - 2a - 3)$

$$= a^2 + 2a\Delta x + (\Delta x)^2 - 2a - 2\Delta x - 3 - a^2 + 2a + 3$$
$$= 2a\Delta x + (\Delta x)^2 - 2\Delta x$$

STEP 2 Divide the result from Step 1 by Δx. Note Δx is a factor of $2a\Delta x + (\Delta x)^2 - 2\Delta x$, so

$$\frac{2a\Delta x + (\Delta x)^2 - 2\Delta x}{\Delta x} = \frac{(2a + \Delta x - 2)\Delta x}{\Delta x}$$
$$= 2a + \Delta x - 2$$

STEP 3 Compute the limit as $\Delta x \to 0$

$$\lim_{\Delta x \to 0} (2a + \Delta x - 2) = 2a - 2$$

There was some messy algebra in Step 1, but now we do have a formula for $f'(a)$: $f'(a) = 2a - 2$.

Substituting the three numbers 0, 1, and 4 into the formula, we have

$$f'(0) = 2 \cdot 0 - 2 = -2$$
$$f'(1) = 0 \qquad \text{and}$$
$$f'(4) = 6$$

Geometrically, Figure 2–22 shows what the tangent lines look like: T_0, T_1, and T_4 are the respective tangent lines. The slope of T_0 is -2; the slope of T_1 is 0 (T_1 is a horizontal line); the slope of T_4 is 6.

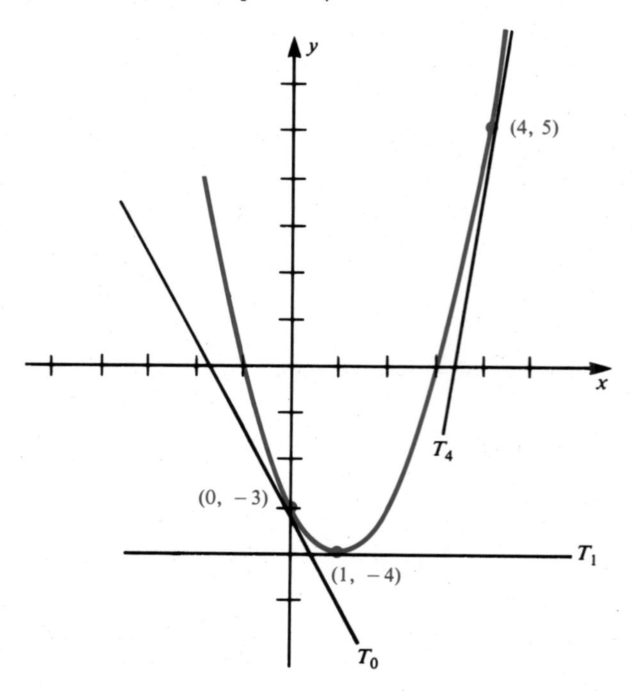

Figure 2–22

Example 4 (*Compare Exercise 23*)

If $f(x) = \frac{1}{x}$, find a formula for $f'(a)$.

Solution We use the three-step process for computing $\lim_{\Delta x \to 0} \frac{f(a + \Delta x) - f(a)}{\Delta x}$.

STEP 1

$$f(a) = \frac{1}{a}; \qquad f(a + \Delta x) = \frac{1}{(a + \Delta x)}$$

$$f(a + \Delta x) - f(a) = \frac{1}{(a + \Delta x)} - \frac{1}{a}$$

To simplify, we combine into one fraction; the common denominator is $(a + \Delta x)a$. Combining, we have

$$= \frac{a}{(a + \Delta x)a} - \frac{a + \Delta x}{(a + \Delta x)a}$$

$$= \frac{a - a - \Delta x}{(a + \Delta x)a}$$

$$= \frac{-\Delta x}{(a + \Delta x)a}$$

Again note that Δx is a factor of the result of Step 1.

STEP 2 Divide the result from Step 1 by Δx. Dividing by Δx is the same as multiplying by $1/\Delta x$.

$$\frac{-\Delta x}{(a + \Delta x)a} \cdot \frac{1}{\Delta x} = \frac{-1}{(a + \Delta x)a}$$

STEP 3 Compute the limit as $\Delta x \to 0$ of the result of Step 2.

$$\lim_{\Delta x \to 0} \frac{-1}{(a + \Delta x)a} = \frac{-1}{(a)a} = \frac{-1}{a^2}$$

Thus, $f'(a) = \frac{-1}{a^2}$.

Example 5 (*Compare Exercise 11*)

Find an equation of the line tangent to the graph $f(x) = 1/x$ at the point $(2, \frac{1}{2})$.

Solution In Example 4, we computed

$$f'(a) = \frac{-1}{a^2}$$

If $a = 2$, then $f'(2) = -\frac{1}{4}$. The tangent line has slope $-\frac{1}{4}$ and goes through

$(2, \frac{1}{2})$ so its point-slope equation is

$$y - \frac{1}{2} = -\frac{1}{4}(x - 2)$$

The graph of $f(x) = 1/x$ and the line tangent to the graph at $(2, \frac{1}{2})$ have been drawn in Figure 2–23.

Notice that if you were to draw any line tangent to this graph, the slope of the tangent line would be negative. This geometric fact is confirmed in the formula for $f'(a)$, the slope. We found in Example 4 that

$$f'(a) = -\frac{1}{a^2}$$

which is always negative.

Other Notation for Derivatives

The rule for evaluating the derivative doesn't have to use the letter a. For example, if

$$f(x) = \frac{1}{x}$$

we now know that

$$f'(a) = -\frac{1}{a^2}$$

We could just as well write this derivative as

$$f'(x) = -\frac{1}{x^2}$$

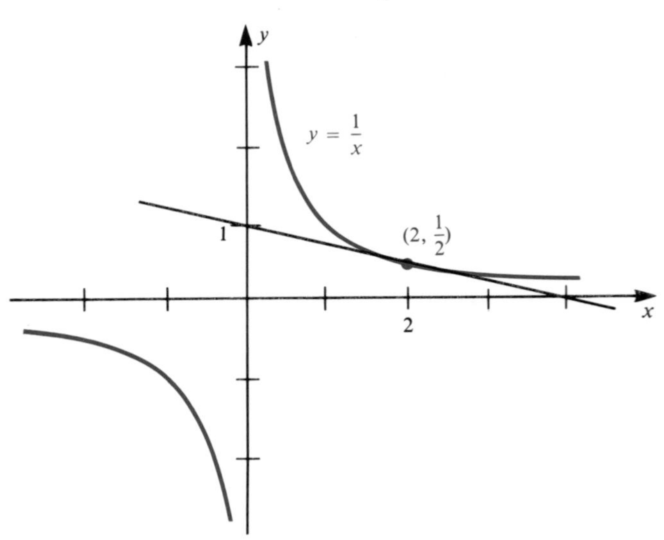

Figure 2–23

Using the result from Example 3, we could also write

$$\text{If } f(x) = x^2 - 2x - 3, \qquad \text{then } f'(x) = 2x - 2$$

In previous sections, we have seen that if $f(x) = mx + b$, then $f'(a) = m$. Thus, we can write

$$\text{If } f(x) = mx + b, \qquad \text{then } f'(x) = m$$

Example 6 (*Compare Exercise 2*)
If $f(x) = 7x - 4$, find $f'(x)$.

Solution

$$f'(x) = 7$$

The derivative is a very important and useful concept in both pure and applied mathematics. During the three hundred years since it was first developed it has been represented by several different notations. Up to now we have used $f'(x)$ to represent the derivative of $f(x)$. Another common notation that we will use is dy/dx (read "the derivative of y with respect to x"). This notation is used to remind us that the derivative is related to slope, and the slope of a line is $\Delta y/\Delta x$. Thus,

$$\frac{dy}{dx} = \lim_{\Delta x \to 0} \frac{\Delta y}{\Delta x}$$

Because we write $y = f(x)$, we also use the notation $\frac{d}{dx} f(x)$, which means the derivative of $f(x)$ with respect to x.

The process of starting with $f(x)$ and then computing the derivative $f'(x)$ is called **differentiation**, or **finding the derivative** of f. Using the result of Example 3 again, we show this terminology in some examples.

Example 7 (*Compare Exercise 1*)

If $y = x^2 - 2x - 3$, what is $\frac{dy}{dx}$?

Solution

$$\frac{dy}{dx} = 2x - 2$$

Example 8 (*Compare Exercise 1*)

If $f(x) = x^2 - 2x - 3$, find $\frac{d}{dx} f(x)$.

Solution

$$\frac{d}{dx} f(x) = 2x - 2$$

Example 9 (*Compare Exercise 1*)
Differentiate $f(x) = x^2 - 2x - 3$.

Solution

$$f'(x) = 2x - 2, \text{ or } \frac{dy}{dx} = 2x - 2, \text{ or } \frac{d}{dx} f(x) = 2x - 2$$

In the next chapter we will show how you can use some computational formulas to write the expression for $f'(x)$ given $f(x)$ without always having to compute a limit.

Examples Where $f'(a)$ Does Not Exist

We conclude this section by looking at the geometry of tangent lines once again, this time to see what happens when the graph does not have a tangent line at a particular point.

Example 10 (*Compare Exercise 17*)
Let $f(x) = |x|$. What is $f'(0)$?

Solution We repeat the three-step process for computing $f'(0)$.

STEP 1

$$f(0) = 0 \qquad f(0 + \Delta x) = f(\Delta x) = |\Delta x|$$

so,

$$f(0 + \Delta x) - f(0) = |\Delta x|$$

STEP 2 Dividing result of Step 1 by Δx we get $|\Delta x|/\Delta x$. We encountered an expression like this in Example 2 of the last section. $|\Delta x|/\Delta x = 1$ if $\Delta x > 0$ but

$$\frac{|\Delta x|}{\Delta x} = -1 \text{ if } \Delta x < 0$$

STEP 3 As we saw in Example 2 of the last section, we should use the left-hand and right-hand limits in this case.

$$\lim_{\Delta x \to 0^+} \frac{|\Delta x|}{\Delta x} = \lim_{\Delta x \to 0^+} 1 = 1, \text{ but}$$

$$\lim_{\Delta x \to 0^-} \frac{|\Delta x|}{\Delta x} = \lim_{\Delta x \to 0^-} -1 = -1. \text{ Therefore,}$$

$$\lim_{\Delta x \to 0} \frac{|\Delta x|}{\Delta x} \quad \text{does not exist.}$$

$f'(0)$ does not exist, and the graph does not have a tangent line at (0, 0). The graph of $f(x) = |x|$ has a "corner" at (0, 0) and so differs, for example, from the graph of $f(x) = x^2$ which is "smooth" at (0, 0). See Figure 2–24.

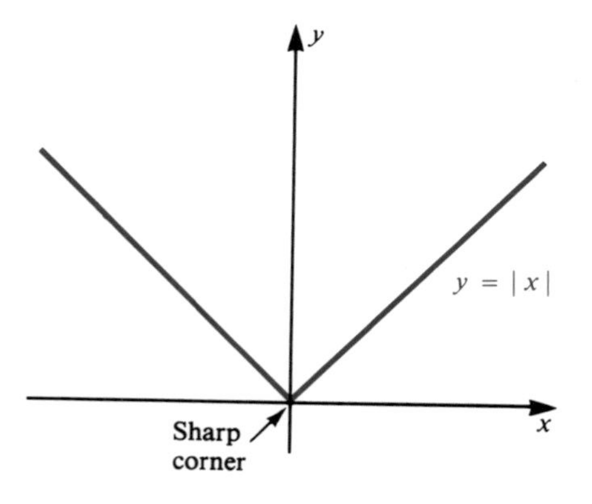

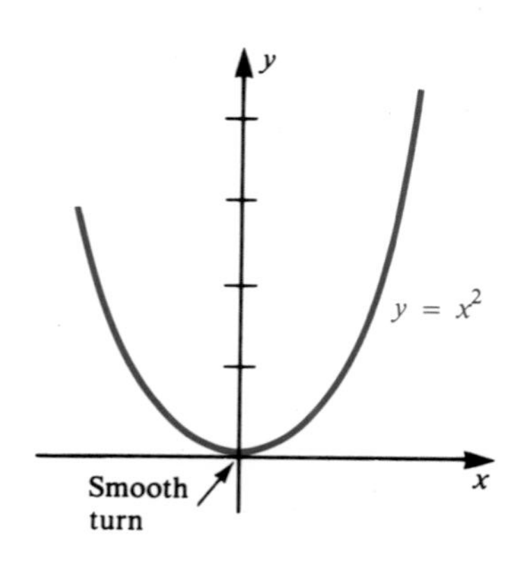

Figure 2–24

In general, the geometric interpretation of $f'(c)$ as the slope of the line tangent to the graph of f at $(c, f(c))$ leads us to three situations when $f'(c)$ is not defined.

Example 11 (*Compare Exercise 15*)
Show the graphical interpretation of the lack of a derivative.

Solution There are three cases.

1. The graph has a corner—a sharp change of direction. See Figure 2–25.
2. The graph has a jump or hole; the function is not continuous. See Figure 2–26.
3. The graph has a tangent line, but the line is the vertical line $x = c$. See Figure 2–27.

In the first two cases, the graph doesn't even have a tangent line, and you can't talk about the slope of something that doesn't even exist! In the third case, remember that a vertical line is the one case of a straight line that doesn't have a slope.

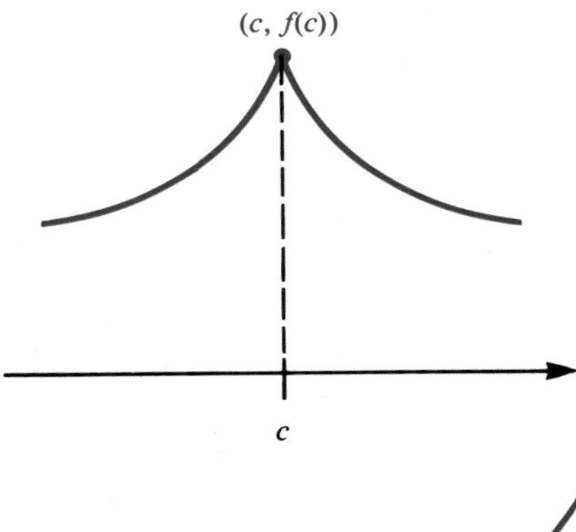

Figure 2–25

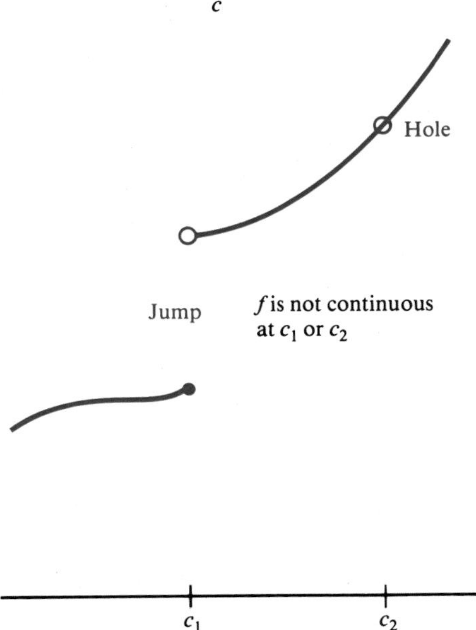

Figure 2–26

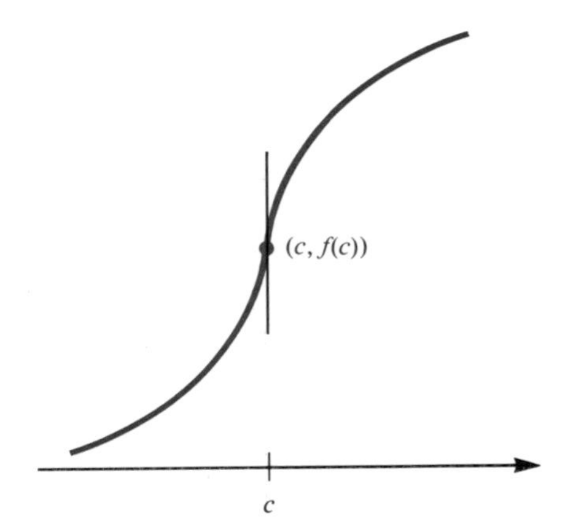

Figure 2–27

2–5 Exercises

I.

1. (*See Examples 4 and 7–9*) If $y = f(x) = \frac{1}{x}$,

   (a) what is $f'(x)$? **(b)** what is $\frac{d}{dx} f(x)$? **(c)** what is $\frac{dy}{dx}$?

   (d) what is the derivative of $f(x)$?

2. (*See Example 6*) If $y = f(x) = 3x + 6$,

   (a) what is $f'(x)$? **(b)** what is $\frac{d}{dx} f(x)$? **(c)** what is $\frac{dy}{dx}$?

   (d) what is the derivative of $f(x)$?

3. (*Compare Example 1*) If $y = x^2 + 1$, then $f'(3) = 6$. Use this information to find an equation of the line tangent to the graph of $f(x)$ at the point $(3, 10)$.
4. If $f(x) = \sqrt{x + 3}$, then $f'(1) = \frac{1}{4}$. Use this information to find an equation of the line tangent to the graph of $f(x)$ at the point $(1, 2)$.
5. If $f(x) = 1(x + 9)$, then $f'(1) = -\frac{1}{100}$. Use this information to find an equation of the line tangent to the curve $y = 1(x + 9)$ at the point $(1, \frac{1}{10})$.
6. If $f(x) = x^3 - 8x$, then $f'(2) = 4$. Use this information to find an equation of the line tangent to the curve $y = x^3 - 8x$ at the point $(2, -8)$.
7. If $f(x) = x^2 - 8x + 6$, then $f'(a) = 2a - 8$.

   (a) What is $f'(x)$?

   (b) What is $\frac{dy}{dx}$ if $y = x^2 - 8x + 6$?

8. If $f(x) = x^3$, then $f'(a) = 3a^2$.

   (a) What is $f'(x)$?

   (b) If $y = x^3$, what is $\frac{dy}{dx}$?

II.

9. (*See Example 1*) If $y = x^2 - 5$, find the slope of the line tangent to the graph at the point $(2, -1)$.

10. If $f(x) = x^2 + 8$, find the slope of the line tangent to the graph at the point $(3, 17)$.

Use formulas established in this section to answer Exercises 11 through 14.

11. (*See Example 5*) Find an equation of the line tangent to the curve $y = 1/x$ at the point whose first coordinate is -2.

12. Find an equation of the line tangent to the curve $y = 1/x$ at the point whose first coordinate is $1/3$.

13. Find an equation of the line tangent to the curve $y = x^2 - 2x - 3$ at the point whose first coordinate is 3.

14. Find an equation of the line tangent to the curve $y = x^2 - 2x - 3$ at the point whose first coordinate is -2.

15. (*See Examples 2 and 11*) Use the graph in Figure 2–28 to answer the following questions.

    (a) For what values of x in the interval $(-3, 6)$ does $f'(x)$ fail to exist?

    (b) At what points on the graph for $-3 < x < 6$ does the graph not have a tangent line?

    (c) What is $f(4)$?

    (d) What is $f'(4)$?

    (e) What is $f(1)$?

    (f) What is $f'(1)$?

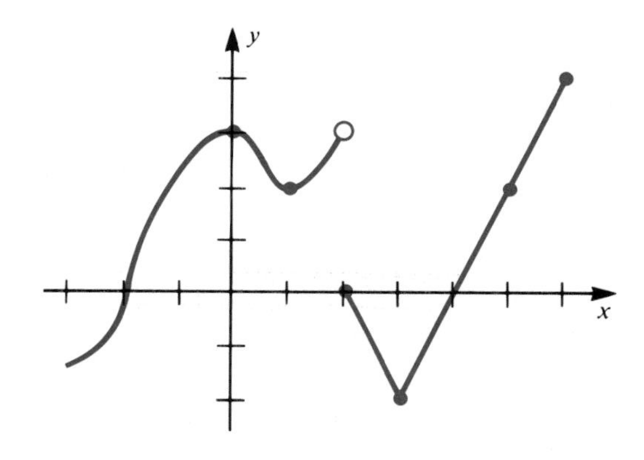

Figure 2–28

16. Use the graph in Figure 2–29 to answer the following questions.
    (a) For what values of x in the interval (0, 10) does $f'(x)$ fail to exist?
    (b) At what points on the graph for $0 < x < 10$ does the graph not have a tangent line?
    (c) What is $f(1)$?
    (d) What is $f'(1)$?
    (e) What is $f(9)$?
    (f) What is $f'(9)$?

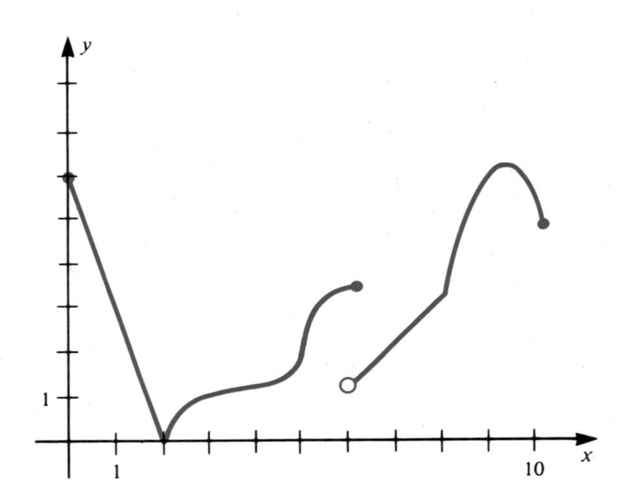

Figure 2–29

III.

17. (*See Example 10*) Use the definition of derivative to show that the curve $y = |x - 3|$ does not have a tangent line at the point (3, 0).
18. Use the definition of derivative to show that the curve $y = |x - 2| + 4$ does not have a tangent line at the point (2, 4).
19. Use the definition of derivative to find the slope of the line tangent to $y = \sqrt{x}$ at the point (9, 3).
20. Use the definition of derivative to find the slope of the line tangent to the curve $y = 1/x^2$ at the point $(4, \frac{1}{16})$.
21. (*See Example 3*) Use the definition of derivative to find the slope of the line tangent to the curve $y = x^2$ at the point (a, a^2).
22. Use the definition of derivative to find the slope of the line tangent to the curve $y = x^2 + 4$ at the point $(a, a^2 + 4)$.
23. (*See Example 4*) Use the definition of derivative to find a formula for $f'(a)$ if $f(x) = 1/(x + 4)$.
24. Use the definition of derivative to find a formula for $f'(a)$ if $f(x) = 3/(5x)$.

IMPORTANT TERMS

2–2	**Average rate of change**	Δx	**Instantaneous rate of change**
	Derivative	**Three-step process for computing the instantaneous rate of change**	
2–3	**Limit**	**Derivative as a limit**	$f'(a)$
	Replacement principle	**Arithmetic rules for limits**	**Limits of polynomials**
	Limit of a quotient of polynomials	**Exponent rule for limits**	**Non-existence of limit**
2–4	**Left-hand limit**	**Right-hand limit**	**Relation between limit and one-sided limits**
	Marginal cost	**Jump in the graph**	**Hole in the graph**
	Continuous function		
2–5	**Slope of tangent line**	**Equation of tangent line**	**Use of definition of derivative to compute the slope of a tangent line**
	Derivative of $1/x$	**Derivative of** $mx + b$	**Non-existence of derivative**

REVIEW EXERCISES

I.

Compute each of the following limits or state that it does not exist:

1. $\lim_{x\to 3} \dfrac{x-3}{x+2}$
2. $\lim_{x\to 3} \dfrac{x+2}{x-3}$
3. $\lim_{x\to 5} \sqrt{31+x}$
4. $\lim_{x\to 5} \dfrac{x^2-9x}{x+4}$
5. $\lim_{x\to 2^+} |x|$
6. $\lim_{x\to 2^-} |x|$
7. $\lim_{x\to 0^-} |x|$
8. $\lim_{x\to -3^-} |x|$
9. If $\dfrac{dy}{dx} = 5x + 8$, what is $f'(x)$?
10. If $y = f(x)$ and $f'(x) = 2x - 2$, what is $\dfrac{dy}{dx}$?
11. If $f(x) = x^2 - 3x + 2$, what is the average rate of change of $f(x)$
    (a) on the interval $[-1, 4]$?
    (b) on the interval $[-1, 2]$?
    (c) on the interval $[2, 4]$?

12. If a new store's daily revenue is given by $R(t) = \sqrt{t + 9}$ in hundreds of dollars t day after it has opened, what is the average change in daily revenue as t increases from 7 to 27?

II.

Use the graph in Figure 2–30 to answer Exercises 13 through 26.

13. Is f continuous on the interval (a) $[-1, 2]$? (b) $[3, 4]$?
14. Is f continuous on the interval (a) $(-3, -1)$? (b) $(0, 4)$?
15. What is

    (a) $\lim_{x \to 1^-} f(x)$? **(b)** $\lim_{x \to 1^+} f(x)$? **(c)** $\lim_{x \to 1} f(x)$?

16. What is

    (a) $\lim_{x \to 0^-} f(x)$? **(b)** $\lim_{x \to 0^+} f(x)$? **(c)** $\lim_{x \to 0} f(x)$?

17. What is

    (a) $\lim_{x \to 2^-} f(x)$? **(b)** $\lim_{x \to 2^+} f(x)$? **(c)** $\lim_{x \to 2} f(x)$?

18. What is

    (a) $\lim_{x \to -1^-} f(x)$? **(b)** $\lim_{x \to -1^+} f(x)$? **(c)** $\lim_{x \to -1} f(x)$?

19. Is the function continuous at

    (a) $x = 0$? **(b)** $x = 1$? **(c)** $x = 2$?

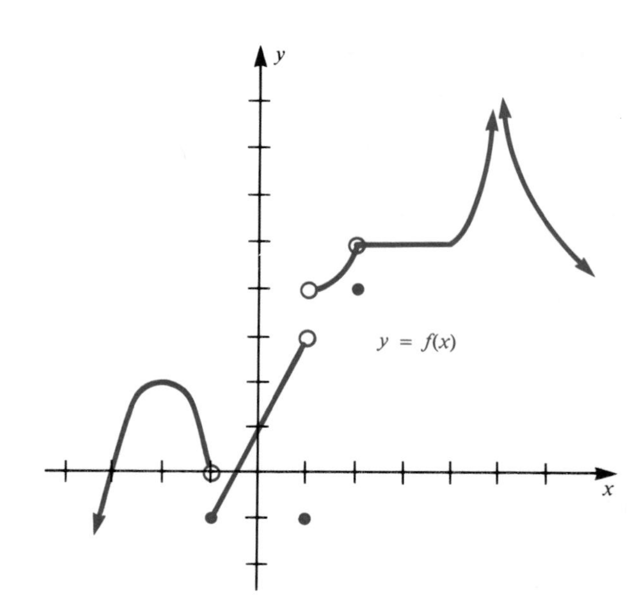

Figure 2–30

20. Is the function continuous at

(a) $x = -1$? **(b)** $x = 4$? **(c)** $x = 5$?

21. Is the function continuous at

(a) $x = -2$? **(b)** $x = 3$? **(c)** $x = -3$?

22. What is

(a) $f(3)$? **(b)** $f'(3)$?

23. What is

(a) $f(0)$? **(b)** $f'(0)$?

24. What is

(a) $f(-2)$? **(b)** $f'(-2)$?

25. What is

(a) $f(1)$? **(b)** $f'(1)$?

26. What is

(a) $f(2)$? **(b)** $f'(2)$?

27. A delivery service charges \$6 to deliver any package weighing less than 4 pounds. If the package weighs 4 pounds or more, the service charges \$$2x$ where x is the package's weight in pounds. If $C(x)$ is the cost in dollars of delivering a package weighing x pounds, what are

(a) $\lim_{x \to 4^-} C(x)$ **(b)** $\lim_{x \to 4^+} C(x)$

(c) $C(4)$ **(d)** Is C continuous at $x = 4$? Explain.

28. Let $f(x) = \begin{cases} 2x + 1, & x \leq 3 \\ x^2 - 2, & x > 3 \end{cases}$. Is f continuous at $x = 3$?

III.

Compute the following limits.

29. $\lim_{x \to 2} \dfrac{x^2 - x - 2}{x - 2}$ **30.** $\lim_{x \to -1} \dfrac{x^2 + 4x + 3}{x + 1}$ **31.** $\lim_{x \to 0} \dfrac{x^3 + 4x^2 - 5x}{x^2 + 2x}$

32. $\lim_{x \to 2} \dfrac{x^2 - 7x + 10}{x^2 - 5x + 6}$ **33.** $\lim_{\Delta x \to 0} \dfrac{\sqrt{4 + \Delta x} - 2}{\Delta x}$ **34.** $\lim_{\Delta x \to 0} \dfrac{(6 + \Delta x)^2 - 36}{\Delta x}$

Use the definition of derivative to compute $f'(a)$ if

35. $f(x) = 3x^2 - 4$ and $a = 1$. **36.** $f(x) = \sqrt{x}$ and $a = 9$.

37. $f(x) = \dfrac{x}{2}$ and $a = 4$. **38.** $f(x) = \dfrac{2}{x}$ and $a = 4$.

39. If $f'(4) = 3$ and $f(4) = 6$, what is an equation of the line tangent to the graph $y = f(x)$ at the point whose first coordinate is 4?

40. If $f(7) = -2$ and $f'(7) = 10$, what is an equation of the line tangent to the graph $y = f(x)$ at the point whose first coordinate is 7?

41. Use the definition of derivative to find $f'(a)$ if $f(x) = \sqrt{x}$.

42. Use the definition of derivative to find $f'(a)$ if $f(x) = x^3$.

43. If $f(x) = \begin{cases} x^2 + 5, & x < 2 \\ -3x + A, & x \geq 2 \end{cases}$, is there a value of A such that f is continuous at $x = 2$? If so what is it?

44. If $f(x) = \begin{cases} 3x + 9, & x < 1 \\ Bx - 6, & x \geq 1 \end{cases}$ is there a value of B such that f is continuous at $x = 1$? If so what is it?

3

Topics in this chapter can be applied to:

Rate of Change in Value ● Revenue, Costs, and Profits ● Rate of Change of Average Cost ● Changes in Rates of Inflation ● Commodity Prices and Demand ● Marginal Costs

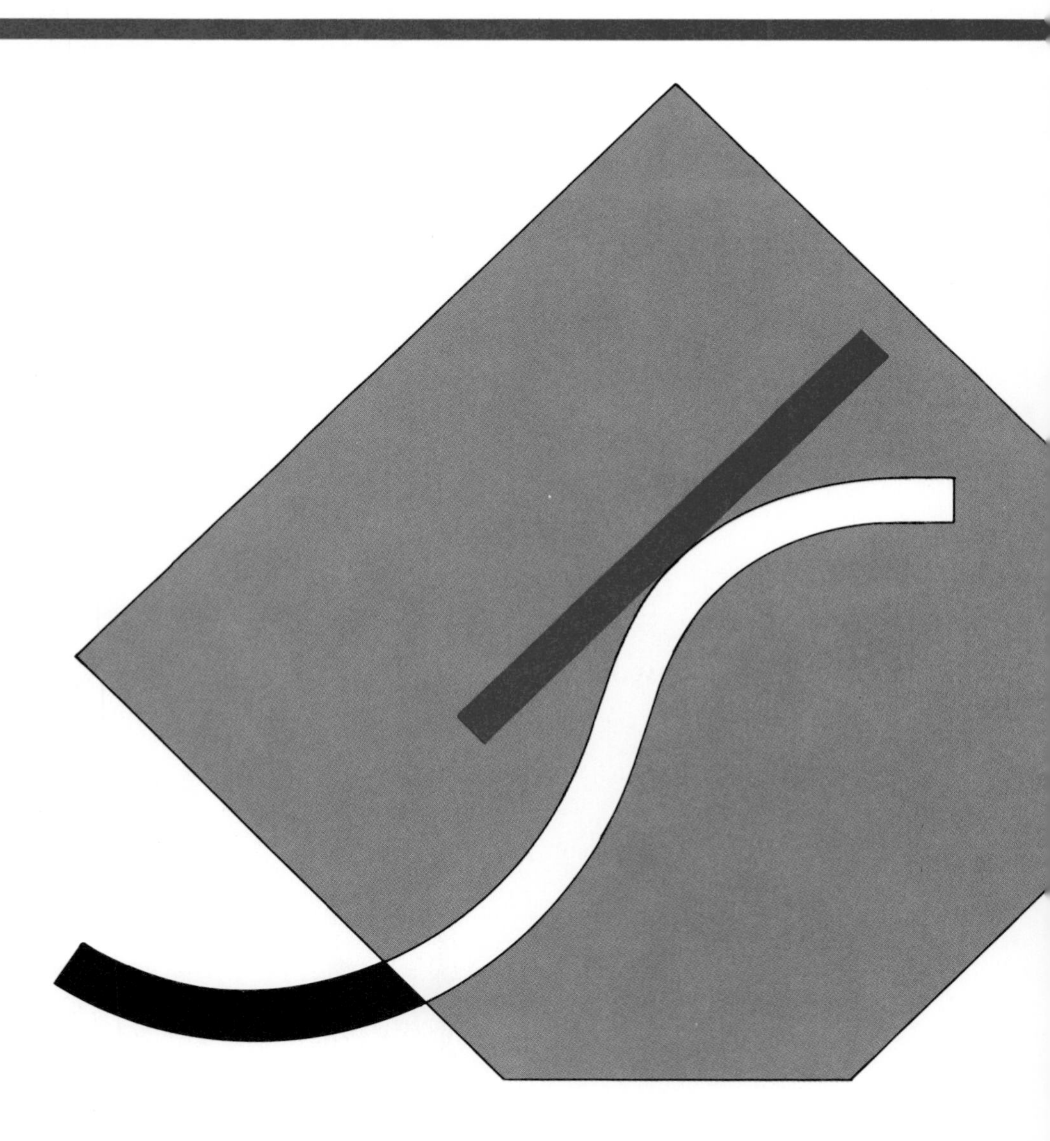

Techniques of Differentiation

- Some Rules for Computing Derivatives
- The Product and Quotient Rules
- The Chain Rule
- Higher Order Derivatives
- Implicit Differentiation
- Related Rates
- The Differential and Linear Approximation

3–1 Some Rules for Computing Derivatives

In the last chapter, we showed how we could use the definition of derivative,

$$\lim_{\Delta x \to 0} \frac{f(x + \Delta x) - f(x)}{\Delta x}$$

to compute $f'(x)$ for some particular functions. We did this to emphasize both the definition of the derivative and the reason that $f'(x)$ can be called **the instantaneous rate of change of** $f(x)$. Using the definition, however, is difficult and time consuming; no one would like having to use the definition every time he or she needed to compute a derivative. Indeed, calculus would not have become a powerful problem-solving tool in so many varied fields without an easier method of computing the derivative. Mathematicians have developed some theorems which will allow us to compute derivatives by following certain rules, and we will show you some of these rules in this and the next two sections.

We found a few special formulas in the examples and exercises of the last chapter. For example, we found that

$$\text{If} \qquad f(x) = mx + b, \qquad \text{then} \qquad f'(x) = m.$$

The derivative of a linear function is the slope of its graph. The derivative gives the instantaneous rate of change (slope) of a function at a point, and a line of the form $y = mx + b$ has the same rate of change (slope) everywhere.

There are two special cases of this formula that we would like to emphasize. We present the first case as our first rule.

Rule 1
The Derivative of a Constant Rule

If $f(x) = c$, then $f'(x) = 0$

In other notation,

If $y = c$, then $\dfrac{dy}{dx} = 0$

In words, **the derivative of a constant is 0.**

To see that this is a special case of the formula for the derivative of a linear function, we point out that if $f(x) = c$ is a constant function, then its graph is a horizontal line. The slope of a horizontal line is zero, and the derivative of a linear function is the slope of the line, so $f'(x) = 0$.

For the second special case, we let $b = 0$ and $m = 1$.

$$\text{If } f(x) = x, \text{ then } f'(x) = 1.$$

We discovered three other similar formulas in Chapter 2.

$$\text{If } f(x) = x^2, \text{ then } f'(x) = 2x.$$
$$\text{If } f(x) = x^3, \text{ then } f'(x) = 3x^2.$$
$$\text{If } f(x) = x^{-1}, \text{ then } f'(x) = -1x^{-2}.$$

These results are all special cases of a theorem called the power rule, which we list as our second rule.

Rule 2
The Power Rule

If $f(x) = x^R$, then $f'(x) = Rx^{R-1}$

In other notation,

If $y = x^R$, then $\dfrac{dy}{dx} = Rx^{R-1}$

Notice how all four of the examples above follow this pattern. Here are some more examples using the power rule.

Example 1 (*Compare Exercise 1*)
If $f(x) = x^7$, what is $f'(x)$?

Solution Here $R = 7$, so $R - 1 = 6$ and $f'(x) = 7x^6$.

Example 2 (*Compare Exercise 7*)

If $f(x) = \sqrt{x^3}$, what is $f'(x)$?

Solution To get the function in the form needed to apply the power rule, we first rewrite $\sqrt{x^3}$ as $(x^3)^{1/2} = x^{3/2}$. Thus, $f(x) = x^{3/2}$, and we can apply the power rule now with $R = \frac{3}{2}$. Then,

$$R - 1 = \frac{3}{2} - 1 = \frac{1}{2}$$

We have

$$f'(x) = \frac{3}{2}x^{1/2}$$

Example 3 (*Compare Exercise 13*)

If $f(x) = \dfrac{1}{x^3}$, what is $f'(x)$?

Solution Again, we must rewrite the function. Rewriting $1/x^3$ as x^{-3}, we see that $f(x) = x^{-3}$, so $R = -3$, $R - 1 = -4$, and finally, that $f'(x) = -3x^{-4} = -3/x^4$

Warning! Remember that you are subtracting one from the exponent; the correct computation in this example is $-3 - 1 = -4$. Do not subtract one from three and then just put a minus sign in front of the result, getting -2 instead of -4.

But, the power rule alone won't get us $f'(x)$ if $f(x) = 6x^3$; the power rule and the remaining computational rules must be followed very literally. The power rule can't be used directly on $f(x) = 6x^3$ because of the constant factor 6. For this function, we need a rule that tells us how to find the derivative of a function of the form $f(x) = cx^R$, for any constant c. The next rule we consider says, more generally, that if a function f has a derivative, and c is a constant, then c times f, written $c \cdot f$, also has a derivative. In fact, the theorem also tells us how to compute that derivative. That formula is our third rule.

Rule 3
The Constant-Times Rule

$$\frac{d}{dx}(c \cdot f(x)) = c \cdot \frac{d}{dx} f(x).$$

In other notation,

If $h(x) = c \cdot f(x)$, then $h'(x) = c \cdot f'(x)$.

In words, **the derivative of a constant times a function is the constant times the derivative of the function.**

Example 4 (*Compare Exercise 9*)
If $f(x) = 6x^3$, what is $f'(x)$?

Solution

$$f'(x) = \frac{d}{dx} f(x) = \frac{d}{dx}(6x^3) = 6\frac{d}{dx}x^3 = 6 \cdot 3x^2 = 18x^2$$

Example 5 Find $\frac{dy}{dx}$ if $y = \frac{7}{x^4}$.

Solution As before, we rewrite the function so we can use the power rule.

$$y = 7x^{-4},$$

so

$$\frac{dy}{dx} = \frac{d}{dx}(7x^{-4}) = 7\frac{d}{dx}x^{-4}$$

$$= 7(-4x^{-5}) = -28x^{-5} = \frac{-28}{x^5}$$

We won't give the following a special name, but we would like to point out to you that we can combine the constant-times rule and the power rule.

$$\text{If} \qquad f(x) = cx^R, \qquad \text{then} \qquad f'(x) = (c \cdot R)x^{R-1}$$

Example 6 (*Compare Exercise 15*)
If $f(x) = 8\sqrt{x}$, what is $f'(x)$?

Solution Rewriting $f(x)$ as $8x^{1/2}$, we have $c = 8$ and $r = \frac{1}{2}$, so

$$f'(x) = 8 \cdot \frac{1}{2}x^{-1/2} = 4x^{-1/2} = \frac{4}{\sqrt{x}}.$$

You do not need to memorize these rules by number, Rule 3 for example, but remember the rule in words:

"The derivative of a constant times a differentiable function is the constant times the derivative of the function."

The following example is to convince you of the reasonableness of this rule.

Example 7 If a company has a fleet of 12 identical cars, and is depreciating each of them at a rate of $3000 per year, what is the rate of depreciation of the whole fleet?

Solution You do not need calculus to figure out what the answer is; simple arithmetic shows the answer is $(3000) \cdot 12 = 36{,}000$ dollars per year. What we want to do here is show how Rule 3 works in a setting where you already know what the answer is. Let $f(t)$ stand for the value of each car when it is t years old. Let $g(t)$ be the value of the entire fleet, so $g(t) = 12 \cdot f(t)$. Because depreciation is the rate of change in value and the derivative is also the rate of change, the depreciation is the derivative. The word depreciation means the value is decreasing, so the derivative is negative. The problem gives us $f'(t) = -3000$, and applying Rule 3 to get the derivative of $g(t) = 12f(t)$, we have $g'(t) = 12 \cdot f'(t)$. Finally,

$$g'(t) = 12(-3000) = -36{,}000$$

Next, what do we do when $f(x)$ is of the form $f(x) = 5x^3 + 9x$? We know how to compute both the derivative of $5x^3$ and the derivative of $9x$. What happens when we add two functions and we already know the derivative of each one? Our fourth rule, the sum rule, answers this question. If f and g are differentiable, so is $f + g$, and you can compute the derivative of $f + g$ by the following rule:

Rule 4
The Sum Rule

$$\frac{d}{dx}(f(x) + g(x)) = \frac{d}{dx}f(x) + \frac{d}{dx}g(x)$$

In other notation,

if $h(x) = f(x) + g(x)$, then $h'(x) = f'(x) + g'(x)$

In words, **the derivative of the sum of differentiable functions is the sum of the derivatives.**

Example 8 (*Compare Exercise 15*)
If $f(x) = 5x^3 + 9x$, what is $f'(x)$?

Solution

$$\begin{aligned} f'(x) &= \frac{d}{dx}f(x) = \frac{d}{dx}(5x^3 + 9x) \\ &= \frac{d}{dx}(5x^3) + \frac{d}{dx}(9x) \\ &= 15x^2 + 9 \end{aligned}$$

The sum rule extends to more than two summands.

Example 9 (*Compare Exercise 25*)
If $f(x) = 9x^2 + 5 + \dfrac{4}{x^2}$, then compute $f'(x)$.

Solution First, rewriting $f(x)$ so that we can apply the power rule to $4/x^2$, we have $f(x) = 9x^2 + 5 + 4x^{-2}$. Now take the derivative of each summand and then add

$$\begin{aligned} f'(x) &= \frac{d}{dx}(9x^2) + \frac{d}{dx}5 + \frac{d}{dx}(4x^{-2}) \\ &= \quad 18x \quad + \quad 0 \quad + \quad -8x^{-3} \\ &= 18x - \frac{8}{x^3} \end{aligned}$$

Example 10 (*Compare Exercise 19*)
What is $f'(x)$ if $f(x) = 6x^3 - 5x^2$?

Solution Rule 4 only deals with the sum of two functions, and here f is the difference of two functions. We can rewrite $f(x)$, however, as $f(x) = 6x^3 + (-5)x^2$ and now apply the sum rule:

$$\begin{aligned} f'(x) &= 18x^2 + (-10)x \\ &= 18x^2 - 10x \end{aligned}$$

Notice that $f'(x) = [(d/dx)6x^3] - [(d/dx)5x^2]$. There was nothing special about this particular example. Because $f(x) - g(x)$ is the same as $f(x) + (-1)g(x)$, we can compute the derivative of the difference of two functions as follows:

$$\frac{d}{dx}[f(x) - g(x)] = \frac{d}{dx}[f(x) + (-1)g(x)]$$

Next, use the sum rule

$$= \frac{d}{dx}f(x) + \frac{d}{dx}[(-1)g(x)]$$

then, use the constant-times rule

$$= \frac{d}{dx}f(x) + (-1)\frac{d}{dx}g(x)$$

$$= \frac{d}{dx}f(x) - \frac{d}{dx}g(x)$$

Thus, we have the following rule.

Rule 5
The Difference Rule

$$\frac{d}{dx}(f(x) - g(x)) = \frac{d}{dx}f(x) - \frac{d}{dx}g(x)$$

In other notation,

If $h(x) = f(x) - g(x)$, then $h'(x) = f'(x) - g'(x)$

In words, **the derivative of the difference of two differentiable functions is the difference of the derivatives.**

Now we can compute the derivative of the difference of two functions directly.

Example 11 Compute $f'(x)$ if $f(x) = 7x^3 - 14x$.

Solution

$$f'(x) = \frac{d}{dx}(7x^3) - \frac{d}{dx}(14x) = 21x^2 - 14$$

The difference rule also extends to more than two differences.

Example 12 (*Compare Exercise 23*)
If $g(x) = 7x^4 - 9x^2 - 10$, find $g'(x)$.

Solution

$$g'(x) = \frac{d}{dx}7x^4 - \frac{d}{dx}9x^2 - \frac{d}{dx}10$$
$$= 28x^3 - 18x - 0$$
$$= 28x^3 - 18x$$

We still must rewrite the function so that each "piece" is in the form cx^R; watch for this in the next two examples.

Example 13 (*Compare Exercise 29*)
If $g(t) = 9t^4 - 6\sqrt{t}$, find $g'(t)$.

Solution We must rewrite $\sqrt{t}$ as $t^{1/2}$ to apply the power rule. Thus,

$$g(t) = 9t^4 - 6\sqrt{t} = 9t^4 - 6t^{1/2}$$

Now we can apply the rules:

$$g'(t) = \frac{d}{dt}(9t^4) - \frac{d}{dt}(6t^{1/2})$$
$$= 9 \cdot 4t^3 - 6 \cdot \frac{1}{2}t^{-1/2}$$
$$= 36t^3 - 3t^{-1/2}$$
$$= 36t^3 - \frac{3}{\sqrt{t}}$$

The next example is to remind you of the other kind of exponents you must rewrite—namely negative exponents.

Warning! Remember that the exponent only goes with the variable. $8x^2 = 8 \cdot x \cdot x$; the eight is not squared. So, when you see something like $1/(3x)$, be careful.

$$\frac{1}{3x} = \frac{1}{3} \cdot \frac{1}{x} = \frac{1}{3}x^{-1} \left[\frac{1}{3x} \neq 3x^{-1}\right]$$

The exponent -1 only goes with the x, just the same way the exponent 2 only goes with the x in $8x^2$.

Example 14 (*Compare Exercise 37*)

If $f(x) = 16x^3 - 4x^2 - \frac{1}{3x}$, find $f'(x)$.

Solution Rewrite $f(x)$ as $f(x) = 16x^3 - 4x^2 - \frac{1}{3}x^{-1}$, and we have

$$f'(x) = 48x^2 - 8x - \frac{1}{3}(-1x^{-2})$$

$$= 48x^2 - 8x + \frac{1}{3x^2}$$

3–1 Exercises

I.

Find $f'(x)$ for each of the following:

1. (*See Example 1*) $f(x) = x^5$
2. $f(x) = x^8$
3. $f(x) = x^{-4}$
4. $f(x) = x^{-7}$
5. $f(x) = x^{4/3}$
6. $f(x) = x^{5/2}$
7. (*See Example 2*) $f(x) = \sqrt{x^5}$
8. $f(x) = \sqrt{x^7}$
9. (*See Example 4*) $f(x) = 10x^4$
10. $f(x) = 7x^{-2}$

Find $\frac{dy}{dx}$ for each of the following:

11. $y = 3x^2$
12. $y = 15x^{1/3}$
13. (*See Example 3*) $y = \frac{9}{x}$
14. $y = \frac{4}{x^2}$
15. (*See Examples 6 and 8*) $y = 10x^3 + 5\sqrt{x}$
16. $y = 8x^2 + 9\sqrt{x}$
17. $y = 15x^{-2} + x^2$
18. $y = 21x + 5x^{-3}$
19. (*See Example 10*) $y = 10x^2 - 4x$
20. $y = 15x - x^{-3}$

II.

Find $f'(x)$ for each of the following:

21. $f(x) = 4x^3 + 5x + 9$
22. $f(x) = 12x^3 + 8 + 4x^{-3}$
23. (*See Example 12*) $f(x) = 6x^2 - 4x + 5$
24. $f(x) = 8x^{3/2} - 3x^2 + 5x - 9$

25. (*See Example 9*)
$f(x) = 9x^3 - 8 + \frac{2}{x^3}$

26. $f(x) = 14 - \frac{4}{x}$

27. $f(x) = 14\sqrt{x} - 9 + \frac{12}{x^4}$

28. $f(x) = 3\sqrt{x} + \frac{5}{x^2}$

Find $\frac{dy}{dx}$ for each of the following:

29. (*See Example 13*)
$y = 3x^3 - 10\sqrt{x}$

30. $y = \frac{8}{x} + 7x^2$

31. $y = 10x^2 - 5 + \frac{8}{x^2}$

32. $y = 6x - 9 + \frac{10}{x^5}$

33. $y = 8\sqrt{x} + 16x - \frac{1}{2}$

34. $y = \frac{8}{x} - \frac{4}{x^2} + \frac{2}{3}$

35. $y = 9\sqrt{x} - \frac{5}{x^2}$

36. $y = \frac{4}{\sqrt{x}} + 8x$

III.

Find the derivative of each of the following:

37. (*See Example 14*)
$f(x) = \frac{1}{4x^2} - \sqrt{x}$

38. $f(x) = \frac{1}{6x} - \frac{8}{\sqrt{x}}$

39. $f(x) = \frac{3}{5x} + \frac{6}{7x^2}$

40. $f(x) = \frac{3}{4x^2} - 8x$

41. $g(t) = \frac{4}{5t^2} - \frac{5t}{3}$

42. $g(t) = \frac{6}{7t^3} - \frac{7}{4t}$

43. $g(x) = 9x - 4\sqrt{x} + \frac{3}{5x^3}$

44. $g(x) = \sqrt{x^3} - \frac{9}{\sqrt{x^3}}$

Do the algebra to rewrite each function so that you can use the rules of this section; then compute the derivative of each of the following:

45. $f(x) = \sqrt{9x}$

46. $f(x) = \frac{1}{\sqrt{16x}}$

47. $f(x) = \left(x + \frac{1}{x}\right)^2$

48. $g(x) = 3x(x^2 + 4)$

49. $g(t) = \frac{t^2 + 6}{3t}$

50. $f(x) = \frac{x + 5}{\sqrt{x}}$

51. The revenue (in dollars) a company gets from installing x sprinkler systems a month is given by $R(x) = 2000x - \frac{1}{4}x^2$. The cost (in dollars) of installing x sprinkler systems per month is given by $C(x) = 720x + 3500$. The profit $P(x)$ is given by $P(x) = R(x) - C(x)$. Compute $R'(x)$, $C'(x)$, and $P'(x)$.

52. If the revenue (in dollars) from producing x clock-radios per week is given by $R(x) = 40x - \frac{1}{10}x^2$, and the cost (in dollars) is given by $C(x) = 25x + 6300$, compute $R'(x)$, $C'(x)$, and $P'(x)$, where $P(x)$, the profit, is given by $P(x) = R(x) - C(x)$.

53. What is an equation of the line tangent to the graph $y = x^2 - 4x + 3$ at the point $(-1, 8)$?

54. What is an equation of the line tangent to the graph $y = \sqrt{x}$ at the point $(9, 3)$?

55. If a rock has moved $-16t^2 + 88t$ feet after t seconds, how fast is it moving when $t = 2$?

56. If a rock is $-16t^2 + 64t + 200$ feet above ground t seconds after it is released, what is the velocity of the rock (a) when $t = 1$? (b) when $t = 2$? (c) when $t = 3$?

3–2 The Product and Quotient Rules

The Product Rule
The Quotient Rule

The Product Rule

We have just seen how to compute the derivative of $f(x) = 9x^2 + 5x$ by recognizing that f is the sum of two simpler functions: $f(x)$ is $9x^2$ plus $5x$.

Just as you can form new functions by adding and subtracting functions, you can combine functions by multiplying or dividing them. Thus, we can recognize that

$$f(x) = (8x^2 - 9)(x^3 - 5x + 6)$$

is the product of $(8x^2 - 9)$ times $(x^3 - 5x + 6)$. Similarly,

$$g(x) = \frac{2x^3}{x^2 + 5}$$

is the quotient of $2x^3$ divided by $(x^2 + 5)$. Mathematicians have developed rules to show how to differentiate products and quotients as well as sums and differences.

First, we show you the rule for the product of two functions. You won't be surprised to learn that this rule is called **the product rule**. If f and g both have derivatives, so does $f \cdot g$, and the rule for computing the derivative of this product follows:

Rule 6
The Product Rule

$$\frac{d}{dx}[f(x) \cdot g(x)] = g(x) \cdot \frac{d}{dx} f(x) + f(x) \cdot \frac{d}{dx} g(x)$$

In other notation,

If $\quad h(x) = f(x) \cdot g(x), \quad$ then $\quad h'(x) = g(x) \cdot f'(x) + f(x) \cdot g'(x)$

The rule in words is a little complicated; it is "**the derivative of the product of two differentiable functions is the second function times the derivative of the first plus the first function times the derivative of the second.**"

Warning! The product rule generally gives students more difficulty than the preceding rules. You may be tempted to compute the product of the derivatives instead of following this rule. Resist the temptation!

Example 1 (*Compare Exercise 13*)
Find $h'(x)$ if $h(x) = (8x^2 - 9)(x^3 - 5x + 6)$.

Solution We apply the product rule with $f(x) = 8x^2 - 9$ and $g(x) = x^3 - 5x + 6$.

$$\begin{aligned} h'(x) &= \frac{d}{dx}[(8x^2 - 9)(x^3 - 5x + 6)] \\ &= (x^3 - 5x + 6)\frac{d}{dx}(8x^2 - 9) + (8x^2 - 9)\frac{d}{dx}(x^3 - 5x + 6) \\ &= (x^3 - 5x + 6)(16x) + (8x^2 - 9)(3x^2 - 5) \end{aligned}$$

Our purpose in this example was to show the mechanics of the product rule, and so we leave the answer in the form the product rule gives us. You may want to check the answer to this example by first multiplying out $(8x^2 - 9)(x^3 - 5x + 6)$ and then computing the derivative.

Our next example uses the product rule in a setting where you already know the answer and shows that writing the product of the derivatives will give the **wrong** answer.

Example 2 (*Compare Exercise 3*)
If $h(x) = 6x^3$, find $h'(x)$.

Solution From previous rules, we know that the answer is $h'(x) = 18x^2$.

If we treat 6 as $f(x)$ and x^3 as $g(x)$ and use the product rule on 6 times x^3, we get

$$\begin{aligned} h'(x) = \frac{d}{dx}(6x^3) &= x^3\frac{d}{dx}6 + 6\frac{d}{dx}x^3 \\ &= x^3 \cdot 0 + 6 \cdot 3x^2 = 0 + 18x^2 \\ &= 18x^2 \end{aligned}$$

You won't do this long method in practice, but if you did you'd get the same answer as you would by the quick method.

Notice what happens if we compute the product of the derivatives:

$$\left(\frac{d}{dx}6\right) \cdot \left(\frac{d}{dx}x^3\right) = 0 \cdot 3x^2 = 0$$

This procedure gives the wrong answer.

Example 3 (*Compare Exercise 1*)
If $f(x) = (3x^2)(4x^3)$, what is $f'(x)$?

Solution Ordinarily, you would probably find $f'(x)$ by doing the multiplication first and then computing the derivative. Again, our goal here is to give an example that lets you check the result of using the product rule correctly.

$$\begin{aligned} f'(x) = \frac{d}{dx}[(3x^2)(4x^3)] &= (4x^3)\frac{d}{dx}(3x^2) + 3x^2\frac{d}{dx}(4x^3) \\ &= (4x^3)(6x) + (3x^2)(12x^2) \\ &= 24x^4 + 36x^4 \\ &= 60x^4 \end{aligned}$$

To see that this answer agrees with our previous rule, let's do the multiplication before computing the derivative.

$$f(x) = (3x^2)(4x^3) = 12x^5$$

so

$$f'(x) = 60x^4$$

Again, the correct answer is **not** simply the product of the derivatives; that computation would give $(6x)(12x^2) = 72x^3$ rather than $60x^4$.

Products of functions often arise in certain applications. For example, the total sales revenue of a company is the product of the number of items it sells and the selling price per item.

Example 4 (*Compare Exercise 36*)
A student service organization wants to sell special sweatshirts for homecoming. The group does some sampling of student opinion and as a result estimates that if they charge $\$p$ per shirt, they can sell $3600 - 100p$ shirts. If R is their total revenue, **(a)** express R as a function of p, and then **(b)** find $R'(p)$.

Solution **(a)** Revenue = (price per shirt)(number of shirts sold)

$$\begin{aligned} &= (p)(3600 - 100p) \\ R(p) &= p(3600 - 100p) \end{aligned}$$

(b) Using the product rule,

$$\begin{aligned} R'(p) &= (3600 - 100p) \cdot 1 + p(-100) = 3600 - 100p - 100p \\ &= 3600 - 200p. \end{aligned}$$

[Check this answer by doing the multiplication first and then computing the derivative of $R(p)$].

In these examples, it may have seemed easier to do the multiplication first and then find the derivative. This is not always true. In fact, for some functions we will encounter later you *must* use the product rule. For now we just look at Example 5, where it's easier to use the product rule first.

Example 5 (*Compare Exercise 13*)
If $f(x) = (x^3 - 4x^2 + 7x - 23)(x^5 + 9x - 10)$, find $f'(x)$.

Solution

$$
\begin{aligned}
f'(x) &= \frac{d}{dx} f(x) \\
&= (x^5 + 9x - 10)\frac{d}{dx}(x^3 - 4x^2 + 7x - 23) \\
&\quad + (x^3 - 4x^2 + 7x - 23)\frac{d}{dx}(x^5 + 9x - 10) \\
&= (x^5 + 9x - 10)(3x^2 - 8x + 7) \\
&\quad + (x^3 - 4x^2 + 7x - 23)(5x^4 + 9)
\end{aligned}
$$

Again, we leave the answer in the form given to us by the product rule.

Reason for the Product Rule. We are presenting these differentiation formulas to you as "rules," but you should not interpret that word to mean that these rules are just made up and can be changed, like the rules of basketball for example. People cannot agree to change the rules for computing derivatives; they are in fact theorems, and represent the way different rates of change interact in the "real world." For example, let's look at the product rule and the way that a rectangle grows.

For purposes of this application, we suppose that both $f(x)$ and $g(x)$ are positive and that both are increasing as x increases. Let $f(x)$ be the width of a rectangle, and let $g(x)$ be the height. Thus, $f(x) \cdot g(x)$ represents the area of the rectangle, and this rectangle is growing as x increases. See Figure 3–1.

Imagine the left and lower boundaries as anchored, so the rectangle is growing only in the direction of the arrows. Let S_1 be the right-hand edge of the rectangle. As S_1 moves to the right, the rate of growth of the area depends on both how long S_1 is and how fast S_1 is moving. The length of S_1 is $g(x)$, the height of the rectangle; how fast S_1 is moving to the right is $f'(x)$, the rate of change of the width. The rate at which the rectangle is growing just by expanding to the right is $g(x) \cdot f'(x)$. A similar discussion shows that the rate at which the rectangle is expanding in the upward direction is $f(x) \cdot g'(x)$. The total rate of expansion is the sum of these two individual rates.

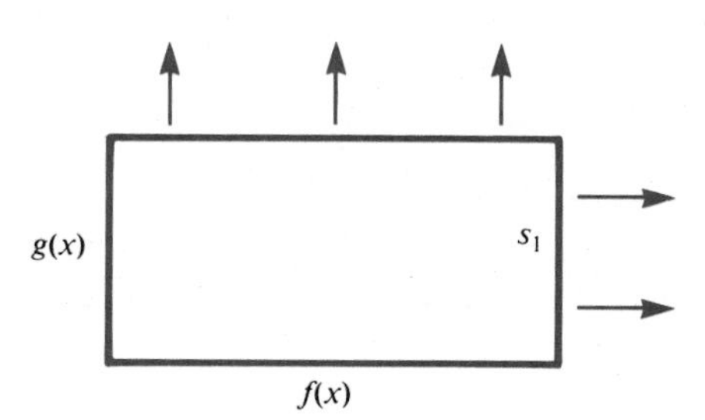

Figure 3–1

The total rate of change of the area $= g(x) \cdot f'(x) + f(x) \cdot g'(x)$. Also, remember that the area is $f(x) \cdot g(x)$, so that the rate of change of the area is the derivative of $f(x) \cdot g(x)$. The rate of change of the area $= (d/dx)[f(x) \cdot g(x)]$.

Setting the two expressions for the rate of change of the area equal to each other gives the product rule:

$$\frac{d}{dx}[f(x) \cdot g(x)] = g(x) \cdot f'(x) + f(x) \cdot g'(x)$$

This discussion does not prove the product rule, but is intended to give you some geometric feeling for why the product rule is what it is. (*Compare Exercise 29*)

The Quotient Rule

You will probably not be surprised that the rule following the product rule is the quotient rule. As in all the rules regarding combinations of functions, we assume that we are starting with two functions that have derivatives.

Rule 7
The Quotient Rule

If $g(x) \neq 0$, and if $h(x) = \dfrac{f(x)}{g(x)}$, then

$$h'(x) = \frac{g(x) \cdot f'(x) - f(x) \cdot g'(x)}{[g(x)]^2}$$

The requirement that $g(x) \neq 0$ insures that $f(x)/g(x)$ makes sense.

When expressed in words, the quotient rule is even more complicated than the product rule:

"The derivative of a quotient of two differentiable functions is equal to the denominator times the derivative of the numerator minus the numerator times the derivative of the denominator all divided by the square of the denominator."

We would like to give you three special warnings about using the quotient rule.

Warning #1! The derivative of the quotient is **not** the quotient of the derivatives.

Example 6 (*Compare Exercise 7*)

If $h(x) = \dfrac{8x^4}{2x}$, find $h'(x)$.

Solution As with our first example following the product rule, if you were to encounter a function like this, the first thing you'd probably do is to simplify and write $h(x) = 4x^3$. Thus, we know the answer is $h'(x) = 12x^2$. Let's use the quotient rule and see what happens. Here, the numerator $f(x) = 8x^4$, and the denominator $g(x) = 2x$. Thus,

$$\frac{d}{dx}\left(\frac{8x^4}{2x}\right) = \frac{2x\left(\frac{d}{dx}8x^4\right) - 8x^4\left(\frac{d}{dx}2x\right)}{(2x)^2}$$

$$= \frac{(2x)(32x^3) - 8x^4(2)}{4x^2}$$

[Note that the new denominator is $(2x)^2$, not $2x^2$]

$$= \frac{64x^4 - 16x^4}{4x^2} = \frac{48x^4}{4x^2} = 12x^2$$

as we expected.

Note that the quotient of the derivatives is $32x^3/2$ or $16x^3$, rather than the correct answer of $12x^2$.

Warning #2! The roles of f and g in the **product** rule are interchangeable—it doesn't matter which comes first because $f(x) \cdot g(x)$ is the same as $g(x) \cdot f(x)$. You have to be more careful with the **quotient** rule; $f(x)/g(x)$ and $g(x)/f(x)$ are generally different expressions; the order in the quotient rule is important.

We can see the importance of both warnings by looking at another example where we know what the answer should be.

We have already dealt with some special quotients. For example, if $f(x) = 5/x^3$, we can find $f'(x)$ by writing $f(x)$ as $5x^{-3}$; using the power rule, we get $f'(x) = -15x^{-4} = -15/x^4$. Let's look at what happens if we use the quotient rule.

Example 7 (*Compare Exercise 9*)

If $f(x) = \frac{5}{x^3}$, find $f'(x)$ using the quotient rule.

Solution

$$f'(x) = \frac{d}{dx}\left(\frac{5}{x^3}\right) = \frac{x^3\left(\frac{d}{dx}5\right) - 5\left(\frac{d}{dx}x^3\right)}{(x^3)^2}$$

$$= \frac{x^3(0) - 5(3x^2)}{x^6}$$

$$= \frac{-15x^2}{x^6} = \frac{-15}{x^4}$$

Notice that the quotient of the derivatives is $0/3x^2 = 0$, and that if we are not careful about subtracting in the right order, we could come up with $15/x^4$ instead of $-15/x^4$.

Example 8 (*Compare Exercise 23*)

Compute $h'(x)$ if $h(x) = \dfrac{x^3 - 2x}{x^2 + 4}$.

Solution $h(x)$ is in the form $\dfrac{f(x)}{g(x)}$, where $f(x) = x^3 - 2x$ and $g(x) = x^2 + 4$.

Thus,

$$h'(x) = \frac{(x^2+4)\dfrac{d}{dx}(x^3-2x) - (x^3-2x)\dfrac{d}{dx}(x^2+4)}{(x^2+4)^2}$$

$$= \frac{(x^2+4)(3x^2-2) - (x^3-2x)(2x)}{(x^2+4)^2}$$

As we did with the product rule, we leave the answer in the form given to us by the quotient rule.

Example 9 (*Compare Exercise 19*)

If $h(x) = \dfrac{x^2}{2x^3 + 9x}$, compute $h'(x)$.

Solution Here $h(x) = \dfrac{f(x)}{g(x)}$, where $f(x) = x^2$, and $g(x) = 2x^3 + 9x$, so

$$h'(x) = \frac{(2x^3+9x)(2x) - x^2(6x^2+9)}{(2x^3+9x)^2} = \frac{4x^4 + 18x^2 - 6x^4 - 9x^2}{(2x^3+9x)^2}$$

$$= \frac{-2x^4 + 9x^2}{(2x^3+9x)^2}$$

Here we were able to simplify the numerator without too many complications, which leads us to our third warning.

Warning # 3! When simplifying the numerator of the derivative, remember that the minus sign goes in front of the whole quantity $f(x) \cdot g'(x)$. For instance, in the preceding example, when we simplified $-x^2(6x^2 + 9)$ we obtained $-6x^4 - 9x^2$. Note the second minus sign. It can be very frustrating to remember to use the quotient rule correctly and then get into trouble by making an arithmetic error. Be careful.

Example 10 (*Compare Exercise* 34)

If the cost C in dollars of producing x clock radios per week is given by $C(x) = \frac{1}{20}x^2 + 15x + 3000$,

(a) what is the average cost per radio?

(b) how is the average cost per radio changing at the production level of 200 radios per week?

Solution **(a)** The average cost per radio is obtained by dividing the total cost of production by the number of radios. If $AC(x)$ stands for the average cost of producing x radios, we have

$$AC(x) = \frac{C(x)}{x}$$

The average cost is

$$AC(x) = \frac{\frac{1}{20}x^2 + 15x + 3000}{x}$$

(b) Next, the rate of change of the average cost is the derivative of the average cost ("rate of change" = "derivative"). We compute $(AC)'(x)$ by using the quotient rule.

$$(AC)'(x) = \frac{x\left[\frac{2}{20}x + 15\right] - 1\left[\frac{1}{20}x^2 + 15x + 3000\right]}{(x)^2}$$

$$= \frac{\frac{2}{20}x^2 + 15x - \frac{1}{20}x^2 - 15x - 3000}{x^2} \qquad \text{[Notice all the sign changes]}$$

$$= \frac{\frac{1}{20}x^2 - 3000}{x^2}$$

Now let $x = 200$.

$$(AC)'(200) = \frac{\left(\frac{1}{20}\right)(200)(200) - 3000}{(200)^2} = \frac{-1000}{40000}$$

$$= \frac{-1}{40}$$

The average cost is decreasing at the rate of \$.025 when the production level is 200 radios per week.

Perhaps you would like to rework the solution to Example 10 by doing the division in $AC(x)$ before computing derivatives. The answer is the same, but which way did you find easier?

The Reason for the Quotient Rule. Why is the quotient rule what it is? To see one way to arrive at this formula, let us start with $h(x) = f(x)/g(x)$. (We will assume that $h'(x)$ does exist.) Multiplying both sides of this equation by $g(x)$, we get $h(x) \cdot g(x) = f(x)$. Now we take the derivative of both sides of the equation

$$h(x) \cdot g(x) = f(x)$$

using the product rule on the left hand side.

$$g(x) \cdot h'(x) + h(x) \cdot g'(x) = f'(x)$$

To solve for $h'(x)$, first subtract $h(x) \cdot g'(x)$ from both sides, obtaining

$$g(x) \cdot h'(x) = f'(x) - h(x) \cdot g'(x)$$

Because by assumption $g(x) \neq 0$, we can divide both sides by $g(x)$.

$$h'(x) = \frac{f'(x) - h(x) \cdot g'(x)}{g(x)}$$

Replace $h(x)$ by its equivalent expression $\frac{f(x)}{g(x)}$.

$$h'(x) = \frac{f'(x) - \left[\frac{f(x)}{g(x)}\right] \cdot g'(x)}{g(x)}$$

Finally, to simplify, multiply numerator and denominator by $g(x)$.

$$h'(x) = \frac{g(x) \cdot f'(x) - f(x) \cdot g'(x)}{(g(x))^2}$$

Again, we emphasize that we have not proven the quotient rule because we haven't proven that $h'(x)$ exists. We have however shown what rule must be used to compute $h'(x)$.

3–2 Exercises

I.

Use the product rule to compute each derivative; then simplify the answer. Check your answer by doing the multiplication first and then computing each derivative.

1. (*See Example 3*) $f(x) = (3x)(7x)$
2. $f(x) = (2x^3)(5x^2)$
3. (*See Example 2*) $f(x) = (9)(x^3)$
4. $f(x) = (x^2 + 3)(x)$
5. $f(x) = (2x + 3)(x - 7)$
6. $f(x) = (3x - 1)(2x + 5)$

Use the quotient rule to compute each derivative; then simplify the answer. Check your answer by doing the division first and then computing each derivative.

7. (*See Example 6*) $f(x) = \frac{3x}{x^2}$
8. $f(x) = \frac{9x^3}{3x^2}$
9. (*See Example 7*) $f(x) = \frac{1}{x}$
10. $f(x) = \frac{6}{x^4}$
11. $f(x) = \frac{x^2}{8}$
12. $f(x) = \frac{2x^3}{3}$

II.

Compute the derivative of each of the following.

13. (*See Examples 1 and 5*) $f(x) = (x^3 - 4x + 1)(x^2 + 9)$

14. $f(x) = (x + 1)(x^2 + x - 10)$

15. $f(x) = \left(x + \frac{1}{x}\right)\left(x - \frac{1}{x}\right)$

16. $f(x) = (x^3 - 9x + 5)(x^4 - 7x + 8)$

17. $f(x) = (x^2 + 3x)\left(x^4 - 7x + \frac{1}{x}\right)$

18. $f(x) = \left(x^2 - \frac{6}{x^2}\right)(x^3 + 8x - 11)$

19. (*See Example 9*) $f(x) = \dfrac{3x + 1}{x - 2}$

20. $f(x) = \dfrac{9x + 5}{x^2 - 8}$

21. $f(x) = \dfrac{x^2 + 4x}{x^3 - 7x}$

22. $f(x) = \dfrac{x^3 - 4x}{\sqrt{x}}$

23. (*See Example 8*) $f(x) = \dfrac{3x^2 - 9x + 2}{x^3 - 5}$

24. $f(x) = \dfrac{7 - 4x^2}{9x^3 + 8x}$

III.

Compute the derivative of each of the following.

25. $f(x) = \dfrac{(x^2 - 8x)(x + 2)}{x^3 - 9}$

26. $f(x) = \dfrac{(x^3 - 5)(6x^2 + 9x)}{x^2 + 1}$

27. $f(x) = \dfrac{3x + 7}{(x + 2)(x^2 + 7)}$

28. $f(x) = \dfrac{7x^2 - 8x}{(14x - 3)(12x^3 - 9x)}$

29. A rectangle has height 16 inches and width 25 inches. The height is increasing at the rate of 3 inches per minute, and the width is increasing at the rate of 2 inches per minute. What is the rate of change of the area of the rectangle? (See Figure 3–1 and discussion.)

30. A rectangle has height 10 inches and width 18 inches. The height is increasing at the rate of $\frac{1}{2}$ inch per second, and the width is increasing at the rate of 2 inches per second. How fast is the area of the rectangle changing?

31. A triangle has base 16 inches and height 9 inches. The base is increasing at the rate of 2 inches per minute, and the height is increasing at the rate of 6 inches per minute. What is the rate of change of the area of the triangle?

32. If a rectangle has height 5 feet and base 10 feet, and if the height is increasing at the rate of 3 feet per minute, and the base is **decreasing** at the rate of 2 feet per minute, what is the rate of change of the area of the rectangle?

33. If $f(2) = 7$, $f'(2) = 3$, $g(2) = 6$, and $g'(2) = 4$, what is $(f \cdot g)'(2)$?

34. (*See Example 10*) A company's weekly cost of producing x hundred gallons of milk is given by

$$C(x) = \frac{1}{50}x^2 + 40x + 6200$$

(a) What is the company's marginal cost function $MC(x)$? (Marginal cost is the rate of change of cost.)

(b) What is the company's average cost function $AC(x)$?

(c) What is the marginal average cost function $MAC(x)$? (Again, "marginal" is the same as "derivative of.")

35. If a company's monthly cost of producing x storm doors is

$$C(x) = \frac{1}{30}x^3 + x^2 + 100x + 9500$$

(a) what is the company's marginal cost function?

(b) what is the company's average cost function?

(c) what is the company's marginal average cost function?

36. (*See Example 4*) A rock group is going to sell T-shirts at their concert. The price p and demand x are related by the equation

$$p = 20 - \frac{x}{120}$$

(a) Express the revenue R as a function of the demand x.

(b) What is the rate of change of $R(x)$?

(c) Express the revenue as a function of the price p.

(d) What is $MR(p)$, the marginal revenue as a function of p?

37. What is an equation of the line tangent to the graph

$$y = \frac{3x + 6}{x - 7}$$

at the point $(4, -6)$?

38. What is an equation of the line tangent to the graph

$$y = (x^2 + 2x + 1)(x^3 - 4x + 9)$$

at the point $(1, 24)$?

3–3 The Chain Rule

Composition of Functions
The Chain Rule
The Generalized Power Rule

We have seen various ways that functions can be combined to build a new function—functions can be added, subtracted, multiplied, or divided. The previous two sections have shown the rules for computing the derivative of the

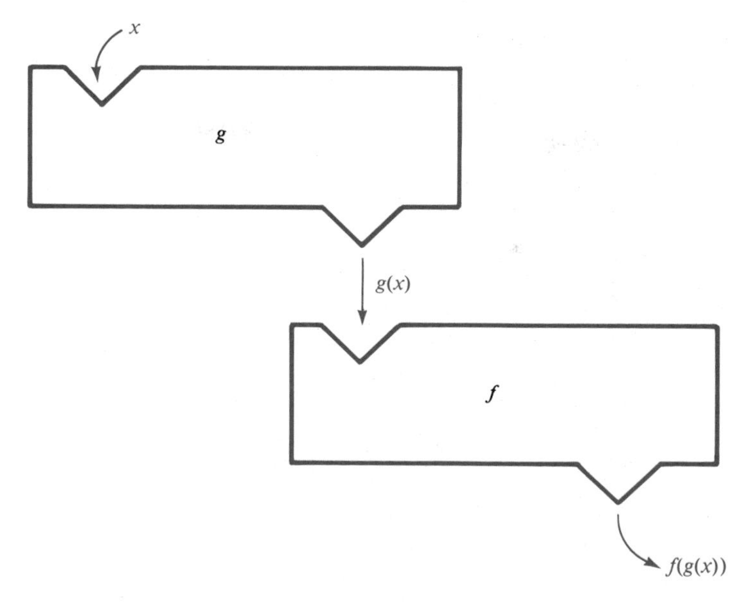

Figure 3–2

new function in terms of the functions that were combined. We will now look at another way of combining two functions to form a new function. This new method is called the **composition of functions.**

Composition of Functions

In Chapter 1, we compared a function to a machine, and we shall use that analogy again to explain composition. Imagine two function machines, f and g, set up as in Figure 3–2. The functions are set up in a sequence, or chain, of evaluations. Given the input x, first evaluate $g(x)$ and then use $g(x)$ as the input for f. The final result of this sequence, or chain, of computations is $f(g(x))$.

Let us look at some specific examples.

Example 1 (*Compare Exercise 1*)
If $g(x) = x^2 + 1$ and $f(x) = 2x + 3$, what is $f(g(x))$?

Solution Remember that $f(x) = 2x + 3$ means that f multiplies any input by 2 and then adds 3 to that result. Any letter can be used to represent the variable. Thus, $f(u) = 2u + 3$ gives the same function as $f(x) = 2x + 3$. If we replace $g(x)$ with the single letter u by setting $u = g(x) = x^2 + 1$, then we can write

$$f(g(x)) = f(u) = 2u + 3$$

To give the final output $f(g(x))$ in terms of the initial input x, simply use the fact that $u = x^2 + 1$.

$$2u + 3 = 2(x^2 + 1) + 3 = 2x^2 + 5$$

$$\text{Thus,} \qquad f(g(x)) = 2x^2 + 5$$

Warning! As in most sequences, the order in which things are done is important. Compare Example 1 with Example 2.

Example 2 If $g(x) = x^2 + 1$ and $f(x) = 2x + 3$, what is $g(f(x))$?

Solution This time $f(x)$ is the first evaluation, so we let the intermediate variable $u = f(x) = 2x + 3$. Then,

$$g(f(x)) = g(u) = u^2 + 1 = (2x + 3)^2 + 1 = 4x^2 + 12x + 10$$

Notice the answers to Examples 1 and 2 are different. Almost always

$$f(g(x)) \neq g(f(x))$$

The order of the composition makes a difference.

Example 3 If $f(x) = x^2 + 9x$ and $g(x) = \sqrt{x}$, compute

(a) $f(g(x))$. **(b)** $g(f(x))$.

Solution **(a)** To evaluate $f(g(x))$, let $u = g(x) = \sqrt{x}$.

$$f(u) = u^2 + 9u$$

Thus, $$f(g(x)) = (\sqrt{x})^2 + 9\sqrt{x} = x + 9\sqrt{x}$$

(b) On the other hand, to evaluate $g(f(x))$ we first treat $f(x)$ as a single variable; let $u = x^2 + 9x$. Then,

$$g(f(x)) = g(u) = \sqrt{u} = \sqrt{x^2 + 9x}$$

The Chain Rule

The function $h(x) = \sqrt{x^2 + 9x}$ is an example of a function whose derivative cannot be computed using only the rules given in the first two sections of this chapter because $h(x)$ cannot be written as the arithmetic combination of functions of the form x^R. The rule we are about to present, the rule used to compute the derivative of the composition of functions, is called the chain rule. Before giving the general statement of the rule, let us show how to compute dy/dx if $y = \sqrt{x^2 + 9x}$. The first function evaluated in the composition is $f(x) = x^2 + 9x$. As above, we introduce a new letter that stands for the output of the f machine. (This letter therefore also stands for the input to the g machine.) Let $u = x^2 + 9x$; u forms the link in the chain between y and x. With this link, we can now state the chain rule:

Rule 8
The Chain Rule

$$\frac{dy}{dx} = \frac{dy}{du} \cdot \frac{du}{dx}$$

In other notation,

If $h(x) = g(f(x))$, then $h'(x) = g'(f(x)) \cdot f'(x)$

Continuing with our example, note that $u = x^2 + 9x$ means that

$$y = \sqrt{u} = u^{1/2}, \text{ so } \frac{dy}{du} = \frac{1}{2}u^{-1/2} = \frac{1}{2\sqrt{u}}$$

Also, $u = x^2 + 9x$ means that $du/dx = 2x + 9$. Substituting these expressions into the chain rule gives

$$\frac{dy}{du} \cdot \frac{du}{dx} = \frac{1}{2\sqrt{u}} \cdot (2x + 9)$$

The last step of the computation is to rewrite dy/dx in terms of x only by substituting back the expression for u in terms of x. Doing this, we obtain

$$\frac{dy}{dx} = \frac{1}{2\sqrt{x^2 + 9x}} \cdot (2x + 9) = \frac{2x + 9}{2\sqrt{x^2 + 9x}}$$

Example 4 (*Compare Exercise 5*)

If $y = 4(x^2 - 8)^3$, find $\dfrac{dy}{dx}$.

Solution Usually, as in this example, the individual functions forming the composition won't be singled out; you must identify them as you identified the individual functions when learning the sum rule or product rule. To help you identify these functions, be on the lookout for parentheses, or polynomials raised to some power, or radical signs. In this example, the polynomial $x^2 - 8$ is cubed. We replace the entire polynomial $x^2 - 8$ by a single letter u. Let $u = x^2 - 8$, and $y = 4(x^2 - 8)^3$ becomes $y = 4u^3$. When we express y in terms of u, you notice that we have a single letter raised to a power, and our previous rules can handle that case.

$$y = 4u^3, \qquad \text{so} \qquad \frac{dy}{du} = 12u^2$$

Also, because

$$u = x^2 - 8, \qquad \frac{du}{dx} = 2x$$

The chain rule requires both $\dfrac{dy}{du}$ and $\dfrac{du}{dx}$.

$$\frac{dy}{dx} = \frac{dy}{du} \cdot \frac{du}{dx} = 12u^2 \cdot 2x$$

The last step is the substitution to get rid of the u. Replace u by $x^2 - 8$:

$$\frac{dy}{dx} = 12(x^2 - 8)^2 \cdot 2x = 24x(x^2 - 8)^2$$

Example 5 If $f(x) = (x^2 - 8x + 9)^4$, find $f'(x)$.

Solution Let $y = (x^2 - 8x + 9)^4$ and $u = x^2 - 8x + 9$. Then $y = u^4$, so

$$\frac{dy}{du} = 4u^3 \qquad \text{and} \qquad \frac{du}{dx} = 2x - 8$$

$$\text{Then,} \qquad \frac{dy}{dx} = \left(\frac{dy}{du}\right)\left(\frac{du}{dx}\right)$$

$$= (4u^3)(2x - 8)$$

$$= 4(x^2 - 8x + 9)^3(2x - 8)$$

Example 6 (*Compare Exercise 13*)
If $f(x) = (x^3 - 8x + 5)^{2/3}$, find $f'(x)$.

Solution Let $y = f(x) = (x^3 - 8x + 5)^{2/3}$ and use the substitution

$$u = x^3 - 8x + 5, \qquad \text{so} \qquad y = u^{2/3}$$

Then,

$$\frac{du}{dx} = 3x^2 - 8 \qquad \text{and} \qquad \frac{dy}{du} = \frac{2}{3}u^{-1/3}$$

Thus,

$$\frac{dy}{dx} = \frac{dy}{du} \cdot \frac{du}{dx} = \left(\frac{2}{3}u^{-1/3}\right)(3x^2 - 8)$$

Obtain the final answer by replacing the u by $x^3 - 8x + 5$.

$$= \frac{2}{3}(x^3 - 8x + 5)^{-1/3}(3x^2 - 8)$$

$$f'(x) = \frac{2(3x^2 - 8)}{3(x^3 - 8x + 5)^{1/3}}$$

Example 7 If $y = \dfrac{3}{x^2 - 4x + 1}$, find $\dfrac{dy}{dx}$.

Solution You already know how to find dy/dx using the quotient rule, but in fact you may find it easier to compute dy/dx using the chain rule. Rewrite the equation as $y = 3(x^2 - 4x + 1)^{-1}$, and let $u = x^2 - 4x + 1$, so $y = 3u^{-1}$.

$$\text{Then,} \qquad \frac{dy}{du} = -3u^{-2} = \frac{-3}{u^2} \qquad \text{and} \qquad \frac{du}{dx} = 2x - 4$$

Thus,

$$\frac{dy}{dx} = \frac{dy}{du} \cdot \frac{du}{dx}$$

$$\frac{dy}{dx} = \frac{-3}{u^2} \cdot (2x - 4) = \frac{-3(2x - 4)}{u^2}$$

$$= \frac{-3(2x - 4)}{(x^2 - 4x + 1)^2}$$

Compute dy/dx using the quotient rule and compare the two procedures.

The Generalized Power Rule

For a function that is the power of another function, the chain rule takes on a special form called the generalized power rule. This rule enables us to compute the derivative more quickly without actually doing the u substitution. To emphasize that this is a particular case of the chain rule, we do not give this rule a new number; instead we call it Rule 8A.

Rule 8A
The Generalized Power Rule

If $y = (f(x))^R$, then $\dfrac{dy}{dx} = R(f(x))^{R-1}f'(x)$

To see that this is a special case of the chain rule, let $u = f(x)$, so $du/dx = f'(x)$. Further, $y = u^R$, so $dy/du = Ru^{R-1}$. Thus,

$$\frac{dy}{dx} = \left(\frac{dy}{du}\right)\left(\frac{du}{dx}\right)$$

$$= (Ru^{R-1})f'(x) = R(f(x))^{R-1}f'(x)$$

Example 8 (*Compare Exercise 9*)

If $y = (5x^4 + 3x)^9$, find $\dfrac{dy}{dx}$.

Solution y is of the form $(f(x))^R$, where $f(x) = 5x^4 + 3x$ and $R = 9$. Computing the derivative of $f(x)$, we have $f'(x) = 20x^3 + 3$. Now, we put the pieces together to form dy/dx.

$$\frac{dy}{dx} = 9(5x^4 + 3x)^8(20x^3 + 3)$$

Example 9 (*Compare Exercise 17*)

If $y = \dfrac{4}{(x^3 + 1)^2}$, find $\dfrac{dy}{dx}$.

Solution Rewrite the function as $y = 4(x^3 + 1)^{-2}$. Then,

$$\frac{dy}{dx} = -8(x^3 + 1)^{-3}(3x^2) = \frac{-24x^2}{(x^3 + 1)^3}$$

Sometimes you have to use the chain rule together with another rule, such as the product rule or quotient rule.

Example 10 (*Compare Exercise 21*)

If $f(x) = x\sqrt{x^3 + 5x}$, find $f'(x)$.

Solution To find $f'(x)$ we must use the product rule because $f(x)$ is x *times* $\sqrt{x^3 + 5x}$.

$$f'(x) = \sqrt{x^3 + 5x}\left(\frac{d}{dx}\,x\right) + x\left(\frac{d}{dx}\sqrt{x^3 + 5x}\right)$$

We have to use the chain rule to compute

$$\frac{d}{dx}\sqrt{x^3 + 5x} = \frac{d}{dx}(x^3 + 5x)^{1/2}$$

Using the generalized power rule with $R = \frac{1}{2}$,

$$\begin{aligned}\frac{d}{dx}(x^3 + 5x)^{1/2} &= \frac{1}{2}(x^3 + 5x)^{-1/2}(3x^2 + 5)\\ &= \frac{3x^2 + 5}{2\sqrt{x^3 + 5x}}\end{aligned}$$

Because $\frac{d}{dx}x = 1$, we have

$$\begin{aligned}f'(x) &= \sqrt{x^3 + 5x} \cdot 1 + \frac{x(3x^2 + 5)}{2\sqrt{x^3 + 5x}}\\ &= \sqrt{x^3 + 5x} + \frac{x(3x^2 + 5)}{2\sqrt{x^3 + 5x}}\end{aligned}$$

Example 11 (*Compare Exercise 25*)
Compute $g'(x)$ if $g(x) = \dfrac{(x^2 + 1)^3}{x^4 - 9x}$.

Solution We must use the quotient rule. While using the quotient rule, we must use the chain rule to compute the derivative of the numerator, $(x^2 + 1)^3$.

$$\begin{aligned}\frac{d}{dx}(x^2 + 1)^3 &= 3(x^2 + 1)^2 2x\\ &= 6x(x^2 + 1)^2\end{aligned}$$

Thus,

$$g'(x) = \frac{(x^4 - 9x)[6x(x^2 + 1)^2] - (x^2 + 1)^3(4x^3 - 9)}{(x^4 - 9x)^2}$$

Sometimes the chain rule must be used more than once.

Example 12 (*Compare Exercise 29*)
If $f(x) = [4x + (3x + 1)^6]^{10}$, find $f'(x)$.

Solution

$$\begin{aligned}f'(x) &= 10[4x + (3x + 1)^6]^9 \cdot \frac{d}{dx}[4x + (3x + 1)^6]\\ &= 10[4x + (3x + 1)^6]^9 \cdot [4 + 6(3x + 1)^5 \cdot 3]\end{aligned}$$

Notice the use of the chain rule in computing the derivative of $(3x + 1)^6$.

3–3 Exercises

I.

1. (*See Example 1*) If $f(x) = x^2 + 3x$ and $g(x) = 2x + 1$, evaluate
   (a) $f(g(x))$ **(b)** $g(f(x))$
2. If $f(x) = 3x - 7$ and $g(x) = 6 - 2x$, evaluate
   (a) $f(g(x))$ **(b)** $g(f(x))$
3. If $f(x) = \sqrt{x}$ and $g(x) = 9x + 25$, evaluate
   (a) $f(g(x))$ **(b)** $g(f(x))$
4. If $f(x) = x^2 + \sqrt{x}$ and $g(x) = 4x$, evaluate
   (a) $f(g(x))$ **(b)** $g(f(x))$

Use the given substitution and the chain rule to compute dy/dx given the following.

5. (*See Example 4*)
   $y = (x^3 - 4x)^5$; $u = x^3 - 4x$
6. $y = (4x^2 - 9x + 3)^3$; $u = 4x^2 - 9x + 3$
7. $y = \sqrt{9x + 16}$; $u = 9x + 16$
8. $y = (25x - 9)^{-1/2}$; $u = 25x - 9$

II.

Find dy/dx given the following.

9. (*See Example 8*)
   $y = (3x^2 + 8)^3$
10. $y = (5x - 3)^4$
11. $y = \sqrt{4x + 9}$
12. $y = (6x - 8)^{-1}$

Find $f'(x)$ given the following.

13. (*See Example 6*)
    $f(x) = (x^2 - 9x + 5)^3$
14. $f(x) = (x^3 - 8)^{1/2}$
15. $f(x) = \sqrt{16x - 4}$
16. $f(x) = 3(x^2 + 4)^{-2}$

Find dy/dx given the following.

17. (*See Example 9*)
    $y = \dfrac{10}{(4x + 1)^3}$
18. $y = \dfrac{1}{(x^2 + 1)^2}$
19. $y = \dfrac{4}{(3x + 1)^3}$
20. $y = \dfrac{6}{(2x^2 + 3)^2}$

III.

Find $f'(x)$ given the following.

21. (*See Example 10*)
$f(x) = x\sqrt{4x + 9}$

22. $f(x) = (3x + 1)(5x^2 + 1)^4$

23. $f(x) = (7x - 8)^3(6x - 4)^2$

24. $f(x) = (x^2 + 1)\sqrt{x^3 - 4x}$

25. (*See Example 11*)
$f(x) = \dfrac{(x^2 + 1)^3}{8x + 10}$

26. $f(x) = \dfrac{(6x - 8)^4}{3x + 9}$

27. $f(x) = \dfrac{3x + 1}{(x^2 - 4)^2}$

28. $f(x) = \dfrac{2x + 9}{(3x - 5)^3}$

29. (*See Example 12*)
$f(x) = (3x + (2x + 1)^3)^5$

30. $f(x) = \sqrt{6x^2 + (5x + 1)^3}$

31. $f(x) = (7x + \sqrt{3x + 1})^3$

32. $f(x) = ((3x + 4)^2 - \sqrt{9x + 4})^3$

33. What is an equation of the line tangent to the graph $y = \sqrt{2x + 17}$ at the point (4, 5)?

34. What is an equation of the line tangent to the graph $y = (2x - 6)^5$ at the point (2, -32)?

35. A company projects that its sales revenue, in millions of dollars, t years after introducing a new model computer will be given by $R(t) = \sqrt{t^2 + 6t}$. What is the rate of change of revenue after 2 years?

36. Management has found that the attentiveness of a new trainee can be measured by $A(t) = (5t^3 + 9)/(t^2 + 3)^2$ after t hours of instruction. What is the rate of change of attentiveness after two hours?

3–4 Higher Order Derivatives

INFLATION SLOWDOWN SPEEDS UP

This headline appeared in the first column on the front page of the *Waco Tribune-Herald* on April 21, 1982, and succinctly sums up the first paragraph, which said:

> The unrelenting recession, driving down some prices for the first time in six years, is dampening inflation at a faster rate than most economists had expected.

"INFLATION" is a measure of the change in the value of the dollar. The word "SLOWDOWN" in the headline then is talking about the rate of change of inflation, while "SPEEDS UP" refers to the rate of change in the slowdown.

The mathematical tool for talking about rate of change is the derivative; the rate of change of the rate of change becomes the derivative of the derivative. If we use f' to denote the derivative of f, then $(f')'$ is the derivative of f'. We can (and will from now on) omit the parentheses and write f'' instead of $(f')'$. We call f'' **the second derivative of f.**

Once we have made this step, we see that we can continue this process (as does the headline). The third derivative of f is the derivative of f'' and is written f''', and so on.

Example 1 (*Compare Exercise 1*)
If $f(x) = x^3$, compute the fourth derivative of f.

Solution

$f(x) = x^3$	the function f
$f'(x) = 3x^2$	the first derivative of f
$f''(x) = 6x$	the second derivative of f
$f'''(x) = 6$	the third derivative of f
$f^{(4)}(x) = 0$	the fourth derivative of f

Observe the notational change for f''''. The notation of repeating the $'$ symbol becomes awkward after a while; it's easy to lose count of the number of marks. We use the symbol $f^{(n)}$ to mean the nth derivative of f when $n \geq 4$.

The alternate notations of dy/dx and y' are also used for $f'(x)$. Using these notations, higher derivatives of f are expressed as:

$$\frac{d^2y}{dx^2}, \quad \text{and} \quad y'' \text{ for } f''(x)$$

$$\frac{d^3y}{dx^3} \quad \text{and} \quad y''' \text{ for } f'''(x)$$

$$\vdots \qquad\qquad \vdots$$

$$\vdots \qquad\qquad \vdots$$

etc.

Example 2 (*Compare Exercise 5*)
If $y = \frac{1}{x}$, find $\frac{d^3y}{dx^3}$.

Solution We write $y = x^{-1}$ in order to use the power rule.

$$\frac{dy}{dx} = -x^{-2}$$

$$\frac{d^2y}{dx^2} = 2x^{-3}$$

$$\frac{d^3y}{dx^3} = -6x^{-4} = -\frac{6}{x^4}.$$

Example 3 (*Compare Exercise 7*)

If $y = (5x + 3)^4$, compute $\dfrac{d^3y}{dx^3}$.

Solution Notice that y is the composition of functions, and so we have to use the chain rule to compute dy/dx. We can use the generalized power rule form of the chain rule with $f(x) = 5x + 3$ and $R = 4$.

$$\frac{dy}{dx} = 4(5x + 3)^3 \cdot 5$$

$$= 20(5x + 3)^3$$

$$\frac{d^2y}{dx^2} = 60(5x + 3)^2 \cdot 5 \qquad \text{[Chain rule again!]}$$

$$= 300(5x + 3)^2$$

$$\frac{d^3y}{dx^3} = 600(5x + 3) \cdot 5$$

$$= 3000(5x + 3)$$

Generally in applications, you won't have to compute more than the first and second derivatives. When the first derivative involves the chain rule, the computation of the second derivative requires some care, as the next example shows.

Example 4 (*Compare Exercise 15*)

If $y = \sqrt{x^4 + 9}$, find y''.

Solution

$$y = (x^4 + 9)^{1/2}, \qquad \text{so}$$

$$y' = \frac{1}{2}(x^4 + 9)^{-1/2} \cdot 4x^3 \qquad \text{[Chain rule]}$$

$$= 2x^3(x^4 + 9)^{-1/2}$$

We had to be careful to use the chain rule when computing y', and we have to continue to use it to compute y''. But something else has snuck in. The chain rule produced a factor, $4x^3$, that is not constant, and so y' is the product of two functions. We must use the product rule as well as the chain rule when computing y''.

$$y'' = (x^4 + 9)^{-1/2}\left[\frac{d}{dx}2x^3\right] + 2x^3\left[\frac{d}{dx}(x^4 + 9)^{-1/2}\right] \qquad \text{(Used the product rule)}$$

$$= (x^4 + 9)^{-1/2}[6x^2] + 2x^3\left[-\frac{1}{2}(x^4 + 9)^{-3/2} \cdot 4x^3\right] \qquad \text{(Used the chain rule)}$$

$$= \frac{6x^2}{\sqrt{x^4 + 9}} - \frac{4x^6}{(\sqrt{x^4 + 9})^3}$$

When computing a particular derivative, you may be called upon to use several rules. Be careful and be patient; don't try to rush to the answer.

Example 5 (*Compare Exercise 13*)
Find $f''(x)$ if $f(x) = (5x^2 - 9)^4$.

Solution The generalized power rule gives

$$f'(x) = 4(5x^2 - 9)^3 \cdot 10x$$
$$= 40x(5x^2 - 9)^3$$
$$f''(x) = (5x^2 - 9)^3 \cdot 40 + 40x \cdot 3(5x^2 - 9)^2 \cdot 10x$$

Don't forget the chain rule at the very end!

$$f''(x) = 40(5x^2 - 9)^3 + 1200x^2(5x^2 - 9)^2$$

Example 6 (*Compare Exercise 17*)
Find y'' if $y = \dfrac{3}{(x^2 + 1)^2}$.

Solution Rewrite y as

$$y = 3(x^2 + 1)^{-2}$$

Then,

$$y' = 3[-2(x^2 + 1)^{-3} \cdot 2x]$$
$$= -12x(x^2 + 1)^{-3}$$
$$y'' = (x^2 + 1)^{-3}(-12) + (-12x)[-3(x^2 + 1)^{-4} \cdot 2x]$$
$$= -12(x^2 + 1)^{-3} + 72x^2(x^2 + 1)^{-4}$$

3–4 Exercises

I.

Find the third derivative of each of the following:

1. (*See Example 1*) $f(x) = 6x + 7$
2. $f(x) = 4x^2 - 9x + 2$
3. $g(t) = t^3 - 8t^2 + 4t - 7$
4. $y = t^4 - 4t^3 + 8t + 1$
5. (*See Example 2*) $f(x) = \dfrac{5}{x^2}$
6. $f(x) = \dfrac{6}{x^4}$

II.

Compute the third derivative of each of the following:

7. (*See Example 3*) $f(x) = (3x + 1)^5$
8. $f(x) = (6x - 4)^3$
9. $f(x) = \sqrt{8x + 4}$

10. $f(x) = (3x - 6)^2$

11. $g(x) = \dfrac{5}{2x + 1}$

12. $h(t) = \dfrac{6}{(3t + 2)^2}$

III.

Compute y'' in Exercises 13 through 20.

13. (*See Example 5*)
$y = (x^2 + 1)^3$

14. $y = (3x^2 - 4)^5$

15. (*See Example 4*)
$y = \sqrt{6x^2 - 8}$

16. $y = \sqrt[3]{27x^3 - 8}$

17. (*See Example 6*)
$y = \dfrac{3}{(x^2 + 1)^3}$

18. $y = -\dfrac{8}{(4x^3 + 8)^2}$

19. $y = \dfrac{4}{(3x^2 + 9)^5}$

20. $y = (4x^2 + 8)^3$

21. If $f(x) = \sqrt{2x^2 + 1}$, find $f''(2)$

22. If $g(t) = (t^3 - 4t)^4$, what is $g''(t)$?

23. A company's daily cost function (in dollars) for producing x units is given by

$$C(x) = \frac{1}{60}x^3 + 5x^2 + 115x + 6200$$

What is the rate of change of the marginal cost when $x = 150$?

24. A sandwich company's weekly cost function (in dollars) for producing x hundred sandwiches per week is given by

$$C(x) = \frac{1}{90}x^3 + x^2 + 84x + 2100$$

What is the rate of change of the marginal cost when the production level is 3000 sandwiches per week?

25. A falling rock is $-16t^2 + 200$ feet above the ground t seconds after it is dropped.

(a) What is the velocity of the rock when $t = 3$?

(b) What is the acceleration of the rock when $t = 3$? (Acceleration is the rate of change of velocity.)

26. An automobile covers

$$\frac{1}{12}t^3 + 15t^2 + 10t \text{ feet}$$

t seconds after it starts. How fast is it accelerating after four seconds?

3–5 Implicit Differentiation

Implicitly Defined Functions

When we write an equation such as $y = 2x^3 - 5x + 7$, we are giving a rule that shows how the value of y depends upon the value of x. For example, if $x = -1$, then we have an explicit formula for computing y: $y = 2(-1)^3 - 5(-1) + 7 = 10$. We say that y is a function of x, and write $y = f(x)$ to show that y depends on x. Now, the equation $y = 2x^3 - 5x + 7$ can be written in other equivalent forms, such as $y - 2x^3 + 5x - 7 = 0$. Remember that in Chapter 1 we saw that a particular line can have several equivalent equations. For example, $2y - 16x + 10 = 0$ is equivalent to $y = 8x - 5$; both equations define the same line. Thus, both equations define the same function but in different forms. When the equation is of the form $y = f(x)$, we say that the function f is given **explicitly**; in any other form, we say that the function is given **implicitly**. Thus, the equation $2y + 16x - 10 = 0$ implicitly defines y as a function of x.

Example 1 (*Compare Exercise 1*)
Each of the following four equations does define y as a function of x. State whether each equation defines y as a function of x **implicitly** or **explicitly**.

(a) $xy = 3x + 2y$
(b) $y = \sqrt{x^4 - 9}$
(c) $y + 6 = 8x$
(d) $y = 4x - y^3$

Solution
(a) Implicitly; the y is not isolated.
(b) Explicitly; the y is isolated.
(c) Implicitly.
(d) Implicitly; note that there is a y term on both sides of the equation.

Sometimes, when an equation implicitly defines y as a function of x, you can write an equivalent equation that gives the function explicitly. The basic idea is to isolate the single term y on one side of the equation and have an expression on the other side that involves only numbers and the variable x. Let's go back to the equations from Example 1 and see what this means in specific cases. We begin with the easiest.

Example 2 (*Compare Exercise 7*)
Given the equation $y + 6 = 8x$, express y explicitly as a function of x.

Solution We can isolate y by subtracting 6 from both sides of the equation: $y = 8x - 6$; $f(x) = 8x - 6$.

Example 3 (*Compare Exercise 11*)
Find the function $y = f(x)$ defined implicitly by the equation $xy = 3x + 2y$.

Solution First, get all the terms that involve y on one side of the equal sign and all the remaining terms on the other side.

$$xy - 2y = 3x$$

Next, factor out y on the left-hand side.

$$y(x - 2) = 3x$$

Now isolate y by dividing both sides by $x - 2$.

$$y = \frac{3x}{x - 2}$$

Sometimes, you may not be able to isolate the y term, as with equation (d) from Example 1. We can rewrite that equation in several forms:

$$y^3 + y = 4x$$
$$y(y^2 + 1) = 4x$$

or

$$y = \frac{4x}{y^2 + 1}$$

but notice even this last form does not isolate y because y also appears on the right-hand side of the equation. Still, the equation $y^3 + y = 4x$ does define y as a function of x even if we cannot write down an explicit formula $y = f(x)$ for this function.

The value of y is dependent on the value of x, and as the value of x changes, the value of y will change also. The derivative of y with respect to x, dy/dx, gives the rate of change of y with respect to x. Knowing this rate of change may be important in an application, but how can you find it if the equation you start with (like $y^3 + y = 4x$) is so complicated that you cannot solve for y explicitly? The technique for finding dy/dx when y and x are related by such complicated equations is called **implicit differentiation.**

Implicit Differentiation

We use equation (d) of Example 1. We begin the process of computing dy/dx by taking the derivative (with respect to x) of both sides of this equation.

$$\frac{d}{dx}y = \frac{d}{dx}(4x - y^3)$$
$$= \frac{d}{dx}4x - \frac{d}{dx}y^3$$

On the right-hand side we used "the derivative of the difference is the difference of the derivatives."

Now, remember that d/dx means to take the derivative with respect to x of the function that follows. Thus on the left, $(d/dx)y$ is the derivative of y with respect to x.

We rewrite $(d/dx)y$ as dy/dx; this is the quantity we are trying to find.

Next, $(d/dx)4x$ is the derivative of $4x$ with respect to x. We rewrite $(d/dx)4x$ as 4.

Finally, we have to compute $(d/dx)y^3$. To see what to do, remember that we are assuming that $y = f(x)$ even though we don't know the explicit formula for $f(x)$. If we replace y by $f(x)$, we have

$$\frac{d}{dx}y^3 = \frac{d}{dx}(f(x))^3$$

Now we can use the generalized power rule.

$$\frac{d}{dx}(f(x))^3 = 3(f(x))^2 \cdot \frac{d}{dx}f(x)$$

If we substitute back the y for $f(x)$, we have

$$\frac{d}{dx}y^3 = 3y^2\frac{d}{dx}y$$

Again, the term $(d/dx)y$ means the derivative of y with respect to x, and can be written as (dy/dx). We have

$$\frac{d}{dx}y^3 = 3y^2\frac{dy}{dx}$$

We can now rewrite our equation

$$\frac{d}{dx}y = \frac{d}{dx}4x - \frac{d}{dx}y^3$$

as

$$\frac{dy}{dx} = 4 - 3y^2\frac{dy}{dx}$$

We use the same method to solve for dy/dx as we used earlier to solve for y. Move all the terms that involve dy/dx to one side of the equal sign and all the other terms to the other side of the equal sign.

$$\frac{dy}{dx} + 3y^2\frac{dy}{dx} = 4$$

Next, factor out dy/dx

$$(1 + 3y^2)\frac{dy}{dx} = 4$$

Finally, divide by the other factor

$$\frac{dy}{dx} = \frac{4}{1 + 3y^2}$$

We could not solve explicitly for y as a function of x, so we can not hope to express dy/dx as a function of x. Generally, we will have dy/dx as a function of both x and y.

Example 4 *(Compare Exercises 15 and 21)*
Given the equation $3y^2 - 2x^3 = y + 5$, find y' in terms of x and y.

Solution We differentiate both sides of the original equation with respect to x.

$$\frac{d}{dx}(3y^2 - 2x^3) = \frac{d}{dx}(y + 5)$$

$$\frac{d}{dx}3y^2 - \frac{d}{dx}2x^3 = \frac{d}{dx}y + \frac{d}{dx}5$$

$$6y\frac{dy}{dx} - 6x^2 = \frac{dy}{dx} + 0$$

$$6y\frac{dy}{dx} - 6x^2 = \frac{dy}{dx}$$

Again, we use the same technique to solve for dy/dx. Get all the terms that involve dy/dx on one side of the equation, and all the remaining terms on the other side.

$$6y\frac{dy}{dx} - \frac{dy}{dx} = 6x^2$$

Factor out dy/dx.

$$(6y - 1)\frac{dy}{dx} = 6x^2$$

Divide by $6y - 1$.

$$\frac{dy}{dx} = \frac{6x^2}{6y - 1}$$

Because we have been able to express y' in terms of x and y, we can find an equation of the line tangent to the curve at a given point, as the next example shows.

Example 5 *(Compare Exercise 27)*
Find an equation of the line tangent to the graph

$$y^3 - 2xy^2 + 6x = 3 \text{ at the point } (2, 3).$$

Solution First, check that (2, 3) is on the curve.

$$3^3 - 2 \cdot 2 \cdot 3^2 + 6 \cdot 2 = 27 - 36 + 12 = 3$$

so $x = 2$, $y = 3$ satisfies the equation and (2, 3) **is** on the curve. Next, the slope of the tangent line is y', so we use implicit differentiation to find an equation involving y', and then solve for y' in terms of x and y. Differentiate both sides with respect to x.

$$\frac{d}{dx}(y^3 - 2xy^2 + 6x) = \frac{d}{dx}(3)$$

$$\frac{d}{dx}(y^3) - \frac{d}{dx}(2xy^2) + \frac{d}{dx}(6x) = 0$$

$$3y^2 \cdot y' - \left[y^2\left(\frac{d}{dx}2x\right) + 2x\left(\frac{d}{dx}y^2\right)\right] + 6 = 0$$

Notice that to compute $(d/dx)(2xy^2)$ we must use the product rule.

$$3y^2 \cdot y' - [y^2 \cdot 2 + 2x \cdot 2y \cdot y'] + 6 = 0$$

$$3y^2y' - 2y^2 - 4xyy' + 6 = 0$$

Now gather all the terms involving y' on one side and all the remaining terms on the other side.

$$3y^2y' - 4xyy' = 2y^2 - 6$$

Factor out y'.

$$y'(3y^2 - 4xy) = 2y^2 - 6$$

Divide by $3y^2 - 4xy$.

$$y' = \frac{2y^2 - 6}{3y^2 - 4xy}$$

Finally, when $x = 2$ and $y = 3$,

$$y' = \frac{2 \cdot 9 - 6}{3 \cdot 9 - 4 \cdot 2 \cdot 3} = \frac{12}{3} = 4$$

The slope of the tangent line is 4.

The point-slope form of the equation of the line tangent to $y^3 - 2xy^2 + 6x = 3$ at the point (2, 3) is $y - 3 = 4(x - 2)$.

3–5 Exercises

I.

State whether each of the following equations defines y as a function of x explicitly or implicitly.

1. (*See Example 1*) $x^2 - 3x = y$
2. $3x + 4y = 8$
3. $6y = 5x^2 - 9x + 2$
4. $y = 8x^2 - y^2$

5. $x^2 + y^2 = 25$

6. $x^2 + 3x = y + 2$

Each of the following equations implicitly defines y as a function of x. Rewrite the equation so that y is given explicitly in the form $y = f(x)$.

7. (*See Example 2*)
$3(y - 1) = 2(x + 5)$

8. $y - 9 = x^2 + 4x + 1$

9. $y(x + 1) = 3x - 7$

10. $4y - 2x = 6y + 5x^2 + 7$

II.

Rewrite each of the following equations so that y is given explicitly as a function of x.

11. (*See Example 3*)
$3y + xy = 10$

12. $y + 5 = x^2y - 9x$

13. $xy + x^2y = 5y - 9x$

14. $y(x^2 + 1) = 6y - 9 + 8x^2$

In Exercises 15 through 20, use implicit differentiation and then solve for y' in terms of x and y.

15. (*See Example 4*)
$x^2 + y^2 = 25$

16. $4y^3 = 5x - 8$

17. $6y^2 + 2y + 5x + 9 = 0$

18. $xy + y^2 = 5$

19. $x^2y + y^3 - 4x^2 = 7$

20. $y^3 - 3x = y + 7$

III.

In Exercises 21 through 26, use implicit differentiation and then solve for dy/dx in terms of x and y.

21. (*See Example 4*)
$3x^2 - 6x^2y + y^2 = 9$

22. $5xy + y^3 = 6x - 8$

23. $4y^3 - 6xy^2 + 3y = 9x + 4$

24. $6x^2y^3 + 4y^2 - 4x = 3y + 8$

25. $\dfrac{x}{y} = y^2 + x$

26. $(x + y)^2 = 3x + 5$

Find an equation of the line tangent to each of the following curves at the indicated point. Check that each point is on the curve.

27. (*See Example 5*)
$y^3 - 2xy + x^2 = 7$; the point is $(2, -1)$

28. $y^3 - x^3 = 28$; the point is $(-1, 3)$

29. Find a value of C so that $(3, -2)$ is on the curve $10y^2 - 4x^2y^2 + x^3 = C$, and then find an equation of the line tangent to that curve at the point $(3, -2)$.

30. If the demand for wheat, x, is related to the price p by the equation $x^2 + 100p^2 = 4100$, what is dp/dx when $p = 4$?

3–6 Related Rates

In the previous section, we saw that the relationship between two variables x and y may be so complicated that we cannot express that relationship explicitly in the form $y = f(x)$. Nevertheless, we can still find dy/dx, the rate of change of y with respect to x.

Frequently in applications, the two variables x and y may depend on a third variable t, usually time. For example, if x is the number of VCR's in use, and y is the number of video tapes being produced, then x and y are related. The more people with VCR's, the more video tapes will be produced; conversely, the more video tapes there are on the market, the more likely someone is to invest in a VCR. Thus, x and y depend on each other. The numbers x and y can also depend on time, both as the number of VCR's grows and also seasonally (with a heavier demand close to Christmas than during the rest of the year). If we could find an equation relating x and y, can we find an equation relating their **rates of change with respect to time**?

The basic principle is this: if two variables x and y are related to each other by some equation, and if both x and y depend on a third variable t, we can find a new equation that gives a relation between dx/dt and dy/dt by differentiating the original equation **with respect to t.**

Let us look at some examples.

Example 1 (*Compare Exercise 1*)
If x and y are related by the equation $x^2 - y^3 = 17$, then find an equation relating x, y, dx/dt, and dy/dt.

Solution We compute the derivative **with respect to t** of both sides of the equation.

$$\frac{d}{dt}(x^2 - y^3) = \frac{d}{dt}17$$

The procedure is very similar to what we did in the previous section. The derivative of the difference is the difference of the derivatives; we use this fact on the left-hand side.

$$\frac{d}{dt}x^2 - \frac{d}{dt}y^3 = \frac{d}{dt}17$$

Now, we must remember to use the generalized power rule *on both members* of the left-hand side

$$\frac{d}{dt}x^2 = 2x \cdot \frac{dx}{dt} \qquad \text{and} \qquad \frac{d}{dt}y^3 = 3y^2 \cdot \frac{dy}{dt}$$

On the right, $(d/dt)(17) = 0$ because the derivative of a constant is 0. So, we have

$$2x\frac{dx}{dt} - 3y^2\frac{dy}{dt} = 0$$

Warning! Remember to take the derivative of **both** sides of the equation, and remember that the derivative of a constant is always 0.

Example 2 (*Compare Exercises 7 and 11*)
Suppose that a runner is jogging on a track and that we can represent the track with the equation $x^2 + 9y^2 = 900$. How is the x-coordinate of the jogger changing when she is at the point $(24, -6)$ if her y-coordinate is decreasing at a rate of 7 feet per second? See Figure 3–3.

Solution Before we do the actual computation, let's try to get some geometric feeling for the situation. If the jogger is at the point $(24, -6)$, and her y-coordinate is decreasing, then she must be moving toward P and away from Q. That means the x-coordinate will be getting smaller; the x-coordinate is decreasing. The rate of change of the x-coordinate is dx/dt; because x is decreasing, we can expect the computation to show that dx/dt is negative. Let's see.

Starting with $x^2 + 9y^2 = 900$, we take the derivative with respect to t of both sides of the equation. Using "the derivative of the sum is the sum of the derivatives," we have

$$\frac{d}{dt}x^2 + \frac{d}{dt}9y^2 = \frac{d}{dt}900$$

Thus,

$$2x\frac{dx}{dt} + 18y\frac{dy}{dt} = 0$$

The problem asks for dx/dt when $x = 24$, $y = -6$, and $dy/dt = -7$. (Again, note that if y is decreasing then dy/dt is negative; dy/dt is -7, not 7.) Substituting these values, we have

$$2 \cdot 24 \cdot \frac{dx}{dt} + 18 \cdot -6 \cdot -7 = 0$$

$$48\frac{dx}{dt} + 756 = 0$$

$$\frac{dx}{dt} = -\frac{756}{48} = -\frac{63}{4}$$

(Notice the expected negative sign). The runner's x-coordinate is decreasing at the rate of $\frac{63}{4}$ feet per second.

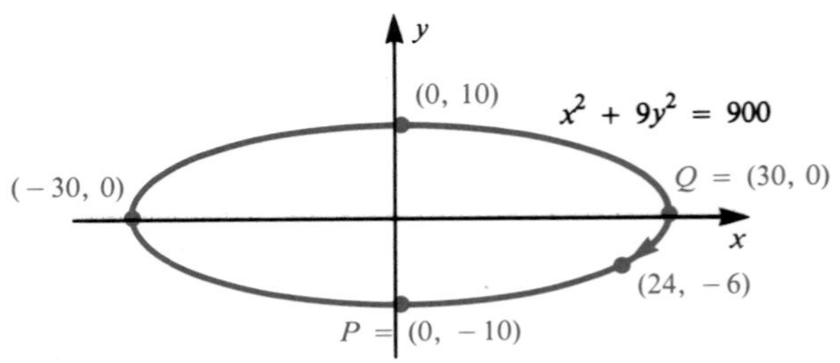

Figure 3–3

Example 3 (*Compare Exercise 21*)
Two cars leave a parking lot at the same time. One travels north at 30 mph, and the other goes east at 40 mph.

(a) Determine how far apart they are after 6 minutes.

(b) Describe how the distance between them is changing.

Solution Let the origin represent the parking lot and, as is usual on maps, let north be up and east to the right. Thus, one car is going up the positive y-axis; denote its position at a given time by $y(t)$. The other car is moving to the right on the positive x-axis; denote its position by $x(t)$. See Figure 3–4.

It would be natural to let d (for **d**istance) represent the distance between the cars, but we shouldn't use d as a variable in calculus with so many other d's being used to mean derivatives. The next letter after x and y is z, so we use z for the distance between the cars. The relation we need is given by the Pythagorean theorem which states that $z^2 = x^2 + y^2$.

One more point about setting up the problem before we start our computations: One reason for stating this problem the way we did is to point out to you the need to use the right units in your computation. The speed of the cars is given in miles per hour, but the time is given in minutes. We must either express time in hours or else tell how fast the cars are going per minute. The time is 6 minutes, which is easy to convert to $\frac{1}{10}$ hour, so we will express time in hours. Now we are set up to answer the questions.

(a) To solve this part, we must find z when $t = \frac{1}{10}$.

Distance travelled = (rate) · (time), so when $t = \frac{1}{10}$

$$x = 40 \cdot \frac{1}{10} = 4$$

and

$$y = 30 \cdot \frac{1}{10} = 3$$

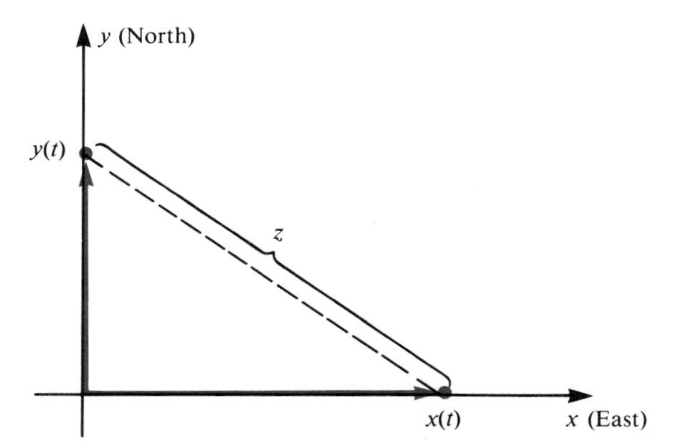

Figure 3–4

Thus, $z^2 = 4^2 + 3^2 = 25$; the distance z is $\sqrt{25}$. After 6 minutes, the cars are 5 miles apart.

(b) To find the rate of change of the distance between the cars, it is important that we start with the equation $z^2 = x^2 + y^2$, take the derivative with respect to t, and **then** substitute numerical values. If we use $z = 5$, $x = 4$, and $y = 3$ before taking derivatives, we won't get a formula relating dz/dt, dx/dt, and dy/dt. Differentiation of both sides of the equation $z^2 = x^2 + y^2$ with respect to t gives us

$$\frac{d}{dt} z^2 = \frac{d}{dt} x^2 + \frac{d}{dt} y^2$$

$$2z \frac{dz}{dt} = 2x \frac{dx}{dt} + 2y \frac{dy}{dt}$$

Now we substitute the values: $z = 5$, $x = 4$, and $y = 3$. Also, the wording of the problem says that x is increasing at the rate of 40; that is, $dx/dt = 40$. Similarly, we have $dy/dt = 30$ (both x and y are increasing so the derivatives are positive). Thus, we have

$$2 \cdot 5 \cdot \frac{dz}{dt} = 2 \cdot 4 \cdot 40 + 2 \cdot 3 \cdot 30$$

$$10 \frac{dz}{dt} = 320 + 180 = 500$$

$$\frac{dz}{dt} = 50$$

dz/dt is positive, so z is increasing. The distance between the cars is increasing at the rate of 50 mph.

Example 4 (*Compare Exercises 17 and 23*)
If x is the number of thousands of running shoes a certain factory manufactures each week, and y is the number of thousands of aerobic shoes it manufactures, then x and y satisfy the equation

$$x^2 - xy + 2y^2 = 22 \qquad \text{(Units are in thousands)}$$

If the current levels are $x = 4$ and $y = 3$, and if the production of running shoes is falling off at the the rate of 100 per week, what is happening to the production of aerobic shoes?

Solution Let t be time measured in weeks. The last sentence tells us that $dx/dt = -0.1$. (Remember units are in thousands; a decrease of 100 shoes means that $\Delta x = -0.1$.) To find dy/dt, we differentiate the equation

$$x^2 - xy + 2y^2 = 22$$

with respect to t.

$$\frac{d}{dt}(x^2 - xy + 2y^2) = \frac{d}{dt}(22)$$

$$\frac{d}{dt}x^2 - \frac{d}{dt}(xy) + \frac{d}{dt}2y^2 = 0$$

$$2x\frac{dx}{dt} - \left[y\frac{dx}{dt} + x\frac{dy}{dt}\right] + 4y\frac{dy}{dt} = 0$$

Now substitute $x = 4$, $y = 3$, and $dx/dt = -0.1$ and then solve for dy/dt.

$$2 \cdot 4(-0.1) - \left[3(-0.1) + 4\frac{dy}{dt}\right] + 4 \cdot 3\frac{dy}{dt} = 0$$

$$-0.8 + 0.3 - 4\frac{dy}{dt} + 12\frac{dy}{dt} = 0$$

$$8\frac{dy}{dt} = 0.5$$

$$\frac{dy}{dt} = 0.0625$$

ANSWER The production of aerobic shoes is increasing at the rate of $62\frac{1}{2}$ pairs per week. (Remember the units: $0.0625 \times 1000 = 62.5$.)

3–6 Exercises

I.

Use the given equation that relates x and y to find another equation that relates x, y, dx/dt, and dy/dt.

1. (*See Example* 1) $x^2 + y^2 = 25$
2. $3x^2 - 4y^3 = 9$
3. $x + 2y = 10$
4. $3x - 4y = 15$
5. $3x^2 - x + 2y^2 = 18$
6. $5x^2 - 3y + 4y^2 = 22$

Solve for dy/dt in Exercises 7 through 11 by using the given values of x, y, and dx/dt.

7. (*See Example 2*) $3x\frac{dy}{dt} - 2\frac{dx}{dt} + y\frac{dy}{dt} = 0$; $x = 3$, $y = -1$, and $\frac{dx}{dt} = 2$
8. $2y^2\frac{dx}{dt} + 4xy\frac{dy}{dt} - 3y^2\frac{dy}{dt} = 0$; $x = -2$, $y = 4$, and $\frac{dx}{dt} = 6$
9. $y\frac{dy}{dt} - x\frac{dx}{dt} = 4y^2\frac{dy}{dt}$; $x = 1$, $y = 2$, and $\frac{dx}{dt} = 3$

10. $3\dfrac{dy}{dt} + 5y\dfrac{dy}{dt} = 7x\dfrac{dx}{dt}$; $x = -2$, $y = 4$, and $\dfrac{dx}{dt} = 5$

II.

Use the given equation to produce another equation that relaxes x, y, dx/dt, and dy/dt. Then use the given values of x, y, and dx/dt to solve for dy/dt.

11. (*See Example 2*)

$x^2 + y^2 = 25$; $x = 3$, $y = -4$, and $\dfrac{dx}{dt} = 2$

12. $3x - 2y^3 = 17$; $x = 5$, $y = -1$, and $\dfrac{dx}{dt} = -2$

13. $2x + 3y = 3$; $x = 3$, $y = -1$, and $\dfrac{dx}{dt} = 4$

14. $5x - 6y = -17$; $x = -1$, $y = 2$, and $\dfrac{dx}{dt} = -3$

15. $x^2 - \sqrt{y} = 7$; $x = 3$, $y = 4$, and $\dfrac{dx}{dt} = 5$

16. $3x + y^2 - \sqrt{x} = 25$; $x = 9$, $y = -1$, and $\dfrac{dx}{dt} = 2$

III.

If x and y are related by the given equation, then find dy/dt for the given values of x, y, and dx/dt.

17. (*See Example 4*)

$6xy = 120$; $x = 2$, $y = 10$, and $\dfrac{dx}{dt} = 3$

18. $x^2 - 2xy + 3y = -34$; $x = 4$, $y = 10$, $\dfrac{dx}{dt} = -2$

19. $3x^2 + 5xy - y^2 = 36$; $x = 2$, $y = 4$, $\dfrac{dx}{dt} = -3$

20. $5x^3y - 6\sqrt{x} + 2y^3 = 20$; $x = 1$, $y = 2$, $\dfrac{dx}{dt} = 10$

21. (*See Example 3*) Two cars leave a parking lot at the same time. One travels north at 20 mph, while the other travels west at 48 mph. How fast is the distance between them changing after 15 minutes?

22. A store has found that the demand, x, for a certain manufactured product and its price, p, are related by $x^2 + xp + p^2 = 172{,}900$. How is the demand changing if the present demand is 400 units, the present price is \$30, and the price is increasing at the rate of \$1.50 per year?

23. (*See Example 4*) A video rental store has projected that if x is the number of tapes that it rents in a certain period, and y is the number of playback machines it rents in the same period, then

$$x - \sqrt{x}\sqrt{y} - y = 2900$$

If the current values are $x = 3600$ and $y = 100$, and if x is increasing at the rate of 120 rentals per week, what is the rate of change of y?

3–7 The Differential and Linear Approximation

Approximation of Change
Approximation of Functional Value
Linear Approximation to $f(x)$
The Differential
Application to Marginal Analysis

Approximation of Change

The central idea in this section is to take a complicated expression $f(x)$ and replace it by a simpler expression $g(x)$, and then use $g(x)$ as an approximation to $f(x)$ near some point a. For example, if $f(x) = \sqrt{x^2 + 7}$, then we can evaluate $f(3)$ easily: $f(3) = \sqrt{9 + 7} = 4$. Now, what if x increases a little bit, say to 3.1? Common sense tells us that $f(3.1) = \sqrt{(3.1)^2 + 7}$ is going to be bigger than $f(3)$ because $(3.1)^2 + 7$ is bigger than 16, so $\sqrt{(3.1)^2 + 7}$ is bigger than $f(3) = \sqrt{16} = 4$. But roughly at least, how much bigger is $f(3.1)$ than $f(3)$? The derivative gives us the rate of change of f, and remembering the analogy between rate of change and speed of a car can help us see how to use the derivative to find an approximate value for $f(3.1)$.

If the speed of a car is a constant, R, then the distance travelled, D, in a given amount of time, T, is given by

$$D = R \cdot T$$

If the rate changes, but not by much, then we can **approximate** the distance travelled by assuming that the rate is constant. For instance, if a car travelled between 59 mph and 61 mph for one minute, then we can approximate the distance travelled by assuming a constant speed of 60 mph. The car travelled approximately one mile.

Returning to our problem of approximating $f(3.1)$, we will proceed in a similar fashion. The number $f'(3)$ gives the instantaneous rate of change of f when $x = 3$; $f'(x)$, the rate of change of $f(x)$, varies as x goes from 3 to 3.1, but we **approximate** the rate of change by the constant $f'(3)$. (This corresponds to assuming constant speed.) The interval from 3 to 3.1 has length 0.1. (This corresponds to the length of time T.) Finally, the change in $f(x)$ over this interval is $f(3.1) - f(3)$. (This corresponds to the distance the car travelled.) We use the symbol $\approx$ to mean "**is approximately equal to**." We have decided that

$$f(3.1) - f(3) \approx f'(3) \cdot (0.1)$$

To carry out these calculations, we first find $f'(x)$.

$$f(x) = (x^2 + 7)^{1/2}$$

Using the generalized power rule,

$$f'(x) = \frac{1}{2}(x^2 + 7)^{-1/2} \cdot 2x = \frac{x}{\sqrt{x^2 + 7}}$$

Now we evaluate $f'(3)$.

$$f'(3) = \frac{3}{\sqrt{16}} = \frac{3}{4} = 0.75$$

Finally, multiply $f'(3)$ by 0.1.

$$f(3.1) - f(3) \approx f'(3) \cdot (0.1) = (0.75)(0.1) = 0.075$$

If x changes from 3 to 3.1, the value of $f(x)$ changes by approximately 0.075. (In this case, the actual change is $f(3.1) - f(3) = 0.07553\ldots$, so our approximation was very good.)

The general statement would be as follows:

> If Δx is small, then $f(a + \Delta x) - f(a) \approx f'(a) \cdot \Delta x$

Using Δy to denote the change in y where $y = f(x)$, we can write

> $\Delta y \approx f'(a) \cdot \Delta x$

In our example above, $a = 3$ and $\Delta x = 0.1$.

Of course, there is no point in approximating the change in $f(x)$ if you can evaluate both $f(a)$ and $f(a + \Delta x)$ fairly easily. In practice, this technique is reserved for functions which cannot be easily evaluated everywhere.

Example 1 (*Compare Exercise 15*)

If $f(x) = x^{1/3}$, what is the approximate change in $f(x)$ as x increases from 8 to 8.5?

Solution Here, it is $f(8)$ that we can evaluate easily, so we let $a = 8$. Next,

$$f'(x) = \frac{1}{3}x^{-2/3}$$

so

$$f'(8) = \frac{1}{3}(8)^{-2/3}$$
$$= \frac{1}{3} \cdot ((8)^{1/3})^{-2}$$
$$= \frac{1}{3} \cdot \frac{1}{4} = \frac{1}{12}$$

Compute the change in x (as before, let "the change in x" be denoted by Δx). Here $\Delta x = 8.5 - 8 = 0.5$. Thus, we have

$$\Delta y \approx f'(a) \cdot \Delta x = \frac{1}{12} \cdot 0.5 = \frac{1}{12} \cdot \frac{1}{2} = \frac{1}{24} = 0.041666\ldots$$

As x increases from 8 to 8.5, the approximate change in $f(x)$ is 0.041666.

We use the same numbers in the next example because we want to emphasize that the same procedure can be used to answer a different question.

Approximation of a Functional Value

Example 2 (*Compare Exercise 19*)
Approximate $\sqrt[3]{8.5}$.

Solution The problem asks for the cube root of a number, so we use the cube root function; let $f(x) = x^{1/3}$. Now the problem is to approximate $f(8.5)$. We look for a number a near 8.5 that we know the cube root of; we choose $a = 8$. The procedure of Example 1 gives us the approximate change of $f(x)$.

$$f(8.5) - f(8) \approx f'(8) \cdot (0.5) = 0.041666\ldots$$

Add $f(8)$ to both sides.

$$f(8.5) \approx f(8) + 0.041666\ldots = 2 + 0.041666$$

Thus, we have

$$\sqrt[3]{8.5} \approx 2.04166\ldots$$

(The actual decimal expansion of $\sqrt[3]{8.5}$ begins 2.0408...)

Before introducing more notation and looking at a graphical interpretation, we give you one more example to show that Δx need not be positive.

Example 3 (*Compare Exercise 21*)
Approximate $\sqrt{15.4}$.

Solution The solution can be broken down into steps.

1. **Choose an appropriate function**. We are asked to approximate the square root of a number, so let $f(x) = \sqrt{x}$.

2. **Choose the number a.** The number a here should be close to 15.4, and we should be able to evaluate $\sqrt{a}$ fairly easily. Here we choose $a = 16$.
3. **Compute Δx.** Remember that Δx is the change in x as x goes from a to $a + \Delta x$. Thus, $\Delta x = (a + \Delta x) - a$ (always subtract a). Here $a = 16$, so $\Delta x = 15.4 - 16 = -0.6$ (Δx can be negative).
4. **Compute $f'(a)$.** First compute $f'(x)$ and then evaluate $f'(a)$. Here

$$f'(x) = \frac{1}{2\sqrt{x}} \qquad \text{so} \qquad f'(16) = \frac{1}{2\sqrt{16}} = \frac{1}{8}$$

5. **Approximate Δy:** $\Delta y \approx f'(a) \cdot \Delta x$. Using the results of steps 3 and 4, we have

$$\Delta y \approx \left(\frac{1}{8}\right)(-0.6) = \left(\frac{1}{8}\right)\left(\frac{-3}{5}\right) = \frac{-3}{40}$$

6. **Approximate $y = f(a + \Delta x)$:** $f(a + \Delta x) \approx f(a) + f'(a)\Delta x$. Here $f(a) = \sqrt{16} = 4$, so

$$\sqrt{15.4} \approx 4 + \frac{-3}{40} = 3\,\frac{37}{40} = 3.925$$

(The actual decimal expansion of $\sqrt{15.4}$ begins 3.9242...)

The process will stop with step 5 if you are asked to approximate **the change in y**. If you are asked to approximate the **value of y**, go on to step 6.

Now we give the general notation and procedure for approximating.

To Find an Approximate Value of $f(b)$

1. Write down the general rule for $f(x)$.
2. Look for a number near b for which you can easily evaluate $f(x)$. Call that number a.
3. Let $\Delta x = b - a$.
4. Compute $f'(x)$ and then evaluate $f'(a)$.
5. Approximate the change in y:

$$\Delta y = f(b) - f(a) \approx f'(a) \cdot \Delta x$$

6. Approximate $f(b)$:

$$f(b) \approx f(a) + f'(a) \cdot \Delta x$$

Linear Approximation to $f(x)$

The quantity $f'(a) \cdot \Delta x$ is the approximate change in $f(x)$ as x goes from a to b; the quantity $f(a) + f'(a) \cdot \Delta x$ is the approximate value of $f(b)$. Since b is any arbitrary number, we could just as well use x in the last approximation formula.

7. $f(x) \approx f(a) + f'(a) \cdot \Delta x = f(a) + f'(a) \cdot (x - a)$

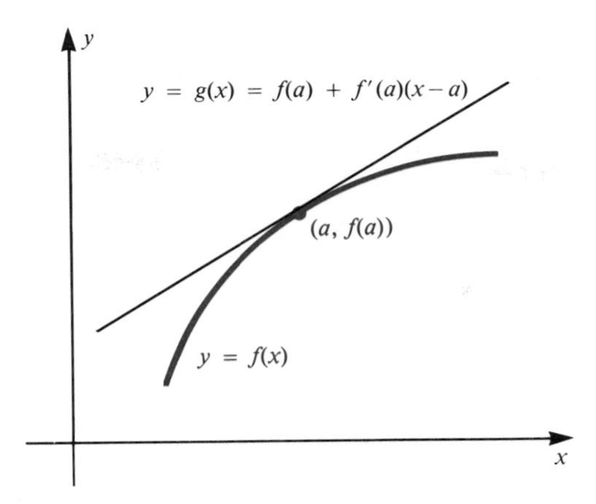

Figure 3–5

The value of $f(x)$ is on the left of the equation; the approximating value is on the right. Let us give a name to the approximating function, say g. Thus,

$$g(x) = f(a) + f'(a)(x - a)$$

In Example 2, we had $f(x) = x^{1/3}$, $a = 8$, $f(a) = 2$ and $f'(a) = \frac{1}{12}$. The function g that approximates $x^{1/3}$ when x is near 8 is given by

$$g(x) = 2 + \frac{1}{12}(x - 8)$$

This is the equation of a straight line. In fact, it is the equation of the line tangent to $y = x^{1/3}$ at the point (8, 2). Looking back at the general equation for $g(x)$,

$$g(x) = f(a) + f'(a)(x - a)$$

we see that the graph of g is always the line tangent to the graph of f at the point $(a, f(a))$. See Figure 3–5.

The Differential

Because there is such a close connection between the derivative of f and the approximation to the change in $f(x)$, we introduce some notation to emphasize this connection. We have already used Δy to mean the actual change in y; now we introduce the notation dy to mean the approximate change in y; $dy = f'(x)\Delta x \approx \Delta y$. See Figure 3–6.

Figure 3–6 also justifies setting $dx = \Delta x$. We give a name to dy in the following definition.

Definition

> dy is called **the differential** of y and is given by
>
> $$dy = f'(x)\,dx$$

You will see this notation, $dy = f'(x)dx$, used again in Chapter 7. Because the differential is so important, we will give you some examples emphasizing its

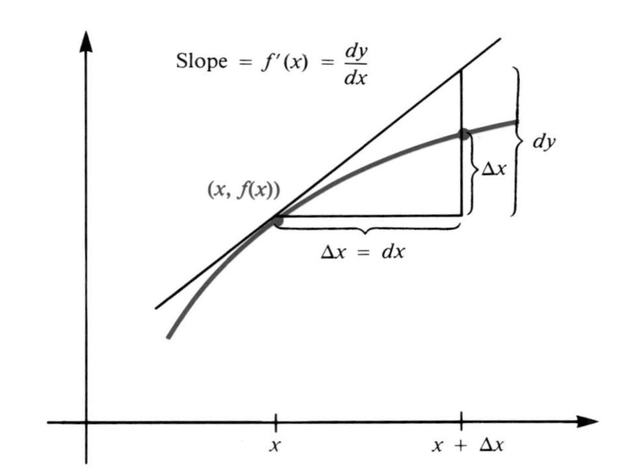

Figure 3–6

definition and then show how this notation can shorten the process of computing an approximation. We will end this section with an important application of the differential in economics.

Example 4 (*Compare Exercise 1*)
Compute dy if $f(x) = x^3 - 4x$.

Solution

$$dy = f'(x)\,dx = (3x^2 - 4)\,dx$$

Example 5 (*Compare Exercise 5*)

Compute dy if $y = \frac{1}{x}$.

Solution Here $f(x) = \frac{1}{x}$ so $f'(x) = \frac{-1}{x^2}$.

$$dy = \frac{-1}{x^2}\,dx$$

In applications, you will have to do more computations after finding dy, so we give you some examples that show these computations.

Example 6 (*Compare Exercise 9*)
Compute dy if $f(x) = \sqrt{x}$, $x = 9$, and $dx = 0.2$.

Solution Here $f(x) = x^{1/2}$, so $f'(x) = \frac{1}{2\sqrt{x}}$

$$f'(9) = \frac{1}{2\sqrt{9}} = \frac{1}{6}$$

$$dx = 0.2 = \frac{1}{5}$$

so

$$dy = f'(x)\,dx$$
$$= \left(\frac{1}{6}\right)\left(\frac{1}{5}\right) = \frac{1}{30}$$

Example 7 (*Compare Exercise 11*)
Compute dy if $y = x^{1/3}$, $x = 64$, and $dx = -6$.

Solution We shorten the computations.

$$dy = \frac{1}{3}x^{-2/3}\,dx$$
$$= \frac{1}{3}[x^{1/3}]^{-2}\,dx$$
$$= \frac{1}{3}[64^{1/3}]^{-2}(-6)$$
$$= \left(\frac{1}{3}\right)(4)^{-2}(-6)$$
$$= \left(\frac{1}{3}\right)\left(\frac{1}{16}\right)(-6)$$
$$= -\frac{1}{8}$$

Example 8 (*Compare Exercise 23*)
Find the approximate value of $\sqrt{22}$.

Solution

1. We have to choose the function; let $f(x) = \sqrt{x}$.
2. We need a number near 22 that we know the square root of; let $x = 25$.
3. Then, dx is the change of x as x goes from 25 to 22; $dx = 22 - 25 = -3$. Note the order of subtraction. The dy notation lets us combine steps 4 and 5.

4. and 5. $dy = f'(x)dx = \dfrac{1}{2\sqrt{x}}dx$. With $x = 25$ and $dx = -3$,

$$dy = \frac{1}{2\sqrt{25}}(-3) = \frac{1}{10}(-3) = -\frac{3}{10} = -0.3$$

6. We have $f(25) = 5$ and $dy = -0.3$, so

$$f(22) \approx f(25) + dy$$
$$f(22) \approx 5 + (-0.3)$$
$$\sqrt{22} \approx 4.7$$

Application to Marginal Analysis

The word "marginal" is widely used in business and economics. You can get a feeling for how important the word is if you look it up in the index of an economics textbook. There will be many entries starting with it. An individual's marginal tax rate, for example, is the tax rate on an additional dollar of income for that individual. We will concentrate on a particular marginal function, marginal cost, but our discussion applies to all marginal functions.

Some definitions of marginal cost are not based on calculus. Rather than defining marginal cost as the derivative of the cost function, marginal cost is sometimes defined as the cost of producing the next item. Letting $C(x)$ be the cost of producing x items, and $MC(x)$ be the marginal cost, this definition would give us

$$MC(x) = C(x + 1) - C(x)$$

Now, using our approximation techniques of this chapter with $\Delta x = (x + 1) - x = 1$, we can write

$$C(x + 1) - C(x) \approx C'(x) \cdot 1 = C'(x)$$

Thus, even with this noncalculus definition of marginal cost, we have $MC(x) \approx C'(x)$.

Example 9 (*Compare Exercise 27*)
A company's cost for producing x refrigerators per week is

$$C(x) = \frac{1}{100}x^2 + 400x + 3000$$

What is the company's marginal cost if it is producing 50 refrigerators per week?

Solution Using calculus, we have

$$C'(x) = \frac{1}{50}x + 400$$

$$C'(50) = 1 + 400 = 401$$

The calculus definition gives an answer quickly: The marginal cost is \$401. For comparison purposes, computing

$$C(51) - C(50)$$

gives

$$\left[\frac{1}{100}(51)^2 + (400)(51) + 3000\right] - \left[\frac{1}{100}(50)^2 + (400)(50) + 3000\right]$$

$$= [26.01 + 20{,}400 + 3000] - [25 + 20{,}000 + 3000]$$

$$= 401.01$$

This lengthier computation gives the answer: The marginal cost is \$401.01.

The difference in the two computations is one penny, or about 0.002%.

Using the derivative as the definition of marginal cost lets us compute the marginal cost more easily. Even more importantly, it also allows us to compute a rule for evaluating marginal cost and so permits us to talk about and analyze the marginal cost *function* more easily. We will continue therefore to identify the marginal cost function as the derivative of the cost function.

3–7 Exercises

I.

Find dy in Exercises 1 through 8.

1. (*See Example 4*) $f(x) = 5x^3 + 4$

2. $f(x) = 3x^2 + 2x - 1$

3. $f(x) = \sqrt{x^2 + 1}$

4. $f(x) = \dfrac{1}{x^2 + 1}$

5. (*See Example 5*) $y = \dfrac{1}{\sqrt{x}}$

6. $y = x^{2/3}$

7. $y = 8x^3 - 4x$

8. $y = (5x + 1)^4$

In Exercises 9 through 14, evaluate dy using the given values of x and dx.

9. (*See Example 6*) $y = x^4 - 3x^2$; $x = 2$, $dx = 0.5$

10. $y = \dfrac{1}{x}$; $x = 5$, $dx = 0.2$

11. (*See Example 7*) $y = \dfrac{1}{x^2 + 1}$; $x = 3$, $dx = -0.4$

12. $y = \sqrt{x}$; $x = 100$, $dx = -3$

13. $y = (2x + 1)^3$; $x = 2$, $dx = 0.2$

14. $y = \dfrac{1}{\sqrt{x}}$; $x = 25$, $dx = -4$

II.

In Exercises 15 through 18, find the approximate change in $f(x)$ as x changes from a to b.

15. (*See Example 1*) $f(x) = x^2 - 3x$; $a = 4$, $b = 4.2$

16. $f(x) = \dfrac{1}{6}x^3 - 9x$; $a = 10$, $b = 10.3$

17. $f(x) = \dfrac{1}{x}$; $a = 20$, $b = 18$

18. $f(x) = \sqrt{x}$; $a = 36$, $b = 32$

19. (*See Example 2*) Use $f(x) = x^{1/3}$ and $a = 27$ to approximate $29^{1/3}$.

20. Use $f(x) = \sqrt{x}$ and $a = 100$ to approximate $\sqrt{104}$.

21. (*See Example 3*) Use $f(x) = \sqrt{x}$ and $a = 25$ to approximate $\sqrt{23}$.

22. Use $f(x) = x^2$ and $a = 3$ to approximate $(2.96)^2$.

III.

Use the techniques of this section to approximate the numbers given in Exercises 23 through 26.

23. (*See Example 8*) $\sqrt{52}$ **24.** $\sqrt{98}$ **25.** $\sqrt[3]{61}$ **26.** $\sqrt[3]{1006}$

27. (*See Example 9*) If a company's cost (in dollars) for producing x garbage disposals per day is given by

$$C(x) = \frac{1}{100}x^2 + 20x + 300$$

what is the approximate cost to the company of producing the 51st garbage disposal? *Hint:* The answer is $MC(50)$.

28. If the cost (in dollars) of producing x subcompact automobiles per hour is given by

$$C(x) = \frac{x^3}{10} - 50x^2 + 8000x + 25{,}000$$

(a) what is the approximate cost of the 101st automobile manufactured?

(b) What is the average cost of manufacturing 100 automobiles?

29. A company's revenue (in dollars) from producing x pairs of shoes per hour is given by

$$R(x) = 50x - \frac{1}{4}x^2$$

Approximately, how will the company's revenue change if production is increased from 40 pairs of shoes per hour to 44 pairs per hour?

IMPORTANT TERMS

3–1	**Rules for computing derivatives**	**Derivative of a constant**	**Power rule**
	Constant-times rule	**Sum rule**	**Difference rule**
3–2	**Product rule**	**Quotient rule**	
3–3	**Composition of functions**	***u*-substitution**	**Chain rule**
	Generalized power rule		
3–4	**Second derivative**	**Acceleration**	**Higher order derivatives**
	Different notations for derivatives		
3–5	**Implicitly defined functions**	**Explicitly defined functions**	**Implicit differentiation**
3–6	**Related rates**		

3–7	**Differential** **dx**	**Linear approximation** **Approximate change in y**	**dy** **Different meanings of marginal and their relation to each other**

REVIEW EXERCISES

Compute $f'(x)$ in Exercises 1 through 8.

1. $f(x) = \sqrt{x^2 + 1}$

2. $f(x) = x\sqrt{x^2 + 1}$

3. $f(x) = \dfrac{x^2 + 1}{x^3 - 4x}$

4. $f(x) = \left(7x^2 - 4 + \dfrac{3}{x}\right)\left(8x^3 - 9x + \dfrac{6}{x^2}\right)$

5. $f(x) = (7x + 4)^2(5x + 1)^3$

6. $f(x) = \dfrac{6x - 4}{9x^2 + 3x}$

7. $f(x) = (3x + \sqrt{5x + 1})^4$

8. $f(x) = [(3x + 1)^2 + (4x - 8)^2]^5$

9. If $y = \dfrac{4}{x^3}$, find y''.

10. If $y = \sqrt{x^3 + 8}$, find $\dfrac{d^2y}{dx^2}$.

11. Find $\dfrac{dy}{dx}$ if $xy + y^2 = 3x$.

12. Find $\dfrac{dy}{dx}$ if $x^3 - 4xy^2 + y^3 = 9$.

13. What is an equation of the line tangent to the graph $x^2 + y^2 = 25$ at the point $(-4, 3)$?

14. What is an equation of the line tangent to the graph $5xy - y^2 = 21$ at the point $(-2, -3)$?

15. If $x^2 + 5y^2 = 129$, what is dy/dt when $x = -7$, $y = -4$, and $dx/dt = 3$?

16. If $6x^2 - 3xy + y^2 = 16$, what is dx/dt when $x = 1$, $y = 5$, and $dy/dt = -3$?

17. Compute the differential dy if $y = \dfrac{3x + 11}{2x - 5}$.

18. Compute the differential dy if $y = \sqrt{x^3 - 4x^2}$.

19. Use the differential to find an approximate value for $\sqrt[3]{10}$.

20. Use the differential to find an approximate value for $\sqrt{31}$.

21. If a company's marginal cost is $6\sqrt{x} + 5$ when production level is x units per week, and if the marginal revenue is $\sqrt{x^2 + 4x}$, what is the company's marginal profit?

22. If $R = x \cdot p$ is the revenue, how is the revenue changing if $x = 10{,}000$, $p = \$4.50$, $dx/dt = -50$ units per week, and $dp/dt = \$0.20$ per week?

23. The cost of producing x items per month is given by

$$C(x) = \tfrac{1}{1000}x^2 + 12x + 6200$$

(a) What is the average cost per item?

(b) What is the marginal average cost?

24. If the cost of producing x items per month is

$$C(x) = \tfrac{1}{600}x^2 + 20x + 5300$$

what is the approximate cost of producing the 101st item?

4

Topics in this chapter can be applied to:
Minimizing Overhead ● Maximizing Productivity ● Marketing Decisions ● Cost Curves ● Minimizing Marginal Costs ● Maximizing Marginal Revenue ● Production Strategies ● Pricing Strategies

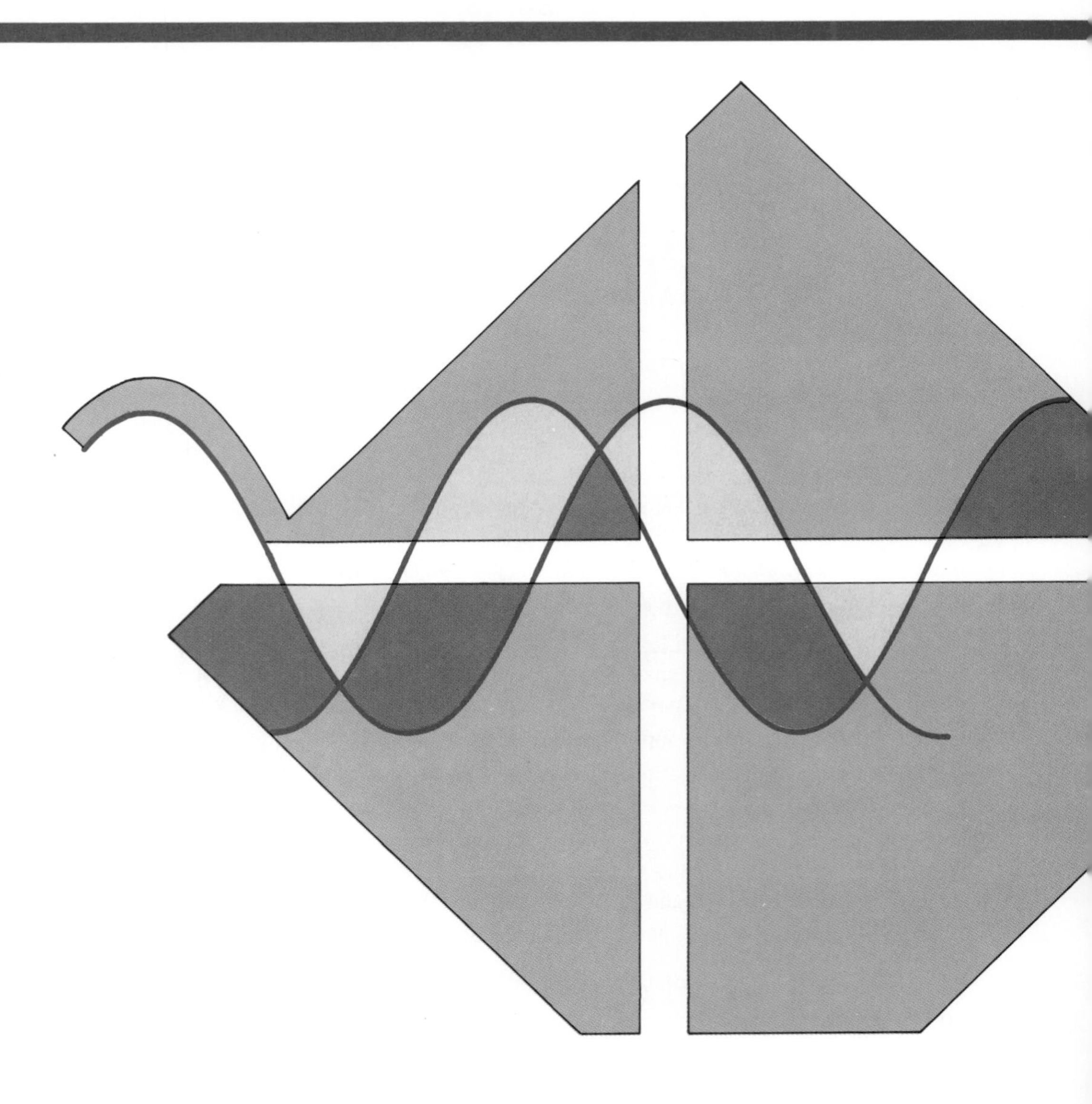

Applications of the Derivative

- The First Derivative Test
- Concavity and the Second Derivative Test
- Extreme Values
- Asymptotes
- Optimization in Applications
- Applications to Business and Economics

One of the major applications of mathematics to real world problems is to find the best, or optimum, way to achieve a goal. "Best" may mean finding a way to minimize a company's overhead or the time required to complete a project. "Best" may mean finding a way to maximize profit or productivity. Some applications are handled by techniques from finite mathematics—the simplex method for example. Some applications are handled with techniques from calculus. In this chapter, you will see how to use the derivative to find the biggest and smallest values of a function. Further, you will see how to use the derivative to provide a graph of a function showing how the functional values rise and fall, and other features of the function important in applications.

4–1 The First Derivative Test

Increasing and Decreasing Functions
Extreme Values
Relative Extreme Values
Critical Numbers

Increasing and Decreasing Functions

The graph in Figure 4–1 gives the temperature in Yourtown, U.S.A. over a period of 24 hours. We will use the temperature function and its graph to establish terminology that will then be used to describe arbitrary functions and graphs.

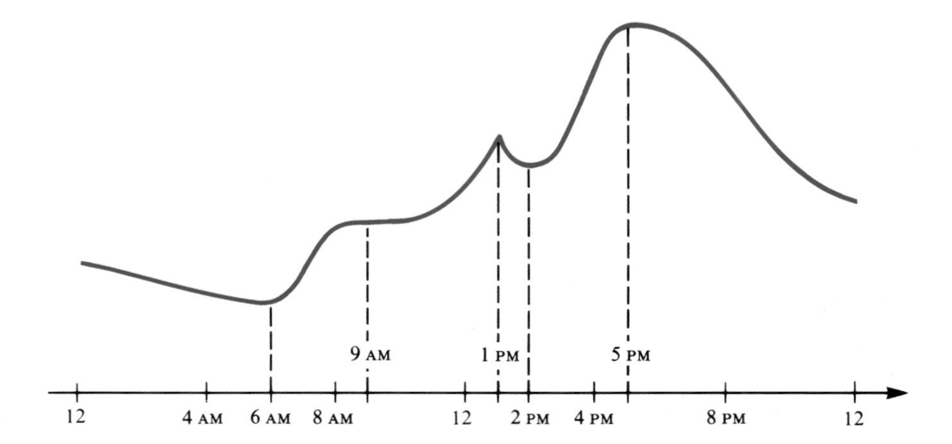

Figure 4–1

By reading the graph, we can describe how the temperature varied. The temperature fell from midnight until 6 AM and started slowly rising. There was a slight period around 9 AM when the temperature held steady, but then it continued to increase until 1 PM. There was a thunderstorm at 1 PM, and the temperature suddenly decreased for a while. When the storm ended, the temperature started climbing again at 2 PM. The highest temperature occurred at 5 PM, and then the temperature decreased for the rest of the day.

We extend this terminology to arbitrary functions like the one whose graph is given in Figure 4–2.

As x goes from a to p, the graph rises. Said another way, as x moves to the right on (a, p), the values of f increase. We say that the function is **increasing** on the open interval (a, p). The function levels off at p, and as x goes from p to q, the graph falls. We say that f is **decreasing** on the open interval (p, q). Similarly, f is increasing on the open interval (q, r), and f is decreasing on the open interval (r, b).

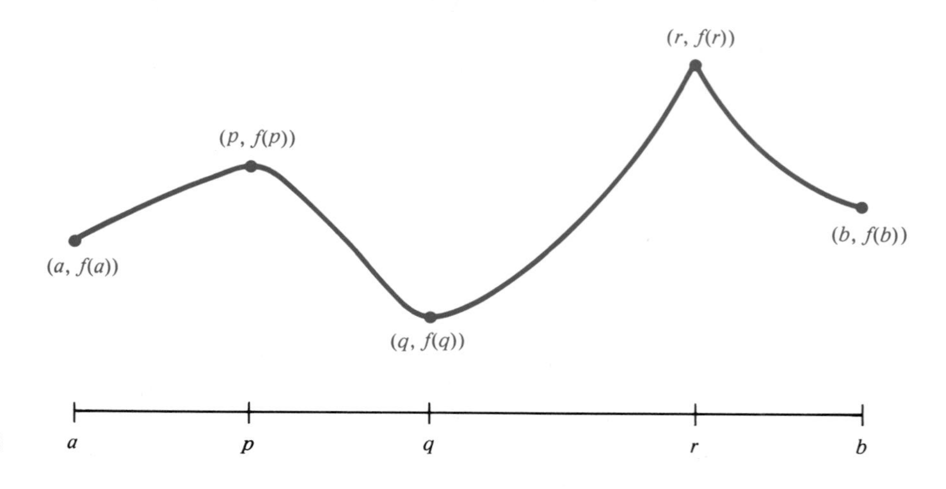

Figure 4–2

Extreme Values

From Figure 4–1, we can see that the temperature reached its maximum value for the day at 5 PM. In Figure 4–2, because $f(r) \geq f(x)$ for all x in (a, b), we say that the function f **has its maximum value on (a, b) when $x = r$**, and that **$f(r)$ is the maximum value of f on (a, b)**. One of the reasons we started with a graph of the temperature rather than just jumping right into the discussion with Figure 4–2 is to help you with the terminology. "What was the highest temperature?" and "When was the temperature the greatest?" are two different questions.

When a marketing researcher wants to use a mathematical model to predict what price will give the biggest profit, the researcher has to distinguish between two different numbers: the one for price and the one for profit. In applications, we will want to know the maximum value of f and what number x produces that maximum value. We must keep straight the distinction between the maximum value of f and the number x that gives the maximum value. A similar distinction must be maintained when talking about least value of the function. The minimum temperature for the day occurred at 6 AM. In an analogous way, we say that **f has its minimum value on (a, b) when $x = q$; the minimum value of f on (a, b) is $f(q)$.** The term **extreme value** of f is used to refer to either a maximum value or a minimum value of f.

Relative Extreme Values

Something interesting happened at 1 PM—the temperature stopped increasing and started decreasing. The same thing happens to the function in Figure 4–2 when $x = p$. The temperature at 1 PM was not the maximum temperature for the whole day, but it was the maximum for a while. Referring to Figure 4–2, we say that f has a **relative maximum** at $x = p$ because $f(p)$ is larger than $f(x)$ for all x near p. (The term **local maximum** at $x = p$ is also used; "local" = "relative.") We say that f has a **relative minimum** at $x = q$ because $f(q)$ is smaller than $f(x)$ for all x near q.

Critical Numbers

Relative extreme values of a function occur when the graph changes direction, and the figures show how a graph can change direction. In Figure 4–2, the graph has a smooth peak at the point $(p, f(p))$ and a sharp peak at the point $(r, f(r))$. We use the derivative to describe the distinction. The graph has a horizontal tangent line at $(p, f(p))$; that is, $f'(p) = 0$. Similarly, the graph shows that $f'(q) = 0$. The graph does not have a tangent line at $(r, f(r))$; $f'(r)$ does not exist. The graph gives a picture of the following important theorem.

Theorem

> If f has a relative extreme value at $x = c$, then either $f'(c) = 0$ or $f'(c)$ does not exist.

This theorem prompts us to make the following definition.

Definition

A number c in the domain of f is called a **critical number** of f if either $f'(c) = 0$ or $f'(c)$ does not exist.

Critical numbers are defined this way because the theorem tells us that when we are looking for the numbers x that produce the extreme values of f, we need only look at these numbers. They are critical in the search for extreme values.

Special caution. If the graph changes direction when $x = c$, then c is a critical number, but not vice-versa; c may be a critical number without the graph changing direction. As the temperature graph shows at 9 AM, the graph may have a horizontal tangent (corresponding to $f'(c) = 0$) without turning around at that point.

Once we've found all the critical numbers, how do we know which ones correspond to points where the graph changes direction and which ones do not? How do we know if we have found a relative maximum, a relative minimum, or neither? Critical numbers are defined in terms of the derivative, and the derivative also tells us how the function is changing near c.

Looking at Figures 4–1 and 4–2, you can see geometrically that when the graph is rising, lines tangent to the graph have positive slope and when the graph is falling, tangent lines have negative slope. From physical applications you know that if the rate of change of a function is positive, the function is increasing; if the rate of change is negative, the function is decreasing.

The First Derivative Test for Increasing and Decreasing

If $f'(x) > 0$ for all x in (a, b), then f is increasing on (a, b).

If $f'(x) < 0$ for all x in (a, b), then f is decreasing on (a, b).

Be careful—these statements refer to two different functions: f and f'. Use their names; avoid ambiguous statements like "it's increasing when it's positive." Even to yourself, say "f is increasing because f' is positive."

If f is continuous and f is increasing just to the left of c and decreasing just to the right of c, then f must have a relative maximum value when $x = c$. Furthermore, the sign of the derivative indicates when f is increasing and when f is decreasing. So a change in the sign of the derivative from positive to negative indicates a relative maximum of f.

We summarize this discussion:

The First Derivative Test for Relative Extreme Values

1. Find the critical numbers of f.
2. Determine the sign of f' on the intervals between the critical numbers.
3. If c is a critical number, then

   (a) $f(c)$ is a relative maximum if f' changes from positive to negative at c.

(b) $f(c)$ is a relative minimum if f' changes from negative to positive at c.

(c) $f(c)$ is not a relative extreme value if f' does not change sign at c.

In determining the sign of f', the critical points play the same role as cut points did in solving inequalities in Chapter 1. The inequalities we are now interested in are $f'(x) > 0$ and $f'(x) < 0$.

We will use $f(x) = 2x^3 - 6x^2 - 18x + 7$ in the next several examples to analyze the first derivative test step by step.

Example 1 (*Compare Exercise 3*)
Find the critical numbers of f if

$$f(x) = 2x^3 - 6x^2 - 18x + 7$$

Solution The function is a polynomial, so $f'(x)$ exists for all x. Thus, the only critical numbers are solutions to $f'(x) = 0$. Computing $f'(x)$, we have $f'(x) = 6x^2 - 12x - 18$. Next, set $f'(x) = 0$ and solve.

$$\begin{aligned} 6x^2 - 12x - 18 &= 0 \qquad \text{Divide by 6} \\ x^2 - 2x - 3 &= 0 \qquad \text{Factor} \\ (x - 3)(x + 1) &= 0 \qquad \text{Solve} \end{aligned}$$

The critical numbers are $x = -1$ and $x = 3$.

Example 2 (*Compare Exercise 7*)
Find the intervals on which f is increasing if

$$f(x) = 2x^3 - 6x^2 - 18x + 7$$

Solution f is increasing if f' is positive, so the problem is equivalent to finding where $f'(x) > 0$. Since $f'(x) = 6x^2 - 12x - 18$, we want to solve $6x^2 - 12x - 18 > 0$. From Example 1, we know the cut points for this inequality are $x = -1$ and $x = 3$.

−1 3

We determine the sign of f' on the intervals between critical points by using test numbers.

(a) From the interval $(-\infty, -1)$, we choose $x = -2$ as a test number.

$$f'(-2) = 6(-2)^2 - 12(-2) - 18 = 30 > 0$$

f' is positive on $(-\infty, -1)$, so f **is increasing on** $(-\infty, -1)$.

(b) From the interval $(-1, 3)$, we choose $x = 0$ as a test number.

$$f'(0) = -18 < 0$$

f' is negative on $(-1, 3)$, so f **is decreasing on** $(-1, 3)$.

(c) From the interval $(3, \infty)$, we choose $x = 4$ as a test number.

$$f'(4) = 6(4)^2 - 12(4) - 18 = 30 > 0$$

f' is positive on $(3, \infty)$, so f **is increasing on** $(3, \infty)$.

ANSWER f is increasing on the intervals $(-\infty, -1)$ and $(3, \infty)$.

Example 3 (*Compare Exercise 11*)
Find and classify the relative extreme values of f if

$$f(x) = 2x^3 - 6x^2 - 18x + 7$$

("Classify" just means to tell whether each relative extreme value is a maximum or a minimum.)

Solution We use "+" for the sign of f' if f' is positive, and "−" for the sign of f' if f' is negative. Using this notation, we can make the following chart with the results of Example 2.

sign of f'	+		−		+
		−1		3	

Notice that f' changes sign from positive to negative at $x = -1$. Thus, the third step in the first derivative test tells us that f has a relative maximum when $x = -1$; the value of the relative maximum is $f(-1) = 17$. Also, because f' changes from negative to positive at $x = 3$, the first derivative test tells us that f has a relative minimum when $x = 3$; the value of the relative minimum is -47.

Example 4 (*Compare Exercise 27*)
Sketch the graph of f if $f(x) = 2x^3 - 6x^2 - 18x + 7$.

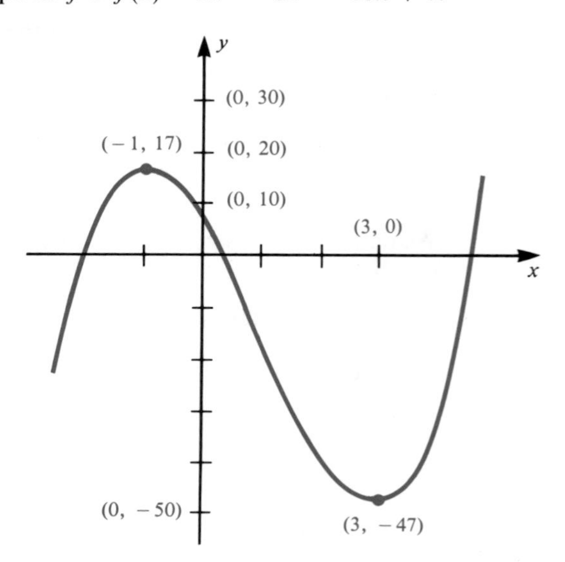

Figure 4–3

Solution We plot the points $(c, f(c))$ where c is a critical number, so here we must plot $(-1, 17)$ and $(3, -47)$. f is increasing on $(-\infty, -1)$, so the graph is rising on this interval. $f'(-1) = 0$, so the graph has a horizontal tangent at $(-1, 17)$. The graph then falls over the interval from $x = -1$ to $x = 3$, turns around smoothly $[f'(3) = 0]$, and then continues to rise indefinitely. See Figure 4–3.

Sometimes you may be asked just to find the critical numbers or the intervals on which a function is increasing, but sometimes you will be asked a question that requires you to put all the steps together.

Example 5 (*Compare Exercise 29*)
Sketch the graph of f if $f(x) = x^3 + 3x^2 - 24x + 2$.

Solution

STEP 1 Find the critical numbers of f.

The function is a polynomial, so $f'(x)$ exists for all x. The only critical numbers are solutions to the equation $f'(x) = 0$. First, compute $f'(x)$.

$$f'(x) = 3x^2 + 6x - 24$$

Next, set $f'(x) = 0$ and solve.

$$3x^2 + 6x - 24 = 0 \qquad \text{Divide by 3}$$
$$x^2 + 2x - 8 = 0 \qquad \text{Factor}$$
$$(x + 4)(x - 2) = 0$$

The critical numbers are $x = -4$ and $x = 2$.

STEP 2 Determine the sign of f' on the intervals between critical numbers.

We begin by drawing the chart.

sign of f' ——|——|——
-4 2

critical numbers

(a) Use a number less than -4 to test the sign of f' on $(-\infty, -4)$.

Using $x = -5$, we have

$$f'(-5) = 3(-5)^2 + 6(-5) - 24 = 21 > 0$$

Now we can start to fill in the chart.

sign of f' —+—|——|——
-4 2

(b) Test the sign of f' between -4 and 2. Using $x = 0$, we have

$$f'(0) = 3(0)^2 + 6(0) - 24 = -24 < 0$$

We can continue to fill in the chart.

sign of f' —+—|—−—|——
-4 2

(c) Finally, test the sign of f' for some $x > 2$. Using $x = 3$, we have

$$f'(3) = 3(3)^2 + 6(3) - 24 = 21 > 0$$

We can now complete our chart.

$$\text{sign of } f' \quad \overset{+}{} \underset{-4}{|} \overset{-}{} \underset{2}{|} \overset{+}{}$$

STEP 3 Sketch the graph of f.

The sign of f' tells us that f is increasing on $(-\infty, -4)$, decreasing on $(-4, 2)$, and then increasing on $(2, \infty)$. Because f is continuous, its graph is one piece. The graph of f must have a general shape like Figure 4–4.

To get a better sketch, we evaluate $f(-4)$ and $f(2)$. Notice that now we are evaluating f; in Step (b) we wanted the sign of f', and we evaluated f' at test points. Now we evaluate f at its critical numbers.

$$f(-4) = (-4)^3 + 3(-4)^2 - 24(-4) + 2 = 82$$
$$f(2) = 2^3 + 3(2)^2 - 24(2) + 2 = -26$$

The graph is shown in Figure 4–5.

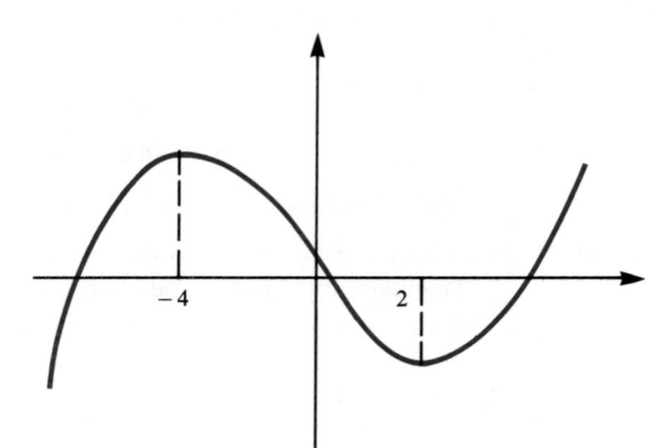

Figure 4–4

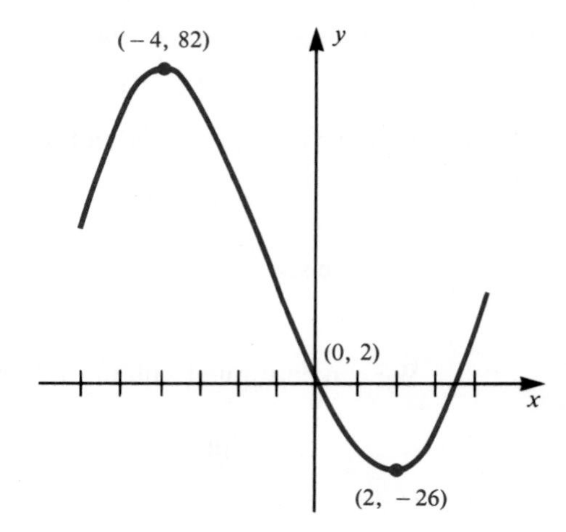

Figure 4–5

Example 6 (*Compare Exercise 35*)
Sketch the graph of f if $f(x) = x^4 - 4x^3 + 2$.

Solution First, compute the derivative of f. $f'(x) = 4x^3 - 12x^2$ exists for all x, so the only critical numbers of f are the solutions to $f'(x) = 0$.

$$4x^3 - 12x^2 = 0$$
$$4x^2(x - 3) = 0$$

The critical numbers are $x = 0$ and $x = 3$. Now we fill in the chart for the sign of f'.

sign of f' ——|——|—— (marks at 0 and 3)

From the interval $(-\infty, 0)$, we choose $x = -1$ as a test number.

$$f'(-1) = 4(-1)^3 - 12(-1)^2 = -16 < 0$$

From the interval $(0, 3)$, we choose $x = 1$.

$$f'(1) = 4(1)^3 - 12(1)^2 = -8 < 0$$

From the interval $(3, \infty)$, we choose $x = 4$.

$$f'(4) = 4(4)^3 - 12(4)^2 = 64 > 0$$

Our completed chart is

sign of f': $-$ | 0 | $-$ | 3 | $+$

Although $x = 0$ is a critical number, f' does not change sign at $x = 0$; therefore, $f(0) = 2$ is not a relative extreme value of f. The graph is falling to the left of 0, then levels off at 0 for an instant, and then falls again to the right of 0.

The sign of f' does change from negative to positive when $x = 3$. Thus, the graph turns around when $x = 3$, and f has a relative minimum when $x = 3$; $f(3) = (3)^4 - 4(3)^3 + 2 = -25$. The graph of f is given in Figure 4–6.

From the graph, we can see that -25 is not only a relative minimum, but in fact is the minimum value of f.

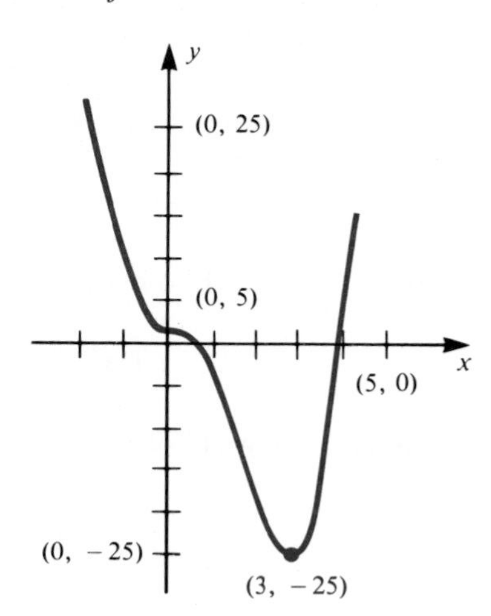

Figure 4–6

Example 7 (*Compare Exercise 33*)
Find the relative extreme values, and sketch the graph of f if

$$f(x) = \frac{x}{4} - 3x^{1/3}$$

Solution First, note that f is continuous for all x, so its graph will be in one piece. Next, compute the derivative of f.

$$f'(x) = \frac{1}{4} - x^{-2/3} = \frac{1}{4} - \frac{1}{x^{2/3}}$$

f' exists for all x, except $x = 0$, so the number $x = 0$ is a critical number. To find the other critical numbers, we need to solve the equation $f'(x) = 0$.

$$\frac{1}{4} - \frac{1}{x^{2/3}} = 0 \qquad \text{Add } \frac{1}{x^{2/3}} \text{ to both sides}$$

$$\frac{1}{4} = \frac{1}{x^{2/3}} \qquad \text{Cross multiply}$$

$$x^{2/3} = 4 \qquad \text{Cube both sides}$$

$$x^2 = 64$$

$$x = \pm 8$$

There are three critical numbers: -8, 0, and 8. We begin our "sign of f'" chart. Fill in this chart as we go.

sign of f' ——|——|——|—— with marks at -8, 0, 8

We have to evaluate $f'(x) = \frac{1}{4} - \frac{1}{x^{2/3}}$, so we choose numbers whose cube root we can compute easily. From $(-\infty, -8)$, we choose $x = -27$.

$$f'(-27) = \frac{1}{4} - \frac{1}{(-27)^{2/3}} = \frac{1}{4} - \frac{1}{(-3)^2} = \frac{1}{4} - \frac{1}{9} > 0$$

From $(-8, 0)$, we choose $x = -1$.

$$f'(-1) = \frac{1}{4} - \frac{1}{(-1)^{2/3}} = \frac{1}{4} - \frac{1}{(-1)^2} = \frac{1}{4} - 1 < 0$$

From $(0, 8)$, we choose $x = 1$.

$$f'(1) = \frac{1}{4} - \frac{1}{(1)^{2/3}} = \frac{1}{4} - \frac{1}{1^2} = \frac{1}{4} - 1 < 0$$

From $(8, \infty)$, we choose $x = 27$.

$$f'(27) = \frac{1}{4} - \frac{1}{(27)^{2/3}} = \frac{1}{4} - \frac{1}{(3)^2} = \frac{1}{4} - \frac{1}{9} > 0$$

Your completed chart should look like

sign of f'	+	−8	−	0	−	8	+

f is increasing on the intervals $(-\infty, -8)$ and $(8, \infty)$.
f is decreasing on the intervals $(-8, 0)$ and $(0, 8)$.
The values of f at the critical numbers are

$$f(-8) = -\frac{8}{4} - 3(-8)^{1/3} = -2 - 3(-2) = 4$$

$$f(0) = \frac{0}{4} - 3(0)^{1/3} = 0 - 0 = 0$$

$$f(8) = \frac{8}{4} - 3(8)^{1/3} = 2 - 3 \cdot 2 = -4$$

We must be careful filling in the curve because $f'(0)$ doesn't exist. In this case, the graph has a vertical tangent at (0, 0), and the graph is shown in Figure 4–7.

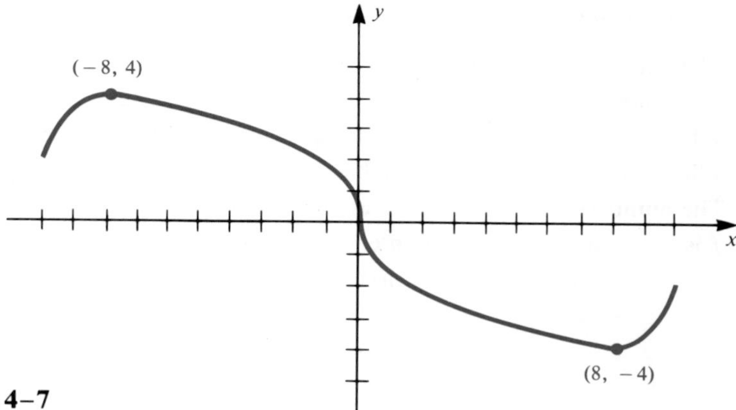

Figure 4–7

4–1 Exercises

I.

1. Use Figure 4–8 to complete the following statements.

   (a) The maximum value of f on $[-3, 5]$ is ______________.
   (b) The minimum value of f on $[-3, 5]$ is ______________.
   (c) f has a relative maximum value when $x =$ ______________.
   (d) f has a relative minimum value when $x =$ ______________.
   (e) The numbers ______________ are the critical numbers of f.

(f) f is increasing on the open interval(s) ______________.

(g) $f'(x) < 0$ on the open interval(s) ______________.

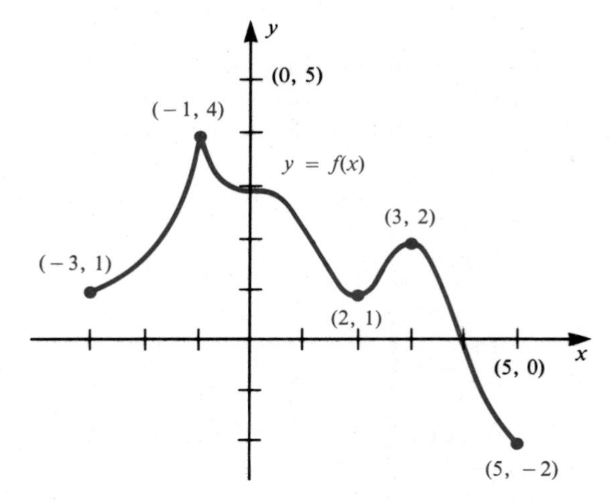

Figure 4–8

2. Use Figure 4–9 to complete the following statements.

(a) The maximum value of f on $[-5, 5]$ is ______________.

(b) The minimum value of f on $[-5, 5]$ is ______________.

(c) When $x =$ ______________, f has a relative maximum value.

(d) When $x =$ ______________, f has a relative minimum value.

(e) The numbers ______________ are critical numbers for f.

(f) f is decreasing on the open interval(s) ______________.

(g) $f'(x) > 0$ on the open interval(s) ______________.

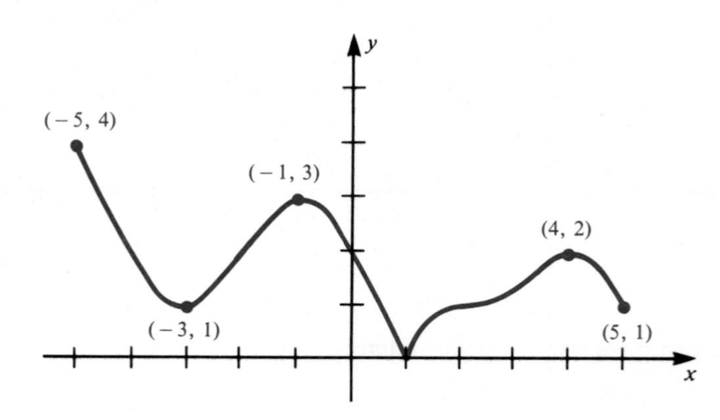

Figure 4–9

Find all the critical numbers of f in Exercises 3 through 6.

3. (*See Example 1*)
$f(x) = 6x^2 - 4x + 8$

4. $f(x) = -3x^2 + 7x - 5$

5. $f(x) = \frac{1}{3}x^3 - \frac{3}{2}x^2 + 2x - 7 \quad - 7$

6. $f(x) = x^3 + \frac{11}{2}x^2 - 4x + 3$

In Exercises 7 through 10, find the intervals on which f is increasing.

7. (*See Example 2*) $f(x) = -5x^2 + 8x - 9$

8. $f(x) = x^2 - 6x + 2$

9. $f(x) = \frac{1}{3}x^3 - x^2 - 8x + 6$

10. $f(x) = -\frac{1}{3}x^3 + 2x^2 + 5x + 1$

Find and classify the relative extreme values of f in Exercises 11 through 14.

11. (*See Example 3*) $f(x) = -5x^2 + 8x - 9$

12. $f(x) = x^2 - 6x + 2$

13. $f(x) = \frac{1}{3}x^3 - x^2 - 8x + 6$

14. $f(x) = -\frac{1}{3}x^3 + 2x^2 + 5x + 1$

II.

Find all the critical numbers of f in Exercises 15 through 18.

15. $f(x) = 2x^3 - x^2 - 4x + 9$

16. $f(x) = 2x^3 + \frac{13}{2}x^2 - 5x + 8$

17. $f(x) = \sqrt{x^2 + 8x + 20}$

18. $f(x) = \sqrt{x^2 - 4x + 15}$

In Exercises 19 through 22, find the intervals on which f is decreasing.

19. $f(x) = \frac{1}{4}x^4 + \frac{1}{3}x^3 - x^2 + 4$

20. $f(x) = \frac{1}{5}x^5 - \frac{4}{3}x^3 + 6$

21. $f(x) = \sqrt{x^2 - 6x + 10}$

22. $f(x) = \sqrt{x^2 - 10x + 30}$

Find the intervals on which f is increasing and the intervals on which f is decreasing, and sketch a graph of f in Exercises 23 through 26.

23. $f(x) = x^2 - 4x$

24. $f(x) = x^2 - 2x - 8$

25. $f(x) = -x^2 + 5x - 4$

26. $f(x) = -x^2 + 7x - 6$

III.

In Exercises 27 through 44, find the intervals on which f is decreasing and the intervals on which f is increasing. Identify relative extrema. Use this information to sketch a graph of the function.

27. (*See Example 4*) $f(x) = x^2 + 5x - 6$

28. $f(x) = -x^2 + 3x + 4$

29. (*See Example 5*) $f(x) = x^3 - 6x^2 + 2$

30. $f(x) = x^3 + 3x^2$

31. $f(x) = x^3 + 12x - 3$

32. $f(x) = -x^3 + 3x^2 + 9x + 1$

33. (*See Example 7*) $f(x) = (x - 8)^{1/3}$

34. $f(x) = (x - 8)^{2/3}$

35. (*See Example 6*)
$f(x) = x^4 + 4x^3 - 1$

36. $f(x) = -3x^4 + 8x^3 + 5$

37. $f(x) = \sqrt{9 - x^2}$

38. $f(x) = \sqrt{x^2 - 4}$

39. $f(x) = (x^2 - 4x)^{1/3}$

40. $f(x) = (x^2 - 2x - 3)^{1/3}$

41. (*See Example 4 of Chapter 1, Section 5*)
$f(x) = x\sqrt{2x + 12}$

42. $f(x) = 2x\sqrt{x + 5}$

43. $f(x) = 4x\sqrt{3 - x}$

44. $f(x) = 2x\sqrt{8 - x}$

45. A consulting firm predicted that t days after a certain new stock started to sell, its price P would be given by

$$P(t) = 140 + 10t - \frac{1}{4}t^2, \ t \geq 0.$$

If the firm is correct,
(a) for what interval(s) will the price of the stock increase?
(b) what will be the largest price of the stock?

46. A book publisher has undertaken market research to arrive at a decision on the number of copies of a certain text to produce. Printing too many copies would mean that many would remain unsold for a long period. Printing too few would imply that the company was not taking full advantage of the available market. The research group arrived at the following equation relating profit P (in dollars) to the number of books published N (in units of one thousand).

$$P = 100(-N^3 + 12N^2 + 60N)$$

Determine the number of books that should be published in order to maximize profit.

47. A baseball is thrown upward. The ball is released with an initial velocity of 96 ft/sec from a height of 6 feet above the ground. Its height in feet above the ground t seconds after it is released is given by

$$h(t) = -16t^2 + 96t + 6$$

(a) For what values of t is the height increasing?
(b) What is the maximum height of the ball?

4–2 Concavity and the Second Derivative Test

Concavity
Points of Inflection

Concavity

Economists typically use a curve like that given in Figure 4–10 to model the cost C as a function of x, where x is the amount of goods produced.

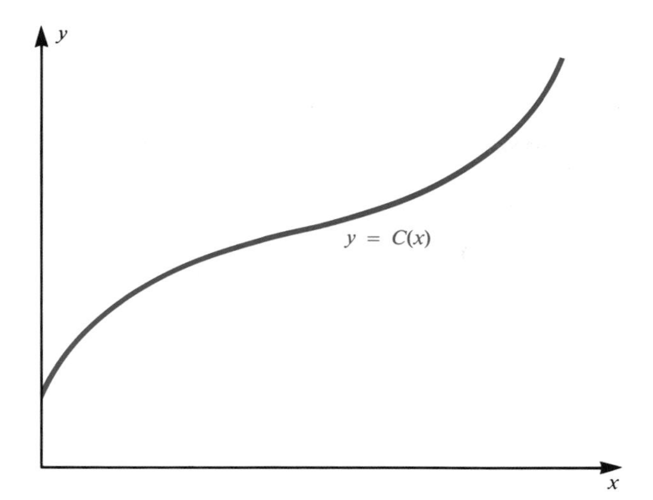

Figure 4–10

Why do they use a curve with this shape? What are the features that a cost curve should have? First, notice that the y-intercept of the curve is positive. This should always be true because the y-intercept is $C(0)$, which represents the fixed costs. Next, notice that the function is increasing, which represents the fact that costs usually rise as the production level rises.

The next feature of the graph that is important to economists is the way the graph "bends" even while it is rising. For small values of x, the slope of the graph is becoming less steep as x increases. For large values of x, however, the slope of the graph is becoming steeper as x increases. This feature of the graph reflects the fact that when x is small, production can be increased efficiently. As production increases however, there comes a point where more workers must be added, maintenance of equipment becomes more expensive, or some other factors make increased production inefficient. The quantitative description of "efficient" is based on the geometry of the graph.

The marginal cost function, MC, is the derivative of the cost function; $MC = C'$. Geometrically therefore, the marginal cost is the slope of the line tangent to the cost curve at $(x, C(x))$. For small values of x, the slopes of these lines decrease as x increases. Said another way, for small values of x, the marginal cost decreases as x increases. See Figure 4–11.

Recall that the sign of the derivative of a function indicates whether that function is increasing or decreasing. Thus, MC will be decreasing if $(MC)'$ is negative. Furthermore, because $MC = C', (MC)' = C''$. The geometric aspect of the cost curve we are trying to describe can now be characterized in terms of the sign of the second derivative of the cost function. When x is small, $C''(x) < 0$.

Similarly, for large values of x, the increasing steepness of the curve reflects the fact that MC is increasing; $C''(x) > 0$.

The first derivative tells us whether the graph is rising or falling; the second derivative tells us which way the graph is bent. The word **concave** is used to describe the informal notion of "bent."

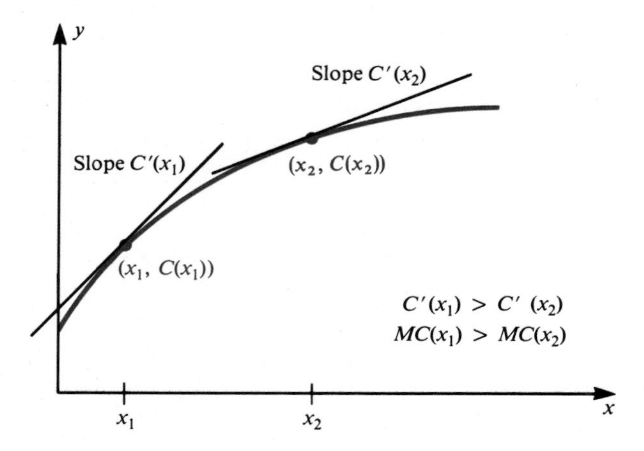

Figure 4–11

Definition

> **(a)** The graph of f is **concave up** on an interval if $f''(x) > 0$ for each x in the interval.
>
> **(b)** The graph of f is **concave down** on an interval if $f''(x) < 0$ for each x in the interval.

Warning! The concavity of the curve has nothing to do with whether the function is increasing or decreasing. There are four possible ways that the sign of f' and the sign of f'' can be paired. Figure 4–12 shows each of these possibilities.

The sign of $f''(c)$ has particular significance if $f'(c) = 0$. Recall that if $f'(c) = 0$, then c is a critical point, and $f(c)$ may be a relative extremum. We saw in the last section that $f(c)$ is a relative extreme value if f' changes sign at c. We now look at the behavior of f'' when f' changes sign. Suppose that $f'(c)=0$, and f' goes from negative to positive; then, the graph of f must look like that in Figure 4–13.

In this situation, as f' goes from negative to positive, f' is increasing. If f' is increasing, then $f''(x) > 0$, and the graph of f is concave up. A similar discussion about what happens at c if $f(c)$ is a relative maximum leads to what is called **the second derivative test.**

The Second Derivative Test for Relative Extreme Values

> If $f'(c) = 0$ and
>
> **(a)** $f''(c) > 0$, then f has a relative minimum at $x = c$.
>
> **(b)** $f''(c) < 0$, then f has a relative maximum at $x = c$.
>
> **(c)** $f''(c) = 0$, then no conclusion can be drawn.

Warning! Condition **(c)** really says that if $f''(c) = 0$, the first derivative test should be used to find out what's going on. See Examples 2, 3, and 4.

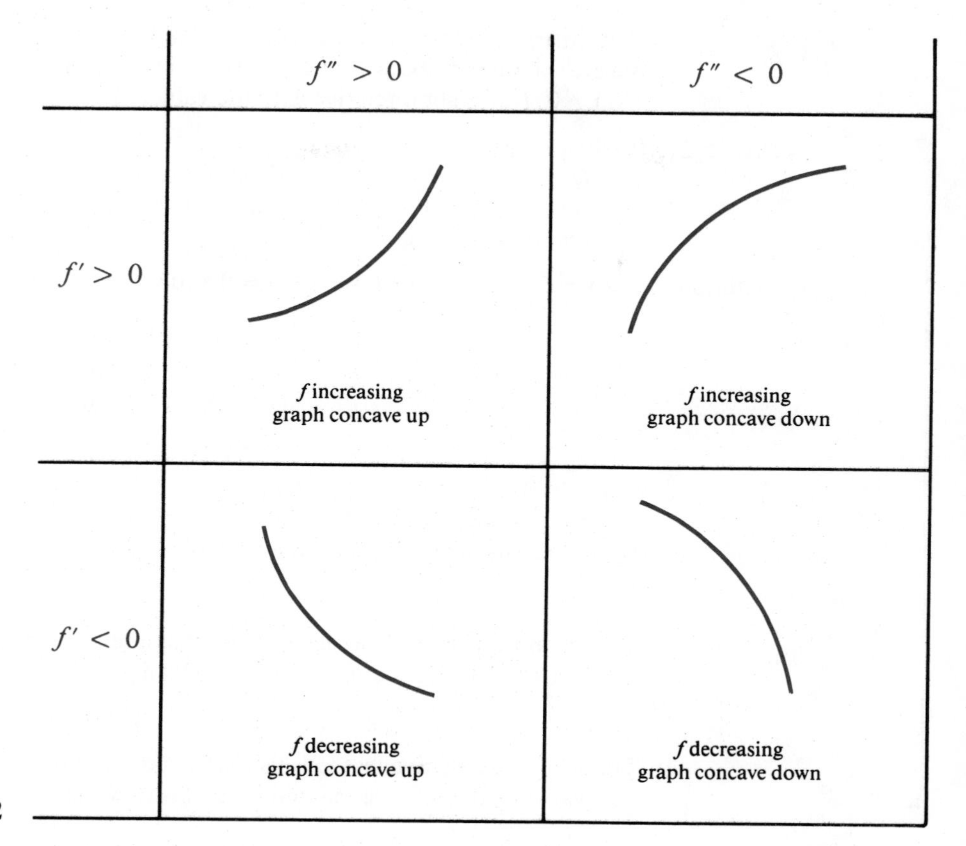

Figure 4–12

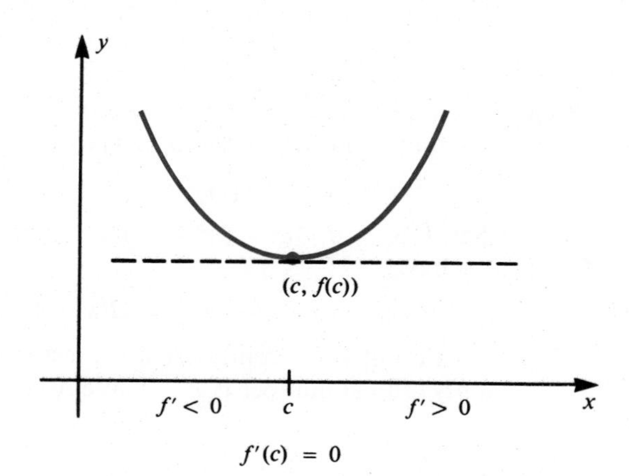

Figure 4–13

Warning! You cannot use the second derivative test at c unless $f'(c) = 0$. If c is a critical number because $f'(c)$ does not exist, then $f''(c)$ doesn't exist either. If $f'(c) \neq 0$, $f(c)$ **is not** a relative extreme value.

Example 1 (*Compare Exercise 5*)
Let $f(x) = 2x^3 - 6x^2 - 18x + 7$. Use the second derivative test to find the relative extreme values of f. (Compare with Example 1 of the previous section. The graph of f is Figure 4–3.)

Solution $f'(x) = 6x^2 - 12x - 18$; set $f'(x) = 0$ and solve as before:

$$6x^2 - 12x - 18 = 0$$
$$x^2 - 2x - 3 = 0$$
$$(x - 3)(x + 1) = 0$$
$$x = 3, x = -1$$

The critical numbers are $x = 3$ and $x = -1$.

Next, $f''(x) = 12x - 12$. (Remember to find the derivative of f' this time.) Now evaluate f'' at the critical numbers.

$$f''(3) = 12 \cdot 3 - 12 = 24 > 0$$

The second derivative test tells us that f has a relative minimum when $x = 3$. The value of the relative minimum is $f(3)$, which is -47.

$$f''(-1) = 12(-1) - 12 = -24 < 0$$

The second derivative test tells us that f has a relative maximim when $x = -1$. The value of the relative maximum is $f(-1)$, which is 17.

Example 2 (*Compare Exercise 11*)
Find the relative extreme values of f if

$$f(x) = x^4 - 4x^3 + 6x^2 - 4x + 3 = (x - 1)^4 + 2$$

Solution $f'(x)$ exists for all x, so the only critical numbers are the solutions to $f'(x) = 0$. We compute $f'(x)$ using the form $f(x) = (x - 1)^4 + 2$.

$$f'(x) = 4(x - 1)^3(1) + 0 = 4(x - 1)^3 \qquad \text{(Note the use of the chain rule)}$$

Set $f'(x) = 0$; $4(x - 1)^3 = 0$ has only one solution, $x = 1$. Computing $f''(x)$, we have

$$f''(x) = 4 \cdot 3(x - 1)^2 1 = 12(x - 1)^2, \qquad \text{so} \qquad f''(1) = 12(0)^2 = 0.$$

Part **(c)** of the second derivative test tells us to use the first derivative test. The only critical number is $x = 1$. We need to complete the following chart:

sign of f' ——————|——————
1

From $(-\infty, 1)$, we choose $x = 0$; $f'(0) = 4(-1)^3 = -4 < 0$
From $(1, \infty)$, we choose $x = 2$; $f'(2) = 4(1)^3 = 4 > 0$

We can now fill in the chart:

$$\text{sign of } f' \quad \overset{-}{}\underset{1}{|}\overset{+}{}$$

By using the first derivative test, we know that f has a relative minimum at $x = 1$; $f(1) = 2$ is the value of the relative minimum. Before drawing the graph of f, we check concavity. Because $f''(x) = 12(x - 1)^2$, $f''(x) > 0$ on $(-\infty, 1)$ and $(1, \infty)$. The graph of f is concave up on both these intervals. The graph of f is given in Figure 4–14.

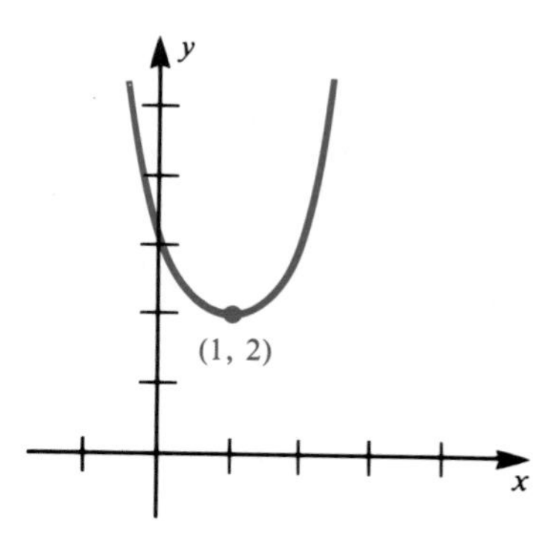

Figure 4–14

Example 3 (*Compare Exercise 11*)
Find the relative extreme values of g if

$$g(x) = -x^4 + 4x^3 - 6x^2 + 4x - 3$$

Solution Notice that $g(x) = -f(x)$, where $f(x)$ is defined as in Example 2. This means that $g'(x) = -f'(x)$ and $g''(x) = -f''(x)$. Thus, $g'(x) = 0$ precisely when $f'(x) = 0$, and $x = 1$ is the only critical number of g. Furthermore, $g''(1) = -f''(1) = 0$, so **(c)** of second derivative test again tells us to use the first derivative test. The sign of g' is the opposite of the sign of f' so we have

$$\text{sign of } g' \quad \overset{+}{}\underset{1}{|}\overset{-}{}$$

The first derivative test tells us that g has a relative maximum at $x = 1$. The concavity of the graph of g is determined by the sign of g''. Here $g''(x) = -12(x - 1)^2$, so $g''(x) < 0$ on $(-\infty, 1)$ and on $(1, \infty)$. The graph of g is concave down on both these intervals. In fact, the graph of g is the reflection in the x-axis of the graph of f, and is given in Figure 4–15.

Example 2 shows that when $f''(c) = 0$, $f(c)$ may be a relative minimum; Example 3 shows that when $f''(c) = 0$, $f(c)$ may be a relative maximum. There is even a third possibility, and we look at this case in Example 4.

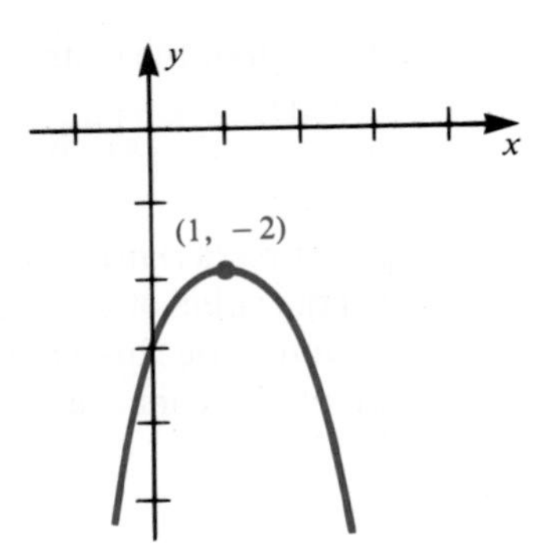

Figure 4–15

Example 4 (*Compare Exercise 11*)
Let $f(x) = x^3 - 3x^2 + 3x + 4$. Find the relative extreme values of f.

Solution $f'(x) = 3x^2 - 6x + 3$ for all x, so the only critical numbers of f are those that satisfy $f'(x) = 0$.

Set

$$3x^2 - 6x + 3 = 0$$
$$x^2 - 2x + 1 = 0$$
$$(x - 1)(x - 1) = 0$$

The function has only one critical number, $x = 1$. Compute $f''(x)$: $f''(x) = 6x - 6$. Thus, $f''(1) = 6 \cdot 1 - 6 = 0$, and **(c)** of the second derivative test tells us nothing. We go back to the first derivative test.

sign of f' ——————|—————— (tick at 1)

Test $f'(x)$ for $x < 1$. We choose $x = 0$;

$$f'(0) = 3 \cdot 0^2 - 6 \cdot 0 + 3 = 3 > 0$$

so f is increasing on $(-\infty, 1)$.

Next, test $f'(x)$ for $x > 1$. We choose $x = 2$;

$$f'(2) = 3(2)^2 - 6 \cdot 2 + 3 = 3 > 0$$

so f is also increasing on $(1, \infty)$.

sign of f' —— + ——|—— + —— (tick at 1)

f' does not change sign at $x = 1$, so f does not have a relative extreme value at $x = 1$. But, $x = 1$ was the only possibility! Therefore, f has no relative extreme values. We discuss its graph below.

We finished Example 4 by discovering that if $f(x) = x^3 - 3x^2 + 3x + 4$, then f has no relative extreme values. But we want to investigate the behavior of f near $x = 1$ more closely. What does the graph of f look like? To draw the graph of f, we compute $f(1) = 1^3 - 3 \cdot 1^2 + 3 \cdot 1 + 4 = 5$. Remember that $f'(1) = 0$, so we must draw a graph that rises up to (1, 5), has a horizontal tangent there, and then rises again on the other side of (1, 5).

The graph rises as x increases, but again we face the question of how the graph bends. We look at f'' again, and analyze the sign of f''. The **procedure** is the same as that for analyzing the sign of f'. Set $f''(x) = 0$ and find the cut points for this equation; then the sign of f'' is constant on intervals between cut points.

Here,

$$f''(x) = 6x - 6$$

so set

$$6x - 6 = 0$$

and solve.

$$x = 1$$

sign of f'' ——————|—————— (cut point at 1)

Test $f''(x)$ for $x < 1$: Choosing $x = 0$, we have

$$f''(0) = 6 \cdot 0 - 6 = -6 < 0$$

Test $f''(x)$ for $x > 1$: Choosing $x = 2$, we have

$$f''(2) = 6 \cdot 2 - 6 = 6 > 0.$$

sign of f'' —— $-$ ——|—— $+$ —— (cut point at 1)

The graph of f is concave down on $(-\infty, 1)$, but concave up on $(1, \infty)$. Such a graph is drawn in Figure 4–16. The line T is the line tangent to the graph at (1, 5).

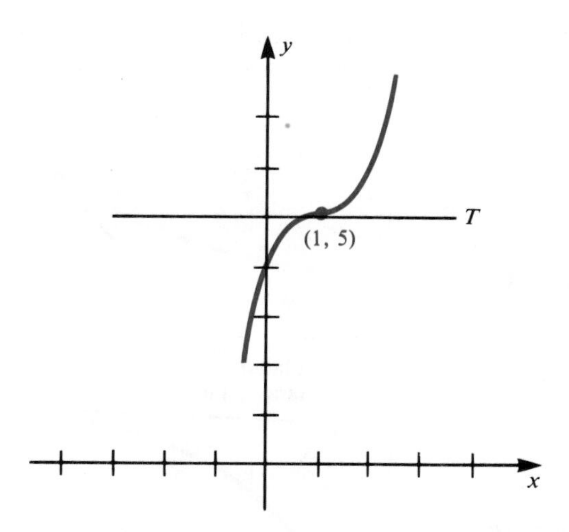

Figure 4–16

Points of Inflection

The graph in Figure 4–16 has a feature at the point (1, 5) that we saw occur in the typical cost curve at the beginning of this section. Now we can describe this feature mathematically; the graph has a change of concavity at (1, 5). In this example and the cost function example, the concavity changes from concave down to concave up. We give a special name to the points on the graph where the concavity changes.

Definition

> The point $(a, f(a))$ on the graph of f is a **point of inflection** if the graph is concave up on one side of $(a, f(a))$ and concave down on the other side. That is, the point $(a, f(a))$ is a point of inflection if the graph changes concavity at $(a, f(a))$.

We are interested in points of inflection because they enable us to sketch a better graph. Furthermore, points of inflection are very important in understanding "the law of diminishing marginal returns" because they help locate the relative maximum values of f', (see Figure 4–17) and relative minimum values of f' (see Figure 4–18).

Remember the definition of a point of inflection; the graph must change concavity at $(a, f(a))$. It is not enough that $f''(a) = 0$. Example 2 shows a function with $f''(1) = 0$; but, $(1, f(1))$ is not a point of inflection. Example 6 will show that $(a, f(a))$ may be a point of inflection even if $f''(a)$ doesn't exist. Points of inflection occur when f'' changes sign, and f'' can change sign at a if either $f''(a) = 0$ or a is not in the domain of f''. Finding the sign of f'' is the same as solving both the inequalities $f''(x) > 0$ and $f''(x) < 0$. We solve these inequalities by using cut points.

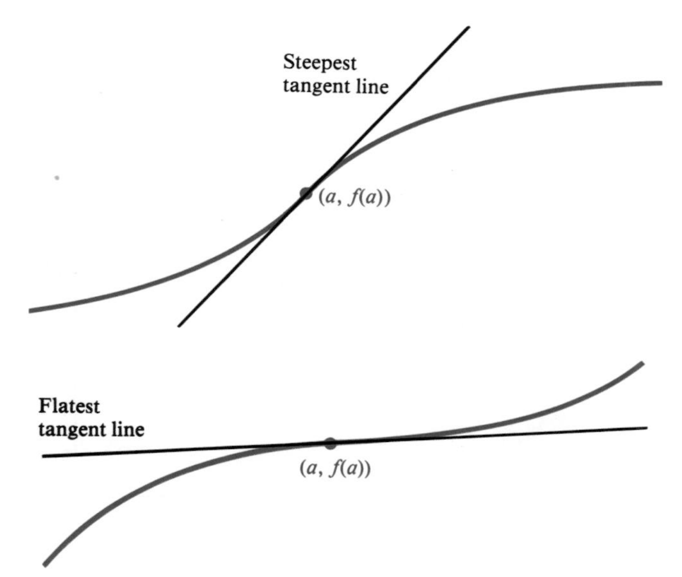

Figure 4–17

Figure 4–18

Example 5 (*Compare Exercise 17*)
Sketch the graph of f if $f(x) = 2 - 9x + 6x^2 - x^3$.

Solution $f'(x) = -9 + 12x - 3x^2$ for all x, so the only critical numbers are the solutions to $f'(x) = 0$. Set

$$-9 + 12x - 3x^2 = 0 \qquad \text{Divide by } -3$$
$$3 - 4x + x^2 = 0$$
$$(x - 3)(x - 1) = 0$$

The critical numbers of f are $x = 1$ and $x = 3$. The chart

sign of f'	$-$	1	$+$	3	$-$

tells us that f is decreasing on both $(-\infty, 1)$ and $(3, \infty)$, and that f is increasing on $(1, 3)$. Evaluating $f(x)$ at the critical numbers, we have $f(1) = -2$ and $f(3) = 2$. To discuss concavity, we look at f''.

$$f''(x) = 12 - 6x$$

The chart

sign of f''	$+$	2	$-$

shows that the graph is concave up on $(-\infty, 2)$ and concave down on $(2, \infty)$. The point $(2, f(2)) = (2, 0)$ is an inflection point.

The graph of f is given in Figure 4–19.

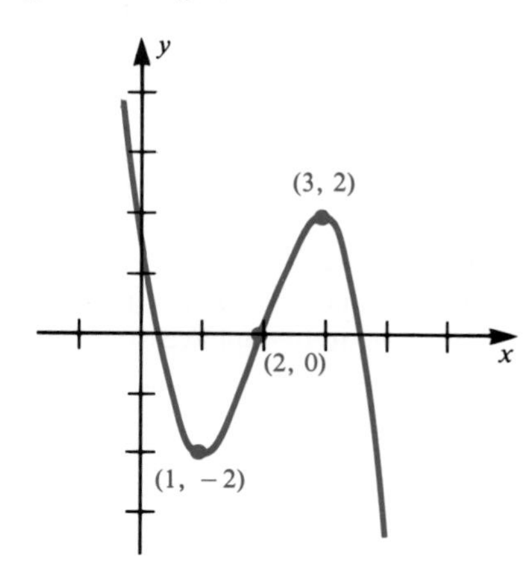

Figure 4–19

Example 6 (*Compare Exercise 27*)
Sketch the graph of $y = (x + 1)^{1/3} + 2$, and label any points of inflection.

Solution If $f(x) = (x + 1)^{1/3} + 2$, then f is continuous for all x, so its graph has just one piece.

$$f'(x) = \frac{1}{3}(x + 1)^{-2/3} = \frac{1}{3(x + 1)^{2/3}}$$

Note that $f'(x)$ does not exist for $x = -1$, so -1 is a critical number. There are no solutions to $f'(x) = 0$. Furthermore, $f'(x) > 0$ if $x \neq -1$, so the sign chart for f' is as follows.

$$\text{sign of } f' \quad \underset{-1}{\overset{+ \qquad\qquad +}{\overline{\qquad\qquad|\qquad\qquad}}}$$

The function f is increasing on $(-\infty, -1)$ and on $(-1, \infty)$. To investigate concavity, we turn to the second derivative.

$$f''(x) = \frac{-2}{9}(x + 1)^{-5/3} = \frac{-2}{9(x + 1)^{5/3}}$$

Next, we analyze the sign of f''. We set $f''(x) = 0$, but $\dfrac{-2}{9(x + 1)^{5/3}} = 0$ has no solution. However, there is a number where $f''(x)$ doesn't exist, namely $x = -1$.

The chart we must fill in is

$$\text{sign of } f'' \quad \underset{-1}{\overline{\qquad\qquad|\qquad\qquad}}$$

To test $f''(x)$ with a number less than -1, we choose $x = -2$:

$$f''(-2) = \frac{-2}{9(-1)^{5/3}} = \frac{-2}{9(-1)} = \frac{2}{9} > 0$$

To test $f''(x)$ with a number greater than -1, we choose $x = 0$:

$$f''(0) = \frac{-2}{9(1)^{5/3}} = \frac{-2}{9 \cdot 1} = \frac{-2}{9} < 0$$

The sign chart for f'' is now complete.

$$\text{sign of } f'' \quad \underset{-1}{\overset{+ \qquad\qquad -}{\overline{\qquad\qquad|\qquad\qquad}}}$$

Evaluating $f(x)$ at $x = -1$, we obtain $f(-1) = 0^{1/3} + 2 = 2$. The sign chart of f'' tells us that $(-1, 2)$ is a point of inflection because the graph of f changes from concave up to concave down at that point. The graph of f is given in Figure 4–20.

The graph of f is concave up over $(-\infty, -1)$ and concave down over $(-1, \infty)$.

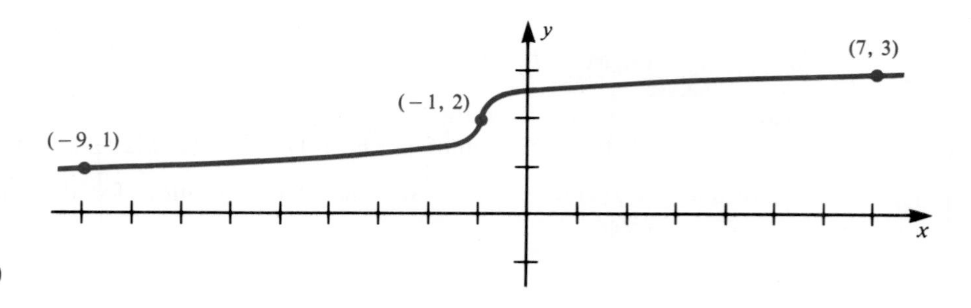

Figure 4–20

Example 7 (*Compare Exercise 23*)
Sketch the graph of $y = 1/x$.

Solution First, since $f(x) = 1/x$, the function f is not continuous for all x; the number $x = 0$ is not in the domain of f. Writing $f(x) = x^{-1}$, we find that

$$f'(x) = -1x^{-2} = \frac{-1}{x^2}$$

and

$$f''(x) = 2x^{-3} = \frac{2}{x^3}$$

Because $f'(x)$ is negative for all x in the domain of f, the sign chart for f' looks like

sign of f'	$-$	0	$-$

f is decreasing on $(-\infty, 0)$ and on $(0, \infty)$.

To analyze the sign of $f''(x) = 2/x^3$, we find the cut points for the equation $2/x^3 = 0$. There are no solutions to this equation, but $x = 0$ is a cut point because 0 is not in the domain of $2/x^3$.

The chart to fill in is

sign of f''		0	

To test $f''(x)$ for $x < 0$, we choose $x = -1$.

$$f''(-1) = \frac{2}{(-1)^3} = \frac{2}{-1}$$
$$= -2 < 0$$

To test $f''(x)$ for $x > 0$, we choose $x = 1$.

$$f''(1) = \frac{2}{(1)^3} = \frac{2}{1}$$
$$= 2 > 0$$

The chart now looks like

sign of f''	$-$	0	$+$

The graph is concave down over $(-\infty, 0)$ and concave up on $(0, \infty)$, but the graph has no point of inflection because the function is not even defined at $x = 0$. The graph of $y = 1/x$ is given in Figure 4–21.

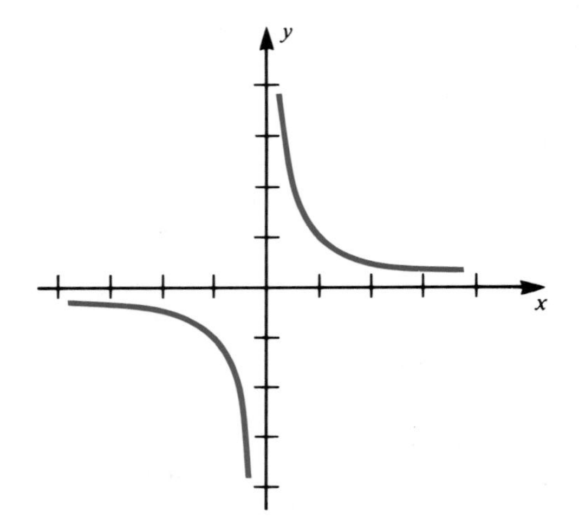

Figure 4–21

4–2 Exercises

I.

Use the graph in Figure 4–22 to answer the four questions that follow. (Some tangent lines have been drawn, and the answers involve only integer values of x.)

1. On what open intervals is the graph concave up?
2. On what open intervals is the graph concave down?
3. What are the points of inflection?
4. What are the numbers in the domain where f has a relative extreme value?

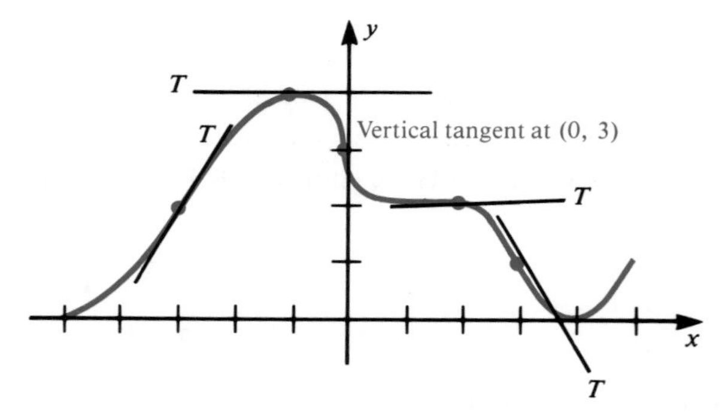

Figure 4–22

II.

In Exercises 5 through 14, use the second derivative test to find the numbers at which f has a relative extreme value, and tell whether f has a relative maximum or minimum at each such number.

5. (*See Example 1*) $f(x) = 7x^2 - 4x - 2$

6. $f(x) = 3x^2 - 6x + 4$

7. $f(x) = -4x^2 + 24x + 5$

8. $f(x) = 6 - 30x - 5x^2$

9. $f(x) = 2x^3 + 9x^2 - 60x + 5$

10. $f(x) = \frac{2}{3}x^3 + x^2 - 24x + 7$

11. (*See Examples 2, 3, and 4*)
$f(x) = 3x^4 - 4x^3 - 12x^2 + 7$

12. $f(x) = x^4 + 4x^3 - 8x^2 - 10$

13. $f(x) = 9x + \frac{1}{x}$

14. $f(x) = 4x + \frac{16}{x}$

In Exercises 15 through 28, identify the intervals on which the graph of f is concave up, the intervals on which the graph of f is concave down, and find the points of inflection.

15. $f(x) = 9x^2 - 14x + 1$

16. $f(x) = 8 - 5x - 7x^2$

17. (*See Example 5*)
$f(x) = x^3 - 6x^2 + 9x - 4$

18. $f(x) = 2x^3 - 24x^2 + 5x - 9$

19. $f(x) = -5x^3 + 10x^2 + 3x - 8$

20. $f(x) = -10x^3 + 20x^2$

21. $f(x) = x^4 + 2x^3 - 36x^2 + 8x - 4$

22. $f(x) = x^4 - 12x^3 + 9x + 10$

23. (*See Example 7*)
$f(x) = 7 - \frac{1}{x}$

24. $f(x) = x + \frac{25}{x}$

25. $f(x) = 8x^2 + \frac{1}{x}$

26. $f(x) = x^2 - \frac{8}{x}$

27. (*See Example 6*)
$f(x) = 2 + (x - 5)^{1/3}$

28. $f(x) = (x + 6)^{2/3} - 4$

III.

In Exercises 29 through 34,

(a) identify the intervals on which f is increasing or decreasing;
(b) identify the intervals on which f is concave up or concave down; and
(c) sketch the graph of f. Plot the points corresponding to relative extrema and points of inflection.

29. $f(x) = (x - 4)^3 - 3$

30. $f(x) = x^3 - 3x - 2$

31. $f(x) = x^3 + 3x^2 - 9x + 1$

32. $f(x) = \frac{1}{4}x^4 + \frac{1}{3}x^3 - x^2 + 2$

33. $f(x) = x - 2\sqrt{x}$

34. $f(x) = 4\sqrt{x} - x + 1$

35. Using preliminary data, a company has decided to describe its cost in manufacturing x units per day by

$$C(x) = \frac{1}{12}x^3 - 100x^2 + 300x + 6100$$

For what value of x is the marginal cost, MC, minimized?

36. A company has determined that if it produces x items per week, then its weekly revenue function R can be modeled by

$$R(x) = \frac{-1}{6}x^3 + 70x^2 + 130x$$

For what value of x is the company's marginal revenue, MR, maximized? This value is called the point of diminishing marginal returns.

4–3 Extreme Values

Existence of a Maximum and a Minimum
Functions Defined on Closed and Bounded Intervals
Functions with Only One Critical Number

In the previous two sections, we have shown how the first and second derivatives determine the shape of a function's graph, and how they can be used to find where a function has its relative extreme values. Frequently however, applications call for not just the relative extreme values of a function, but for the function's maximum or minimum value over the whole domain. This section gives methods of finding the maximum and minimum value of a function. We concentrate in this section on developing the mathematical techniques that are used in applications. Thus, when we deal with applications in Section 4–5, we can then concentrate on how to apply these techniques.

Existence of a Maximum and a Minimum

The function whose graph is drawn in Figure 4–23 has a relative minimum when $x = -2$ and a relative maximum when $x = 3$, but does not have either a minimum value or a maximum value.

Certain conditions often arise in applications, however, which guarantee that a function satisfying one of these conditions will have an extreme value. We will deal with two of these conditions, the first of which restricts the domain of the function to an interval of form $[a, b]$.

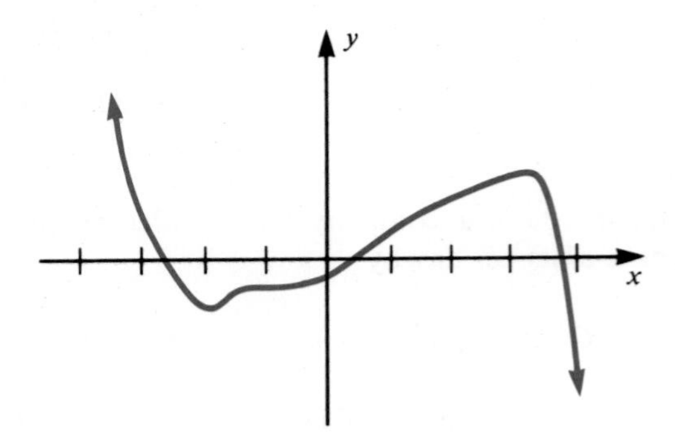

Figure 4–23

Functions Defined on Closed and Bounded Intervals

A closed and bounded interval is an interval of form $[a, b]$. There is a theorem that guarantees that every continuous function whose domain is a closed and bounded interval has both a maximum value and a minimum value on that interval. We can take advantage of this theorem to simplify finding extreme values. To see how, suppose that f is continuous on $[a, b]$ and has its maximum value at c. If c is in the open interval (a, b), then $f(c)$ is a relative maximum and so c is a critical number. If c is in $[a, b]$, but not in the open interval (a, b), then c is either a or b. Thus, f has its maximum value either at a critical number or at one of the two endpoints a and b. Example 1 shows how these observations can quicken our computations.

Example 1 (*Compare Exercise 1*)
If $f(x) = x^2 - 6x + 4$ for $-2 \leq x \leq 5$, find the maximum value of f.

Solution The restriction $-2 \leq x \leq 5$ means that the domain of the function is the interval $[-2, 5]$. The function is continuous, so the theorem assures us that both maximum and minimum values of f exist. Furthermore, the derivative of f exists for all x in $(-2, 5)$; we have $f'(x) = 2x - 6$. Setting $f'(x) = 0$, we find that the only critical number is $x = 3$.

The maximum value of f is either $f(3)$, $f(-2)$, or $f(5)$. To find out which one, we simply perform these three evaluations.

$$f(3) = -5; \qquad f(-2) = 20; \qquad f(5) = -1$$

The largest of these three numbers is 20, so the maximum value of f must be $f(-2) = 20$.

The same reasoning we used to find the maximum value of f also works to find the minimum value. Looking back at the end of Example 1, we can also say that the minimum value of f on $[-2, 5]$ is $f(3) = -5$. (*See Exercise 1*).

Theorem

> If f is continuous on $[a, b]$, then both the maximum value of f and the minimum value of f occur when x is either a critical number or an endpoint.

This type of search for extreme values is sometimes referred to as **the candidates test**.

The Candidates Test

> If f is continuous on an interval of the form $[a, b]$, find all the places where f could possibly have an extreme value (the critical numbers and the endpoints), and then evaluate the function at each of these points. These values are the candidates for the maximum and minimum. The greatest candidate is the maximum value of f on $[a, b]$, and the least candidate is the minimum.

Warning! Remember the distinction between the maximum value of f and the number in the domain where f achieves its maximum.

Example 2 *(Compare Exercise 11)*
Find the maximum and minimum values of f on the interval $[-2, 4]$ if $f(x) = 2x^3 - 3x^2 - 12x + 10$.

Solution The function is continuous on $[-2, 4]$ and $f'(x) = 6x^2 - 6x - 12$. Because $f'(x)$ exists for all x in $(-2, 4)$, the only critical numbers are solutions to $f'(x) = 0$.

Set

$$6x^2 - 6x - 12 = 0$$

and solve

$$x^2 - x - 2 = 0$$

$$(x - 2)(x + 1) = 0$$

The only critical numbers are $x = -1$ and $x = 2$. Now, all we have to do is evaluate $f(-1)$ and $f(2)$, and then $f(-2)$ and $f(4)$, and compare these numbers. All the relevant information is displayed in the following table.

	Critical Numbers		Endpoints	
x	-1	2	-2	4
$f(x)$	17	-10	6	42

The maximum value of f is 42 when $x = 4$. The minimum value of f is -10 when $x = 2$.

Example 3 *(Compare Exercise 21)*
Find the maximum and minimum values of f on the interval $-1 \le x \le 7$ if $f(x) = (2x - 6)^{2/3} + 5$.

Solution

$$f'(x) = \frac{2}{3}(2x - 6)^{-1/3}(2)$$

$$= \frac{4}{3(2x - 6)^{1/3}}$$

There are no solutions to $\dfrac{4}{3(2x - 6)^{1/3}} = 0$. The only critical number is $x = 3$ because $f'(3)$ is not defined. The candidates for maximum and minimum are $f(3)$, $f(-1)$, and $f(7)$. $f(3) = 5$; $f(-1) = (-8)^{2/3} + 5 = (-2)^2 + 5 = 9$; and $f(7) = (8)^{2/3} + 5 = 9$. The maximum value of f on $[-1, 7]$ is 9, and the minimum value is 5.

Example 4 (*Compare Exercise 25*)
Find the extreme values of f on the interval $[0, 6]$ if $f(x) = x\sqrt{36 - x^2}$.

Solution $f(x) = x(36 - x^2)^{1/2}$, so

$$f'(x) = (36 - x^2)^{1/2} + x\left(\frac{1}{2}\right)(36 - x^2)^{-1/2}(-2x)$$
$$= \sqrt{36 - x^2} - \frac{x^2}{\sqrt{36 - x^2}}$$

Thus, $f'(x)$ exists for all x in $(0, 6)$. Next, $f'(x) = 0$ when

$$\sqrt{36 - x^2} = \frac{x^2}{\sqrt{36 - x^2}}$$

Cross multiply to get

$$36 - x^2 = x^2$$
$$36 = 2x^2$$
$$18 = x^2$$
$$x = \pm\sqrt{18}$$

Reject $x = -\sqrt{18}$ because this number is not in the domain of f. Finally, we use the candidates test. The critical number is $\sqrt{18}$; $f(\sqrt{18}) = \sqrt{18}\sqrt{36 - 18} = \sqrt{18}\sqrt{18} = 18$. Evaluating f at the endpoints 0 and 6, we have

$$f(0) = 0 \text{ and } f(6) = 0$$

If $f(x) = x\sqrt{36 - x^2}$ for $0 \leq x \leq 6$, the maximum value of f is 18; the minimum value of f is 0.

A model that assumes a linear relationship between price and demand lends itself to the candidates test.

Example 5 (*Compare Example 33*)
Find the production level that will give a company its maximum revenue if the relation between the price p and production level x is given by

$$p = -\frac{1}{40}x + 10$$

Solution Since x is production level, the application makes us assume that $x \geq 0$. And since p is price, we would also assume that $p \geq 0$. Since

$$p = -\frac{1}{40}x + 10$$

the inequality $p \geq 0$ is the same as

$$-\frac{1}{40}x + 10 \geq 0$$

Solving this inequality for x, we get $x \leq 400$. Thus, x must satisfy $0 \leq x \leq 400$. The general expression for revenue is $R = x \cdot p$. In this case, we can express R as a function of x by

$$R(x) = x\left(-\frac{1}{40}x + 10\right) = -\frac{1}{40}x^2 + 10x, \qquad 0 \leq x \leq 400$$

R is continuous on the closed and bounded interval $[0, 400]$, so we can use the candidates test.

$$R'(x) = -\frac{1}{20}x + 10$$

is defined for all x in $(0, 400)$, so the only critical numbers will be solutions to

$$-\frac{1}{20}x + 10 = 0$$

$x = 200$ is the only critical number.

	Critical Number	Endpoints	
x	200	0	400
$R(x)$	1000	0	0

The maximum revenue is 1000, and the best production level is $x = 200$.

Functions with Only One Critical Number

Sometimes, an application forces you to try to find the extreme value of a function whose domain does not include both endpoints or whose domain is unbounded. Example 8 will show a typical problem of this sort. To explain how to handle this situation, we still assume that the function we are dealing with is continuous and that its domain is an interval. In this instance, no further conditions are assumed about the interval. Instead, we assume that the function has only one critical number. The following theorem tells how to handle this case.

The One-Critical-Number Test

Suppose f is continuous on an interval and has only one critical number c in the interval

1. If $f(c)$ is a relative minimum, then $f(c)$ is the minimum value of f.
2. If $f(c)$ is a relative maximum, then $f(c)$ is the maximum value of f.

Warning! Be careful applying this theorem—it requires that f have **exactly one** critical number. The theorem does **not** say that if $f(c)$ is the only relative maximum, then $f(c)$ is the maximum value of f. In fact, Example 2 gives a situation where f has only one relative maximum, namely $f(-1)$, but $f(-1)$ is not the maximum. Notice however that $x = -1$ is not the only critical number; $x = 2$ is another critical number for f. To use the one-critical-number test, you must make sure that f has a **total of only one critical number**.

Example 6 (*Compare Example 13*)
Find the maximum value of f on $[-1, 5]$ if $f(x) = 14 + 24x + 3x^2 - x^3$

Solution $f'(x) = 24 + 6x - 3x^2$. Set $f'(x) = 0$ and solve

$$24 + 6x - 3x^2 = 0$$
$$-8 - 2x + x^2 = 0$$
$$x^2 - 2x - 8 = 0$$
$$(x - 4)(x + 2) = 0$$

Careful! The equation $24 + 6x - 3x^2 = 0$ has two solutions: $x = 4$ and $x = -2$, but -2 is not in the domain of the function; -2 is not in $[-1, 5]$. The function has only one critical number: $x = 4$. Also, $f''(x) = 6 - 6x$, so $f''(4) = 6 - 24 = -18 < 0$. Thus, f has a relative maximum at $x = 4$, and $x = 4$ is the **only** critical number. Therefore, f has its maximum value on $[-1, 5]$ when $x = 4$. The maximum value of f is $f(4) = 94$.

Frequently, there is more than one way to do a particular problem. For instance, you can check the answer to Example 6 by using the candidates test. The test that relies on f having only one critical number is especially useful when the domain of the function is not a closed and bounded interval and when you can easily compute $f''(c)$. We give two more examples in this section. The first is a straightforward computation, and the second is an application.

Example 7 (*Compare Exercise 5*)
Find the extreme value of f on the interval $(0, \infty)$ if

$$f(x) = x + \frac{32}{x^2}$$

Solution $f(x) = x + 32x^{-2}$, so $f'(x) = 1 - 64x^{-3} = 1 - \dfrac{64}{x^3}$; $f'(x)$ exists for all $x > 0$.

Set $f'(x) = 0$ and solve

$$1 - \frac{64}{x^3} = 0$$

$$1 = \frac{64}{x^3}$$

$$x^3 = 64$$

$x = 4$ is the only critical number.

Next, $f''(x) = 192x^{-4}$, so $f''(4) = \frac{192}{256} > 0$, and f has a relative minimum when $x = 4$. Since 4 is the only critical number, $f(4) = 6$ is the minimum value of f. Since there are no endpoints and no other critical number, f does not have a maximum value.

Example 8 (*Compare Exercise 35*)
A trucking company has data which indicates that on a 300-mile trip, the cost of operating a truck that averages V miles per hour is $0.40 + (V/80)$ dollars per mile. The company pays its drivers \$20 per hour. In order to minimize costs, at what speed should the trucks travel?

Solution The time spent travelling is $(300/V)$ hours, so the labor cost (in dollars) to the company is

$$(20)\left(\frac{300}{V}\right) = \frac{6000}{V}$$

The mileage cost (in dollars) of operating the truck is

$$300\left(0.40 + \frac{V}{80}\right) = 120 + \frac{300V}{80}$$

The total cost in dollars in terms of V is the sum of these two costs:

$$C(V) = \frac{6000}{V} + 120 + \frac{300V}{80}$$

and the domain is all $V > 0$ (We will rule out excessive speeds later, if necessary). Note that we **cannot** use the candidate test to solve this problem because the domain is not of the form $[a, b]$.

$$C'(V) = -\frac{6000}{V^2} + \frac{300}{80} = -\frac{6000}{V^2} + \frac{15}{4}$$

$C'(V)$ exists for all V in the domain of C. To find the critical numbers, set

$$-\frac{6000}{V^2} + \frac{15}{4} = 0$$

and solve

$$\frac{15}{4} = \frac{6000}{V^2}$$

$$V^2 = 1600$$

$V = \pm 40$; reject $V = -40$ because the domain requires $V > 0$.
There is only one critical number: $V = 40$.

$$C''(V) = \frac{12000}{V^3}, \qquad \text{so} \qquad C''(40) = \frac{12000}{(40)^3} > 0$$

$C''(40)$ is positive, so C has a relative minimum when $V = 40$. The cost is a relative minimum at 40 mph, and there is only one critical number, so the cost is minimized if the trucks average 40 mph.

4–3 Exercises

I.

In Exercises 1 through 4, use the candidates test to find the maximum and minimum values of f on the given interval.

1. (*See Example 1*) $f(x) = x^2 - 4x + 1, \ -3 \le x \le 3$
2. $f(x) = x^2 + 2x - 3, \ -2 \le x \le 2$
3. $f(x) = -3x^2 + 6x - 5, \ -2 \le x \le 3$
4. $f(x) = -2x^2 + 8x + 3, \ -4 \le x \le 5$

In Exercises 5 through 8, find the minimum value of f on the given interval.

5. (*See Example 7*) $f(x) = x + \frac{9}{x}$, for $x > 0$
6. $f(x) = x + \frac{4}{x^2}$, for $x > 0$
7. $f(x) = 2x^2 + \frac{32}{x}$, for $x > 0$
8. $f(x) = x^2 + \frac{2}{x}$, for $x > 0$

II.

In Exercises 9 through 14, find the maximum and minimum values of f on the given interval.

9. $f(x) = x^2 - 8x + 2, \quad -2 \le x \le 3$
10. $f(x) = x^2 + 6x - 2, \quad -1 \le x \le 4$
11. (*See Example 2*) $f(x) = x^3 - 6x^2 + 9x - 4, \quad -1 \le x \le 2$
12. $f(x) = 4x^3 + 6x^2 - 24x + 5, \quad -3 \le x \le 3$
13. (*See Example 6*) $f(x) = 6 + 24x + 3x^2 - x^3, \quad 0 \le x \le 5$
14. $f(x) = 2 + 21x + 9x^2 - x^3, \quad -2 \le x \le 4$

In Exercises 15 through 20, find the maximum value of f on the indicated interval.

15. $f(x) = \frac{8}{2 + x^2}, \ -\infty < x < \infty$
16. $f(x) = \frac{6}{3 + x^2}, \ -\infty < x < \infty$
17. $f(x) = 3x + \frac{12}{x}$, for $x < 0$
18. $f(x) = -x^2 - \frac{16}{x}$, for $x > 0$
19. $f(x) = 1 - x^{2/3}, \ -\infty < x < \infty$
20. $f(x) = 7 - 6x^{2/3}, \ -\infty < x < \infty$

III.

In Exercises 21 through 26, find maximum and minimum values of f on the given interval.

21. (*See Example 3*) $f(x) = (x^2 - 9)^{1/3}, \ -1 \le x \le 6$
22. $f(x) = (x^2 - 9)^{1/3}, \ -6 \le x \le 1$
23. $f(x) = (x^2 - 9)^{2/3}, \ -1 \le x \le 6$
24. $f(x) = (x^2 - 9)^{2/3}, \ -6 \le x \le 1$

25. (*See Example 4*)
$f(x) = x\sqrt{16 - x^2}, 0 \le x \le 4$

26. $f(x) = 2x\sqrt{25 - x^2}, 0 \le x \le 5$

In Exercises 27 through 32, find an extreme value of f on the given interval, and identify whether the value is the maximum or minimum.

27. $f(x) = 4x + \dfrac{100}{x}, x > 0$

28. $f(x) = 4x + \dfrac{100}{x}, x < 0$

29. $f(x) = \dfrac{x^2 + 1}{x^2 + 4}, \quad -\infty < x < \infty$

30. $f(x) = \dfrac{x^2 + 4}{x^2 + 1}, \quad -\infty < x < \infty$

31. $f(x) = x - 3x^{1/3}$, for $x > 0$

32. $f(x) = x - 18\sqrt{x}$, for $x > 0$

33. (*See Example 5*) Find the production level that will give a company its maximum revenue if the relation between the price p and production level x is given by

$$p = -\frac{1}{50}x + 30$$

What is the corresponding price?

34. Find the production level that will give a company its maximum revenue if the relation between the price p and production level x is determined by the equation $x^2 + 25p^2 = 7200$

35. (*See Example 8*) Suppose that the trucking company in Example 8 finds its dollar cost per mile is $0.40 + (V/80)$, no matter how long the trip. What is the most economical speed if the length of the trip is 500 miles (rather than 300 miles)?

36. The same trucking company as in Example 8 finds that the dollar cost per mile for a new truck is $0.45 + (V/80)$. What is the most economical average speed for a new truck?

37. A different trucking company finds that its costs for operating a truck at V miles per hour is $0.35 + (V/100)$ dollars per mile, and this firm pays its drivers \$25 per hour. What is the most economical average speed for this form on a 300-mile trip?

4–4 Asymptotes

Average-Cost Function
Geometry of Asymptotes
Vertical Asymptotes
Horizontal Asymptotes

Average-Cost Function

A certain firm has modeled its cost by a linear function C, where $C(x)$, the cost of producing x items, is given by $C(x) = 1.5x + 4$; as we have seen, the 4 represents fixed costs, and the 1.5 is the cost per item. The firm is assuming that its marginal cost is constant; $MC(x) = C'(x) = 1.5$. Using this model, what does the average cost curve look like? In general, we let $AC(x)$ be the **a**verage **c**ost of producing x items, so

$$AC(x) = \frac{C(x)}{x}$$

In this particular example then,

$$AC(x) = \frac{1.5x + 4}{x} = 1.5 + \frac{4}{x}$$

We sketch the graph of $AC(x)$ for $x > 0$, since those are the only values of x that make sense in our application.

$AC(x) = 1.5 + (4/x)$ is continuous for all $x > 0$, so the graph has one piece. Because $(AC)'(x) = -4/x^2$, $(AC)'(x) < 0$ for all x. (Note that x^2, the denominator, is always positive.) The negative derivative indicates that the average cost is always decreasing. The second derivative is $(AC)''(x) = 8/x^3$. $(AC)''(x) > 0$ for all x. (We're assuming that x is positive, so x^3 is positive.) Therefore, the graph of the average-cost function is always concave up.

From the first and second derivative, we know that the general shape of the graph is like that given in Figure 4–24. But there are many graphs with this general shape.

We have seen that we can graph functions by plotting only a few points if the points we plot are carefully chosen. If the domain of the function is an interval of the form $[a, b]$, then certainly we must always choose to plot $(a, f(a))$ and $(b, f(b))$. But what if the domain of the function is some other type of interval? For instance, in this example $AC(x) = 1.5 + (4/x)$, and the domain is $(0, \infty)$. The question for this function is, "What happens to $AC(x)$ when x is near 0 or when x is very large?"

Geometry of Asymptotes

What happens to $AC(x) = 1.5 + (4/x)$ if x is a large number? The term $4/x$ is then a number close to 0, so $1.5 + (4/x)$ is close to 1.5. In fact, because $4/x$ is positive, we can even say that $1.5 + (4/x)$ is a little bit bigger than 1.5. Furthermore, the larger the value of x, the closer $1.5 + (4/x)$ is to 1.5. We again use the

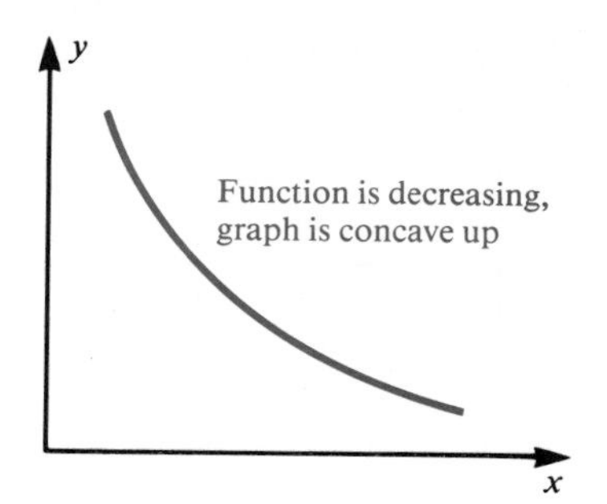

Figure 4–24

language of limits to describe the notion of "close to." This time, we write

$$\lim_{x \to \infty} \left(1.5 + \frac{4}{x}\right) = 1.5$$

This is the mathematical shorthand for saying "if x is a large number, then $1.5 + (4/x)$ is close to 1.5." Geometrically, this means the second coordinates of points on the graph are very close to 1.5 when x is big. The points are very close to those of height 1.5. Figure 4–24 gave us the general shape of the graph, and now if we draw the horizontal line $y = 1.5$ with a dotted line, we can see that the tail of the graph on the right must look like Figure 4–25.

The graph is getting close to the line $y = 1.5$. When this happens, we say that the line $y = 1.5$ is an **asymptote** of the graph. Because the line $y = 1.5$ is horizontal, we call the line $y = 1.5$ a **horizontal asymptote** of the graph.

Warning! The asymptote is the **line** $y = 1.5$. Asymptotes are lines, not numbers.

Definition

If $\lim_{x \to \infty} f(x) = b$, then the line whose equation is $y = b$ is called a **horizontal asymptote** of the graph of f.

Now, we return to our example of the average cost $AC(x) = 1.5 + (4/x)$. The term $4/x$ is the portion of the fixed cost shared by each of the x items produced; the more items, the smaller this share is for each individual item. Conversely, the fewer items produced, the larger the share per item. We assume the units are such that x can take on values between 0 and 1 (for example, the item could be gallons of gasoline, and x could be measured in units of 10,000). What happens to $1.5 + (4/x)$ if x is near 0? If x is near 0, then $1/x$ is a large number—if x is 0.00001, then $1/x$ is 100,000. The smaller x is, the larger $1/x$ is, and in fact, $1/x$ gets arbitrarily large. We write this as

$$\lim_{x \to 0^+} \frac{1}{x} = \infty$$

Note that $x \to 0^+$ means the right-hand limit at 0. The graph of $1.5 + (4/x)$ is given in Figure 4–26.

When x is close to 0, the graph is nearly vertical; it is very close to the vertical line $x = 0$ (the y-axis). We again use the word asymptote to describe the situation

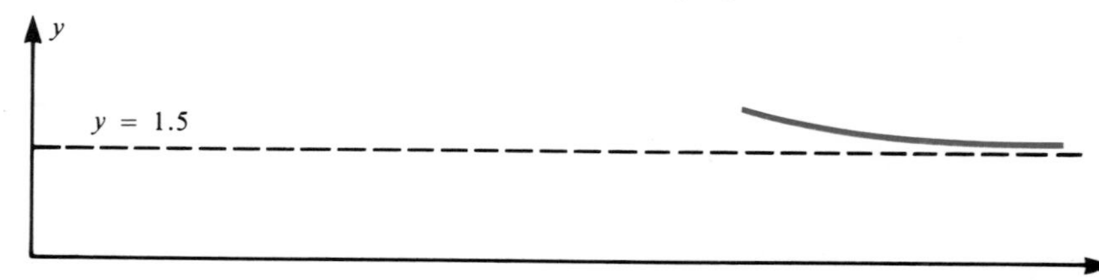

Figure 4–25

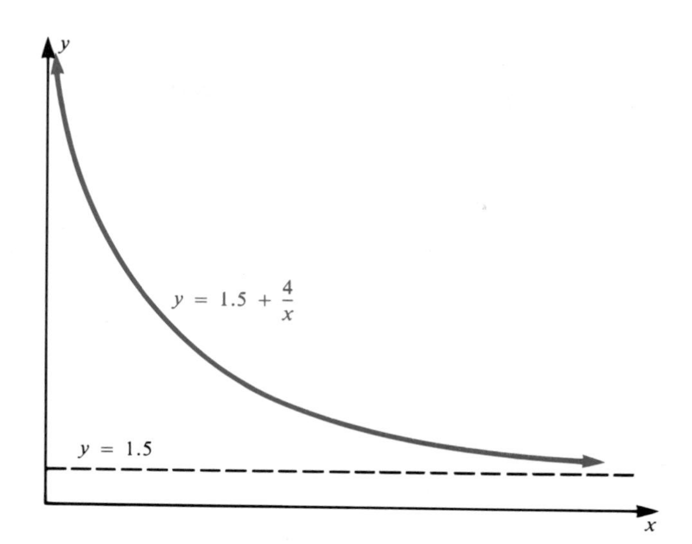

Figure 4–26

when the graph is close to a line. In this instance, because the line is vertical, we say that the line $x = 0$ is a **vertical asymptote** of the graph of $AC(x)$.

Definition

> If $\lim_{x \to a+} f(x) = \infty$, then the vertical line whose equation is $x = a$ is called a **vertical asymptote** of the graph of f.

Remember that the general idea behind the definition of asymptote is this: If the graph of f is very close to a line L whenever $|x|$ or $|f(x)|$ is large, then the line L is an asymptote of the graph. We will now examine asymptotes more closely, starting with vertical asymptotes.

Vertical Asymptotes

There are four pictures that could be described by saying "the line $x = a$ is a vertical asymptote of the graph $y = f(x)$."

In Figure 4–27(i), the picture is further described by saying

$$\lim_{x \to a^+} f(x) = \infty$$

In Figure 4–27(iii), as x approaches a from the right, the values of f drop off steeply; $f(x)$ is negative but very large in absolute value. We describe this situation by saying $\lim_{x \to a^+} f(x) = -\infty$. Figure 4–27(ii) and Figure 4–27(iv) are described by left-hand limits. We now expand the definition of vertical asymptote given above.

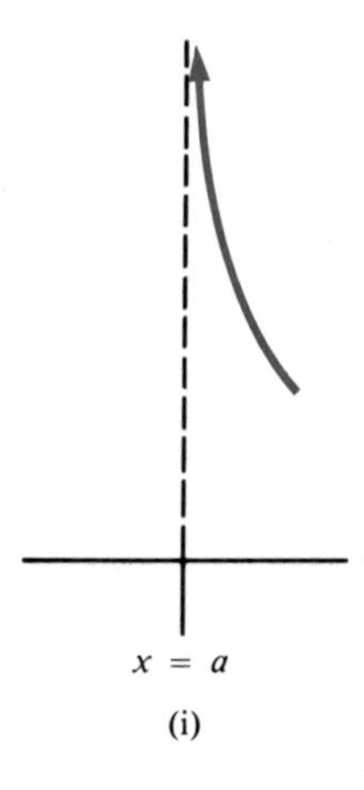

(i)

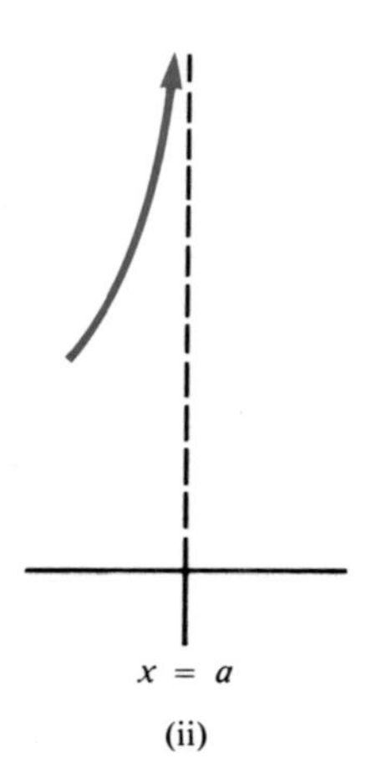

(ii)

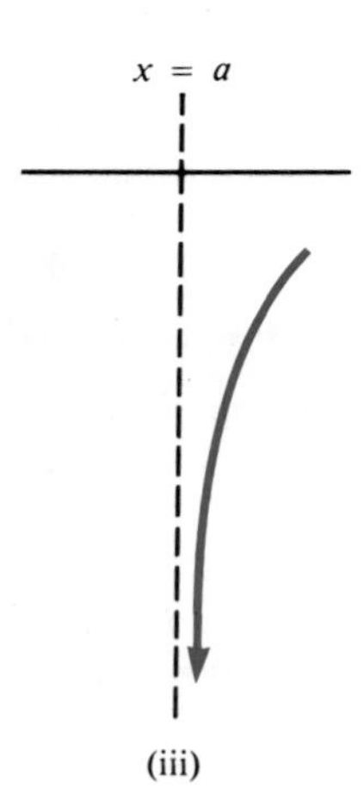

(iii)

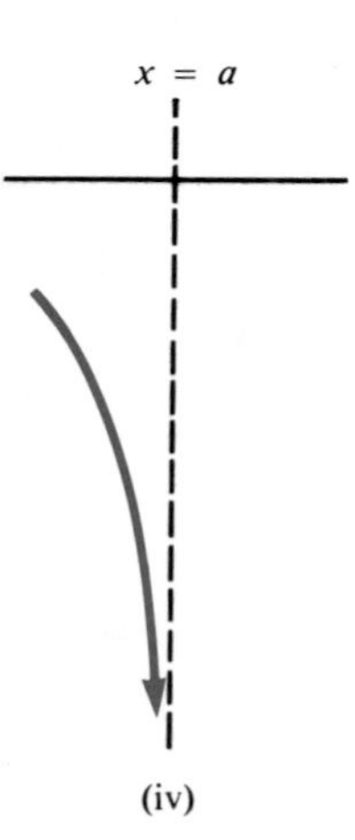

(iv)

Figure 4–27

Definition

The line $x = a$ is a **vertical asymptote** of the graph of f if at least one of the following four conditions is met:

(i) $\lim_{x\to a^+} f(x) = \infty$,

(ii) $\lim_{x\to a^-} f(x) = \infty$,

(iii) $\lim_{x\to a^+} f(x) = -\infty$, or

(iv) $\lim_{x\to a^-} f(x) = -\infty$.

How do you tell if a graph has a vertical asymptote? Like most general questions, this one doesn't have a short answer other than "check the definition."

There is one situation however which will indicate that a vertical asymptote exists; you should then use limits to see which of the four types of vertical asymptotes the graph has.

Theorem If $f(x) = P(x)/Q(x)$ where $P(x)$ and $Q(x)$ are polynomials, with $Q(a) = 0$ but $P(a) \neq 0$, then the line $x = a$ is a vertical asymptote of the graph of f.

Example 1 (*Compare Exercises 1 and 19*)

Let $f(x) = \dfrac{x+4}{x-2}$. Discuss the vertical asymptotes of the graph of f.

Solution $f(x) = \dfrac{P(x)}{Q(x)}$, where $P(x) = x + 4$ and $Q(x) = x - 2$. Now, $Q(x) = 0$ if $x = 2$. Furthermore, $P(2) = 6 \neq 0$. The theorem quoted above guarantees that the line $x = 2$ is a vertical asymptote.

We now use one-sided limits to get a better understanding of the graph. Remember that Figure 4–27 presented four pictures of what the graph can look like near a vertical asymptote.

First, we investigate

$$\lim_{x \to 2^+} \frac{x+4}{x-2}$$

If x is close to 2 and bigger than 2, then both $x + 4$ and $x - 2$ are positive. Thus,

$$f(x) = \frac{x+4}{x-2}$$

is the quotient of positive numbers, so $f(x)$ is positive. Just to the right of $x = 2$, the graph must be above the x-axis, and so looks like Figure 4–27(i). Thus,

$$\lim_{x \to 2^+} \frac{x+4}{x-2} = \infty$$

Next, we examine

$$\lim_{x \to 2^-} \frac{x+4}{x-2}$$

If x is a little bit less than 2, then $x + 4$ is positive and $x - 2$ is negative. Thus,

$$f(x) = \frac{x+4}{x-2}$$

is of the form

$$\frac{\text{positive number}}{\text{negative number}}$$

and is therefore negative. Thus, $\lim_{x \to 2^-} f(x) = -\infty$, and the graph must look like Figure 4–27(iv).

Example 2 (*Compare Exercises 11 and 23*)
Let

$$f(x) = \frac{x^2 - 6x + 5}{x^2 - 4x + 3}$$

Discuss the vertical asymptotes of the graph of f.

Solution Here $P(x) = x^2 - 6x + 5 = (x - 5)(x - 1)$ and $Q(x) = x^2 - 4x + 3 = (x - 3)(x - 1)$. The denominator of $f(x)$ is zero when $x = 3$ and when $x = 1$. These give two **possible** lines for vertical asymptotes. Note however that $P(1) = 0$ as well as $Q(1) = 0$. The theorem does not apply in this situation. We must look at the limit of $f(x)$ at $x = 1$.

$$\lim_{x \to 1} f(x) = \lim_{x \to 1} \frac{(x - 5)(x - 1)}{(x - 3)(x - 1)} = \lim_{x \to 1} \frac{x - 5}{x - 3} = \frac{-4}{-2} = 2$$

None of the four limit conditions for an asymptote is true, so the line $x = 1$ is **not** a vertical asymptote. Figure 4–28 shows the graph of f for x close to 1.

Next, check the line $x = 3$; $P(3) = -2 \neq 0$, so the line $x = 3$ is a vertical asymptote.

Again, we look at the right- and left-hand limits to see what type of infinite limits we have in this case.

We have seen that we can rewrite $f(x)$ as $\dfrac{x - 5}{x - 3}$ if $x \neq 1$.

We investigate the right-hand limit, namely

$$\lim_{x \to 3^+} f(x) = \lim_{x \to 3^+} \frac{x - 5}{x - 3}$$

1. The fact that $\lim_{x \to 3^+} x - 5 = -2 \neq 0$ and $\lim_{x \to 3^+} x - 3 = 0$ tells us that

$$\lim_{x \to 3^+} \frac{x - 5}{x - 3}$$

will be infinite. We must decide on the sign.

2. If x is a little bit bigger than 3, then $x - 5$ is close to -2, and so the numerator is negative. Because $x > 3$, $x - 3$ is positive so the denominator is positive. A negative number divided by a positive number is negative. Thus, when x is a little bit bigger than 3, $f(x)$ is negative.
3. Putting (1) and (2) together, we have

$$\lim_{x \to 3^+} \frac{x - 5}{x - 3} = -\infty$$

We now know that the line $x = 3$ is a vertical asymptote of the graph of f, and that just to the right of $x = 3$, the graph looks like Figure 4–27(iii).

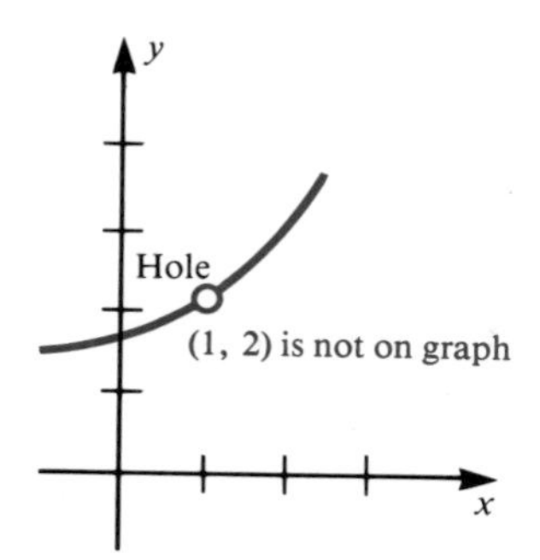

Figure 4–28

Example 3 (*Compare Exercise 27*)
Discuss the vertical asymptotes of the graph of f if

$$f(x) = \frac{6}{(x+1)(x-4)}$$

Solution $f(x)$ is of the form $P(x)/Q(x)$, with $P(x) = 6$ and $Q(x) = (x+1)(x-4)$; $Q(x) = 0$ when $x = -1$ and when $x = 4$. So, we investigate the behavior of $f(x)$ near $x = -1$ and $x = 4$.

A complete investigation of the situation requires us to look at four one-sided limits: $\lim_{x \to -1^+} f(x)$, $\lim_{x \to -1^-} f(x)$, $\lim_{x \to 4^+} f(x)$, and $\lim_{x \to 4^-} f(x)$. We find two of these limits now.

First,

$$\lim_{x \to -1^+} f(x) = \lim_{x \to -1^+} \frac{6}{(x+1)(x-4)}$$

1. The limit of the numerator is not 0, while the limit of the denominator is 0. Therefore, the limit of the quotient is infinite, and we must determine the sign.
2. We assume that x is just a little bit bigger than -1. The numerator is always 6, which is positive. Now look at the denominator. $x + 1$ is positive and $x - 4$ is negative, so the denominator, $(x+1)(x-4)$, is negative. (Remember that x is a little bit bigger than -1.)

   Thus, $f(x)$ is of the form $\dfrac{\text{positive number}}{\text{negative number}}$, and so $f(x)$ is negative.
3. Steps (1) and (2) imply $\lim_{x \to -1^+} f(x) = -\infty$. To the right of $x = -1$, the graph looks like Figure 4–29.

Second, we look at $\lim_{x \to 4^+} f(x)$.

1. $\lim_{x \to 4^+} \dfrac{6}{(x+1)(x-4)}$ is of the form $\dfrac{\text{nonzero limit}}{\text{zero limit}}$, so the limit is not a number. We investigate the sign of the infinite limit.

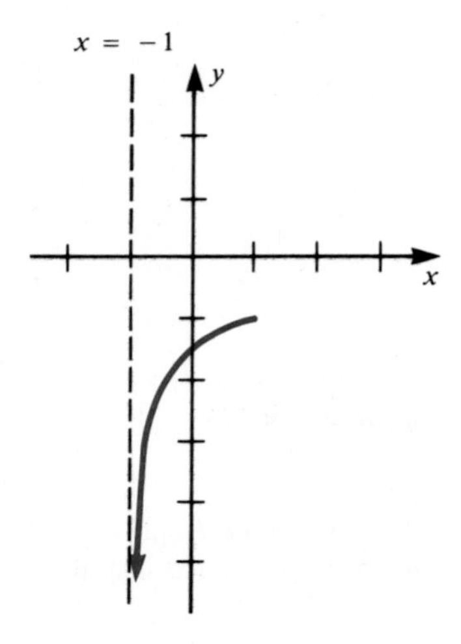

Figure 4–29

2. If x is a little bit bigger than 4, then

   6 is positive
   $x + 1$ is positive
   $x - 4$ is positive, so

$$\frac{6}{(x+1)(x-4)}$$

is positive. Thus,

$$\lim_{x \to 4^+} \frac{6}{(x+1)(x-4)} = \infty$$

To the right of $x = 4$, the graph looks like Figure 4–30.

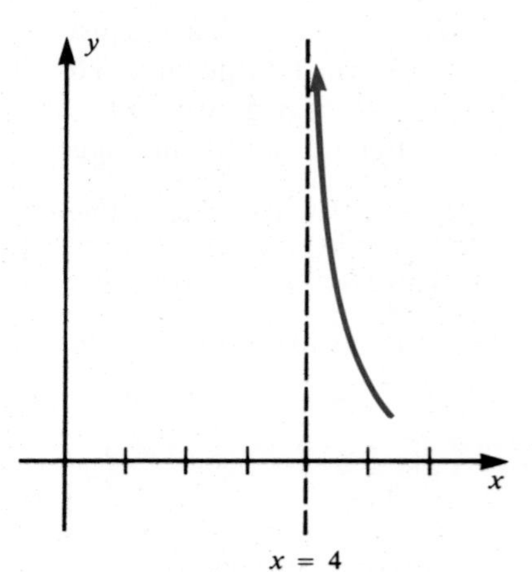

Figure 4–30

Warning! Example 2 shows that you cannot determine whether there is an asymptote just by looking at the denominator.

Example 3 shows that you cannot evaluate $\frac{6}{0}$ as ∞. You cannot claim that a positive number divided by 0 is positive infinity. To compute the correct sign, you must look at the sign of $f(x)$ on each side of the asymptote.

Horizontal Asymptotes

Vertical asymptotes describe the behavior of a function near a point in its domain where the function "blows up." That is, vertical asymptotes help us to understand what the graph looks like when $|f(x)|$ is arbitrarily large for values of x near some fixed number. Horizontal asymptotes describe the long term behavior of the function. That is, horizontal asymptotes help us understand what the graph looks like when $|x|$ is arbitrarily large, and $f(x)$ is near some fixed number. To describe the behavior of f to the far right of its graph, we write $\lim_{x \to \infty} f(x)$; to describe the behavior of f to the far left of its graph, we write $\lim_{x \to -\infty} f(x)$.

A graph may have no horizontal asymptotes, as in Figure 4–31(i); one horizontal asymptote, as in Figure 4–31(ii); or two horizontal asymptotes, as in Figure 4–31(iii).

Many functions that give rise to horizontal asymptotes in applications are of the form $f(x) = P(x)/Q(x)$, where $P(x)$ and $Q(x)$ are polynomials. There is a nice theorem that describes the situation completely in three cases.

Theorem

Let $P(x) = a_n x^n + \cdots + a_0$, $a_n \neq 0$ (the degree of $P(x)$ is n)
Let $Q(x) = b_m x^m + \cdots + b_0$, $b_m \neq 0$ (the degree of $Q(x)$ is m)

Case 1. If $m > n$,

$$\lim_{x \to \infty} \frac{P(x)}{Q(x)} = 0 \qquad \text{and} \qquad \lim_{x \to -\infty} \frac{P(x)}{Q(x)} = 0$$

Thus, the line $y = 0$ is the horizontal asymptote.

Case 2. If $m = n$,

$$\lim_{x \to \infty} \frac{P(x)}{Q(x)} = \frac{a_n}{b_m} = \lim_{x \to -\infty} \frac{P(x)}{Q(x)}$$

Thus, the line

$$y = \frac{a_n}{b_m}$$

is the horizontal asymptote.

Case 3. If $m < n$, then

$$\lim_{x \to \infty} \frac{P(x)}{Q(x)} \quad \text{and} \quad \lim_{x \to -\infty} \frac{P(x)}{Q(x)}$$

are not finite.

Thus, the graph does not have a horizontal asymptote.

We will not pursue Case 3 further.

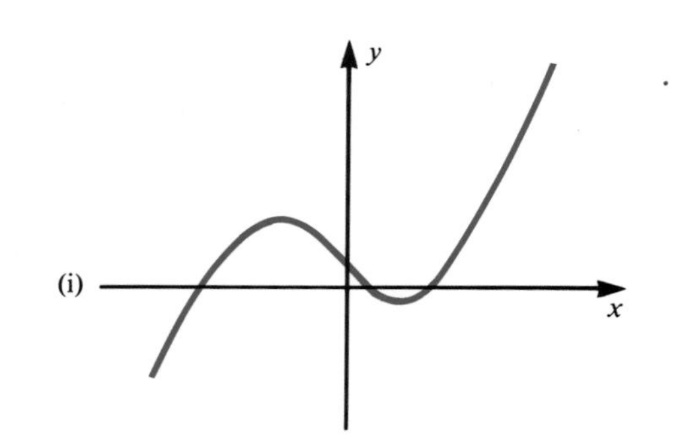

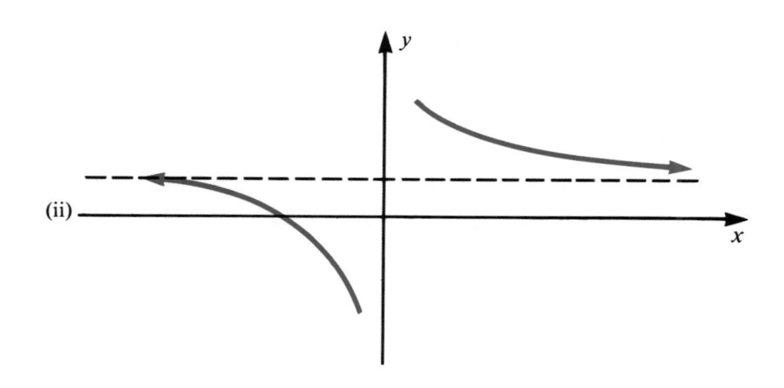

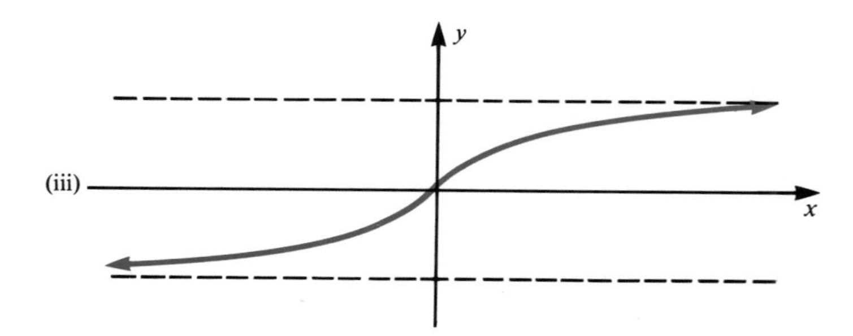

Figure 4–31

Example 4 (*Compare Exercise 9*)
Discuss the horizontal asymptotes of the graph of f if

$$f(x) = \frac{3x + 5}{4x^2 - 9x + 8}$$

Solution Here $f(x) = \frac{P(x)}{Q(x)}$, where $P(x) = 3x + 5$ has degree 1, and $Q(x) = 4x^2 - 9x + 8$ has degree 2. [The degree of the denominator] > [the degree of the numerator], so this example falls into Case 1.

$$\lim_{x \to \infty} \frac{3x + 5}{4x^2 - 9x + 8} = 0$$

and the x-axis is a horizontal asymptote (to the far right).

$$\lim_{x \to -\infty} \frac{3x + 5}{4x^2 - 9x + 8} = 0$$

and the x-axis is a horizontal asymptote (to the far left).

Example 5 (*Compare Exercise 17*)
Discuss the horizontal asymptotes of the graph of f if

$$f(x) = \frac{3x^2 + 9x - 5}{5x^2 - x + 4}$$

Solution Here $f(x) = \frac{P(x)}{Q(x)}$, where $P(x) = 3x^2 + 9x - 5$ has degree 2, and $Q(x) = 5x^2 - x + 4$ also has degree 2. [The degree of the denominator] = [the degree of the numerator], so this example falls into Case 2. Thus, $\lim_{x \to \infty} \frac{P(x)}{Q(x)}$ = the ratio of the coefficients of the highest degree terms = 3/5.

The line $y = \frac{3}{5}$ is a horizontal asymptote (to the far right). Similarly,

$$\lim_{x \to -\infty} \frac{3x^2 + 9x - 5}{5x^2 - x + 4} = \frac{3}{5}$$

so the line $y = \frac{3}{5}$ is a horizontal asymptote (to the far left).

Example 6 (*Compare Exercise 3*)
Discuss the horizontal asymptotes of the graph of f if

$$f(x) = \frac{(x - 3)(x + 1)}{x - 2}$$

Solution Here $f(x) = \frac{P(x)}{Q(x)}$, where $P(x) = (x - 3)(x + 1)$ has degree 2, and $Q(x) = x - 2$ has degree 1. This function falls into Case 3, and the graph has no horizontal asymptotes.

Example 7 (*Compare Exercise 43*)
If a firm's average-cost function for producing x items is given by

$$AC(x) = \frac{3x + 42}{x}$$

what is $\lim_{x \to \infty} AC(x)$? Sketch a graph of $AC(x)$ for $x > 0$.

Solution We can apply the theorem with $P(x) = 3x + 42$ (degree 1) and $Q(x) = x$ (also degree 1).

The degrees of the denominator and the numerator are the same, so

$$\lim_{x \to \infty} \frac{3x + 42}{x} = 3$$

When a large amount is being produced, the average cost is close to 3.

The line $x = 0$, which is the y-axis, is a vertical asymptote of AC. Since

$$\frac{3x + 42}{x} > 0$$

when x is just a little bit bigger than 0, $\lim_{x \to 0^+} AC(x) = \infty$.

Next, writing $AC(x) = 3 + (42/x)$, we have $AC'(x) = -42/x^2 < 0$. Thus, AC is decreasing on the whole interval $(0, \infty)$. Finally, $AC''(x) = 84/x^3 > 0$ when $x > 0$, so the graph is concave up on $(0, \infty)$. The graph of AC is given in Figure 4–32.

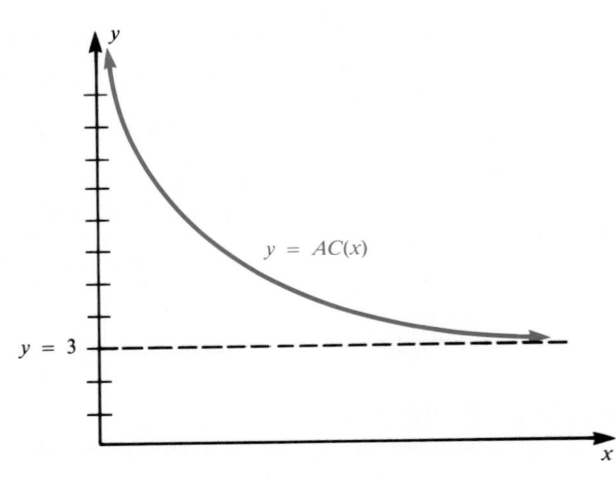

Figure 4–32

For the last example in this section, we show how to put all these pieces together.

Example 8 (*Compare Exercise 35*)
Determine any vertical and horizontal asymptotes, and sketch the graph of f if

$$f(x) = \frac{3x^2 + 6}{x^2 - 4}$$

Solution $f(x) = P(x)/Q(x)$ where $P(x) = 3x^2 + 6$ and $Q(x) = x^2 - 4$. The lines $x = 2$ and $x = -2$ are vertical asymptotes because $Q(2) = 0$ while $P(2) \neq 0$, and $Q(-2) = 0$ while $P(-2) \neq 0$. Note that $3x^2 + 6$ is always positive. We have

$$\lim_{x \to 2^+} \frac{3x^2 + 6}{x^2 - 4} = \infty$$

and

$$\lim_{x \to 2^-} \frac{3x^2 + 6}{x^2 - 4} = -\infty$$

Furthermore, $\lim_{x \to -2^+} f(x) = -\infty$ and $\lim_{x \to -2^-} f(x) = \infty$. The line $y = 3$ is a horizontal asymptote because the degree of $P(x) = 2 =$ the degree of $Q(x)$, and the ratio of the coefficients of x^2 is $\frac{3}{1} = 3$.

To sketch the graph of f we look at $f'(x)$.

$$f'(x) = \frac{(x^2 - 4) \cdot 6x - (3x^2 + 6) \cdot 2x}{(x^2 - 4)^2}$$

$$= \frac{-36x}{(x^2 - 4)^2}$$

$f'(2)$ and $f'(-2)$ do not exist, and $f'(x) = 0$ when $x = 0$. The sign chart for f' is as follows:

sign of f'	+	-2	+	0	−	2	−

$f(0) = -3/2$ is a relative maximum.

Notice, the fact that f is increasing on $(-\infty, -2)$ must agree with the asymptotic behavior of f near $x = 2$. The graph of f is given in Figure 4–33.

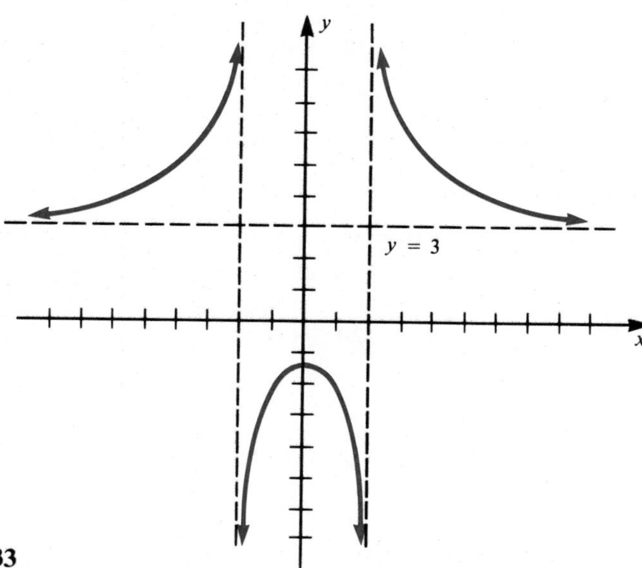

Figure 4–33

4–4 Exercises

I.

Use the theorems from this section on the existence of asymptotes to give the equation of any vertical or horizontal asymptotes that the following functions may have.

1. (*See Example 1*) $f(x) = \dfrac{x+5}{x-6}$

2. $f(x) = \dfrac{2x-4}{x-2}$

3. (*See Example 6*) $f(x) = \dfrac{x+1}{x^2-9}$

4. $f(x) = \dfrac{x-8}{x^2-16}$

5. $f(x) = \dfrac{3x+7}{(2x+6)(x-4)}$

6. $f(x) = \dfrac{x+4}{(2x+10)(x-7)}$

7. $f(x) = \dfrac{4x-7}{5x+10}$

8. $f(x) = \dfrac{3x^2+8}{x^2+5}$

9. (*See Example 4*) $f(x) = \dfrac{4x-9}{x^2+1}$

10. $f(x) = \dfrac{x^2-9}{x^3+8}$

Compute the following limits.

11. (*See Example 2*) $\displaystyle\lim_{x\to 3^-} \frac{x-5}{x-3}$

12. $\displaystyle\lim_{x\to 2^+} \frac{x-7}{x-2}$

13. $\displaystyle\lim_{x\to 5^+} \frac{6-2x}{x-5}$

14. $\displaystyle\lim_{x\to -3^+} \frac{x+1}{x+3}$

15. $\displaystyle\lim_{x\to \infty} \frac{3x^2-9}{5x^2+8x+2}$

16. $\displaystyle\lim_{x\to -\infty} \frac{3x+9}{x^2+10x-8}$

17. (*See Example 5*) $\displaystyle\lim_{x\to -\infty} \frac{5x+7}{10x+4}$

18. $\displaystyle\lim_{x\to \infty} \frac{4x^2+3x-9}{3x^2+2x-5}$

II.

Identify the vertical and horizontal asymptotes for each of the following functions, and compute the appropriate limits.

19. (*See Example 1*) $f(x) = \dfrac{3x+2}{x+5}$

20. $f(x) = \dfrac{6}{x-4}$

21. $f(x) = \dfrac{3x+2}{x^2+4}$

22. $f(x) = \dfrac{x+1}{2x+10}$

23. (*See Example 2*) $f(x) = \dfrac{x-1}{x^2-3x+2}$

24. $f(x) = \dfrac{x}{x^2+3x}$

25. $f(x) = \dfrac{x^2 - 5x + 4}{x^2 - 3x - 4}$

26. $f(x) = \dfrac{x^2 - x - 6}{x^2 - 3x - 10}$

Compute each of the following limits.

27. (*See Example 3*)
$\lim_{x \to -3^+} \dfrac{1}{(x + 5)(x + 3)}$

28. $\lim_{x \to 2^-} \dfrac{1}{(x + 3)(x - 2)}$

29. $\lim_{x \to -4^-} \dfrac{x - 2}{(x + 4)(x - 1)}$

30. $\lim_{x \to -1^-} \dfrac{x}{(x + 1)(x + 3)}$

31. $\lim_{x \to 2^-} \dfrac{x - 3}{(x - 2)(x + 5)}$

32. $\lim_{x \to 6^-} \dfrac{x - 2}{(x - 6)(x + 1)}$

III.

Sketch the graphs of the following functions. Draw any vertical and horizontal asymptotes using dotted lines.

33. $f(x) = \dfrac{1}{x^2}$

34. $f(x) = \dfrac{5}{x + 2}$

35. (*See Example 8*)
$f(x) = \dfrac{2x - 8}{x - 3}$

36. $f(x) = \dfrac{2x - 6}{x + 1}$

37. $f(x) = 1 + \dfrac{4}{x^2 + 1}$

38. $f(x) = \dfrac{6}{x^2 - 9}$

39. $f(x) = \dfrac{x^2 - 6x + 5}{x^2 - 3x + 2}$

40. $f(x) = \dfrac{1}{x^2 + x - 6}$

41. $f(x) = \dfrac{3}{x^2 + 4x}$

42. $f(x) = \dfrac{3x^2 - 16}{x^2 - 16}$

43. (*See Example 7*) A manufacturing company has a daily cost in dollars given by $C(x) = 8x + 1200$, where x is the number of units manufactured each day.
(a) What is the approximate average cost per item for large scale production?
(b) Sketch the graph of the average cost function.

4–5 Optimization in Applications

Getting Started
A Six-Step Process

The techniques of Section 1, 2, and 3 of this chapter showed you how to find the maximum and minimum values of a function when you already know the function. One of the major difficulties in applying mathematics, however, is the construction of the mathematical function that adequately represents, or models, the real world problem to be solved. In short, a major difficulty is "getting started." These real world problems go by various names—applications, models, word problems, and even, in an attempt to reduce students' "math anxiety," story problems. Whatever they are called, they are important because life is a word problem; one seldom meets a function just sitting around waiting for someone to find its maximum value.

Getting Started

The key to doing these problems is getting your paper dirty. You may get the impression by reading texts or listening to lectures that when the presentation of the solution begins, the writer or lecturer already knows what the answer is. Don't read or listen to the solutions as if they are finished products. Instead, think of a solution as a process; it is the way the problem is being attacked. The final answer is only one part of the solution. You shouldn't expect to know what the answer is before you write anything down. Some students don't give themselves a fair chance to do word problems because they don't view the solution process in a step-by-step manner. We have broken down the solution process into six steps. So that the steps themselves don't become too abstract, we begin by showing them in a concrete setting as we solve the following problem. We will give the general step and then show how it works in this particular problem.

The Six-Step Process

Example 1 (*Compare Exercise 1*)
A university wants to build a rectangular patio next to the student union building. As part of its beautification program, the university will transplant some bushes and landscape three sides of the patio with these bushes. The student union building will serve as the fourth side of the rectangle. The university already has enough bushes to allow 800 feet for the three landscaped sides. What is the area of the largest patio the university can build without having to purchase more bushes?

Solution

STEP 1 **Read through the whole problem. Then go back and find out what quantity the problem is asking you to maximize or minimize**. (Be on the look out for question marks, or words like "find" or "determine.") Write down what it is you are trying to find.

There are a lot of words in this problem (on purpose—we want to show you how to pick out essentials and how to effectively ignore the non-essentials). The "?" follows the last sentence. The question is "What is the area of the largest patio....?" Therefore the answer will be in the form "The maximum area of the patio is...."

STEP 2 **Write down a general formula or equation that represents the quantity to be maximized or minimized**.

Here the question involves area, so we need some formula involving area. Read the problem again—do we want the area of a circle, a rectangle, a triangle, or what? The patio is to be a rectangle!

area of rectangle = length × width
(or base × height, or some equivalent statement).

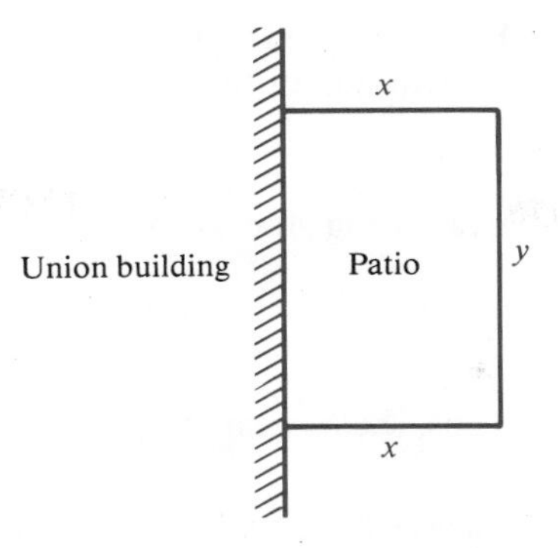

Figure 4–34

STEP 3 **Try to draw a picture or a chart representing the data in the problem.** Introduce some letters that represent the variables in your formula from Step 2. Make sure the units used in the problem are consistent with each other.

For this problem we draw a rectangle. One side of the rectangle will be a building, so we indicate that in our drawing. Next, we use A to represent the area of the rectangle, x to represent the width of the rectangle, and y for the length. The unit of measurement in the problem was feet, so x and y are in feet. All units are feet or square feet. We now have $A = x \cdot y$, and we want to find the largest value of A. See Figure 4–34.

STEP 4 **Use information from the problem to express the quantity you're looking for in terms of only one variable.** (If you do this, you have then framed the problem in such a way that you can use the techniques of the first three sections of this chapter to solve the problem.)

Here, we know we have 800 feet of bushes to put around the three sides labeled x, y, and x. Thus, $2x + y = 800$. We use this relation to solve for y: $y = 800 - 2x$. Next, we substitute this expression for y into our formula from Step 3.

$$\begin{aligned} A = x \cdot y &= x(800 - 2x) \\ &= 800x - 2x^2 \end{aligned}$$

We have a rule that expresses A in terms of x. Now we need the domain of the function. Because x represents width, we must have $x \geq 0$. Similarly, $y \geq 0$. Using $y = 800 - 2x$, we must have $800 - 2x \geq 0$ or $400 \geq x$. Thus, the domain of the function is $0 \leq x \leq 400$.

Now we have completely determined a function of one variable, and we wish to maximize that function.

STEP 5 **Use calculus to maximize or minimize** (whichever you're after) **the function from Step 4.**

Here, $A(x) = 800x - 2x^2$, so $A'(x)$ exists for all x, and $A'(x) = 800 - 4x$. Set

$$A'(x) = 800 - 4x = 0$$

and solve

$$4x = 800$$

$$x = 200 \text{ is the only critical number.}$$

Because the domain of A is the closed and bounded interval $[0, 400]$, we can use the candidates test. Evaluating $A(x)$ at the endpoints, we have

$$A(0) = 0 \qquad \text{and} \qquad A(400) = 0$$

Evaluating $A(x)$ at the critical number, we have

$$\begin{aligned} A(200) &= 800 \cdot 200 - 2(200)^2 \\ &= 160{,}000 - 80{,}000 = 80{,}000 \end{aligned}$$

Thus, the maximum value of A occurs when $x = 200$.

STEP 6 **Answer the question. Use correct units.**

The question does not ask for the dimensions of the patio; it asks for the area of the patio.

$$A(200) = 80{,}000$$

ANSWER The maximum area of the patio is 80,000 sq. ft.

We will go through this six-step process more briskly in the next example.

Example 2 (*Compare Exercise 3*)
An apartment complex has 120 units. The manager can rent all 120 units if the rent is \$280 per month. Some preliminary data indicates that for every \$20 increase in the monthly rent, 5 units will become vacant. What rent should the manager charge to achieve the greatest revenue?

Solution

STEP 1 We want to maximize revenue.

STEP 2 Revenue = (number of units rented) · (rent per apartment)

STEP 3 Let R = revenue, x = number of units rented, and p = rental price per unit. All the units in the problem are dollars and dollars per month, so the units match. It's hard to draw a meaningful picture for this problem, but here is a table showing some data.

x	120	115	110
p	280	300	320

$$R = x \cdot p$$

STEP 4 The problem states that if $\Delta p = 20$, then $\Delta x = -5$. Thus, $\dfrac{\Delta p}{\Delta x} = \dfrac{20}{-5} = -4$ is a constant; x and p are linearly related. If we treat p as a linear function of x, then the slope of the graph is -4. The problem also states that if $x = 120$, then $p = 280$. We use the point-slope equation of the line and then solve for p.

$$p - 280 = -4(x - 120)$$
$$p - 280 = -4x + 480$$
$$p = -4x + 760$$

(See remark immediately following this example.)

$$R = x \cdot p \text{ now becomes } R = x(-4x + 760) = -4x^2 + 760x$$

x is the number of units rented, so $0 \leq x \leq 120$.

STEP 5 Maximize $R = -4x^2 + 760x$, $0 \leq x \leq 120$. R is differentiable for all x in $(0, 120)$, and

$$R'(x) = -8x + 760$$

Set

$$-8x + 760 = 0$$

and solve.

$$-8x = -760$$
$$x = 95$$

The only critical value is $x = 95$.

Because there is only one critical number, we can use the second derivative test. $R''(x) = -8$, so $R''(95) = -8 < 0$. Thus, $x = 95$ gives a relative maximum value to R. Since $x = 95$ is the only critical point, $x = 95$ gives the maximum value to R.

STEP 6 The problem asked for the value of p that maximizes R; we have found the value of x that maximizes R. In Step 4, we found that

$$p = -4x + 760$$

so when $x = 95$,

$$p = -4(95) + 760 = -380 + 760 = 380$$

ANSWER Profit will be maximized when the manager charges $380 per month.

We would like to make some remarks about Step 4. You don't need to do this step exactly as we did. For instance, you could use two points from the chart and write

$$\frac{p - 280}{x - 120} = \frac{300 - 280}{115 - 120} = \frac{20}{-5} = -4$$

Then simplify

$$\frac{p - 280}{x - 120} = -4$$

$$p - 280 = -4(x - 120)$$
$$p = -4x + 760$$

As another instance, you could compute $\frac{\Delta x}{\Delta p}$. Just remember to use the same variable on the top in both fractions.

$$\frac{120 - 115}{280 - 300} = \frac{x - 115}{p - 300}$$

$$\frac{-1}{4} = \frac{x - 115}{p - 300}$$

$$-1(p - 300) = 4(x - 115) = 4x - 460$$

$$p = -4x + 760$$

Or, you could solve for x in terms of p, and write the revenue as a function of p.

You may also have noticed that Examples 1 and 2 presented us with the same situation in Step 5. In both cases, there is only one critical number, and the domain is of the form $[a, b]$. In Example 1, we used the candidates test, which depends on the form of the interval. In Example 2, we found the maximum by a test that depends on the function having only one critical point. We also could have used the first derivative test in either example. Do not put yourself in a bind by thinking that there is only one correct way to do a problem.

Example 3 (*Compare Exercise 7*)

The design department wants a certain company to adopt a new company logo. The new logo has essentially a square shape, so they want the company to design new boxes with a square front. The volume of the box must be 108 cubic inches. Also, the boxes must have a bottom that is better made than the sides and top. The bottom costs 10¢ per square inch, whereas the top and sides only cost 2¢ per square inch. What are the dimensions of the box that minimize its cost?

Solution

STEP 1 We want to minimize the cost of the box.

STEP 2 Volume of box = (length)(height)(depth); cost of a side = (cost per square inch)(area in square inches); total cost = (cost of top) + (cost of bottom) + (cost of four sides).

STEP 3 Draw a picture of a box. Let x = the length of the box, y = its height, z = its depth. See Figure 4–35. All the cost units are cents per square inch, and all length units are inches, so the units match.

STEP 4 We need a cost function.

The top and bottom of the box are the same size; they are x inches by z inches. The cost of the top is $2xz$ cents. The cost of the bottom is $10xz$ cents. The total cost of the top and bottom is $10xz + 2xz = 12xz$ cents.

The front and back of the box are both x inches by y inches; each costs $2xy$ cents. The total cost for the front and back is $2xy + 2xy = 4xy$ cents.

The other two sides of the box are both y inches by z inches; each costs $2yz$ cents. The total cost for these two sides is $4yz$ cents.

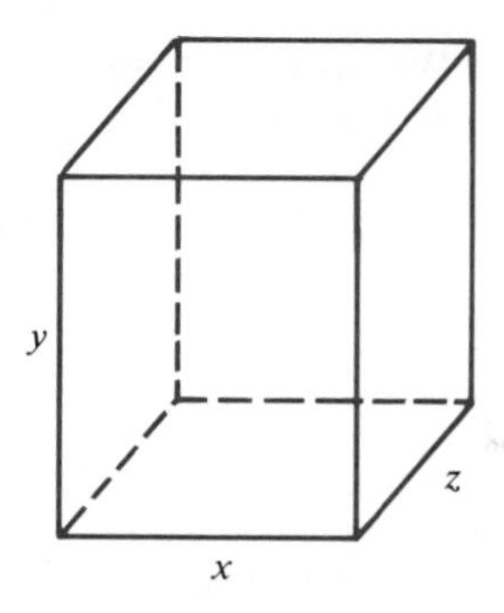

Figure 4–35

Our first expression for the cost is

$$C = 12xz + 4xy + 4yz$$

Next, we need to use some information in this particular problem to write C as a function of one variable. The design department wants a square front, so $x = y$. Now we can eliminate y and write $C = 12xz + 4xx + 4xz = 16xz + 4x^2$.

We have used one piece of information (a square front) to get rid of one variable. But we still have C expressed in terms of x and z. We must use another piece of information to solve for one of these variables in terms of the other. The volume must be 108, so $xyz = 108$. Since $x = y$, we can write

$$xxz = 108$$
$$x^2 z = 108$$

or

$$z = \frac{108}{x^2}$$

Now we can write C as a function of x.

$$C(x) = 16x\left(\frac{108}{x^2}\right) + 4x^2 = \frac{1728}{x} + 4x^2$$

The domain of C is all $x > 0$.

STEP 5 Minimize C.

$C'(x) = \dfrac{-1728}{x^2} + 8x$, so $C'(x)$ exists for all $x > 0$. To find the critical numbers, set $C'(x) = 0$ and solve.

$$\frac{-1728}{x^2} + 8x = 0$$

$$-1728 + 8x^3 = 0$$
$$8x^3 = 1728$$
$$x^3 = 216$$
$$x = 6$$

We only have one critical number, and we use the second derivative test at $x = 6$. Note that we cannot use the candidates test.

$$C''(x) = \frac{3456}{x^3} + 8$$

so

$$C''(6) = \frac{3456}{216} + 8 > 0$$

C has a relative minimum at $x = 6$. Furthermore, C is continuous for $x > 0$ with only one critical number, which gives a relative minimum. Therefore, the minimum of C occurs when $x = 6$.

STEP 6 The problem asks for the dimensions of the least expensive box: $y = x$, so the sides are 6 inches by 6 inches.

$$z = \frac{108}{x^2} = \frac{108}{36} = 3$$

The depth of the box is 3 inches.

4–5 Exercises

1. (*See Example 1*) A farmer wishes to enclose a rectangular field with a fence. One side of the field is bounded by a straight river; this side of the field does not require fencing. What is the maximum area the farmer can enclose with 4000 feet of fencing?
2. A rancher wishes to construct a pen to keep his pigs and chickens. The pen will be in the shape of a rectangle with a dividing fence down the middle parallel to one of the sides. What is the maximum total area the rancher can enclose with 600 feet of fencing?
3. (*See Example 2*) An airline company charges $400 regular one-way fare between New York and London in May. Its jumbo jets, with a capacity of 380 passengers, fly with an average of 300 passengers. It has been estimated that each $20 fare reduction attracts 10 more passengers. What fare should be charged to maximize revenue?
4. An orange grove is being planted. The aim is to plant the number of trees that will yield the maximum number of oranges per acre. As the number of trees per acre increases, the average yield per tree decreases. Existing statistics for the area imply that a grove with 30 trees per acre will yield, on an average, 400 oranges per tree. The yield per tree will be reduced by approximately 10 oranges for every additional tree per acre. Find the number of trees per acre that should be planted to give a maximum total yield per acre.

5. A restaurant is presently selling its special hamburger plate for \$3.20. The restaurant sells 240 plates per day. For every 20¢ decrease in price, the restaurant can sell 30 more plates. What price should the restaurant charge to maximize its revenue?
6. A video rental store is now charging \$3 per night to rent a movie and is renting 100 movies per night. The manger believes that for every decrease of 10¢ in the rental price, the store will rent 5 more movies. What rental price should the store charge to maximize its revenue?
7. (*See Example 3*) A closed box must have a volume of 9000 cubic inches. The bottom of the box will be twice as wide as it is long. Find the dimensions that will minimize the surface area of such a box.
8. An open box is to be made from a square piece of cardboard that is 12 inches on the side. Four equal squares will be cut from each corner of the cardboard, and then the sides will be folded up to form the box. What is the volume of the largest box that can be made this way?
9. A company wants to install machinery that will occupy 1600 square feet of floor space. The machinery is to be arranged in a rectangle. Furthermore, there must be a 1-foot margin of free space on all four sides of the machinery. The company wants to use the least amount of floor space possible. What dimensions will require the least amount of floor space?
10. A man wishes to enclose 1000 square feet with fencing. He wants the enclosed area to be in the shape of a rectangle. The fencing along the side of the rectangle that faces the street will cost \$4 per foot, and the fencing on the other three sides costs \$1 per foot. What are the dimensions of the rectangle that minimize the man's total cost?

4–6 Applications to Business and Economics

Relation Between Average Cost and Marginal Cost
Marginal Analysis of Profit
Marginal Analysis and Changes in Cost

In this section, we introduce some general principles of economics and show how they may be justified by the use of calculus.

Relation Between Average Cost and Marginal Cost

We saw in Section 2 that some cost functions are best modeled by a curve whose shape is shown in Figure 4–36.

Such a curve shows that, while costs increase with increased production, for small levels of production the cost curve is concave down; the marginal cost is decreasing due to efficiency. There is a point where the concavity of the graph

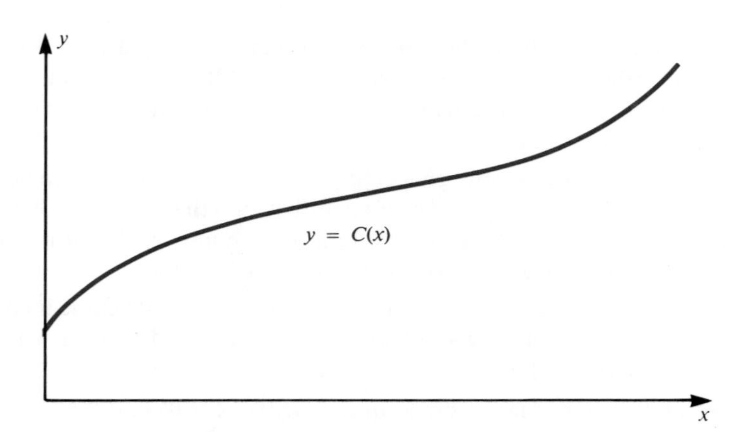

Figure 4–36

changes, and the cost curve becomes concave up, reflecting the lack of efficiency. The marginal cost is increasing; more and more effort, and hence money, is needed to increase production. Such a cost curve is usually modeled by a third-degree polynomial. (This is why many of the cost functions we use in this text are cubics.) That means that the marginal cost curve is a quadratic—it is the derivative of a cubic; $MC = C'$. The graph of the marginal cost then is a parabola. Since the marginal cost decreases and then increases, the parabola must open up. A typical marginal cost function is drawn in Figure 4–37.

Now we look at the general shape of a typical average-cost function AC, where $AC(x) = \dfrac{C(x)}{x}$. The average-cost function has the line $x = 0$ as a vertical asymptote (the average cost is very large when x is close to 0). Furthermore, because $C(x)$ is a cubic, for large x the average-cost function looks very much like a parabola. We now draw the average cost curve, AC, into Figure 4–37; we get Figure 4–38.

The figure is drawn so it looks as if the two graphs cross where the average cost turns around. We look at a specific example.

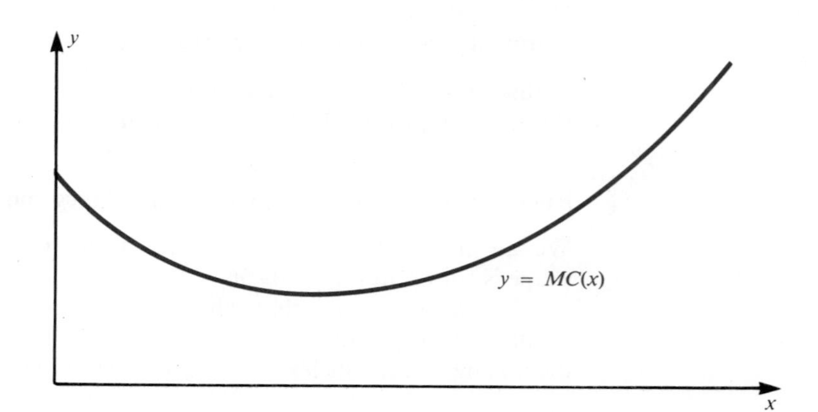

Figure 4–37

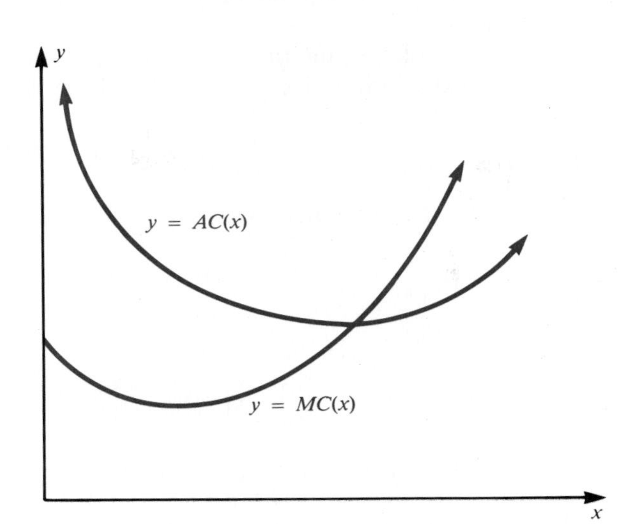

Figure 4–38

Example 1 (*Compare Exercise 1*)
A firm has modeled its cost C at different levels of production x by

$$C(x) = \frac{1}{2}x^3 - 8x^2 + 50x + 200 \qquad \text{for } x \geq 0$$

Find the production level that gives the smallest average cost, and compare $MC(x)$ and $AC(x)$ at that level.

Solution

$$AC(x) = \frac{C(x)}{x}$$

$$= \frac{1}{2}x^2 - 8x + 50 + \frac{200}{x} \qquad \text{for } x > 0$$

To minimize AC, we first find AC'.

$$(AC)'(x) = x - 8 - \frac{200}{x^2}$$

$(AC)'(x)$ exists for all $x > 0$, so the only critical numbers for AC are where $(AC)'(x) = 0$. Set

$$x - 8 - \frac{200}{x^2} = 0$$

$$x^3 - 8x^2 - 200 = 0$$

This factors. (Don't panic! You will be given the factors when you need them in the exercises.)

$$x^3 - 8x^2 - 200 = (x - 10)(x^2 + 2x + 20)$$

Use of the quadratic formula shows that $x^2 + 2x + 20 = 0$ does not have any real solutions. Hence, the only critical number for AC is $x = 10$.

$$AC(10) = \frac{1}{2}(10)^2 - 8(10) + 50 + \frac{200}{10} = 40$$

Now we compute $MC(10)$.

$$MC(x) = C'(x) = \frac{3}{2}x^2 - 16x + 50$$

and so

$$MC(10) = \frac{3}{2}(10)^2 - 16(10) + 50 = 40$$

Thus, $AC(10) = MC(10)$. Figure 4–38 was accurate in this case.

In fact, Figure 4–38 represents a general law of economics.

A Law of Economics

The marginal-cost curve cuts the average-cost curve at the lowest point of the average-cost curve.

Said another way,

Theorem

When AC is minimized, $MC(x) = AC(x)$.

To see why this theorem is true, we start with

$$AC(x) = \frac{C(x)}{x}$$

Next, look at the derivative of both sides, using the quotient rule on the right.

$$(AC)'(x) = \frac{x \cdot C'(x) - C(x) \cdot 1}{x^2}$$

$$= \frac{x \cdot C'(x) - C(x)}{x^2}$$

When AC has its minimum value, $(AC)'(x) = 0$. If $(AC)'(x) = 0$, then the numerator of $(AC)'(x) = 0$, so

$$x \cdot C'(x) - C(x) = 0$$

$$x \cdot C'(x) = C(x)$$

or, since $x \neq 0$,

$$C'(x) = \frac{C(x)}{x}$$

Because $C'(x) = MC(x)$ and $\dfrac{C(x)}{x} = AC(x)$, we have

$$MC(x) = AC(x)$$

Under the assumption that the cost curve has the general shape shown in Figure 4–36, AC has only one critical number. That critical number gives the minimum of AC and satisfies the equation $MC(x) = AC(x)$.

Marginal Analysis of Profit

Again, we refer to the typical curve used to describe cost as given in Figure 4–36. Because revenue is frequently the product of two numbers, (units sold) (price per unit), the curve used to model revenue is frequently a piece of a parabola as shown in Figure 4–39.

We now draw these two curves simultaneously; see Figure 4–40. Notice the distinctive geometric feature that occurs at x_1 and x_2; the tangent lines to the two curves are parallel. Two lines are parallel if they have the same slope, and the slope of the tangent line is the derivative. Thus, we must have $R'(x_1) = C'(x_1)$ and $R'(x_2) = C'(x_2)$.

The quantity $R(x) - C(x)$ is the profit, $P(x)$. (For some x, $C(x) > R(x)$, and the profit is negative; a negative profit is the same thing as a loss.) To find the maximum profit, you first find the critical numbers for P.

$$P(x) = R(x) - C(x)$$

so

$$P'(x) = R'(x) - C'(x)$$

The marginal profit equals the marginal revenue minus the marginal cost; $MP = MR - MC$. The critical numbers for P are where $P'(x) = 0$, but

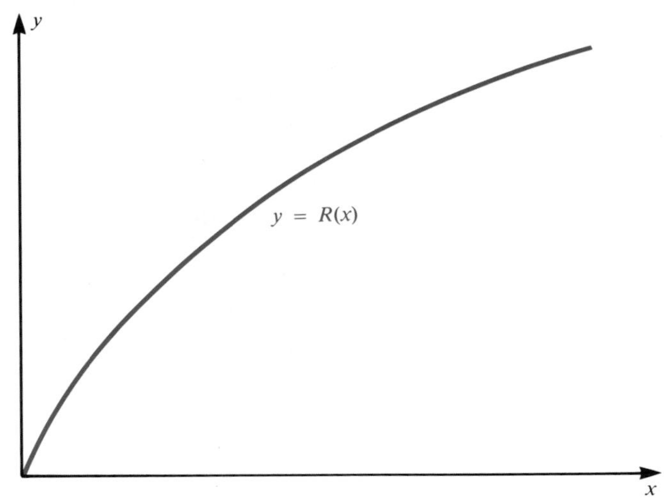

Figure 4–39

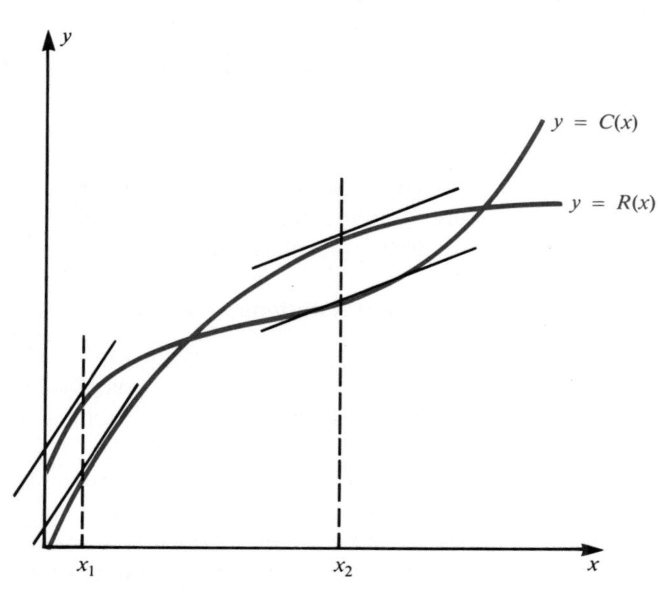

Figure 4–40

$P'(x)=0$ precisely when $R'(x)=C'(x)$. Thus, the maximum profit occurs when $MR(x) = MC(x)$.

Law of Economics

> When the maximum profit occurs the marginal cost and the marginal revenue are equal.

In other words,

Theorem

> When P has its maximum value, $MR(x) = MC(x)$.

Unfortunately, the minimum profit (that is, the maximum loss) also occurs when $MR = MC$. Figure 4–40 shows that $MR = MC$ at both x_1 and x_2. But with curves of this general shape, $P(x_1)$ is negative while $P(x_2)$ is positive. This leads to the following statement, often given as a general rule for optimizing profit: *To maximize profit, select the largest output at which marginal revenue equals marginal cost.* We do not present this statement as a law because it is not true for all cost and revenue functions. It is true under the general assumptions that led to Figure 4–40.

The following example is typical.

Example 2 (*Compare Exercise 5*)
A company produces x units of a certain product per day. The company models its daily cost by

$$C(x) = \frac{1}{150}x^3 - 3x^2 + 525x + 6000$$

The daily revenue function the company uses is given by

$$R(x) = -x^2 + 453x$$

Find the production level that maximizes the daily profit.

Solution

$$MC(x) = \frac{1}{50}x^2 - 6x + 525$$

$$MR(x) = -2x + 453$$

Set

$$MC(x) = MR(x)$$

and solve.

$$\frac{1}{50}x^2 - 6x + 525 = -2x + 453$$

$$\frac{1}{50}x^2 - 4x + 72 = 0$$

Using the quadratic formula and a calculator, we get

$$x = \frac{-b \pm \sqrt{b^2 - 4ac}}{2a} = \frac{4 \pm \sqrt{16 - (288/50)}}{2(1/50)} = \frac{4 \pm 3.2}{1/25}$$

$$= 25[4 \pm 3.2] = 100 \pm 80 = 20 \text{ and } 180$$

Selecting the larger value of x tells us that the production level that maximizes profit for the company should be 180 units per day. We can check this conclusion by evaluating $P(x)$. We find that $P(20) = -6693\frac{1}{3}$ and $P(180) = 6960$. The company should indeed choose the production level of 180 units per day.

Economists classify markets into various types. A **perfectly competitive market** is one where there are many companies operating with the same product.* The price of the product is some constant determined by the open competitive market. No one buyer or seller is able to influence the price of the product. For example, products such as paper towels, nails, dishwashing detergent, refrigerators, and washing machines fall into this type of market.

Let p be the price of the product as determined by the open market. The revenue obtained from producing x items is $p \cdot x$. Thus, in this market the revenue function is given by

$$R(x) = p \cdot x, \text{ where } p \text{ is a constant}$$

With P a constant, the marginal revenue is

$$R'(x) = p$$

* For discussions of the other types of markets, we refer you to a book such as Mansfield, Edwin. *Economics, Principles, Problems and Decisions* (W. W. Norton and Company, 1977).

The optimum output in such a market is then ordinarily given by the largest value of x that satisfies

$$C'(x) = p$$

The following example illustrates this situation.

Example 3 (*Compare Exercise 9*)
A company is manufacturing and selling refrigerators in a perfectly competitive market where the price of a refrigerator is \$600. The company is currently manufacturing 270 refrigerators per day. A management consulting firm has been given the task of determining whether an increase or decrease in production would lead to increased profits. The firm has determined the cost function to be given by

$$C(x) = \frac{x^3}{120} - 4x^2 + 880x + 4250$$

where x is the number of refrigerators produced per day. What should the consulting firm recommend?

Solution The marginal cost is

$$C'(x) = \frac{x^2}{40} - 8x + 880$$

The market price is $p = 600$.
The optimum output will satisfy the equation

$$C'(x) = p$$

$$\frac{x^2}{40} - 8x + 880 = 600$$

$$\frac{x^2}{40} - 8x + 280 = 0$$

Solving for x by using the quadratic formula,

$$x = \frac{-b \pm \sqrt{b^2 - 4ac}}{2a} = \frac{8 \pm \sqrt{(-8)^2 - 4(1/40)(280)}}{2(1/40)} = \frac{8 \pm 6}{(1/20)}$$

$$= 20[8 \pm 6] = 280 \text{ or } 40$$

Selecting the larger number, we see that the optimum output should be 280 refrigerators per day. The management consulting firm should recommend that the company increase output from the present 270 refrigerators per day to 280 refrigerators per day.

Marginal Analysis and Changes in Cost

How should a company react when its costs increase? There are two entirely different situations.

Case 1. **Changes that affect fixed cost.** A company may find that the taxes on its property have increased, or that a new labor contract will increase its fixed labor cost. Can the company pass any of these increased costs on to the consumer? **All other things being equal**, the surprising answer is **NO**! We have just seen that the optimum production strategy for a company is determined by analysis of its marginal revenue and marginal cost. The fixed cost is the constant term in the cost function. When you take the derivative of the cost function, the derivative of the fixed cost is 0. **Fixed cost does not affect the marginal cost function**. The solutions to $MR(x) = MC(x)$ are not affected by a change in fixed cost. Profits will be lowered, but the company's optimum production strategy does not change.

Of course, one of the things that makes economics so difficult is that "all other things being equal" never happens. The price-demand curve, and hence the formula for $R(x)$, will probably change due to inflation and other factors. The company may even decide to increase its advertising, thereby increasing the demand (it hopes), which in turn will allow the company to increase prices.

Case 2. **Changes that affect variable cost.** Various plans for special taxes on gasoline have been proposed, and they continue to be debated. Generally the proposals have two aims in mind: first, to collect more money for the government; second, to drive up the price of gasoline, thereby reducing the demand and so reducing the amount of oil that the United States imports. The general name for a tax levied on the producer in the form of a fixed amount per item produced is **excise tax**. We have seen that taxes affecting the fixed cost of a company do not affect production strategy. How does an excise tax, which increases the variable cost per item, affect optimum production strategy (**all other things being equal**)? Such a tax affects the whole industry. Our functions, or models, should not be the same in this situation as the ones we use for one small producer in a perfectly competitive market. Since the whole industry is affected, we treat the industry as if it were just one producer—a situation called a **monopoly**.

Example 4 (*Compare Exercise 13*)
A company's revenue function is given by

$$R(x) = -\frac{1}{50}x^2 + 19x$$

and its cost function by

$$C(x) = 11x + 100$$

Determine the optimal output and the maximum profit.

Example 5 If fixed costs increase from 100 to 120, what should the company's new production strategy be, and what will be its new profit?

Example 6 (*Compare Exercise 13*)
If a tax of 2 per unit is imposed, what should the company's new production strategy be, and what will be its new profit?

Solution to Example 4

$$MR(x) = R'(x) = -\frac{1}{25}x + 19;$$

$$MC(x) = C'(x) = 11$$

Set $\quad MR(x) = MC(x) \quad$ and solve.

$$-\frac{1}{25}x + 19 = 11$$

$$-\frac{1}{25}x = -8$$

$x = 200$ is the best production level.

$$P(200) = R(200) - C(200)$$

$$= -\frac{1}{50}(200)^2 + 19(200) - [11(200) + 100] = 700$$

The maximum profit is 700.

Solution to Example 5

The new cost function is given by $C(x) = 11x + 120$. But $MC(x)$ still equals 11, and $MR(x)$ is also unaffected. The equation $MR(x) = MC(x)$ is still

$$-\frac{1}{25}x + 19 = 11$$

with solution

$$x = 200$$

The best production level for the company remains $x = 200$. Now $P(200) = 680$. The profit is lowered by the amount of the increase in the fixed cost, but the optimum production strategy is unchanged.

Solution to Example 6

$$MR \text{ is still the same; } MR(x) = -\frac{1}{25}x + 19$$

$$MC \text{ has been increased; } MC(x) = 13$$

Solving $(-1/25)x + 19 = 13$, we get $x = 150$. The optimum production strategy is now $x = 150$, and the company should cut back on production. The new profit is

$$P(150) = -\frac{1}{50}(150)^2 + 19(150) - [13(150) + 100]$$

$$= 350$$

In Example 6, production must be cut back, and the profit will be reduced. If the production level was kept at $x = 200$ however, the new profit would be $P(200) = 300$. Staying with the old strategy would yield even less profit.

The last three examples dealt with the changes in production that an industry should make (or not make) in response to increased costs. As consumers, we are generally more interested in the pricing strategy of the company rather than its production strategy.

Example 7 (*Compare Exercise 15*)
How is the price of the commodity in Examples 5 and 6 affected?

Solution We assumed that

$$R(x) = -\frac{1}{50}x^2 + 19x$$
$$= x\left(-\frac{1}{50}x + 19\right)$$
$$= x \cdot p, \text{ where } p \text{ is price.}$$

Hence, we are assuming that the price is given by

$$p = -\frac{1}{50}x + 19$$

Thus in Example 4, when $x = 200$, the price $p = 15$. The change in fixed cost discussed in Example 5 did not affect the production level, x, and so the price is also unchanged. The change in variable costs in Example 6 caused the best production strategy for the industry to change from 200 to 150. At the new production level of 150, the new price is

$$p = -\frac{1}{50}(150) + 19 = 16 \text{ per unit}$$

The price increased by 1.

Example 7 shows that the best strategy for the company is to pass half of the excise tax on to the consumer, but to absorb half of the tax itself. This analysis runs counter to the belief that companies never have to bear the brunt of taxes, and that they can always pass any tax on to the consumer. In some cases they may be able to do so, but in some cases they cannot.

4–6 Exercises

1. (*See Example 1*) A company has calculated that the total cost of manufacturing x automobiles per day is

$$C(x) = \frac{x^3}{20} - 15x^2 + 3000x + 25{,}600$$

(a) What is the average-cost function for the company? **(b)** Find the production level that minimizes the average cost of each car. **(c)** What is that minimum average cost? **(d)** Compare the marginal cost at this production level.
[*Hint:* Use $x^3 - 150x^2 - 256{,}000 = (x - 160)(x^2 + 10x + 16{,}000)$]

2. The total cost of producing x units per week is

$$\frac{x^3}{8} - 29x^2 + 4000x + 14{,}400.$$

(a) Determine the average-cost function. **(b)** How many units per week would minimize the average cost of each item? **(c)** Show that this occurs at the production level where average cost is equal to marginal cost.
[*Hint:* Use $x^3 - 116x^2 - 57{,}600 = (x - 120)(x^2 + 4x + 480)$]

3. The total cost of manufacturing x items per day is

$$\frac{x^3}{4} - 49x^2 + 4{,}000x + 10{,}000$$

(a) Compute the average-cost and marginal-cost functions. **(b)** Find the daily production that minimizes average cost per item. **(c)** Show that when this occurs, the average cost is equal to marginal cost.
[*Hint:* $x^3 - 98x^2 - 20{,}000 = (x - 100)(x^2 + 2x + 200)$]

4. The total cost of manufacturing x units per day is

$$\frac{x^3}{200} - 2x^2 + 1000x + 9680$$

(a) Determine the average-cost and marginal-cost functions. **(b)** Find the daily production that minimizes cost per unit. **(c)** Show that this occurs when average cost is equal to marginal cost.
[*Hint:* $x^3 - 200x^2 - 968{,}000 = (x - 220)(x^2 + 20x + 4400)$]

5. (*See Example 2*) The cost and revenue functions of a manufacturing company are given by

$$C(x) = \frac{x^3}{180} - 4x^2 + 1000x + 4000$$

and

$$R(x) = -x^2 + 700x$$

Here, x is the number of units produced in one day. Determine the output that leads to maximum profit and also the maximum daily profit that can be realized.

6. A company is operating with the cost and revenue functions given by

$$C(x) = \frac{x^3}{150} - 4x^2 + 2150x + 2700$$

and

$$R(x) = -x^2 + 1900x$$

Here, x is the number of units produced in one week. Determine the output that leads to the maximum weekly profit and the size of that profit.

7. The cost and revenue functions of a manufacturing company are given by

$$C(x) = \frac{x^3}{180} - 3x^2 + 1305x + 3500$$

and

$$R(x) = -x^2 + 1200x$$

where x is the number of units produced daily. The current daily production is 195 units. Will a change in production rate lead to an increase in profits?

8. A company that manufactures motor boats is operating with weekly cost and revenue functions given by

$$C(x) = \frac{x^3}{120} - 6x^2 + 2360x + 4500$$

and

$$R(x) = -x^2 + 2000x$$

The current weekly output is 370 boats. Would a change in this output lead to increased profits?

9. (*See Example 3*) A company is manufacturing and selling a certain type of microcomputer for \$400 in a perfectly competitive market. Its cost function is given by

$$C(x) = \frac{x^3}{270} - 2x^2 + 600x + 2420$$

where x is the number of computers manufactured in one week. The company is currently manufacturing 325 computers per week. Should it change its production rate?

10. A company manufactures a cooling fan that sells for \$125 in a perfectly competitive market. The cost of manufacturing x fans per day is given by

$$C(x) = \frac{x^3}{240} - 4x^2 + 1000x + 3200$$

How many fans per day should be manufactured to realize maximum profit? What will that maximum profit be?

11. A company manufactures television sets in a perfectly competitive market. The price of a television set is \$350. The cost for producing x sets per day is

$$C(x) = \frac{x^3}{150} - 3x^2 + 600x + 2750$$

How many sets per day should be manufactured to maximize profits?

12. A company operating in a perfectly competitive market has a daily cost for producing x units given by

$$C(x) = \frac{x^3}{180} - 2x^2 + 600x + 4500$$

The price of a unit, as determined by the market, is \$495. **(a)** What should the daily output be to maximize profits? **(b)** What is the maximum daily profit that can be attained?

13. (*See Examples 4 and 6*) The weekly revenue and weekly cost of a manufacturing company are

$$R(x) = -2x^2 + 1620x$$

and

$$C(x) = 18x^2 + 60x + 2000$$

Here, x is the weekly output. A tax of \$120 per unit is to be levied. **(a)** Determine the optimal output before the tax is imposed. **(b)** What is the optimal output after the tax has been imposed?

14. The weekly revenue and weekly cost functions of a manufacturing company are given by

$$R(x) = -x^2 + 1400x$$

and

$$C(x) = 14x^2 + 200x + 1000$$

Here, x is weekly output. A tax of \$90 per unit is to be imposed. **(a)** Determine the optimal output before the tax is imposed. **(b)** What is the optimal output after the tax is imposed?

15. (*See Example 7*) Use the fact that the price p is given by

$$p = \frac{R(x)}{x}$$

to determine how the company in Exercise 13 will change its pricing strategy.

16. Use

$$p = \frac{R(x)}{x}$$

to determine how the company in Exercise 14 will change its pricing structure.

IMPORTANT TERMS

4–1	**Increasing functions**	**Decreasing functions**	**Maximum value**
	Minimum value	**Relative maximum value**	**Relative minimum value**

	Extreme value	**Relative extreme values**	**Critical numbers**
	First derivative test		
4–2	**Concave up**	**Concave down**	**Point of inflection**
	Second derivative test		
4–3	**Closed and bounded interval**	**Candidates test**	**Test involving only one critical number**
4–4	**Vertical asymptote**	**Horizontal asymptote**	**Limit at infinity**
	$\lim_{x \to \infty} f(x)$	$\lim_{x \to -\infty} f(x)$	**Infinite limit**
	$\lim_{x \to a^+} f(x) = \infty$	$\lim_{x \to a^-} f(x) = \infty$	$\lim_{x \to a^+} f(x) = -\infty$
	$\lim_{x \to a^-} f(x) = -\infty$		
4–5	**Six-step process for solving word problems**		
4–6	**Average cost**	**Marginal cost**	**Marginal revenue**
	Marginal profit	**Fixed cost**	**Variable cost**

REVIEW EXERCISES

Find the maximum and minimum of each of the functions given in Exercises 1 through 10. Determine the intervals on which f is increasing. Indicate any points of inflection. Sketch a graph of f.

1. $f(x) = \dfrac{1}{x}, \quad 2 \le x \le 5$

2. $f(x) = \dfrac{1}{x}, \quad -6 \le x \le -\dfrac{1}{2}$

3. $f(x) = x^2 - 4x + 5, \quad 0 \le x \le 5$

4. $f(x) = -2x^2 - 8x + 3, \quad -1 \le x \le 3$

5. $f(x) = x + \dfrac{32}{x^2}, \quad 1 \le x \le 8$

6. $f(x) = 2x + \dfrac{27}{x^2}, \quad 1 \le x \le 6$

7. $f(x) = x^5 - 5x^4 + 2, \quad -1 \le x \le 5$

8. $f(x) = x^4 - 4x^3 + 6, \quad -1 \le x \le 4$

9. $f(x) = x + \sqrt{36 - x^2}, \quad -6 \le x \le 0$

10. $f(x) = x + \sqrt{36 - x^2}, \quad 0 \le x \le 6$

For each of the functions given in Exercises 11 through 18, find the intervals on which f is decreasing. Indicate all asymptotes. Sketch the graph of f, indicating any points of inflection.

11. $f(x) = x\sqrt{36 - x^2}$

12. $f(x) = (x - 8)^{1/3}$

13. $f(x) = \dfrac{1}{x^2 - 1}$

14. $f(x) = \dfrac{x}{x^2 - 1}$

15. $f(x) = \dfrac{x^2}{x^2 - 1}$

16. $f(x) = \dfrac{1}{x^2 + 6}$

17. $f(x) = \dfrac{x^2 - 1}{x - 1}$

18. $f(x) = \dfrac{x^2 - 9}{x + 3}$

19. If the sum of two positive numbers is 16, what is the maximum value of their product?

20. If the product of two positive numbers is 16, what is the minimum value of their sum?

21. **(a)** What is the minimum value of f if $f(x) = \dfrac{3x}{x^2 + 4}$?

(b) What is the maximum value of f if $f(x) = \dfrac{3x}{x^2 + 4}$?

22. A company's cost function for producing x items per day is given by

$$C(x) = \frac{x^2}{100} + 2x + 400$$

What level of production minimizes the average cost per item?

23. A ball is thrown straight up; its height (in feet) after t seconds is given by $h(t) = -16t^2 + 64t + 5$. What is the maximum height the ball reaches?

24. A company's price-supply equation is given by $4x + 3p = 120$. What price should the company charge to maximize its revenue?

25. A company's cost function is given by

$$C(x) = \frac{1}{3}x^3 + x^2 + 15x + 2$$

and its revenue is given by

$$R(x) = 4x^2 + 10x$$

What value of x maximizes the company's profit?

26. A rental company finds that the price it charges per day to rent an automobile is linearly related to the number of cars that it rents. If the company rents 40 cars when it charges \$30 and rents 20 cars when it charges \$35 per day, what rental fee will maximize the company's revenue?

27. A company manufacturing refrigerators is operating in a perfectly competitive market. The refrigerators are selling for \$400, and the company's cost for producing x of these per week is

$$C(x) = \frac{x^3}{300} - x^2 + 436x + 3000$$

How many refrigerators should the company manufacture each week to maximize its profit?

28. A company is currently producing x items per week. Its revenue is given by

$$R(x) = 300x - x^2$$

and its cost is

$$C(x) = \frac{x^2}{10} + 80x + 200$$

What is the company's present optimal production strategy? How should the company change its production level if a new tax of 11 per unit is levied?

5

Topics in this chapter can be applied to:

Inflation ● Population Growth ● Radioactive Decay ● Compounding Interest ● Investment Planning ● City Planning ● Budget Management ● Managing Resource Consumption

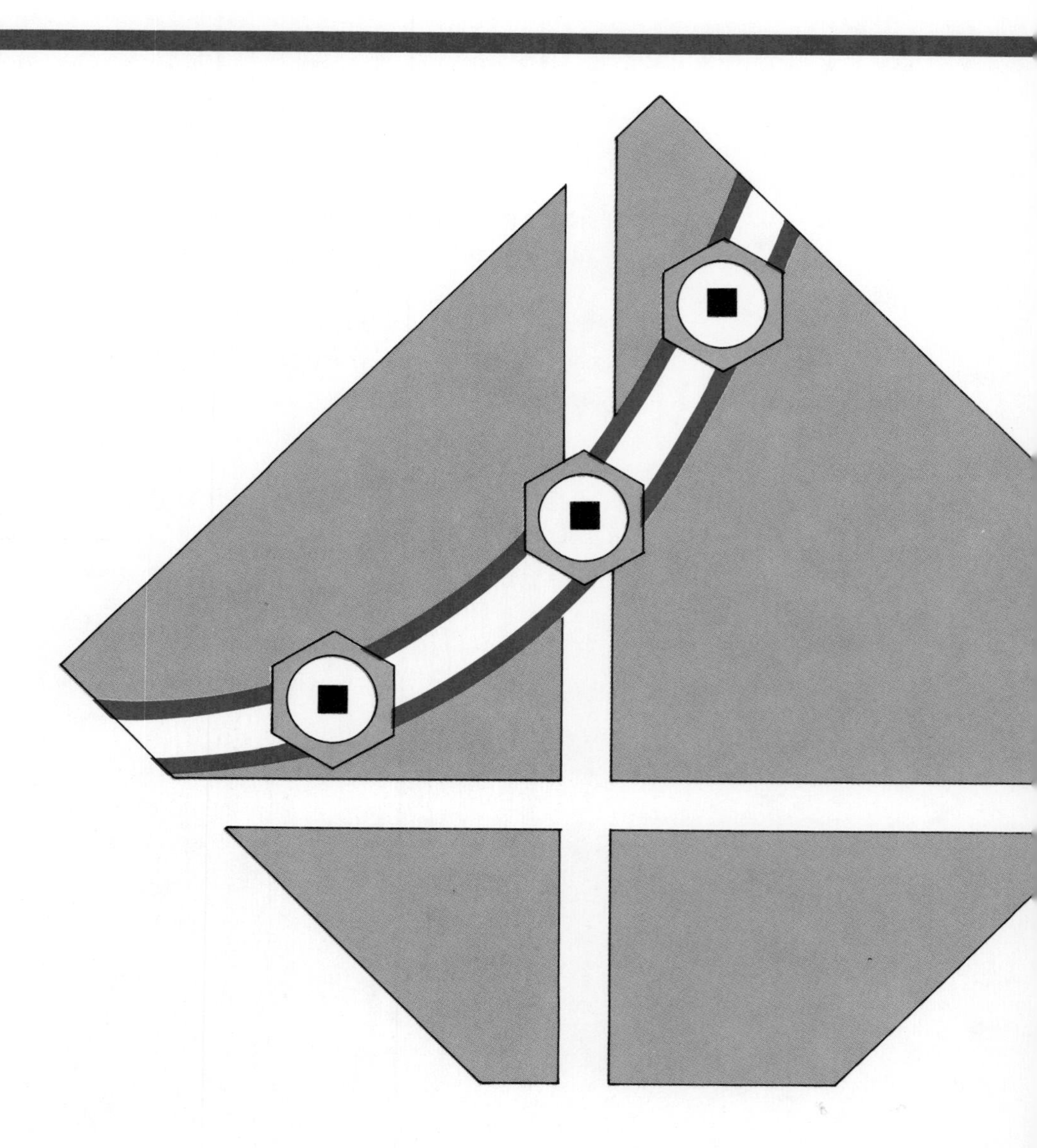

Exponential and Logarithm Functions

- The Exponential Function
- The Natural Logarithm Function
- The Derivative of e^x
- The Derivative of $\ln x$
- Applications of Exponential Functions
- Bases Other than e

5–1 The Exponential Function

Different Ways of Describing Change
The Effect of Compounding
Continuous Compounding and the Number e
Exponential Functions and Their Graphs

Different Ways of Describing Change

We have seen that the derivative is the mathematical tool used to describe the rate of change of a function. For example, if $A(t)$ is the amount of money an investment is worth t years after the investment is made, $A'(t)$ describes the rate of change of $A(t)$ in dollars-per-year. Similarly, if $P(t)$ is the population of a country at time t years, then $P'(t)$ describes the rate of change of the population in people per year. In general, if $y = f(x)$, then $f'(x)$ gives the rate of change of y per unit of x.

Often however, changes in the amount of an investment or changes in the population of a country are described not in terms of the actual **amount** of the change, but in terms of the **percentage** of the change. To see the distinction, suppose that two people, A and B, both invest money in an account that pays interest at the rate of 6% per year. Suppose that A invests \$1000, and B invests \$10,000. Let $A(t)$ be the amount of A's investment after t years, and similarly, let $B(t)$ be the amount of B's investment. After one year, A's investment is worth \$1060, while B's is worth \$10,600. Although the **percentage change** in the two amounts is the same, the **amount of change** in the two accounts is different. B's account has increased more than A's account; $B'(t) > A'(t)$. People use the word "interest" to mean both the actual amount of interest ($600 for B and $60 for A), and the rate of interest (6% for both A and B); you must determine by the context which is meant.

Many other real-world situations lend themselves to talking about change in terms of percentage rates—inflation, population growth, and radioactive decay are but a few. This chapter will develop the mathematical tools needed to deal with these situations, and one of the tools is new. We will encounter new types of functions—ones that are not built by using polynomials or radicals. To show how these functions arise from real-world situations, we begin with a specific discussion about savings accounts.

The Effect of Compounding

We develop the theory in an example.

Example 1 (*Compare Exercise 1*)
A person deposits $1000 in a savings account that pays an interest rate of 5%, compounded annually. Assuming no deposits or withdrawals, how much money will be in the account after four years?

Solution The compounding period, here one year, tells how often the interest amount is added to the savings account. Here, the account is worth a constant $1000 for a year, and then the bank adds 5% of $1000 (which is $50) to the account. [5% of $1000 = (.05)\cdot(1000)$.]

At the beginning of the second year, the account is worth $1000 + 50 = 1050$ dollars. At the end of the second year, the bank adds 5% of $1050 to the account—during the second year the bank pays interest on the interest earned during the first year. This is what is meant by compounding. The interest added to the account at the end of the second year is 5% of $\$1050 = (.05)(\$1050) = \$52.50$, so at the beginning of the third year the account is worth $1050 + 52.50 = 1102.50$ dollars. We could continue in this manner to solve this particular problem, but we are going to want to solve other problems with different beginning investment amounts (here $1000), different interest rates (here 5%), and different compounding periods (here 1 year).

We leave this particular example now by replacing our particular numbers with letters. We hope to see more easily what is happening, and at the same time, to derive a formula for the more general case.

Let P = the beginning investment—the beginning **p**rincipal.
Let r = the annual interest **r**ate.
Let n = the **n**umber of compounding periods.
Let $A(n)$ = the **a**mount in the account after ***n*** compounding periods.

The amount after 1 year is

$$A(1) = \underbrace{P}_{\text{principal}} + \underbrace{rP}_{\text{interest}} = P(1 + r)$$

The amount after 2 years is

$$A(2) = \underset{\text{principal}+}{P(1+r)} + \underset{\text{interest}}{r(P(1+r))} = P(1+r)^2$$

The formula $A(2) = P(1+r)^2$ can be obtained by factoring $P(1+r)$ from the expression $P(1+r) + r(P(1+r)) = P(1+r)(1+r) = P(1+r)^2$. It is easier, and more useful, to avoid this algebra by thinking for a bit about the first formula: $A(1) = P(1+r)$. This formula says that you find the amount after 1 year by multiplying the amount you started with by the number $(1+r)$. Now, regard $P(1+r)$ as the amount at the beginning of the second year; you find how much the account is worth 1 year from then by multiplying $P(1+r)$ by the number $(1+r)$. If you start the year with $P(1+r)$ dollars, you will have $[P(1+r)] \cdot (1+r) = P(1+r)^2$ when the year is over. Now, regard $P(1+r)^2$ as the beginning amount during the third year; the amount at the end of this year is $P(1+r)^2$ multiplied by $(1+r)$, or $P(1+r)^3$.

$$A(3) = P(1+r)^3$$

We can continue in this manner, establishing the formula

$$A(n) = P(1+r)^n$$

Returning to our particular example,the amount after 4 years is

$$A(4) = 1000(1+.05)^4 = 1000(1.05)^4 = (1000)(1.21551) = 1215.51 \text{ dollars.}$$

What happens for different compounding periods? For example, what if interest was added every six months (semi-annually) rather than every year? Specifically, how much will be added after a six-month period if \$1000 is invested at a 5% annual rate compounded semi-annually? The bank does not pay 5% of the amount after only six months. In fact, what the bank does is pay the proportionate amount; after one half of a year, the bank pays one half of 5%, or $2\frac{1}{2}$%. The amount of interest paid is $2\frac{1}{2}$% of \$1000 = (.025)(1000) = \$25.00. Remember that n is the number of compounding periods, so after six months, n is 1, and we have

$$A(1) = 1000 + \left(\frac{.05}{2}\right)(1000) = 1000\left(1 + \frac{.05}{2}\right) = 1025 \text{ dollars}$$

We can go through the same argument we went through in our first example to show that

$$A(2) = 1000\left(1 + \frac{.05}{2}\right)^2 = 1050.62 \text{ dollars}$$

After one year, the amount in the account would be \$1050.62. After four years, the amount would be (remember that there are now eight compounding periods)

$$A(8) = 1000\left(1 + \frac{.05}{2}\right)^8 = 1218.40 \text{ dollars}$$

More frequent compounding has meant that the account is worth more after four years; \$1218.40 compared to \$1215.51.

If the compounding is done monthly, then the interest rate used in the formula is $\frac{.05}{12}$. The amount after four years would be

$$A(48) = 1000\left(1 + \frac{.05}{12}\right)^{48} = 1220.90 \text{ dollars}$$

We are now ready to state the general formula.

If an initial principal P is invested in an account that pays interest at an annual rate r, compounded m times per year, the amount in the account after n compounding periods will be

Formula I

$$A(n) = P\left(1 + \frac{r}{m}\right)^{n}$$

We can introduce the real variable t (the time expressed in years) in place of n (the number of compounding periods) by noticing that the (number of compounding periods) = (number of compounding periods per year) · (number of years). In symbols,

$$n = m \cdot t$$

We can now express $A(t)$ by the formula

Formula II

$$A(t) = P\left(1 + \frac{r}{m}\right)^{mt}$$

Continuous Compounding and the Number *e*

Some popular compounding periods are annually ($m = 1$), semi-annually ($m = 2$), quarterly ($m = 4$), monthly ($m = 12$), and daily (m usually taken to be 360 or 365). You may have seen advertisements for "interest compounded continuously." What does that mean? There is no value of m that will give compounding instant-by-instant. We can however imagine more frequent compounding than daily—the compounding period could be an hour ($m = 24 \cdot 365 = 8760$), a minute ($m = 8760 \cdot 60 = 525{,}600$), or a second ($m = 31{,}536{,}000$). We would then have to compute numbers of the form $(1 + (r/m))^{mt}$ for very large values of m.

We will avoid these horrendous computations because mathematicians have been able to prove an amazing theorem. Before presenting the general theorem, however, we need to examine a specific example, the case when $r = 1$ and $t = 1$. Thus, we want to look at numbers of the form $(1 + (1/m))^m$.

The following table evaluates the expression $(1 + (1/m))^m$ for various values of m.

m	1	10	100	1000	10,000	100,000
$\left(1+\frac{1}{m}\right)^m$	2	2.59374	2.70481	2.71692	2.71815	2.71827

As m gets bigger and bigger, the expression $(1 + (1/m))^m$ gets closer and closer to a fixed number whose decimal expansion begins 2.7182818284590.... Because this decimal expansion does not repeat, we cannot write this number as a fraction. Like π, this number is irrational. Thus, we need special notation for this number. The letter people use for this number is the English letter ***e***. This is similar to using the Greek letter π to denote the number 3.14159265.... [This notation is in honor of a distinguished mathematician, Leonhard Euler (1707–1783), who obtained many results concerning this number and whose last name begins with the letter e.]

The number e occurs in so many applications that many calculators have a special button that enables you to compute e^r for various values of r.

The amazing theorem we referred to above is this:

Theorem

> No matter what the value of r, as m gets bigger and bigger, the expression $(1 + (r/m))^m$ gets closer and closer to e^r.

(For $r = 1$, note that $e^1 = e$.) This theorem allows us to derive the formula to use for continuous compounding. By inserting a new pair of parentheses, we can rewrite Formula II as

$$A(t) = P\left[\left(1 + \frac{r}{m}\right)^m\right]^t$$

We get closer and closer to the idea of continuous compounding by using more and more compounding periods; that is, by letting m get larger and larger. When m is a very large number, $(1 + (r/m))^m$ is very close to e^r. Thus, when m is very large

$$P\left[\left(1 + \frac{r}{m}\right)^m\right]^t \text{ is close to } P[e^r]^t$$

All this lets us now write the formula we use to model continuous compounding.

Formula for Continuous Compounding

> If P dollars is invested at an annual interest rate of r, compounded continuously, then after t years the amount A is given by
>
> ***Formula III*** $\qquad A(t) = P \cdot e^{rt}$

The next example compares the difference between continuous compounding and the annual and semi-annual compounding discussed above.

Example 2 (*Compare Exercise 17*)
If \$1000 is invested at 5%, compounded continuously, how much is the investment worth after four years?

Solution We let $P = 1000$, $r = .05$, and $t = 4$ in Formula III, $A(t) = Pe^{rt}$. Then, $A(4) = 1000e^{.05(4)} = 1000e^{.2}$. Using a calculator, we find $e^{.2} \approx 1.2214$, so $A(4) \approx 1000(1.2214) = 1221.40$ dollars ($\approx$ means approximately equal).

The graphs in Figure 5–1 show the growth of A for $m = 2$, $m = 4$, $m = 12$, and lastly, for continuous compounding. Notice how the first three graphs "approach" the last graph as m gets bigger.

Looking at Formula III again, $A(t) = Pe^{rt}$, we see that we have just encountered a new type of function; the variable t is in the **exponent**. Functions with a variable exponent are called exponential functions, and they are used in a wide variety of applications. The discussion so far has shown how one such function arises in a financial application; but before we go on to other applications, we will give you a quick review of exponential functions and their graphs. You might also want to review the material on exponents in Chapter 0.

Exponential Functions and Their Graphs

The exponential functions we will look at are those of the form $f(x) = kb^{cx}$, where k and c are any two real numbers, and b is a positive number. This is the form in Formula III, with $k = P$, $c = r$, and $b = e$. *The variable x is in the exponent.* The

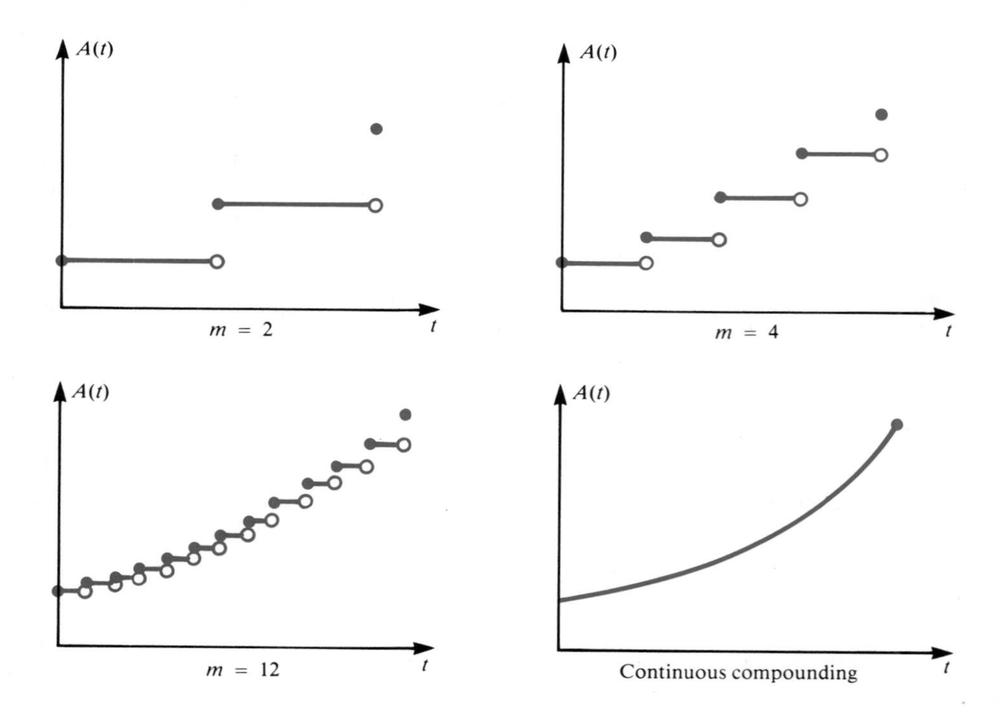

Figure 5–1

number b is called the **base**, and we require b to be positive so that we don't encounter complex numbers like $(-1)^{1/2} = \sqrt{-1}$.

We begin by letting k and c both equal 1, so that we are looking at functions of the form $f(x) = b^x$. We start with values of b that are more familiar than the number e.

Example 3 Let $b = 1$. Graph $y = 1^x$.

Solution The case when $b = 1$ is not very interesting, for then $f(x) = 1^x = 1$ for all values of x.

The graph of 1^x is drawn in Figure 5–2.

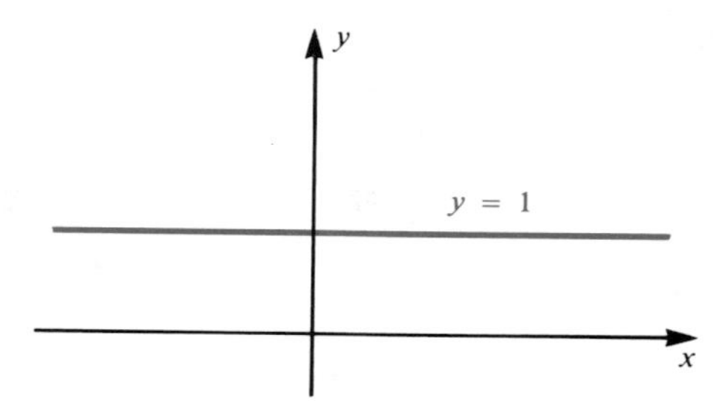

Figure 5–2

Example 4 (*Compare Exercise 37*)
Let $b = 2$. Graph $y = 2^x$.

Solution To see what the graph looks like, we computed the following table of values, plotted those points, and then drew a smooth curve through those points. The result is in Figure 5–3.

x	-5	-4	-3	-2	-1	0	1	2	3	4	5
$f(x) = 2^x$	.03125	.0625	.125	.25	.5	1	2	4	8	16	32

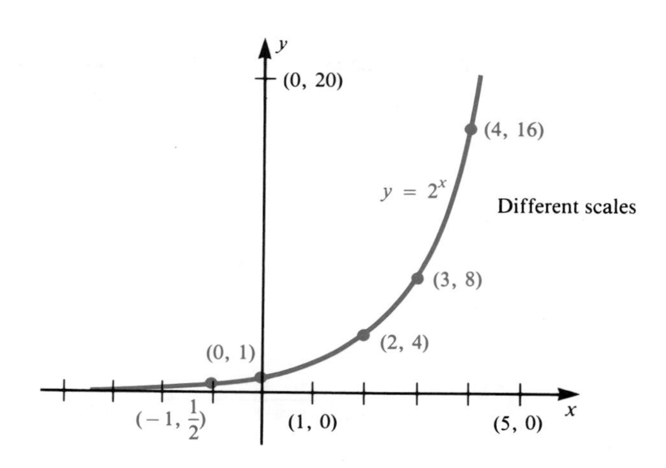

Figure 5–3

Notice the different scales on the x- and y-axes—the function 2^x grows very rapidly. (See Exercise 45.) Also notice that the entire graph lies above the x-axis. For negative values of x, the values of 2^x are positive but small. When $x = -10$,

$$f(-10) = 2^{-10} = \frac{1}{2^{10}} \approx .00098$$

Our next example is closely related to $y = 2^x$.

Example 5 Let $b = \frac{1}{2}$. Graph $f(x) = \left(\frac{1}{2}\right)^x$.

Solution Again, we construct a table of values.

x	−5	−4	−3	−2	−1	0	1	2	3	4	5
$\left(\frac{1}{2}\right)^x$	32	16	8	4	2	1	.5	.25	.125	.0625	.03125

The graph is given in Figure 5–4.

The tables in the last two examples are very similar; the ordering is just reversed. The reason for this can be seen by rewriting $\frac{1}{2}$ as 2^{-1}. Then,

$$\left(\frac{1}{2}\right)^x = (2^{-1})^x = 2^{-x}$$

Imagine the graph of $y = 2^x$ as a piece of wire you are able to rotate around the y-axis [so the graph is like a weather vane that can pivot around the point (0, 1)]. If you rotate the graph of $y = 2^x$ through a turn of 180 degrees, you get the graph of $y = (\frac{1}{2})^x = 2^{-x}$. The graphs are mirror images of each other.

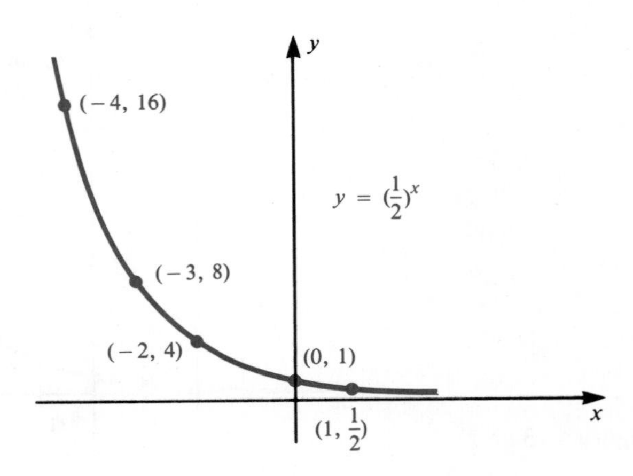

Figure 5–4

These two graphs indicate the two general shapes of the graphs of exponential functions. If $f(x) = b^x$, then the graph of f goes through the point (0, 1). If $b > 1$, then f is increasing, and its graph is similar to Figure 5–3. The larger the value of b, the more steep the shape of the curve. If $0 < b < 1$, then f is decreasing, and its graph is similar to Figure 5–4. The graphs of $y = (1/b)^x$ and $y = b^x$ are mirror images of each other.

Values of k and c different from 1 don't change the general appearance of the graphs of exponential functions. The one exception is $k = 0$. If $k = 0$, then the function reduces to $y = 0 \cdot b^{cx} = 0$. The graph is just the line $y = 0$, the x-axis. If $k > 0$, $y = kb^{cx}$ will still have one of the two general shapes shown in Figure 5–5.

If $k < 0$, then the values of y are negative, and the graphs are like those in Figure 5–5, except they are upside down. See Figure 5–6.

Notice that the graph of $y = kb^{cx}$ goes through the point $(0, k)$ and does not meet the x-axis (unless $k = 0$).

Here is another specific example.

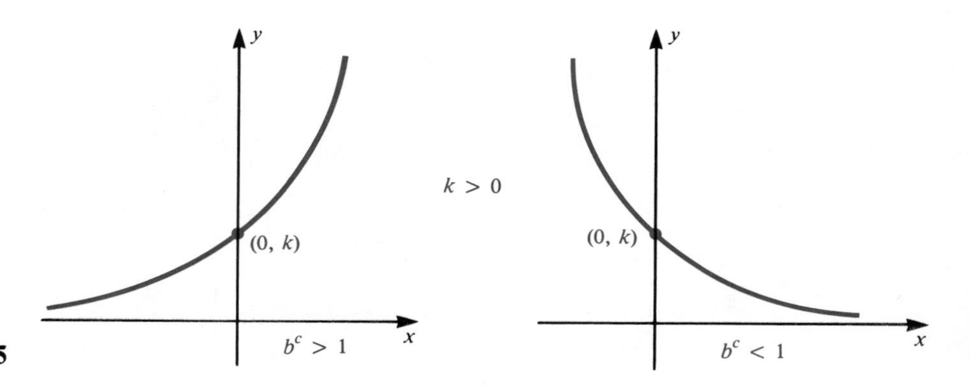

Figure 5–5

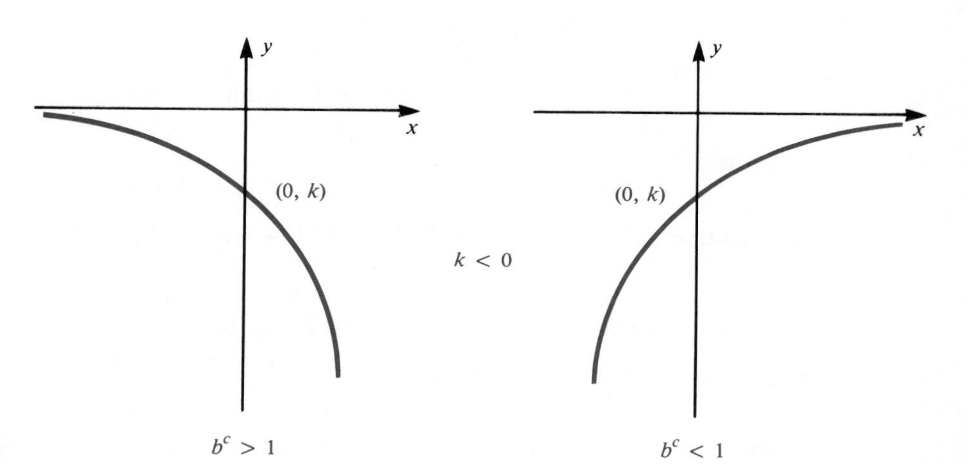

Figure 5–6

Example 6 (*Compare Exercise 41*)

Let $k = 5$, $b = 2$, and $c = \frac{1}{4}$. Graph $y = 5 \cdot 2^{x/4}$

Solution The following table was constructed using a calculator, and the values of y were rounded to the nearest tenth.

x	−5	−4	−3	−2	−1	0	1	2	3	4	5
y	2.1	2.5	3.0	3.5	4.2	5	5.9	7.1	8.4	10	11.9

The graph is shown in Figure 5–7.

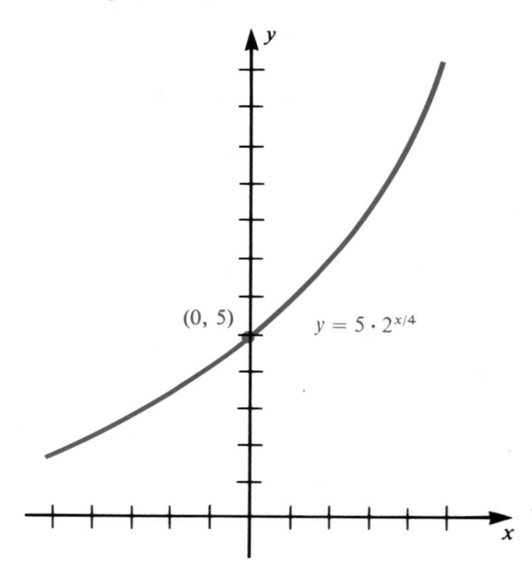

Figure 5–7

Example 7 (*Compare Exercises 23 and 27*)
Compare the graphs of $y = 2^{3x}$ and $y = 8^x$.

Solution Because $2^{3x} = (2^3)^x = 8^x$, the graphs are identical. This example is to point out that the same function may have different forms.

Example 8 (*Compare Exercise 31*)
Rewrite the function $f(x) = 3^{x+2}$ in the form $f(x) = k3^x$.

Solution $3^{x+2} = 3^x \cdot 3^2 = 9 \cdot 3^x$. Thus $k = 9$, and we have $3^{x+2} = 9 \cdot 3^x$.

We conclude this section by returning to the function that made us look at exponential functions in the first place.

Example 9 (*Compare Exercise 39*)
Graph $y = e^x$.

Solution Remember that $e = 2.71828\ldots$, so this graph will be very close to $y = (2.7)^x$. It will be "between" the graphs of $y = 2^x$ and $y = 3^x$.

x	−5	−4	−3	−2	−1	0	1	2	3	4	5
$y = e^x$	.007	.02	.05	.14	.37	1	2.7	7.4	20.1	54.6	148.4

The graph is shown in Figure 5–8.

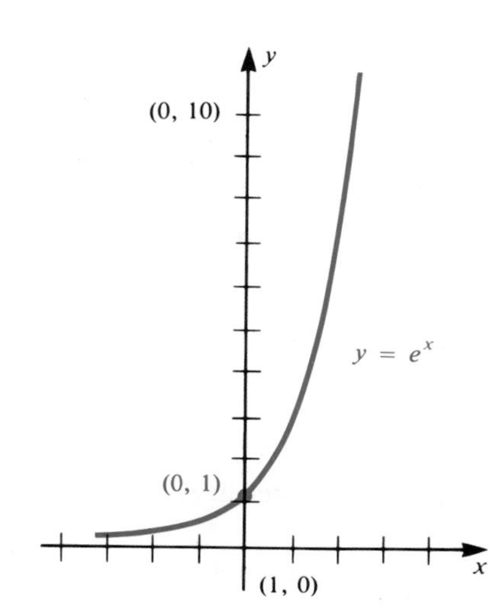

Figure 5–8

5–1 Exercises

I.

1. (*See Example 1*) \$1000 is invested at 6% interest, compounded annually.

   (a) How much is the investment worth after one year?
   (b) How much is the investment worth after two years?
   (c) How much is the investment worth after four years?

2. \$1000 is invested at $7\frac{1}{2}$% interest, compounded annually.

   (a) How much is the investment worth after one year?
   (b) How much is the investment worth after three years?
   (c) How much is the investment worth after six years?

3. An investment is made at 8% interest, compounded annually. How much is the investment worth after three years if the original investment was

   (a) \$1000?
   (b) \$2000?
   (c) \$10,000?

4. \$10,000 is invested at 8% interest. How much is the investment worth after one year if the compounding period is
   (a) one year?
   (b) six months?
   (c) three months?
5. \$10,000 is invested at 9% interest. How much is the investment worth after two years if the compounding period is
   (a) one year?
   (b) three months?
   (c) one month?

Evaluate and simplify each of the following.

6. $f(0)$ if $f(x) = e^x$
7. $f(1)$ if $f(x) = 16^{x/2}$
8. $g(0)$ if $g(x) = e^{-3x}$
9. $g(3)$ if $g(x) = 4^{x/2}$
10. $f(-1)$ if $f(x) = 3^x$
11. $f(-2)$ if $f(x) = 8^{x/3}$
12. $f(-3)$ if $f(x) = 16^{x/2}$

II.

13. After how many years is \$500 invested at 8% interest, compounded quarterly, worth $\$500(1.02)^{12}$?
14. What is the interest rate, compounded quarterly, if after one year a \$1000 investment is worth $\$1000(1.03)^4$?
15. After four years, what is the value of a \$1000 investment made at 6% interest, compounded semi-annually? [*Hint:* the answer is in the form \$1000(______)$^{\_\_\_\_}$.]
16. After five years, what is the value of a \$3000 investment made at 10% interest, compounded quarterly? [\$3000(______)$^{\_\_\_\_}$.]
17. (*See Example 2*) After how many years is \$2000 invested at 6% interest, compounded continuously, worth $\$2000e^{.12}$?
18. What is \$2000 invested at 7% interest, compounded continuously, worth after 8 years? (\$2000$e^{____}$).

III.

Evaluate and simplify each of the following.

19. $f(2)$ if $f(x) = 5 \cdot 3^x$
20. $f(-1)$ if $f(x) = 6 \cdot 2^{3x}$
21. $f(0)$ if $f(x) = 7 \cdot 3^{x/2}$
22. $f(6)$ if $f(x) = 6 \cdot 4^{x/3}$

Find b in the following four exercises.

23. (*See Example 7*) $5^{2x} = b^x$
24. $9^{x/2} = b^x$
25. $3^{-2x} = b^x$
26. $16^{-x/2} = b^x$

Find c in the following four exercises.

27. (*See Example 7*) $2^{cx} = 8^x$
28. $9^{cx} = 3^x$
29. $5^{cx} = (\frac{1}{5})^x$
30. $16^{cx} = (\frac{1}{2})^x$

Find k in the following four exercises.

31. (*See Example 8*) $k4^x = 4^{x+2}$

32. $k2^x = 2^{x+5}$

33. $k3^x = 3^{x-2}$

34. $k5^x = 5^{x-3}$

In Exercises 35 through 42, sketch the graph of f.

35. $f(x) = 3^x;\ -3 \le x \le 3$

36. $f(x) = 3^{-x};\ -3 \le x \le 3$

37. (*See Example 4*) $f(x) = (\frac{1}{2})2^x;\ -3 \le x \le 3$

38. $f(x) = 2 \cdot 2^x;\ -3 \le x \le 3$

39. (*See Example 9*) $f(x) = \frac{1}{2}e^x;\ -3 \le x \le 3$

40. $f(x) = e^{-x};\ -3 \le x \le 3$

41. (*See Example 6*) $f(x) = 2 \cdot 3^{x/2}$

42. $f(x) = -1 \cdot 2^x$

43. **(a)** If \$1000 is invested at 10% interest compounded quarterly, how much is the investment worth six years after it is made?
    (b) If the interest is compounded continuously, how much is the investment worth after six years?

44. If \$1000 is invested at 7% interest, compounded continuously, how much is the investment worth three years after it is made?

45. There is a folk tale that the game of chess was invented as an entry in a contest sponsored by a bored king who was looking for new amusements. The king was so pleased with the game that he offered the inventor anything the inventor would choose. The inventor asked for one grain of wheat in the square in the lower left-hand corner of the chess board; two grains in the square to the right of the first one; four grains in the next square, and so on (see Figure 5–9). The king thought the inventor should ask for more; after all there are thousands of grains of wheat in a single bushel. The king's country produced one trillion grains of wheat that year. Was the king able to give the inventor his wish?

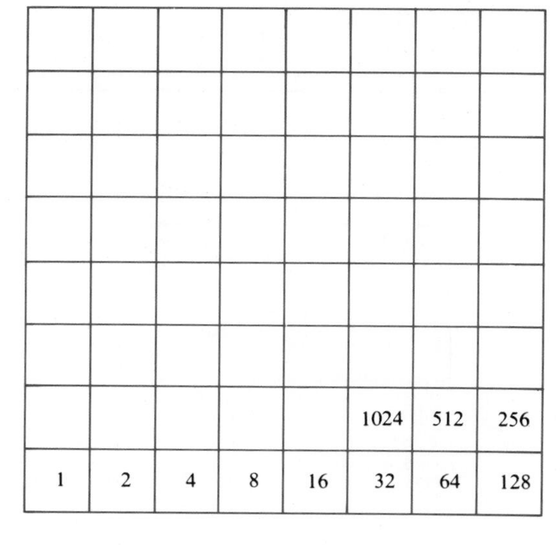

Figure 5–9

5–2 The Natural Logarithm Function

Doubling An Investment

We return to our investigation of the situation where someone invests \$1000 at 5%, compounded continuously. From the previous section, we know that $A(t)$, the amount the investment is worth after t years, is given by $A(t) = 1000\, e^{.05t}$, and that the graph of A looks like Figure 5–10.

Up to now we have been using t as our independent variable and then determining $A(t)$. Now we would like to ask a slightly different question about this investment. We change our point of view and ask, "How long must we wait for the investment to be worth a certain amount?" To be specific, how long will it take for the investment to double; for what value of t is $A(t) = 2000$? In terms of the function machine, we have the situation shown in Figure 5–11.

If the output is 2000, what is the input? We look at how this question can be interpreted graphically.

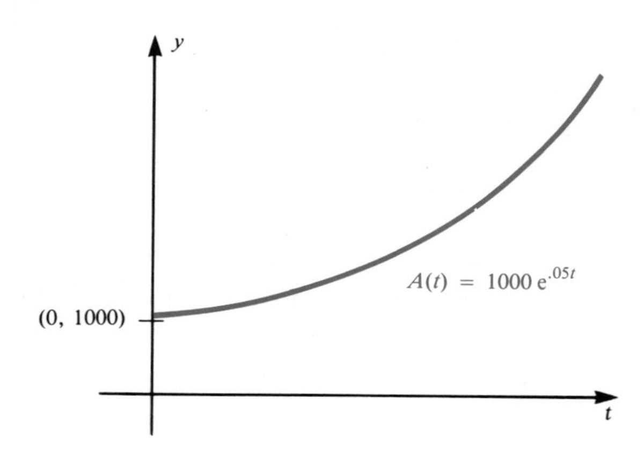

Figure 5–10

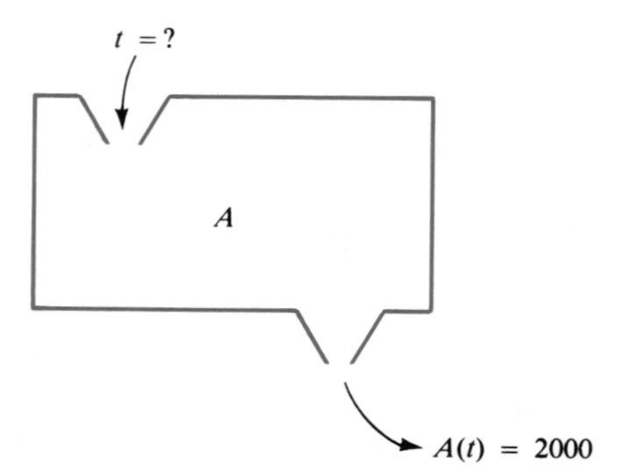

Figure 5–11

Normally, you would graph the function A by thinking of starting on the horizontal axis (the t-axis) and computing $A(t)$ (the height), and then plotting the point $(t, A(t))$. This time we're asking, "If you know the height of the point, what is its first coordinate?" Specifically, if the point is $(t, 2000)$, what is t?

To answer this question, we need to solve for t in the equation $1000e^{.05t} = 2000$. Dividing both sides by 1000, we want to solve $e^{.05t} = 2$. As an intermediate step, we change variables by letting $x = .05t$. Now, if we solve $e^x = 2$, then we can determine t by

$$t = \frac{x}{.05} = 20x$$

The question can now be phrased, "What is the first coordinate of the point $(x, 2)$ on the graph $y = e^x$?"

We know that $e^0 = 1$ (which is below 2) and that $e^1 = e\,(= 2.71\ldots$, which is above 2), so x is somewhere between 0 and 1. See Figure 5–12.

To get a better idea of the value of x, we do some more evaluations. $e^{.5} = 1.64\ldots$, which is less than 2, so x is between .5 and 1.

We can continue in this manner, getting x to be about .7 ($e^{.7} = 2.01\ldots$). See Figure 5–13.

Because $t = 20x$, we have $t \approx (20)(.7) = 14$. Here is the answer to our question; it would take about 14 years for the investment to grow to \$2000.

The Natural Logarithm Function

In the above example, we wanted to solve the equation $e^x = 2$. If we ask how long it would take for the investment to triple, we would want to solve the equation $e^x = 3$. The general problem is to solve the equation $e^x = y$ for x when you know the value of y. If we know the output of $f(x) = e^x$, can we find the input? We want a function that "undoes" the exponential function; given the number e^x as input, this new function will return x to us. This function is so important that it has a special name—the **natural logarithm function**. (It is also called the logarithm

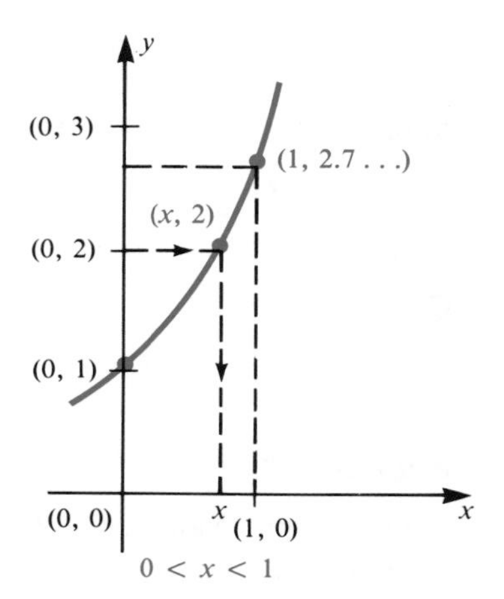

Figure 5–12

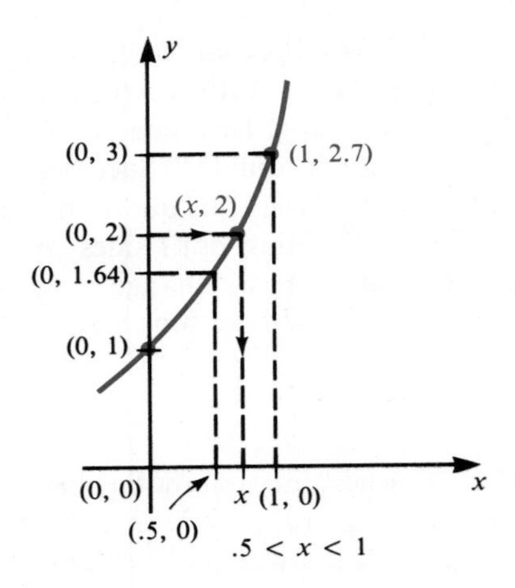

Figure 5–13

function to the base e.) The term "natural" is used because the number e is forced on us by nature. Furthermore, nature does us a favor here because the calculus formulas for derivatives are simpler using base e than any other base.

Warning! It is not generally true that if you know the output of a function machine, then you can figure out what the input must have been. For example, if $f(x) = x^2$, and we know that the output is 9, there is no way of determining whether the input was 3 or -3. The key fact we are using is that the **graph of the exponential function is never at the same height twice**. A given value of y is the result of a unique input. Thus, if $e^c = e^d$, then $c = d$.

Definition

> The natural logarithm function is written **ln**, and is defined by
>
> $\ln B = A$ if and only if $e^A = B$

The definition of the natural logarithm involves two equations that are equivalent. To help us understand more fully the relationship between the natural logarithm function and the exponential function, we use these two equations to produce a third.

Equation (1) $\ln B = A$

is equivalent to

Equation (2) $e^A = B$

If we use Equation (2) to replace B by e^A in Equation (1), then Equation (1) becomes

Equation (3) $\ln e^A = A$

We give some examples to show you how to use Equation (3) to perform some evaluations.

Example 1 (*Compare Exercise 1*)
Compute $\ln e^2$.

Solution With $A = 2$, Equation (3) tells us that $\ln e^2 = 2$.

Example 2 (*Compare Exercise 5*)
Compute $\ln \frac{1}{e}$.

Solution Rewrite $\frac{1}{e}$ as e^{-1}. Thus, $$\ln \frac{1}{e} = \ln e^{-1} = -1$$

Using x instead of A, Equation (3) becomes $\ln e^x = x$. Thus, ln does exactly what we wanted it to do; given e^x as its input, ln gives us x as its output.

We use Equations (1) and (2) to produce another important equation. This time we use Equation (1) to replace A by $\ln B$ in Equation (2); we get

Equation (4) $$e^{\ln B} = B$$

Replacing B by x, Equation (4) becomes $e^{\ln x} = x$. Given $\ln x$ as input, the exponential function gives us x as output. The exponential and logarithm functions undo each other. See Figure 5–14.

In general, functions that "undo" each other are called inverse functions, but we will not pursue the general theory here. We do use this special relationship, however, to see what the graph of $y = \ln x$ looks like.

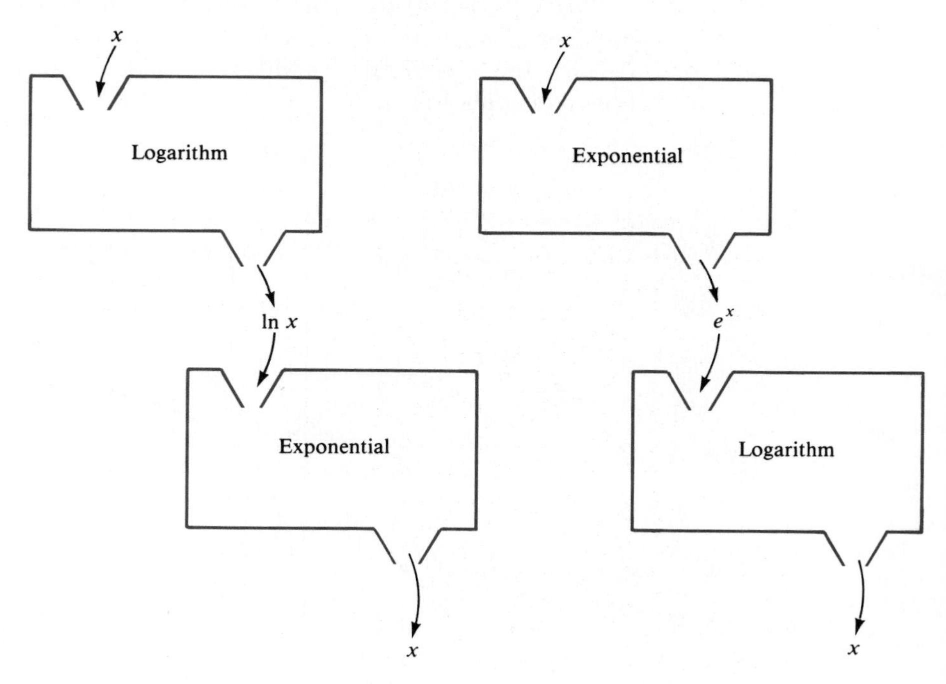

Figure 5–14

We proceed through a sequence of equivalent statements.

The point (B, A) is on the graph of $y = \ln x$

if and only if

$$\ln B = A$$

if and only if

$$e^A = B$$

if and only if

the point (A, B) is on the graph of $y = e^x$

In summary, (B, A) is on the graph of $y = \ln x$ precisely when (A, B) is on the graph of $y = e^x$. See Figure 5–15.

If you rotate the plane about the dotted line $y = x$, the graph of $y = e^x$ will be rotated so that it lies exactly on top of the graph $y = \ln x$.

There are some important observations we need to make about $\ln x$.

1. **The domain of the logarithm function is $(0, \infty)$.** The fact that e^x is never zero or negative means that $\ln x$ is only defined if x is positive.

Two special values of $\ln x$ are

2. **$\ln 1 = 0$**, and
3. **$\ln e = 1$.**

We were led to the natural logarithm function by trying to answer the question of when an exponential function would achieve a certain value. We return to this type of question with some examples that show how the ln function is used in applications.

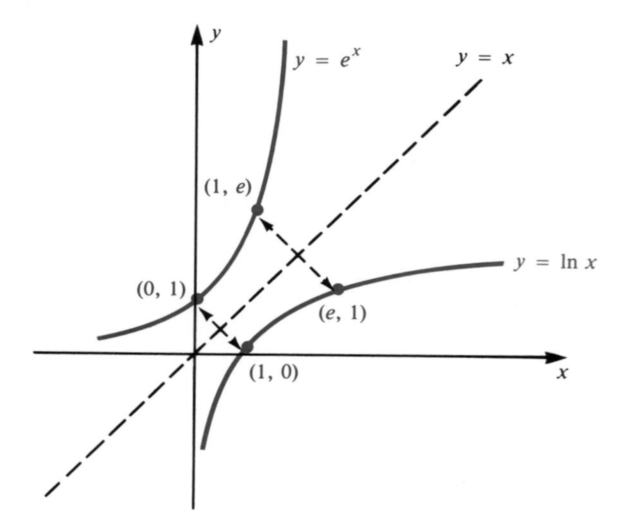

Figure 5–15

Example 3 (*Compare Exercise 35*)
Solve $e^x = 5$.

Solution Apply ln to both sides of the equation $e^x = 5$.

$$\ln e^x = \ln 5$$

Now we use $\ln e^x = x$; $x = \ln 5$

Example 4 (*Compare Exercise 47*)
How long will it take a \$1000 investment to become worth \$2000 at 8% interest, compounded continuously?

Solution The initial amount is \$1000, and $r = 8\% = .08$, so

$$P(t) = 1000\, e^{.08t}$$

We need to solve for t in the equation

$$2000 = 1000\, e^{.08t}$$

Dividing by 1000, we get $\quad 2 = e^{.08t}$

Taking ln of both sides, $\quad \ln 2 = \ln e^{.08t}$

Now, $\ln e^{.08t} = .08t$, so

$$\ln 2 = .08t$$

$$8t = 100 \ln 2$$

$$t = \frac{100 \ln 2}{8}$$

Using a calculator to approximate t, we have

$$t \approx \frac{(100)(.6931)}{8} = 8.66\ldots$$

ANSWER The investment will be worth \$2000 in about $8\frac{2}{3}$ years, or 8 years and 8 months.

We change the beginning principal in Example 5, but still ask how long it will take the investment to double.

Example 5 How long will it take a \$6000 investment to become worth \$12,000 at 8% interest, compounded continuously?

Solution Solve for t in the equation

$$12{,}000 = 6000\, e^{.08t} \qquad \text{Divide by 6000}$$

$$2 = e^{.08t}$$

We just solved this equation in Example 4.

ANSWER The investment will be worth \$12,000 in approximately 8 years and 8 months.

Notice that the two examples have the same solution because

$$\frac{2000}{1000} = \frac{12{,}000}{6{,}000} = 2$$

We can state a general principle: an amount invested at 8% interest, compounded continuously, will double in value in about 8 years and 8 months. Different interest rates will of course have different doubling periods, but all the different periods are independent of the size of the investment.

Example 6 (*Compare Exercise 49*)
How long will it take an investment to double if the interest is 6%, compounded continuously?

Solution Let P be the initial amount invested. Then, our model for $A(t)$ gives us

$$A(t) = Pe^{.06t}$$

We want to know when $A(t) = 2P$, so we solve

$$2P = Pe^{.06t} \qquad \text{Divide by } P$$

$$2 = e^{.06t} \qquad \text{Take ln of both sides}$$

$$\ln 2 = \ln e^{.06t} = .06t$$

$$t = \frac{\ln 2}{.06} \approx 11.55$$

ANSWER The investment will double in a little more than $11\frac{1}{2}$ years.

Properties of ln *x*

We have stressed the special relationship between the ln function and the exponential function and have shown how the ln function enables us to solve equations that have a variable in an exponent.

Historically, logarithms were first introduced in the 17th century in a different setting. Their use became widespread because of the following properties that can be used to simplify calculations. We list these properties without proofs.

Arithmetic Properties of ln

If p and q are positive, and m is any number, then

1. $\ln (p \cdot q) = \ln p + \ln q$
2. $\ln \left(\frac{p}{q}\right) = \ln p - \ln q$
3. $\ln (p^m) = m \ln p$

Log tables became very important aids to computation, especially in navigation. It is ironic that the computational needs of modern navigation have made log tables obsolete. We now need to navigate at high speed in an ocean of space, rather than at low speed in an ocean of water. Modern navigational computations need to be done quickly and on board the space craft. These needs prompted the invention and miniaturization of calculators. If the only importance of logarithms were for calculations, they would be a thing of the past (like the slide-rule). It is because logarithms and exponentials are so important **as functions in applications** using calculus that they still deserve our attention. We need them as functions to describe the behavior of real-world phenomena.

We conclude this section with some more examples using ln to solve exponential equations.

Solving Exponential Equations

We need to be able to solve equations, such as $18 = 2^x$, that involve a base other than e. To solve equations of this type, we establish the following property:

Theorem

$$\text{If } b > 0, \qquad \ln b^c = c \ln b$$

To see this, we rewrite b as $b = e^{\ln b}$.

Then, $b^c = (e^{\ln b})^c = e^{c \ln b}$. Now take ln of both sides of the equation $b^c = e^{c \ln b}$.

$$\ln b^c = \ln e^{c \ln b} = c \ln b.$$

Note that if $b = e$, then $\ln e^c = c \ln e$ and $\ln e = 1$ means as before, $\ln e^c = c$.

Example 7 (*Compare Exercise 39*)
Solve for x if $20 = 2^{3x}$.

Solution Take ln of both sides:

$$\ln 20 = \ln 2^{3x} = 3x \ln 2 \qquad \text{(Here } c \text{ is the expression } 3x\text{)}$$

Thus,

$$\frac{\ln 20}{\ln 2} = 3x$$

$$\frac{\ln 20}{3 \ln 2} = x$$

Example 8 (*Compare Exercise 43*)
Solve for x if $42 = 3 \cdot 5^{2x}$

Solution The simplest thing to do is to divide both sides by 3 as the first step.

$$14 = 5^{2x} \qquad \text{Now take ln of both sides}$$

$$\ln 14 = \ln 5^{2x}$$

$$\ln 14 = 2x \ln 5 \qquad \text{Finally, divide by } 2 \ln 5$$

$$\frac{\ln 14}{2 \ln 5} = x$$

Using a calculator, we can approximate x by

$$\frac{2.639}{2(1.609)} \approx x$$

$$x \approx .8199$$

5–2 Exercises

I.

Simplify each of the following expressions.

1. (*See Example 1*) $\ln e^3$
2. $\ln e^{1/2}$
3. $\ln e^{-4}$
4. $\ln e^{x/2}$
5. (*See Example 2*) $\ln \frac{1}{e^2}$
6. $\ln \frac{1}{e^3}$
7. $\ln 1$
8. $\ln \sqrt{e}$
9. $e^{\ln 4}$
10. $e^{\ln (1/3)}$
11. $e^{\ln 5}$
12. $e^{\ln (x/3)}$

In the next four examples, we use approximate values of the functions involved to emphasize that e^2, for example, is a number.

13. (*See Figure 5–15*) If (2, 7.4) is on graph of $y = e^x$, then what corresponding point is on the graph of $y = \ln x$?
14. If (.5, 1.6) is on the graph of $y = e^x$, then what corresponding point is on the graph of $y = \ln x$?
15. If (4, 1.4) is on the graph of $y = \ln x$, then what corresponding point is on the graph of $y = e^x$?
16. If (.3, − 1.20) is on the graph of $y = \ln x$, then what corresponding point is on the graph of $y = e^x$?

II.

Simplify each of the following expressions.

17. $e^{2 \ln 3}$ *Hint:* First rewrite as $(e^{\ln 3})^2$
18. $e^{2 \ln 4}$
19. $e^{(1/2) \ln 9}$
20. $e^{-\ln 8}$
21. $e^{-\ln 2}$
22. $\ln(e^2 \cdot e^3)$
23. $(\ln e^2)(\ln e^3)$
24. $\ln \left(\frac{1}{e^2}\right)$
25. $\ln(e^4 \cdot e)$
26. $(\ln e^4)(\ln e)$

Solve for x in the following equations; write the answer as an integer.

27. $40 = 5 \cdot 2^x$
28. $1 = 8 \cdot 2^x$
29. $16 = 2^x$
30. $\frac{1}{9} = 3^x$
31. $48 = 3 \cdot 2^{-x}$
32. $6 = 96 \cdot 2^{-x}$
33. $5 = 160 \cdot 2^{-x}$
34. $4 = 36 \cdot 3^{-x}$

III.

Solve for x in the following equations; write the answer using the natural logarithm function.

35. (*See Example 3*) $16 = e^{.8x}$
36. $3 = e^{x/5}$
37. $10 = 40e^{-.02x}$
38. $12 = 4e^{-.3x}$
39. (*See Example 7*) $18 = 2^x$
40. $21 = 3^x$
41. $105 = 5^{2x}$
42. $37 = 4^{x/3}$
43. (*See Example 8*) $20 = 5 \cdot 3^x$
44. $18 = 2 \cdot 4^x$
45. $36 = 4 \cdot 3^{2x}$
46. $40 = 10 \cdot 2^{x/3}$
47. (*See Example 4*) \$1000 is invested at 6% interest, compounded continuously. When will the investment be worth \$3000?
48. How long will it take an investment earning 7% interest, compounded continuously, to double?
49. (*See Example 6*) How long will it take an investment earning 14% interest, compounded continuously, to double?
50. How much must be invested at 8%, compounded continuously, so that the investment will be worth \$10,000 after 6 years?
51. How much must be invested at 5%, compounded continuously, so that the investment is worth \$20,000 after 30 years?
52. If an investment doubles every ten years, what is the interest rate (continuous compounding)?
53. If the population of the earth is growing at the rate of 2%, compounded continuously, how long will it take the population to double?

5–3 The Derivative of e^x

The Derivative of $f(x) = e^x$
Use of the Chain Rule with Exponential Functions

The Derivative of $f(x) = e^x$

We begin by recalling the definition of derivative as given in Chapter 2. The definition is

$$f'(x) = \lim_{\Delta x \to 0} \frac{f(x + \Delta x) - f(x)}{\Delta x}$$

We now use this definition to find the derivative of $f(x) = e^x$.

Theorem

If $f(x) = e^x$, then $f'(x) = e^x$.

The function $f(x) = e^x$ is its own derivative!

To see how this formula is obtained, we repeat the three-step process that we used in Chapter 2.

Step 1. Compute and simplify $f(x + \Delta x) - f(x)$.

$$f(x) = e^x, \text{ so } f(x + \Delta x) = e^{x+\Delta x}$$

Because $e^{a+b} = e^a e^b$, we can write $e^{x+\Delta x} = e^x e^{\Delta x}$. Thus,

$$\begin{aligned} f(x + \Delta x) - f(x) &= e^{x+\Delta x} - e^x \\ &= e^x e^{\Delta x} - e^x \\ &= e^x(e^{\Delta x} - 1) \end{aligned}$$

Step 2. Divide the result of Step 1 by Δx. There's not much we can do in this step except write

$$\frac{e^x(e^{\Delta x} - 1)}{\Delta x} = e^x\left[\frac{e^{\Delta x} - 1}{\Delta x}\right]$$

Step 3. Compute the limit of the result from Step 2 as $\Delta x \to 0$. Notice that the factor e^x is independent of Δx. Thus,

$$\lim_{\Delta x \to 0}\left(e^x \frac{e^{\Delta x} - 1}{\Delta x}\right) = e^x \lim_{\Delta x \to 0} \frac{e^{\Delta x} - 1}{\Delta x}$$

We need to know the value of

$$\lim_{\Delta x \to 0} \frac{e^{\Delta x} - 1}{\Delta x}$$

We shall not prove the following result, but if b is any positive number, then

$$\lim_{\Delta x \to 0} \frac{b^{\Delta x} - 1}{\Delta x} = \ln b$$

As a special case, we have

$$\lim_{\Delta x \to 0} \frac{e^{\Delta x} - 1}{\Delta x} = \ln e = 1$$

Putting the three steps together,

If $f(x) = e^x$, then

$$\begin{aligned} f'(x) &= e^x \lim_{\Delta x \to 0} \frac{e^{\Delta x} - 1}{\Delta x} \\ &= e^x \cdot 1 \\ &= e^x \end{aligned}$$

which establishes the theorem.

Because the derivative gives the slope of a tangent line, the geometric interpretation of this formula is as follows: the slope of the line tangent to the graph of $y = e^x$ at any point P is the same as the second coordinate of P. More succinctly,

$$\text{if } y = e^x, \text{ then } y' = y.$$

Warning! This formula is very different from the formulas for derivatives you have been using up to this point. In the previous chapters, every exponent was a number. When a function involves an exponent that is a fixed number, you use the power rule to compute the derivative. When the exponent is a variable, you must use a different formula. Be careful to keep the distinction between a constant exponent and a variable exponent. See Figure 5–16.

x^b ← Constant exponent $\qquad$ b^x ← Variable exponent

Figure 5–16

Example 1 (*Compare Exercise 3*)
Find $f'(x)$ if $f(x) = x^3 + e^x$.

Solution $f'(x) = 3x^2 + e^x$
We used two distinct formulas, the power rule formula for $\dfrac{d}{dx}x^3$ and the exponential rule formula for $\dfrac{d}{dx}e^x$.

Notice the use of different formulas in the following examples.

Example 2 (*Compare Exercise 5*)
Find $f'(x)$ if $f(x) = x^2e^x$.

Solution We must use the product rule:

$$\frac{d}{dx}(x^2e^x) = e^x\left(\frac{d}{dx}x^2\right) + x^2\left(\frac{d}{dx}e^x\right)$$
$$= e^x \cdot 2x + x^2 \cdot e^x$$
$$f'(x) = 2xe^x + x^2e^x$$

Example 3 (*Compare Exercise 11*)
Find $f'(x)$ if $f(x) = \dfrac{e^x}{4x^3}$.

Solution Here we use the quotient rule:

$$f'(x) = \frac{4x^3\left(\dfrac{d}{dx}e^x\right) - e^x\left(\dfrac{d}{dx}4x^3\right)}{(4x^3)^2}$$
$$= \frac{4x^3 \cdot e^x - e^x \cdot 12x^2}{16x^6}$$
$$= \frac{4x^3e^x - 12x^2e^x}{16x^6}$$

Now simplifying,

$$= \frac{4x^2e^x(x-3)}{16x^6}$$

$$= \frac{e^x(x-3)}{4x^4}$$

Example 4 (*Compare Exercise 15*)
Find $f'(x)$ if $f(x) = (e^x - 4x)^3$.

Solution We can use the generalized power rule:

$$f(x) = (g(x))^R, \text{ with } g(x) = e^x - 4x \text{ and } R = 3$$
$$f'(x) = R(g(x))^{R-1} \cdot g'(x), \text{ and } g'(x) = e^x - 4$$

Hence,

$$f'(x) = 3(e^x - 4x)^2(e^x - 4)$$

Use of the Chain Rule with Exponential Functions

But what about an exponent other than x? For example, what is the derivative of e^{5x}? When the exponent is a function, you will have to use the chain rule, but first a reminder:

The Chain Rule Revisited. The chain rule says that

$$\frac{dy}{dx} = \frac{dy}{du} \cdot \frac{du}{dx}$$

We now show how the chain rule works with exponential functions.

Example 5 (*Compare Exercise 7*)
If $f(x) = e^{5x}$, what is $f'(x)$?

Solution We let $y = e^{5x}$ and define the intermediary variable u by setting $u = 5x$ so that $y = e^u$. Then,

$$\frac{dy}{du} = e^u \qquad \text{and} \qquad \frac{du}{dx} = 5$$

so

$$\frac{dy}{dx} = \frac{dy}{du} \cdot \frac{du}{dx}$$
$$= e^u \cdot 5$$
$$= e^{5x} \cdot 5 = 5e^{5x}$$
$$f'(x) = 5e^{5x}$$

The generalized power rule is a special form of the chain rule. Here is another special form of the chain rule to help you compute $f'(x)$ for the case $f(x) = e^{g(x)}$.

Theorem

If $f(x) = e^{g(x)}$, then $f'(x) = e^{g(x)} \cdot g'(x)$.

To see this, we apply the chain rule to $y = e^{g(x)}$ by letting $u = g(x)$ so that $y = e^u$. Then,

$$\frac{dy}{du} = e^u \qquad \text{and} \qquad \frac{du}{dx} = g'(x)$$

$$\begin{aligned} \frac{dy}{dx} &= \frac{dy}{du} \cdot \frac{du}{dx} \\ &= e^u \cdot g'(x) \\ &= e^{g(x)} \cdot g'(x) \end{aligned}$$

In Example 4, $g(x) = 5x$ and $g'(x) = 5$, so $f'(x) = e^{5x} \cdot 5 = 5e^{5x}$.

Example 6 (*Compare Exercise 27*)
Find an equation of the line tangent to the curve $y = 4e^{3x}$ when $x = 2$.

Solution To find the slope, we compute $\frac{dy}{dx}$. Since 4 is a constant,

$$\begin{aligned} \frac{d}{dx}(4e^{3x}) &= 4\frac{d}{dx}e^{3x} \\ &= 4(e^{3x} \cdot 3) \\ &= 12e^{3x} \end{aligned}$$

With $x = 2$, the slope of the tangent line is $12e^6$, and the value of y is $4e^6$, so the line has the equation

$$y - 4e^6 = 12e^6(x - 2)$$

The next example gives the general formula for $f'(x)$ if $f(x) = ke^{cx}$.

Example 7 If $y = ke^{cx}$, find the formula for y'.

Solution If $y = ke^{cx}$,

$$\begin{aligned} y' &= \frac{d}{dx}(ke^{cx}) \\ &= k\frac{d}{dx}(e^{cx}) \\ &= ke^{cx} \cdot c \\ &= y \cdot c \\ &= cy \end{aligned}$$

Thus,

$$y' = cy.$$

This is an important formula in applications, and we will investigate some of these applications in Section 5–5. The equation $y' = cy$ says that the rate of change of y is directly proportional to y.

Example 8 If $f(x) = e^{4x}$, what is $f''(x)$?

Solution

$$f'(x) = e^{4x} \cdot 4 = 4e^{4x}$$

so

$$f''(x) = 4e^{4x} \cdot 4 = 16e^{4x}$$

Example 9 (*Compare Exercises 19 and 29*)
If $f(x) = e^{x^2}$, what is $f''(x)$?

Solution

$$f'(x) = e^{x^2} \cdot 2x = 2xe^{x^2}$$

To compute $f''(x)$, we must use both the product rule and the chain rule. We use the product rule on $(2x)(e^{x^2})$ and, along the way, must use the chain rule on e^{x^2}.

$$f''(x) = e^{x^2}(2) + 2x(e^{x^2} \cdot 2x) = 2e^{x^2} + 4x^2e^{x^2}.$$

Sometimes you must use the chain rule more than once.

Example 10 (*Compare Exercise 33*)
Find $f'(x)$ if $f(x) = (e^{x^2} + x)^3$.

Solution First, use the generalized power rule form of the chain rule.

$$\begin{aligned} f'(x) &= 3(e^{x^2} + x)^2 \cdot \frac{d}{dx}(e^{x^2} + x) \\ &= 3(e^{x^2} + x)^2(e^{x^2} \cdot 2x + 1) \end{aligned}$$

Notice the use of the chain rule in computing $\frac{d}{dx} e^{x^2}$:

$$= 3(e^{x^2} + x)^2(2xe^{x^2} + 1)$$

We can again use the first and second derivatives as tools to investigate how these new functions behave.

Example 11 (*Compare Exercise 47*)
Sketch the graph of $f(x) = e^{2x}$.

Solution First, remember that e^{2x} is positive for all values of x, so the graph will be above the x-axis. Next, $f'(x) = e^{2x} \cdot 2 = 2e^{2x}$ is also positive for all x. Thus, f is increasing on $(-\infty, \infty)$. Finally, $f''(x) = 2e^{2x} \cdot 2 = 4e^{2x} > 0$, so the graph of f is concave up on $(-\infty, \infty)$.

To find the y-intercept, we compute $f(0) = e^{(2)(0)} = e^0 = 1$. See Figure 5–17 for the graph of $f(x) = e^{2x}$.

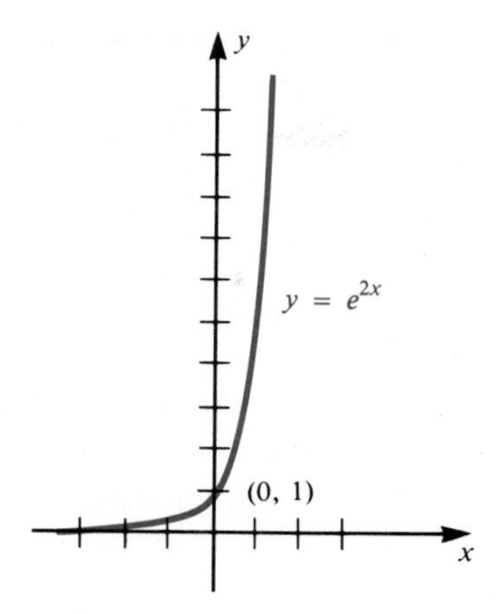

Figure 5–17

Example 12 (*Compare Exercise 45*)
Find the maximum and minimum values of f if $f(x) = xe^{-2x}$ on the interval $[-5, 5]$.

Solution To compute $f'(x)$, we use the product rule on $(x)(e^{-2x})$ and must remember to use the chain rule when computing $\frac{d}{dx}e^{-2x}$. We have

$$\begin{aligned} f'(x) &= e^{-2x}(1) + x(e^{-2x}(-2)) \\ &= e^{-2x} - 2xe^{-2x} \qquad \text{Factor } e^{-2x} \\ &= e^{-2x}(1 - 2x) \end{aligned}$$

$f'(x)$ exists for all x, so the only critical numbers are solutions to $e^{-2x}(1 - 2x) = 0$. Since $e^{-2x} = 0$ has no solution, $x = \frac{1}{2}$ is the only critical number of f.

The domain of f is a closed and bounded interval, so we use the candidates test:

$$\begin{aligned} f(5) &= 5e^{-10} \\ f(-5) &= -5e^{10} \\ f\left(\frac{1}{2}\right) &= \frac{1}{2}e^{-1} = \frac{1}{2e} \end{aligned}$$

$f(-5) = -5e^{10}$ is the only negative number, so the minimum value of f is $-5e^{10}$. Comparing the other two numbers,

$$\frac{1}{2e} > \frac{5}{e^{10}}$$

so the maximum value of f is

$$f\left(\frac{1}{2}\right) = \frac{1}{2e}$$

5–3 Exercises

I.

Compute $f'(x)$ for each of the following.

1. $f(x) = 4e^x$
2. $f(x) = -7e^x$
3. (*See Example 1*) $f(x) = x^3 - 2e^x$
4. $f(x) = 4x^2 + 5e^x$
5. (*See Example 2*) $f(x) = x^3e^x$
6. $f(x) = 2xe^x$
7. (*See Example 5*) $f(x) = 6e^{3x}$
8. $f(x) = 3e^{-2x}$
9. $f(x) = 8e^{x/2}$
10. $f(x) = 12e^{x/4}$
11. (*See Example 3*) $f(x) = \dfrac{x^3 + 2}{e^x}$
12. $f(x) = \dfrac{x^2 - 1}{3e^x}$
13. $f(x) = \dfrac{e^x + 2}{x^2}$
14. $f(x) = \dfrac{2e^x - 5}{x^3}$
15. (*See Example 4*) $f(x) = (e^x - 7)^4$
16. $f(x) = (e^x + 7x^2)^3$
17. $f(x) = \sqrt{e^x + 1}$
18. $f(x) = \dfrac{1}{\sqrt{2e^x + 3}}$

II.

Compute $f'(x)$ for each of the following.

19. (*See Example 9*) $f(x) = 2x^3e^{5x}$
20. $f(x) = 7x^2e^{3x}$
21. $f(x) = (e^{4x} + 1)^3$
22. $f(x) = \sqrt{e^{2x} + 5}$
23. $f(x) = 3x^2 + xe^{-x}$
24. $f(x) = x^2e^x - 4x$
25. $f(x) = (\sqrt{x^2 + 1})e^{4x}$
26. $f(x) = (3x + 1)^2e^{5x}$
27. (*See Example 6*) What is an equation of the line tangent to the curve $y = e^{x/3}$ at the point whose first coordinate is 6?
28. What is an equation of the line tangent to the curve $y = e^{-2x}$ at the point whose first coordinate is 4?

III.

Find $f''(x)$ for each of the following.

29. (*See Example 9*) $f(x) = e^{x^2+x}$
30. $f(x) = e^{3x^2-x}$
31. $f(x) = x^2e^{4x}$
32. $f(x) = 2x^3e^{-x}$

Find $f'(x)$ for each of the following.

33. (*See Example 10*) $f(x) = \sqrt{e^{4x} + 5x}$
34. $f(x) = (e^{x^2} - 6x^3)^4$
35. $f(x) = (x^2 + 1)e^{\sqrt{3x+1}}$

36. $\dfrac{5e^{(4x-1)^2}}{(x^2+1)^3}$
37. What is an equation of the line tangent to the curve $y = e^{-x}$ at the point whose first coordinate is ln 4?
38. What is an equation of the line tangent to the curve $y = e^{x/3}$ at the point whose first coordinate is ln 8?
39. (*See Example 12*) What are the critical numbers of $f(x) = x^2e^{-x}$?
40. What are the critical numbers of $f(x) = xe^{x/3}$?
41. On what intervals is f decreasing if $f(x) = \dfrac{e^x}{x}$?
42. On what intervals is f decreasing if $f(x) = xe^x$?
43. What is the point of inflection on the curve $y = xe^{-x}$?
44. What is the point of inflection on the curve $y = xe^{x/2}$?
45. (*See Example 12*) Find the extreme values of $f(x) = xe^{x/2}$, $-4 \leq x \leq 4$.
46. Find the extreme values of $f(x) = x^2e^x$, $-1 \leq x \leq 4$.
47. (*See Example 11*) Sketch the graph of $f(x) = xe^x$, $-5 \leq x \leq 5$.
48. Sketch the graph of $f(x) = xe^{-x}$, $-5 \leq x \leq 5$.

5–4 The Derivative of ln *x*

The Derivative of ln *x*
Use of the Chain Rule with the Logarithm Function

We will not establish the formula for the derivative of $f(x) = \ln x$ by using the definition of the derivative. Instead, we use the relation between the functions $\ln x$ and e^x to first show that the natural logarithm function does indeed have a derivative, and then to find the formula for this derivative.

Recall from Section 5–2 that if we let $g(x) = \ln x$ and $f(x) = e^x$, then g and f "undo" each other in the sense that $g(f(x)) = \ln e^x = x$ for all x in the domain of f, and $f(g(x)) = e^{\ln x} = x$ for all x in the domain of g. There is a general theorem that states that whenever f and g are two functions that have this relationship and f' exists and doesn't equal 0, then g' also exists. Because $f'(x) = e^x$, which is never equal to 0, this general theorem guarantees that the derivative of the natural logarithm exists. We now use the chain rule to establish the formula for this derivative.

The Derivative of ln *x*

Theorem

If $g(x) = \ln x$, then $g'(x) = \dfrac{1}{x}$.

Remember that the domain of ln is $(0, \infty)$, so this formula is valid for $x > 0$.

To establish this theorem, we use a special case of the form of the chain rule given in the previous section. The formula we wish to use is

$$\text{if } h(x) = e^{g(x)}, \qquad \text{then} \qquad h'(x) = e^{g(x)} \cdot g'(x)$$

In particular, let $g(x) = \ln x$ so that $h(x) = e^{\ln x} = x$. We have

$$x = h(x) = e^{g(x)}$$

Now differentiate,

$$1 = h'(x) = e^{g(x)} \cdot g'(x)$$

Again, use $e^{g(x)} = e^{\ln x} = x$ so

$$1 = x \cdot g'(x)$$

Solving for $g'(x)$, we have

$$g'(x) = \frac{1}{x}$$

Example 1 (*Compare Exercise 3*)
Compute $f'(x)$ if $f(x) = x^3 \ln x$

Solution Using the product rule, we have

$$\begin{aligned} f'(x) &= (\ln x)(3x^2) + (x^3)\left(\frac{1}{x}\right) \\ &= 3x^2(\ln x) + x^2 \end{aligned}$$

Example 2 (*Compare Exercise 13*)
Find $f'(x)$ if $f(x) = \dfrac{\ln x}{x^4}$.

Solution We must use the quotient rule.

$$\begin{aligned} f'(x) &= \frac{x^4(1/x) - (\ln x)4x^3}{(x^4)^2} \\ &= \frac{x^3 - (\ln x)4x^3}{x^8} = \frac{x^3(1 - 4\ln x)}{x^8} \\ &= \frac{1 - 4\ln x}{x^5} \end{aligned}$$

Next, we look at the composition of various functions with $\ln x$ and see how the chain rule applies in these cases.

Example 3 (*Compare Exercise 5*)
Find $f'(x)$ if $f(x) = \ln(3x)$.

Solution **Warning!** The formula only gives the derivative of $\ln x$. You cannot conclude that $f'(x) = 1/(3x)$. Computing the derivative of f in the form written above requires the use of the chain rule.

We will remind you of the chain rule and show how it applies here, but first we do this example another way, by rewriting $f(x)$. We use the fact that

$$\ln(p \cdot q) = \ln p + \ln q$$

so with $p = 3$ and $q = x$,

$$\ln(3x) = \ln 3 + \ln x$$

Now we take the derivative of f.

$$\begin{aligned} \frac{d}{dx}(\ln 3x) &= \frac{d}{dx}(\ln 3 + \ln x) \\ &= \frac{d}{dx}(\ln 3) + \frac{d}{dx}(\ln x) \\ &= 0 + \frac{1}{x} \\ &= \frac{1}{x} \end{aligned}$$

ANSWER $f'(x) = \dfrac{1}{x}$

There are a few surprises in this computation (which is the real reason we did it). First, please observe that $\ln 3$ is a constant, and so its derivative is 0. (Sometimes when first learning to find derivatives involving ln, students will write $\frac{1}{3}$ for the derivative of ln 3). Second, the final result may also be surprising; $\ln x$ and $\ln 3x$ have the same derivative. This is due to the property of ln that allowed us to write $\ln 3x$ as $\ln 3 + \ln x$. You know that $\ln x$ and $(4 + \ln x)$ have the same derivative because the derivative of the constant 4 is 0. Remember that ln 3 is also a number, just as 4 is a number ($\ln 3 = 1.09861\ldots$).

Example 4 (*Compare Exercise 11*)
If $f(x) = \ln(3x)$, compute $f'(x)$.

Solution This is the same function as in Example 3, but this time we do use the chain rule. The intermediate evaluation is $3x$; we let $u = 3x$ so that $y = \ln u$;

$$\frac{dy}{du} = \frac{1}{u} \qquad \text{and} \qquad \frac{du}{dx} = 3$$

Thus,

$$\begin{aligned} f'(x) = \frac{dy}{dx} &= \frac{dy}{du} \cdot \frac{du}{dx} \\ &= \frac{1}{u} \cdot (3) \end{aligned}$$

Replace u by $3x$ so that the final answer is expressed in terms of x:

$$= \frac{1}{3x} \cdot 3 = \frac{1}{x}$$

Remember that you cannot get different answers using different techniques; there may be more than one way to proceed through the computation, but the end results must be the same.

Example 5 (*Compare Exercise 15*)

If $y = \ln(x^2 - 3x)$, find $\frac{dy}{dx}$.

Solution In the process of evaluating $\ln(x^2 - 3x)$, you first evaluate $x^2 - 3x$ and then take the logarithm of $x^2 - 3x$. As we did in Example 4, let u be an intermediary variable; here, $u = x^2 - 3x$. Then, $y = \ln u$, so

$$\frac{dy}{du} = \frac{1}{u} \qquad \text{and} \qquad \frac{du}{dx} = 2x - 3$$

$$\frac{dy}{dx} = \frac{dy}{du} \cdot \frac{du}{dx}$$

$$= \left(\frac{1}{u}\right)(2x - 3) = \frac{2x - 3}{u}$$

Finally, replace u by $x^2 - 3x$ so that the final answer is expressed in terms of x.

$$\frac{dy}{dx} = \frac{2x - 3}{x^2 - 3x}$$

Example 6 (*Compare Exercise 37*)
If $f(x) = \ln x^3$, find $f'(x)$.

Solution This derivative can also be computed two different ways.

1. Use the chain rule with $u = x^3$ and $y = \ln u$;

$$\frac{dy}{du} = \frac{1}{u} \qquad \text{and} \qquad \frac{du}{dx} = 3x^2$$

$$f'(x) = \frac{1}{u} \cdot 3x^2 = \frac{1}{x^3} \cdot 3x^2 = \frac{3}{x}$$

$$f'(x) = \frac{3}{x}$$

2. We compute the derivative again, this time using a property of ln to rewrite $f(x) = \ln(x^3) = 3(\ln x)$. Thus,

$$\frac{d}{dx} f(x) = \frac{d}{dx} 3(\ln x) = 3 \frac{d}{dx} \ln x = 3 \cdot \frac{1}{x} = \frac{3}{x}$$

$$f'(x) = \frac{3}{x}$$

Once again, we have used two different techniques for computing $f'(x)$, and once again, the two techniques yield the same result.

Example 7 (*Compare Exercise 23*)
If $g(x) = (\ln x)^3$, find $g'(x)$.

Solution You should contrast this example with Example 6; the order of the operations is very important. For instance, if we let $x = 2$, for the function in Example 6, we have $f(2) = \ln 2^3 = \ln 8 = 2.0794\ldots$; while for the function in this example, we have $g(2) = (\ln 2)^3 = (0.6931\ldots)^3 = 0.3330\ldots$.

To compute $g'(x)$, we again have to use the chain rule, but now in the form of the generalized power rule—$g(x)$ is $(\ln x)$ raised to a power. Let $u = \ln x$, then $y = u^3$ and

$$\begin{aligned} g'(x) = \frac{dy}{dx} &= \frac{dy}{du} \cdot \frac{du}{dx} \\ &= 3u^2 \cdot \frac{1}{x} \\ &= 3(\ln x)^2 \cdot \frac{1}{x} \\ &= \frac{3(\ln x)^2}{x} \end{aligned}$$

Use of the Chain Rule with the Logarithm Function

We were able to give special forms of the chain rule for the generalized power rule and the exponential function, and we can again give a special case of the chain rule when f is of the form $f(x) = \ln(g(x))$.

Special Form of Chain Rule

If $f(x) = \ln(g(x))$, then

$$\begin{aligned} f'(x) &= \frac{1}{g(x)} \cdot g'(x) \\ &= \frac{g'(x)}{g(x)} \end{aligned}$$

This formula results from using the chain rule, with $y = f(x) = \ln(g(x))$, and setting $u = g(x)$. Then, $y = \ln u$ so

$$\frac{dy}{du} = \frac{1}{u} \qquad \text{and} \qquad \frac{du}{dx} = g'(x)$$

$$\begin{aligned} f'(x) = \frac{dy}{dx} &= \frac{dy}{du} \cdot \frac{du}{dx} \\ &= \frac{1}{u} \cdot g'(x) \\ &= \frac{1}{g(x)} \cdot g'(x) = \frac{g'(x)}{g(x)} \end{aligned}$$

Example 8 (*Compare Exercise 29*)
If $f(x) = \ln(x^3 + 4x)$, find $f'(x)$.

Solution We can use the special form of the chain rule, with $g(x) = x^3 + 4x$. Thus, $g'(x) = 3x^2 + 4$, and

$$f'(x) = \frac{1}{x^3 + 4x} \cdot (3x^2 + 4)$$
$$= \frac{3x^2 + 4}{x^3 + 4x}$$

Use of this special form of the chain rule allows you to compute the derivative of $\ln(g(x))$ without having to go through the u substitution, and so quickens your computation. To see this, go back and do Example 5 using this form. Use of this form also makes it easier to compute more complicated derivatives.

Example 9 (*Compare Exercise 41*)
Find $f'(x)$ if $f(x) = [\ln(x^2 + x)]^3$.

Solution $f(x)$ is of the form $(h(x))^3$, so first, use the generalized power rule.

$$f'(x) = 3(h(x))^2 h'(x)$$

Now, compute $h'(x)$.

$$h(x) = \ln(g(x)) = \ln(x^2 + x)$$
$$h'(x) = \frac{2x + 1}{x^2 + x}$$

Putting the pieces together, we have

$$f'(x) = 3[\ln(x^2 + x)]^2\left(\frac{2x + 1}{x^2 + x}\right)$$

We conclude with another example that also uses the chain rule twice.

Example 10 (*Compare Exercise 45*)
If $f(x) = \ln(x + (x^2 + 1)^3)$, find $f'(x)$.

Solution Let $g(x) = x + (x^2 + 1)^3$. Then, $g'(x) = 1 + 3(x^2 + 1)^2 \cdot 2x$. Note that the $2x$ is due to the chain rule used on $(x^2 + 1)^3$.

$$g'(x) = 1 + 6x(x^2 + 1)^2$$

Thus,

$$f'(x) = \frac{g'(x)}{g(x)}$$
$$= \frac{1 + 6x(x^2 + 1)^2}{x + (x^2 + 1)^3}$$

5–4 Exercises

I.

Compute $f'(x)$ in each of the following exercises.

1. $f(x) = x^2 + \ln x$
2. $f(x) = 7x^2 - 3\ln x$
3. (*See Example 1*) $f(x) = x \ln x$
4. $f(x) = (x^2 + 1)\ln x$
5. (*See Example 3*) $f(x) = \ln x + \ln 4$
6. $f(x) = (\ln 5)(\ln x)$
7. $f(x) = \dfrac{\ln x}{\ln 10}$
8. $f(x) = \dfrac{\ln x}{\ln 7}$
9. $f(x) = e^x \ln x$
10. $f(x) = x^2 \ln x$
11. (*See Example 4*) $f(x) = \ln(6x)$
12. $f(x) = \ln(x/2)$
13. (*See Example 2*) $f(x) = \dfrac{\ln x}{x}$
14. $f(x) = \dfrac{x}{\ln x}$

II.

Compute $f'(x)$ in each of the following exercises.

15. (*See Example 5*) $f(x) = \ln(x^2 + 1)$
16. $f(x) = \ln(5x - 3)$
17. $f(x) = [\ln(x + 1)]^3$
18. $f(x) = \ln(x^2 + 3x)$
19. $f(x) = x\ln(3x - 8)$
20. $f(x) = \dfrac{\ln(5x - 1)}{3x^2}$
21. $f(x) = \dfrac{\ln(x^2 + 1)}{6x}$
22. $f(x) = 4x\ln(6x + 11)$
23. (*See Example 7*) $f(x) = (\ln x)^4$
24. $f(x) = \ln x^5$
25. $f(x) = (\ln x)e^{5x}$
26. $f(x) = \ln(e^{2x} + 5)$
27. $f(x) = x^2(\ln 6)$
28. $f(x) = (\ln 4)^2$
29. (*See Example 8*) $f(x) = \ln(x^2 - 3x + 1)$
30. $f(x) = \ln(x^3 - 9x)$
31. $f(x) = \ln(4x + 8)$
32. $f(x) = \ln(x + e^x)$

III.

Compute $f'(x)$ in each of the following exercises.

33. $f(x) = x\ln e^2$
34. $f(x) = e^{(\ln 4)x}$
35. $f(x) = e^{\ln x}$
36. $f(x) = x^{\ln e}$
37. (*See Example 6*) $f(x) = \ln\sqrt{x}$
38. $f(x) = \ln(x^3 + 5x)$
39. $f(x) = \ln(3x + 1)^4$
40. $f(x) = \ln\sqrt{5x + 1}$
41. (*See Example 9*) $f(x) = \sqrt{\ln(x^2 + 1)}$
42. $f(x) = [\ln(x^3 - 8)]^2$
43. $f(x) = [\ln(5x + 1)]^2$
44. $f(x) = e^{x^2}\ln(x^3 - 8x + 9)$

45. (*See Example 10*) $f(x) = \ln(e^{2x} + \sqrt{4x + 10})$

46. $f(x) = \ln(e^{2x} + e^{4x})$

47. $f(x) = [\ln(x + e^{2x})]^3$

48. $f(x) = [\ln(x^2 + \sqrt{3x + 5})]^2$

Compute $f''(x)$ in the following exercises.

49. $f(x) = x^2 \ln x$

50. $f(x) = x(\ln x)^2$

51. $f(x) = x \ln \sqrt{x + 1}$

52. $f(x) = x^2 e^{4x}$

5–5 Applications of Exponential Functions

We began this chapter by considering the problem of describing a rate of change when that rate is given in terms of the percentage of change rather than the amount of change. The particular situation we dealt with, continuously compounding interest, led us to look at the number e and exponential functions with base e. The natural logarithm was introduced to help solve equations involving such exponential functions. Next, we found the formulas for the derivatives of these functions. In particular, we found that

$$\text{if } y = ke^{cx}, \qquad \text{then} \qquad y' = cy$$

This formula allows us to answer the original question about rates of change given as percentages. For example, letting y represent the size of the population, we can translate the statement

"The population is growing at the rate of 2%" to

$$y' \text{ is } 2\% \text{ of } y$$

to

$$y' = .02y$$

Here $c = .02$.

An exponential function is the function to use when the rate of change is given in terms of percentage, or said another way, when the rate of change of the function is directly proportional to the value of the function. Thus, when you want to describe population growth, inflation, unemployment statistics, the national debt, or any quantity for which the rate of change is given in terms of percentage, you should be using an exponential function as your model. We will do some examples showing these applications, and then discuss further the implication of "exponential growth."

Example 1 (*Compare Exercise 1*)

A certain city is growing at the continuous rate of 5%. If the population was 100,000 in 1980, what will it be in the year 2000?

Solution Let us use $P(t)$ to denote the population at time t. The rate of change (growth) is 5% of the population (amount).

$$P'(t) = .05P(t)$$

We are using t instead of x, and $P(t)$ instead of $f(x)$, but this equation tells us that $P(t)$ is of the form $P(t) = ke^{ct}$.

Notice that in the expression $y' = cy$, the function was $y = ke^{cx}$. In this example, $c = .05$; thus, we have

$$P(t) = ke^{.05t}$$

The number c in the expression $y' = cy$ is the coefficient of x in the expression $y = ke^{cx}$.

To find k, we must use information about a particular value of y. Here, we need to use information about the size of the population at a particular time. When we are given data in this form, it is convenient to choose $t = 0$ to correspond to the year 1980. The convenience comes from the fact that if $P(t) = ke^{ct}$, then

$$P(0) = ke^{c0} = ke^{0} = k \cdot 1 = k$$

The number k is the value of y when $x = 0$.

If $t = 0$ corresponds to 1980, then the problem gives us the information that $P(0) = 100{,}000$, so $k = 100{,}000$. We can now write

$$P(t) = 100{,}000e^{.05t}$$

The problem asks for the population in the year 2000, which is 20 years after 1980. The year 2000 corresponds to $t = 20$:

$$\begin{aligned} P(20) = 100{,}000e^{(.05)(20)} &= 100{,}000e^{1} \\ &= 100{,}000e \\ &\approx 271{,}828 \end{aligned}$$

It would be silly to calculate the population using more decimal places of e. In fact, it is not wise, in practice, to say much more than that this model predicts a population of a little over 270,000 in the year 2000.

Example 2 (*Compare Exercise 5*)

If \$1000 is invested in a trust fund in the year 2000, and the fund pays $5\frac{3}{4}\%$ interest, compounded continuously, how much will be in the fund **(a)** after 100 years? **(b)** after 200 years?

Solution First, we invite you to estimate the two answers; you may be surprised by the following computation. Let $A(t)$ be the amount the fund is worth t years after it is established; the continuous percentage growth rate tells us that $A(t)$ is of the form

$$A(t) = ke^{ct}$$

As in Example 1, the growth rate is c, and k is the initial amount. Here $c = 5\frac{3}{4}\% = .0575$ and $k = 1000$. So, $A(t) = 1000e^{.0575t}$

(a) Setting $t = 100$,

$$\begin{aligned} A(100) &= 1000e^{(.0575)(100)} \\ &= 1000e^{5.75} \\ &\approx (1000)(314.19) \\ &= 314{,}190 \end{aligned}$$

After 100 years, the fund is worth \$314,190.

Now that you know the answer to the first question, would you like to guess again at the second answer?

(b) We need to compute $A(200)$.

$$\begin{aligned} A(200) &= 1000e^{(.0575)(200)} \\ &= 1000e^{11.5} \\ &\approx (1000)(98715.77) \\ &= 98{,}715{,}770 \end{aligned}$$

After 200 years, the fund is worth \$98,715,770!

Exponential functions can grow very rapidly, as this example shows very dramatically. The amount of growth in the first 100 years is sizable, and we could all wish one of our great-grandparents had created such a trust fund for us. The growth during the second hundred years is almost too much to believe. An investment of \$1000 has grown to nearly \$100 million. Imagine the social impact an investment like that can have. Benjamin Franklin did, and he established such a fund that now, approximately 200 years after his death, provides scholarships for medical students. We should note that historically, a rate of $5\frac{3}{4}\%$ is a high rate of interest, and in the exercises you will see how small changes in the interest rate can affect the total amount, especially over a long period of time.

In the next few examples, we explore different ways people can present the same data; growth at a constant percentage rate can lead to some interesting discussions.

Suppose the federal budget grows at the steady rate of 7% under two presidents, X and Y. Assume that the budget starts at \$1000 billion under President X, and X is president for four years.

Example 3 (*Compare Exercise 6*)
What will the budget be after four years?

Solution Let $B(t)$ = the budget (in billions of dollars) after t years. Percentage growth rate means we should write

$$B(t) = ke^{ct}$$

Here, $k = 1000$ (units are billions of dollars) and $c = .07$. Using $t = 4$, we have

$$\begin{aligned} B(4) = 1000e^{(.07)4} &= 1000e^{.28} \\ &\approx 1323 \end{aligned}$$

After four years, the budget is approximately \$1323 billion.

Now President Y takes over.

Example 4 (*Compare Exercise 7*)
What will the budget be **(a)** after three years of Y's term; **(b)** after four years?

Solution **(a)** We continue letting $t = 0$ correspond to the beginning of X's term, so to answer **(a)** we use $t = 7$.

$$B(7) = 1000e^{(.07)7} = 1000e^{.49} \approx 1632$$

(b) With $t = 8$, we have

$$B(8) = 1000e^{(.07)8} = 1000e^{.56} \approx 1751$$

Now consider how these numbers can be debated. Under X, the budget grew (in amount) a total of \$323 billion, for an average growth of about \$80 billion per year. Under Y, the budget grew by \$(1751 − 1632) billion = \$119 billion during the last year. Since $119 \approx (1\frac{1}{2})(80)$, some will claim that the budget is growing $1\frac{1}{2}$ times as fast under Y as it did under X! Yet remember that the percentage growth was a steady 7% under both presidents. Opponents of Y will talk about amounts; supporters of Y will talk about percentages. The effects of percentage growth rate on magnitudes are not well understood by the general public.

Sometimes, growth rates are given directly in terms of percentage; sometimes, the notion of doubling period (or half-life) is used. The presence of a doubling period also indicates that you should use an exponential function to model the real world situation. We have seen that the number c represents the instantaneous percentage growth rate, and we now show how to find c when you are given a doubling period.

Example 5 (*Compare Exercise 8*)
A certain bacteria colony doubles in size every 20 minutes. If there are 1000 bacteria present at 1 PM, how many will there be at 2:30 PM on the same day?

Solution Because the doubling time is given in minutes, we will measure time in minutes. We know how many bacteria are in the colony at 1 PM, so we let $t = 0$ correspond to 1 PM. We use $P(t)$ for the number of bacteria t minutes after 1 PM.

$$P(t) = ke^{ct}$$

and we must solve for k and c.

$$P(0) = 1000, \qquad \text{so} \qquad k = 1000.$$

$P(t)$ is of the form

$$P(t) = 1000e^{ct}$$

Furthermore, we are given that the population doubles in 20 minutes, so $P(20) = 2000$. By letting $t = 20$ in the expression $P(t) = 1000e^{ct}$, we have

$$P(20) = 1000e^{c(20)}$$

We set these two expressions for $P(20)$ equal to each other and then solve for c:

$$2000 = 1000e^{20c} \qquad \text{Divide by 1000}$$

$$2 = e^{20c} \qquad \text{Take ln of both sides}$$

$$\ln 2 = \ln e^{20c} = 20c \qquad \text{Divide by 20}$$

$$\frac{\ln 2}{20} = c \qquad (c \approx .0347)$$

We now have both k and c.

$$P(t) = 1000e^{[(\ln 2)/20]t}$$

To find the population at 2:30, we need to calculate $P(90)$.

$$\begin{aligned} P(90) &= 1000e^{[(\ln 2)/20]90} \\ &= 1000e^{[(9 \ln 2)/2]} \\ &\approx 1000e^{3.119} \approx 22{,}627 \end{aligned}$$

The colony is over 22 times its original size.

Some applications deal not with growth, but rather with diminishing quantities. For example, psychologists may model retention rate of knowledge in terms of the percentage of the knowledge forgotten each week. The application we will deal with here has to do with radioactive decay. Certain materials are called radioactive because they give off radiant energy. This is called radioactive decay. One such element that exists in our atmosphere is Carbon-14 (C-14). Living animals and plants absorb the C-14 from the atmosphere and, because the supply is being replenished, the level of C-14 in living organisms remains constant. When the organism dies, however, the intake stops and the amount of C-14 diminishes due to radioactive decay. We will let $R(t)$ be the measure of C-14 t years after the death of the organism. But we do not want to get too technical, so we will not worry about the units of $R(t)$. Scientists tell us that $R(0) = 15.3$ and that the half-life of C-14 is about 5730 years.

Example 6 Find k and c for $R(t) = ke^{ct}$, where $R(t)$ is the amount of C-14 left t years after an organism dies.

Solution $R(t) = ke^{ct}$ and $R(0) = 15.3$ tells us that $k = 15.3$, so

$$R(t) = 15.3e^{ct}$$

Now, a half-life of 5730 years means that

$$R(5730) = \frac{1}{2}R(0) = \frac{1}{2}(15.3) = 7.65$$

Also from the formula for $R(t)$,

$$R(5730) = 15.3e^{c5730}$$

We put these two expressions for $R(5730)$ equal to each other and solve for c.

$$7.65 = 15.3e^{c5730} \qquad \text{Divide by 15.3}$$

$$\frac{1}{2} = e^{5730c} \qquad \text{Take ln of both sides}$$

$$\ln\frac{1}{2} = 5730c \qquad \text{Divide by 5730}$$

$$c = \frac{\ln\left(\frac{1}{2}\right)}{5730} \qquad (\approx -.000121)$$

We can use the fact that $\ln(\frac{1}{2}) = -\ln 2$ to write

$$R(t) = 15.3e^{-[(\ln 2)/5730]t}$$

Note that because R is a decreasing function, c must be negative. Notice also the similarity of the steps in Examples 5 and 6.

Example 7 *(Compare Exercise 11)*

A piece of wood from the funeral boat of an Egyptian pharaoh was measured for C-14 radioactivity. The level was found to be 9.75. How long ago did this pharaoh die?

Solution The measurement tells us that $R(t) = 9.75$; we need to solve for t.

$$9.75 = 15.3e^{[-(\ln 2)/5730]t} \quad \text{Divide by 15.3}$$

$$.63725 = e^{[-(\ln 2)/5730]t} \quad \text{Take ln of both sides}$$

$$\ln .63725 = [-(\ln 2)/5730]t$$

$$t = \frac{(5730)(\ln .63725)}{-\ln 2} \approx 3725$$

The pharaoh died about 3725 years ago.

This method of dating fossils was developed by Dr. Willard F. Libby of the University of Chicago in the late 1940's. He received a Nobel prize for this work in 1960. If you are interested in reading further about this method of dating fossils, we recommend *Radiocarbon Dating*, by Willard F. Libby, University of Chicago Press, 1955; and *Before Civilization*, by Colin Renfrew, Alfred A. Knopf, New York, 1973. The first book is Libby's original work; the second has a very readable discussion of carbon dating, including an appendix which discusses the assumptions made in the theory.

5–5 Exercises

1. (*See Example 1*) Property values in a certain suburb have been growing at a continuous rate of 5%. If the average house cost \$80,000 in 1980, what would the average cost be in the year 2000?
2. If the average cost of a new automobile increases at the continuous rate of 6%, and if in 1985 this average cost was \$10,500, what will the average cost be in 1995?
3. If the inflation rate in a certain country is 4%, continuously compounded, what would a person have to earn in 1990 just to stay even if the person's salary in 1985 was \$20,000?
4. If \$500 is invested in a trust fund that pays interest at a continuous rate of 3%, how much will the fund be worth **(a)** after 100 years? **(b)** after 200 years?
5. (*See Example 2*) If \$500 is invested in a trust fund paying interest at a continuous rate of 4%, how much will the fund be worth, **(a)** in 100 years? **(b)** in 200 years?

6. (*See Example 3*) Suppose that a state government's budget grows at a continuous rate of 10%. If the budget is $20 billion when Governor A takes office, what will it be four years later?

7. (*See Example 4*) For the same state as in Example 6, if Governor B replaces A after four years, compare the growth of the budget during B's last year to the total growth under A.

8. (*See Example 5*) The population in a certain country is doubling every 20 years. If the country had 100 million people in 1950, how many will it have **(a)** in 2000? **(b)** in 2050?

9. If the U.S. consumption of coal doubles every 30 years, and the U.S. consumed 600 million tons in 1980, how much will the U.S. consume in 2020?

10. The per-capita GNP of country A was $2000 in 1980 and was growing at a continuous rate of 8% per year. [GNP = Gross National Product]

    (a) When will the per-capita GNP of country A reach $6000?

    (b) The per-capita GNP of country B was $4400 in 1980 and was growing at a continuous rate of 5%. When will the per-capita GNP of country B reach $6000?

    (c) When will the per-capita GNP of country A equal that of country B?

The following results were obtained by measuring the radioactivity of samples. Determine the age of each sample; use 5730 years as the half-life of C-14.

11. (*See Example 7*) Wood from the foundation cribbing for a fortification wall in a mound at Alisher Huyuh, Turkey. $R(t) = 10.26$.

12. Wood from the floor of a central room in a large Hilani (palace) of the Syro-Hittite period in the city of Tayinat in northwest Syria. $R(t) = 11.17$.

13. Slab of wood from a roof beam of the tomb of Vizier Hemaka, contemporaneous with King Udimu, First Dynasty, in Sakkara, Egypt. $R(t) = 8.33$.

14. Lake mud from Neasham near Darlington in the extreme north of England. Correlated directly with the last glacial stage. $R(t) = 3.96$.

5–6 Bases Other than *e*

Rewriting b^x
The Derivative of $y = b^x$
Logarithms with Bases Other than e
Rewriting $y = \log_b x$
The Derivative of $y = \log_b x$

In Section 5–1, we wrote the general form of an exponential function as kb^{cx}. Then we specialized to the case when the base b was the special number e. In this section, we show that in fact, every exponential function can be written using base e. Further, we introduce logarithms to bases other than e, and show that they too can be written using only the natural logarithm. Thus, the only exponential or logarithm functions you really need are those whose base is e. Because it may be more convenient, however, in certain applications to use other bases, we derive the formulas for the derivatives of b^x and $\log_b x$.

Rewriting b^x

Let b be an arbitrary positive number. We write the equation $b^x = e^{cx}$ and solve for c. Take ln of both sides.

$$\ln b^x = \ln e^{cx}$$
$$x \ln b = cx$$
$$\ln b = c$$

And we have found c. Now that we know what c is, we can rewrite b^x more directly. Since $b = e^{\ln b}$, we have

$$b^x = (e^{\ln b})^x = e^{(\ln b)x}$$

Example 1 (*Compare Exercise 1*)
Express $f(x) = 2^x$ using base e.

Solution $2 = e^{\ln 2}$, so $f(x) = 2^x = e^{(\ln 2)x} = e^{x \ln 2}$. It is customary to write $(\ln 2)x$ as $x \ln 2$ to avoid parentheses.

Example 2 (*Compare Exercise 5*)
Express $f(x) = 2^{x/3}$ using base e.

Solution Again, $2 = e^{\ln 2}$, so $2^{x/3} = e^{(\ln 2)(x/3)}$. Thus $f(x) = e^{(x \ln 2)/3}$.

Example 3 (*Compare Exercise 7*)
If $f(x) = 4 \cdot 3^x$, rewrite $f(x)$ using base e.

Solution We have to rewrite the base; here $b = 3$ and $3 = e^{\ln 3}$.

Thus

$$\begin{aligned} 4(3)^x &= 4(e^{\ln 3})^x \\ &= 4e^{x \ln 3} && \text{(Exact form)} \\ &\approx 4e^{1.0986x} && \text{(Using the decimal approximation to ln 3)} \end{aligned}$$

The Derivative of $y = b^x$

To find the derivative of $y = b^x$, we again rewrite the base. We have just seen that $b = e^{\ln b}$, so we write $y = (e^{\ln b})^x = e^{x \ln b}$. Now, we use the chain rule to compute the derivative. The special form of the chain rule for an exponential function with base e was obtained in Section 5–3.

$$\text{If } f(x) = e^{g(x)}, \qquad \text{then} \qquad f'(x) = e^{g(x)} \cdot g'(x)$$

In this case, $g(x) = x \cdot \ln b$, and so $g'(x) = \ln b$. Remember that $\ln b$ is just a constant. Using this formula,

$$\text{if } f(x) = e^{x \ln b}, \qquad \text{then} \qquad f'(x) = e^{x \ln b} \cdot \ln b$$

One further step can be made by going back to the original expression for $e^{x \ln b}$, namely b^x.

Theorem If $f(x) = b^x$, then $f'(x) = b^x \ln b$

Example 4 (*Compare Exercise 31*)
If $f(x) = 6^x$, what is $f'(x)$?

Solution Using $b = 6$, $f'(x) = 6^x \ln 6$.

For a more general exponent, if $f(x) = b^{g(x)}$, then $f(x) = e^{(\ln b)g(x)}$, and we have the following theorem.

Theorem If $f(x) = b^{g(x)}$, then

$$\begin{aligned} f'(x) &= e^{(\ln b)g(x)} \cdot \ln b \cdot g'(x) \\ &= b^{g(x)} \cdot \ln b \cdot g'(x) \end{aligned}$$

Notice that if $b = e$, then $\ln b = 1$, and we have the formulas for the derivative of e^x and $e^{g(x)}$, as before.

Example 5 (*Compare Exercise 35*)
If $f(x) = 4^{x^2}$, what is $f'(x)$?

Solution Here, $b = 4$ and $g(x) = x^2$. Thus, $g'(x) = 2x$, and we have $f'(x) = 4^{x^2}(\ln 4)2x$.

Example 6 (*Compare Exercise 39*)
If $f(x) = 3 \cdot 5^{x^2+x}$, find $f'(x)$.

Solution Here, $b = 5$ and $g(x) = x^2 + x$. The number 3 is simply a constant times 5^{x^2+x}.

$$f'(x) = (3 \cdot 5^{x^2+x})(\ln 5)(2x + 1)$$

Logarithms with Bases Other than *e*

If $b > 0\,(b \neq 1)$, we can now define the logarithm to the base b as the function that "undoes" b^x. Thus, the function $\log_b x$ is related to b^x as the function $\ln x$ is related to e^x. Specifically,

$$\log_b M = N \qquad \text{if and only if} \qquad b^N = M$$

We now give you some examples to stress this definition and the relation between logarithms and exponentials. The examples will be done by using the definition of the logarithm.

Example 7 (*Compare Exercise 11*)
What is $\log_2 8$?

Solution If we let $N = \log_2 8$, then $2^N = 8$, and we recognize that $N = 3$. Thus, $\log_2 8 = 3$.

Example 8 (*Compare Exercise 15*)
What is $\log_2(\frac{1}{8})$?

Solution We want B so that $2^B = \frac{1}{8}$; by "inspection," or "trial and error," $B = -3$. Thus, $\log_2(\frac{1}{8}) = -3$.

Example 9 (*Compare Exercise 23*)
What is $\log_2(-8)$?

Solution This expression is undefined; 2^N is always positive. The equation $2^N = -8$ has no solution.

Warning! As with the natural logarithm, the domain of any logarithm function is all positive numbers; $\log_b x$ may be negative, but x cannot be negative. Be sure you understand the difference between Examples 8 and 9.

Rewriting $y = \log_b x$

As we mentioned, the traditional use of logarithms has been to aid in some calculations, and so a very commonly used logarithm is $\log_{10}$ (in fact, it is called the common logarithm). Often, all that is written is "log"; the base 10 is understood. With $\log_b$ notation, the natural logarithm is the same as $\log_e$, but the natural logarithm is almost always written as "ln." Many calculators feature both a **log** and a **ln** button. We will show you how to compute $\log_b p$ for any base b using only the ln function.

$$N = \log_b M \qquad \text{means} \qquad b^N = M$$

Take ln of both sides of the equation $b^N = M$.

$$\ln b^N = \ln M$$
$$N \ln b = \ln M$$

Divide both sides by $\ln b$.

$$N = \frac{\ln M}{\ln b}$$

Remember that we began with $N = \log_b M$, so these two expressions for N give us the following theorem:

Theorem

$$\log_b M = \frac{\ln M}{\ln b}$$

Using the more common variable x in place of M, we have

$$\log_b x = \frac{\ln x}{\ln b}$$

Example 10 (*Compare Exercise 17*)
What is $\log_3 81$?

Solution We compute $\log_3 81$ in two ways. First, we use the definition $\log_3 81 = N$, where $3^N = 81$. The number 81 was chosen for this example so that we can find the answer as an integer; $N = 4$.

Second, we use the conversion formula and compare results;

$$\log_3 81 = \frac{\ln 81}{\ln 3}$$

Our calculator gives $\ln 81 \approx 4.3944\ldots$ and $\ln 3 \approx 1.0986\ldots$, and when we perform the division $(\ln 81)/(\ln 3)$ on our calculator, it gives the answer 4.

Of course, once we go to the calculator we are dealing with approximations, and we were "lucky" to get exactly 4 as a result of our computations.

Example 11 (*Compare Exercise 21*)
Find a numerical approximation to $\log_2 12$.

Solution $\log_2 12 = N$ means $2^N = 12$; but we cannot solve this by inspection. We know that N is some number between 3 and 4 because $2^3 = 8$ is less than 12, but $2^4 = 16$ is larger than 12. The conversion formula gives us

$$\log_2 12 = \frac{\ln 12}{\ln 2}$$

Now, using the ln function on our calculator we obtain

$$\frac{\ln 12}{\ln 2} \approx \frac{2.4849}{.6931} \approx 3.585 \qquad \text{(Rounded to 3 places)}$$

The Derivative of $y = \log_b x$

To compute the derivative of $\log_b x$, rewrite $\log_b x$ as

$$\frac{\ln x}{\ln b} = \frac{1}{\ln b} \ln x$$

and remember that $\dfrac{1}{\ln b}$ is a constant.

The derivative of $\ln x$ is simply $1/x$. Thus, if

$$f(x) = \left(\frac{1}{\ln b}\right) \ln x, \qquad \text{then} \qquad f'(x) = \left(\frac{1}{\ln b}\right)\left(\frac{1}{x}\right) = \frac{1}{x \ln b}$$

Theorem

If $f(x) = \log_b x$, then $f'(x) = \dfrac{1}{x \ln b}$.

Example 12 (*Compare Exercise 27*)
If $f(x) = \log_3 x$, find $f'(x)$.

Solution Using $b = 3$, we have

$$f'(x) = \frac{1}{x \ln 3}$$

The corresponding chain rule formula is

Theorem

If $f(x) = \log_b g(x)$, then $f'(x) = \dfrac{1}{\ln b} \dfrac{1}{g(x)} \cdot g'(x)$

You can either learn this form directly or just remember to write

$$\log_b g(x) = \frac{\ln g(x)}{\ln b}$$

and use the chain rule when computing the derivative of $\ln g(x)$.

Example 13 (*Compare Exercise 43*)
If $f(x) = \log_2(8x + 7)$, find $f'(x)$.

Solution We use the theorem with $b = 2$ and $g(x) = 8x + 7$. Thus, $g'(x) = 8$, and

$$f'(x) = \left(\frac{1}{\ln 2}\right)\left(\frac{1}{8x + 7}\right) \cdot 8$$

$$= \frac{8}{(\ln 2)(8x + 7)}$$

5–6 Exercises

I.

Rewrite each of the following in the form $f(x) = ke^{cx}$ for some c and k.

1. (*See Example 1*) $f(x) = 3^x$
2. $f(x) = 10^x$
3. $f(x) = 9^{-x}$
4. $f(x) = \left(\frac{1}{2}\right)^x$
5. (*See Example 2*) $f(x) = 7^{x/2}$
6. $f(x) = 9^{x/3}$
7. (*See Example 3*) $f(x) = 5 \cdot 8^x$
8. $f(x) = 2 \cdot 10^x$
9. $f(x) = 7 \cdot 4^{x/3}$
10. $f(x) = 32^{x/6}$

II.

What integer or fraction is represented by each of the following?

11. (*See Example 7*) $\log_3 81$

12. $\log_5 25$

13. $\log_8 2$

14. $\log_{16} 4$

15. (*See Example 8*) $\log_3\left(\frac{1}{3}\right)$

16. $\log_2\left(\frac{1}{32}\right)$

Rewrite each of the following expressions using the natural logarithm. Use a calculator with an ln button to find a decimal approximation.

17. (*See Example 10*) $\log_6 36$

18. $\log_{16} 2$

19. $\log_{25} 5$

20. $\log_8\left(\frac{1}{2}\right)$

21. (*See Example 11*) $\log_{10} 14$

22. $\log_5 30$

23. (*See Example 9*) $\log_2(-16)$

24. $\log_3(-9)$

25. $\log_{-3} 10$

26. $\log_{-4}(-8)$

Find the derivative of each of the following.

27. (*See Example 12*) $f(x) = \log_2 x$

28. $f(x) = \log_{10} x$

29. $f(x) = x^2 \log_4 x$

30. $f(x) = 3x^5 \log_{10} x$

31. (*See Example 4*) $f(x) = 10^x$

32. $f(x) = 8^x$

33. $f(x) = 7^{x/2}$

34. $f(x) = 2^{x/3}$

III.

Find $f'(x)$ for each of the following.

35. (*See Example 5*) $f(x) = 16^{x^2}$

36. $f(x) = 9^{\sqrt{x}}$

37. $f(x) = 10^{\sqrt{4x+1}}$

38. $f(x) = 3^{2x^2+5x}$

39. (*See Example 6*) $f(x) = 10 \cdot 3^{x^2-2x}$

40. $f(x) = 3 \cdot 2^{3x^2-1}$

41. $f(x) = 2^{2^x}$

42. $f(x) = 2^{x^2}$

43. (*See Example 13*) $f(x) = \log_3(4x - 2)$

44. $f(x) = \log_9(x^2 + 1)$

45. $f(x) = \log_4(2x^2 - 8)^3$

46. $f(x) = 3\log_4(2x^2 - 8)$

47. $f(x) = [\log_4(x^2 - 9)]^3$

48. $f(x) = \sqrt{\log_{10}(3x + 1)}$

49. $f(x) = \log_x e$

50. $f(x) = \log_x 10$

IMPORTANT TERMS

5–1 **Compound interest** — **Compounding period** — **Continuous compounding** — **Number e** — **Exponential functions and their graphs**

5–2	**Doubling period**	**Natural logarithm function**	**Relation between logarithm function and exponential function**
	Solving exponential equations		
5–3	**Derivative of $f(x) = e^x$**	**Use of the chain rule with the exponential function**	
5–4	**Derivative of $f(x) = \ln x$**	**Use of the chain rule with the logarithm function**	
5–5	**Percentage growth rate**	**Half-life**	**Carbon dating**
5–6	**$b^x = e^{x \ln b}$**	**$\log_b x = (\ln x / \ln b)$**	**Derivative of $f(x) = b^x$**
	Derivative of $f(x) = \log_b x$	**Use of the chain rule with general exponential and logarithmic functions**	

REVIEW EXERCISES

In Exercises 1 through 4, find $f'(x)$

1. $f(x) = x^2e^{3x}$ **2.** $f(x) = e^x \ln x$ **3.** $f(x) = \ln(x^2 + 1)$ **4.** $f(x) = \ln\sqrt{6x + 8}$

5. On what interval is f increasing if $f(x) = x \ln x$?

6. On what interval is the graph of $y = \dfrac{x}{e^x}$ concave down?

7. If a certain colony of bacteria doubles in size every minute, and if after 30 minutes a test tube containing these bacteria is full, when was the test tube only half-full?

8. A colony of a certain bacteria doubles in size every 17 minutes. How long will it take a colony to grow to

(a) 20 times its original size?

(b) 100 times its original size?

(c) 1000 times its original size?

9. Samples of basketry were taken from Lovelock Cave in Nevada. They had a C-14 reading of 12.4. Approximately how old were the baskets?

10. The population of a certain country is growing at the continuous rate of 3% per year, and the present population of the country is 25 million.

(a) In how many years will the country have 100 million people?

(b) How many people will the country have after 100 years?

11. If $10,000 is invested at 8% interest, how much is the investment worth after 10 years if the compounding is done

(a) quarterly? **(b)** continuously?

12. If an investment pays 7%, compounded continuously, then what rate would a second account have to pay to yield the same amount of interest if the second account is compounded annually?

13. A certain bank card charges interest at the rate of $1\frac{1}{2}\%$ per month. How much interest would a holder of the card have to pay for the use of \$1000 for one year? (Assume that no payments were made during the year.)

14. What is an approximate decimal equivalent for

(a) $\left(1 + \dfrac{1}{1{,}000{,}000}\right)^{1{,}000{,}000}$

(b) $\left(1 - \dfrac{1}{1{,}000{,}000}\right)^{1{,}000{,}000}$

6

Topics in this chapter can be applied to problems such as:
Manufacturing Costs and Cost Functions • Market Research and Pricing Policy • Production Levels and Cost Management • Price-Demand Equations • Production-Profit Relationships • Consumers' Surplus • Resource Flows and Capital Accumulations • Gini Index

The Integral

- The Indefinite Integral
- Introduction to Differential Equations
- The Definite Integral
- The Definite Integral and Area
- Applications of Area
- Summations and Definite Integrals

6–1 The Indefinite Integral

From Marginal Cost to Total Cost
Antiderivatives and Integral Notation
Checking Answers
Integration Formulas
Fractional and Negative Exponents

In the previous four chapters, we have studied one of the two major topics in calculus—the derivative. We turn now to the other major calculus topic—the integral.

From Marginal Cost to Total Cost

We have seen how to compute the derivative of a function f and how to use the derivative to analyze properties of f. For instance, if $P(x)$ represents the profit from selling x watches, we have seen how marginal analysis can show how many watches should be produced to maximize profit. Often, however, an application arises where the function isn't known to begin with, but what is known is the rate of change of the function. For example, if a company knows that it costs \$17 to

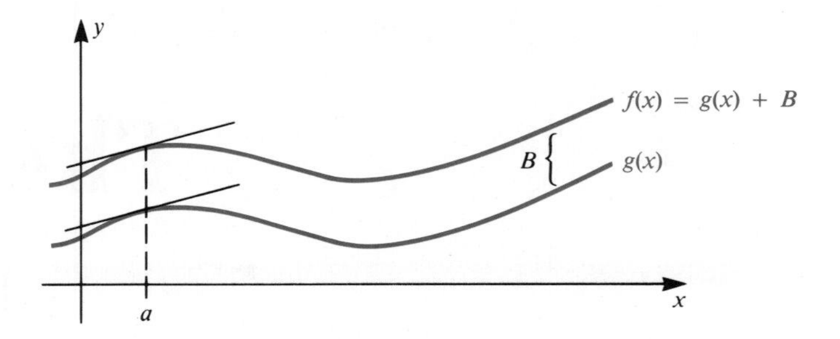

Figure 6–1

manufacture each wristwatch, then that knowledge implies that the cost function is defined by $C(x) = 17x + B$, where B represents the fixed costs. Using the language of marginal analysis, if $MC(x) = 17$, then $C(x) = 17x + B$. This example gives a simple application of a more general principle. If we know the marginal cost function, we know what the cost function looks like, except for an undetermined added constant term representing the fixed costs. The particular cost function is not completely known until the fixed costs are known. The two cost functions $C_1(x) = 17x + 109$ and $C_2(x) = 17x + 420$ have the same marginal cost function; $MC_1(x) = MC_2(x) = 17$. Furthermore, if f is any function with $f'(x) = 17$, then $f(x) = 17x + B$ for some constant B. This fact is part of a more general mathematical statement.

Theorem

> If two functions f and g have the same derivative on some interval, then there is a constant B so that $f(x) = g(x) + B$ for all x in the interval.

In other words, if $f'(x) = g'(x)$ on some interval, then $f(x) - g(x)$ is a constant on that interval.

A geometric interpretation may help in understanding this theorem. We know from Chapter 4 that f' gives us knowledge about the shape of the graph of f. Thus, if $f'(x) = g'(x)$ on some interval, the graphs of f and g have the same shape on this interval. The theorem says that these graphs are "parallel." In Figure 6–1, the graph of f is obtained by sliding the graph of g up B units:

$$f(x) = g(x) + B$$

Observe that $f'(x) = g'(x)$ means that the lines tangent to the graphs of f and g are parallel for every value of x (these tangents are drawn for $x = a$).

If $f'(x) = 17$, we are able to say that $f(x) = 17x + B$. We now address the problem of finding $f(x)$ if we start with a more complicated $f'(x)$.

Example 1 (*Compare Exercise 1*)
If $f'(x) = 2x$, what is $f(x)$?

Solution We are looking for a function whose derivative is $2x$; one such function is $g(x) = x^2$. The theorem tells us that $f(x)$ must be of the form $f(x) = g(x) + B$. Thus, $f(x)$ is of the form $f(x) = x^2 + B$. Note that we cannot determine a particular f; we can only determine $f(x)$ up to an additive constant.

Antiderivatives and Integral Notation

Definition

> F is called an **antiderivative** of f if $F'(x) = f(x)$.

Referring to the previous example, we can now say that x^2 is an antiderivative of $2x$ and also, with $B = -9$, that $x^2 - 9$ is an antiderivative of $2x$. Furthermore, all antiderivatives of $2x$ are of the form $x^2 + B$. (The "anti" indicates that the process is going in the opposite direction from computing derivatives.) There is special mathematical notation for antiderivatives, just as the notations $f'(x)$ and dy/dx are used for derivatives. The notation will not single out a particular antiderivative, since B can be any number. The notation stands for *all* antiderivatives.

Definition

> The collection of all antiderivatives of f is written
>
> $$\int f(x)\,dx$$

Thus, $\int f(x)\,dx$ means the collection of all functions of the form $F(x) + B$, where $F'(x) = f(x)$. We write

$$\int f(x)\,dx = F(x) + B$$

We will also use another name for the collection of all antiderivatives of f.

Definition

> $\int f(x)\,dx$ is called the **indefinite integral** of f.

We can now rephrase the result of Example 1 in the following way:

$$\int 2x\,dx = x^2 + B$$

There are historical reasons for this notation. The symbol $\int$ is called an **integral sign**. This symbol is an elongated S, which is the first letter in the word "summation." We shall see later that there is a close connection between integrals and summations. The dx is called **the differential**, and is also there for historical reasons. Furthermore, it proves very useful in actually computing antiderivatives, as we will see in Chapter 7. The $f(x)$ is called the **integrand**; the constant B is called the **constant of integration**. We have been using B to denote the constant term because our discussion began with a cost function C, and we didn't want to confuse the issue with several C's floating around. Generally, mathematicians use C, the first letter of the word constant, as the constant of integration. So, $\int 2x\,dx = x^2 + C$. Finally, antidifferentiation is also called **integration**.

We summarize:
The indefinite integral of f, written $\int f(x)\,dx$, is the collection of functions of the form $F(x) + C$, where $F'(x) = f(x)$.

Warning! Before showing you some techniques that you can use to find antiderivatives, we wish to emphasize that we are now trying to find an **antiderivative** for a given function. For some time now, it has been almost automatic that when you saw a function, you computed its derivative. In Chapter 3, we would give you a function and ask for the derivative. Here it is as if you're being handed the answer and are being asked, "What is the question?" Your skills in computing derivatives will still be very useful, and you can use these skills to make sure that you have found the correct antiderivative.

Checking Answers

Example 2 Find $\int (15x^2 - 8)\,dx$.

Solution $\int (15x^2 - 8)\,dx = 5x^3 - 8x + C$.

At first glance, you may not understand this example because we didn't explain how to work the problem; we just presented an answer. We did this on purpose! We're just about to give some antidifferentiation rules that will let you see how we arrived at our answer. But first, the point of this example is to emphasize that even if you don't know how we arrived at our answer, you can check to see if it is correct—simply compute the derivative of $5x^3 - 8x$. If the derivative of $5x^3 - 8x$ is $15x^2 - 8$, then the answer to the example is correct; otherwise, the answer is wrong. The answer is correct.

Example 3 (*Compare Exercise 5 for Examples 3 through 6*)

True or false: $\displaystyle\int \left(8x - \frac{6}{x^3}\right) dx = 4x^2 + \frac{3}{x^2} + C$?

Solution Here, $f(x) = 8x - \dfrac{6}{x^3}$ and $F(x) = 4x^2 + \dfrac{3}{x^2}$. Compute $F'(x)$. The derivative of $4x^2 + \dfrac{3}{x^2}$ is $8x - \dfrac{6}{x^3}$. **True.**

Example 4 True or false: $\int e^{3x}\,dx = \frac{1}{3}e^{3x} + C$?

Solution The derivative of $\frac{1}{3}e^{3x}$ is $\frac{1}{3}e^{3x} \cdot 3$ (remember the chain rule), which is e^{3x}. **True.**

Example 5 True or false: $\int (8x - 7)^3\,dx = \frac{1}{4}(8x - 7)^4 + C$?

Solution The derivative of $\frac{1}{4}(8x-7)^4$ is $(\frac{1}{4})4(8x-7)^3 \cdot 8$ (chain rule again) $= 8(8x - 7)^3$. But $(8x - 7)^3 \neq 8(8x - 7)^3$. **False.**

Example 6 True or false: $\int 6x^2\,dx = 2x^3$?

Solution Although $2x^3$ is an antiderivative of $6x^2$, the answer to this example is **false**. The correct answer is $2x^3 + C$.

Integration Formulas

Now we give some rules to help you actually compute the correct answers.

Power Rule for Integration

If p is a real number, $p \neq -1$, then

$$\int x^p\,dx = \frac{1}{p+1}x^{p+1} + C$$

Our first rule for computing antiderivatives is a power rule; compare this rule with our first rule for computing derivatives. The power rule for derivatives is

$$\text{If } f(x) = x^R, \qquad \text{then} \qquad f'(x) = Rx^{R-1}$$

To verify the power rule for integration, apply the power rule for derivatives to $\frac{1}{p+1}x^{p+1}$; $\frac{d}{dx}\left(\frac{1}{p+1}x^{p+1}\right) = \left[\frac{1}{p+1}\right](p+1)x^p = x^p$.

Example 7 (*Compare Exercise 11*)
Find $\int x^3\,dx$.

Solution Here $p = 3$.

$$\int x^3\,dx = \frac{1}{3+1}x^{3+1} + C$$

$$= \frac{1}{4}x^4 + C$$

Warning! The case $\int 1/x\,dx = \int x^{-1}\,dx$ (here $p = -1$) is not covered by this rule. This rule can't work for $p = -1$ because then $1/(p+1)$ would be $1/[(-1)+1] = \frac{1}{0}$ and $\frac{1}{0}$ doesn't make sense. We will discuss $\int 1/x\,dx$ in the next section.

Our second rule for differentiation told us how to handle a constant times a function.

$$(c \cdot f)'(x) = cf'(x)$$

This rule gives us a corresponding integration rule:

Constant-times Rule

$$\int c \cdot f(x)\,dx = c \cdot \int f(x)\,dx$$

Example 8 (*Compare Exercise 17*)
Find $\int 3x\,dx$.

Solution

$$\int 3x\,dx = 3\int x\,dx \qquad \text{(Constant-times rule)}$$

$$= 3\left(\frac{1}{2}x^2\right) + C \qquad \text{(Power rule)}$$

$$= \frac{3}{2}x^2 + C$$

Just as you learned to combine these two rules into one step when computing derivatives, you will also be able to combine them when computing integrals.

Example 9 (*Compare Exercise 17*)
Find $\int 12x^2\,dx$.

Solution

$$\int 12x^2\,dx = 12 \cdot \frac{1}{3}x^3 + C$$

$$= 4x^3 + C$$

The next two rules for finding derivatives had to do with sums and differences of functions; we follow that pattern again.

Sum Rule

$$\int [f(x) + g(x)]\,dx = \int f(x)\,dx + \int g(x)\,dx$$

The integral of the sum is the sum of the integrals, and

Difference Rule

$$\int [f(x) - g(x)]\,dx = \int f(x)\,dx - \int g(x)\,dx$$

The integral of the difference is the difference of the integrals.

Example 10 (*Compare Exercise 19*)
Find $\int (9x^2 - 6)\,dx$.

Solution

$$\int (9x^2 - 6)\,dx = \int 9x^2\,dx - \int 6\,dx$$

$$= \left(9 \cdot \frac{1}{3}x^3 + C_1\right) - (6x + C_2)$$

$$= 3x^3 - 6x + C_1 - C_2$$

where C_1 is the constant of integration associated with $\int 9x^2\,dx$, and C_2 is the constant associated with $\int 6\,dx$. The difference of two constants is constant, so we just write one constant of integration.

$$\int (9x^2 - 6)\,dx = 3x^3 - 6x + C$$

You need not write down constants like C_1 and C_2 when finding the antiderivative of a sum or difference. Just remember to include one constant at the end of the antiderivative.

Example 11 (*Compare Exercise 21*)
Find $\int (5x^2 - 8x + 9)\,dx$.

Solution

$$\int (5x^2 - 8x + 9)\,dx = \frac{5}{3}x^3 - 4x^2 + 9x + C$$

Fractional and Negative Exponents

Of course, the power rule can be used with exponents that are fractions or that are negative, and we would like to make some helpful remarks about handling these types of exponents.

You may find it easier when dealing with fractions to notice that dividing by a fraction is the same as multiplying by its reciprocal; this will allow you to avoid an awkward computation.

Example 12 (*Compare Exercise 27*)
Find $\displaystyle\int x^{1/3}\,dx$.

Solution Here $p = \frac{1}{3}$, so $p + 1 = \frac{4}{3}$; and the reciprocal of $\frac{4}{3}$ is $\frac{3}{4}$.

$$\int x^{1/3}\,dx = \frac{3}{4}x^{4/3} + C$$

Example 13 (*Compare Exercise 25*)
Find $\displaystyle\int \frac{1}{\sqrt{x}}\,dx$.

Solution

$$\int \frac{1}{\sqrt{x}}\,dx = \int x^{-1/2}\,dx$$

$p = -\dfrac{1}{2}$, so $p + 1 = \dfrac{1}{2}$ and the reciprocal of $\frac{1}{2}$ is 2.

$$\begin{aligned}\int x^{-1/2}\,dx &= 2x^{1/2} + C\\ &= 2\sqrt{x} + C\end{aligned}$$

We have two cautions if p is negative. First, be careful to add 1 to p.

Example 14 (*Compare Exercise 23*)
Find $\displaystyle\int \frac{1}{x^4}\,dx$.

Solution We have to rewrite the integrand so that we can apply the power rule:

$$\int \frac{1}{x^4}\,dx = \int x^{-4}\,dx$$

Here $p = -4$ so $p + 1 = -3$ and $\dfrac{1}{p+1} = -\dfrac{1}{3}$

$$\int x^{-4}\,dx = -\frac{1}{3}x^{-3} + C$$

$$= -\frac{1}{3x^3} + C$$

Warning! *Don't* see the 4 and add the 1 to 4 and then paste on a minus sign.

$$\int x^{-4}\,dx \neq -\frac{1}{5}x^{-5} + C$$

Our second caution has to do with the coefficient of x^p.

Example 15 (*Compare Exercise 29*)

Find $\displaystyle\int \frac{1}{7x^3}\,dx$.

Solution Note, $\dfrac{1}{7x^3} = \dfrac{1}{7}x^{-3}$, not $7x^{-3}$. $\left(7x^{-3} = \dfrac{7}{x^3}\right)$

So,
$$\int \frac{1}{7x^3}\,dx = \int \frac{1}{7}x^{-3}\,dx$$

$$= \frac{1}{7}\left(-\frac{1}{2}x^{-2}\right) + C = -\frac{1}{14}x^{-2} + C$$

$$= -\frac{1}{14x^2} + C$$

Sometimes, you must do some arithmetic before using these rules.

Example 16 (*Compare Exercise 33*)

Find $\int x^2(x + 5)\,dx$.

Solution We cannot apply the power rule to both factors; remember the derivative of the product is **not** the product of the derivatives. Do the multiplication first, and then do the integration.

$$\int x^2(x+5)\,dx = \int (x^3 + 5x^2)\,dx$$

$$= \frac{1}{4}x^4 + \frac{5}{3}x^3 + C$$

Example 17 (*Compare Exercise 37*)

Find $\displaystyle\int \frac{x^3 - 9}{x^2}\,dx$.

Solution Again, do the indicated algebra first so that the integrand has the right form.

$$\int \frac{x^3 - 9}{x^2}\,dx = \int \left(\frac{x^3}{x^2} - \frac{9}{x^2}\right) dx$$
$$= \int (x - 9x^{-2})\,dx$$
$$= \frac{1}{2}x^2 - \left(\frac{9}{-1}\right)x^{-1} + C$$
$$= \frac{1}{2}x^2 + \frac{9}{x} + C$$

6–1 Exercises

I.

Find the general expression for $f(x)$ if

1. *(See Example 1)* $f'(x) = x^3 - x + 5.$

2. $f'(x) = \sqrt{x} - 6x.$

3. $f'(x) = \dfrac{1}{x^5}.$

4. $f'(x) = \dfrac{2}{3x^4}.$

In Exercises 5 through 10, determine whether the equations are true or false.

5. *(See Examples 3 through 6)* $\displaystyle\int (2x + 1)^5\,dx = \frac{1}{6}(2x + 1)^6 + C$

6. $\displaystyle\int \sqrt{3x + 1}\,dx = \frac{2}{9}(3x + 1)^{3/2} + C$

7. $\displaystyle\int x(x + 1)^2\,dx = \left(\frac{x^2}{2}\right)\frac{1}{3}(x + 1)^3 + C$

8. $\displaystyle\int \frac{1}{x^3}\,dx = \frac{1}{\frac{1}{4}x^4} + C$

9. $\displaystyle\int (x^2 - 4x + 3)\,dx = \frac{1}{3}x^3 - 2x^2 + 3x$

10. $\displaystyle\int e^{5x}\,dx = e^{5x} + C$

Compute the following indefinite integrals.

11. *(See Example 7)* $\displaystyle\int x^4\,dx$

12. $\displaystyle\int x^{-3}\,dx$

13. $\displaystyle\int x^{2/3}\,dx$

14. $\displaystyle\int \sqrt{x}\,dx$

15. $\displaystyle\int 14\,dx$

16. $\displaystyle\int (4x^5 - 8)\,dx$

17. *(See Examples 8 and 9)* $\displaystyle\int 3(x - 2)\,dx$

18. $\displaystyle\int 10x^4\,dx$

19. *(See Example 10)* $\displaystyle\int (6x^3 - 8x)\,dx$

20. $\int (9x^2 - 6x + 5)\,dx$

21. (*See Example 11*) $\int (x^3 - x)\,dx$

22. $\int (x^5 - 4x^2 + x - 7)\,dx$

II.

Compute the following indefinite integrals.

23. (*See Example 14*) $\int \frac{1}{x^5}\,dx$

24. $\int \frac{6}{x^3}\,dx$

25. (*See Example 13*) $\int \left(\sqrt{x} - \frac{2}{\sqrt{x}}\right)dx$

26. $\int x^{1/5}\,dx$

27. (*See Example 12*) $\int \frac{4}{x^{1/3}}\,dx$

28. $\int 5\sqrt{x^3}\,dx$

29. (*See Example 15*) $\int \left(\frac{1}{4x^3} - \frac{4}{x^2}\right)dx$

30. $\int \left(\frac{2}{3x^2} - \frac{1}{x^3}\right)dx$

31. $\int \left(\frac{1}{3\sqrt{x}} - \frac{x}{5}\right)dx$

32. $\int \left(\frac{9}{x^4} - \frac{1}{6\sqrt{x}}\right)dx$

III.

Find the following indefinite integrals.

33. (*See Example 16*) $\int x(x - 4)\,dx$

34. $\int x^2(x - 3)\,dx$

35. $\int \sqrt{x}(x + 3)\,dx$

36. $\int \sqrt{x}\left(x^2 - \frac{1}{x}\right)dx$

37. (*See Example 17*) $\int \frac{x^2 - 9}{x^2}\,dx$

38. $\int \frac{3x^2 + 4x}{x^5}\,dx$

39. $\int (x + 1)(x + 3)\,dx$

40. $\int (x^2 + 5)(x - 2)\,dx$

41. $\int \frac{x^2 - 5x}{x}\,dx$

42. $\int \frac{x^3 - 8x + 6}{x^3}\,dx$

43. $\int (\sqrt{x} + 1)^2\,dx$

44. $\left(x^2 + \frac{1}{x}\right)^2 dx$

45. If a company's marginal revenue when producing x items is given by $MR(x) = -6x + 200$, what is the company's revenue function? ($x = 0$ implies that $R = 0$.)

46. The marginal cost function of a company manufacturing x units per week is given by

$$C'(x) = \frac{9x^2}{16} - 20x + 8000$$

The fixed cost is $120,000 per week. What is the company's total cost function?

47. A company's marginal cost function is given by

$$MC(x) = \frac{1}{2}x^2 - x + 2$$

and the company's fixed costs are 11. What is the company's cost function?

48. The marginal revenue of a company manufacturing x units per day is $R'(x) = -2x + 500$. What is the company's revenue function?

6–2 Introduction to Differential Equations

Differential Equations
Particular Solutions to Differential Equations
Solution of $y' = \frac{1}{x}$
Antiderivatives of Exponential Functions

Differential Equations

A **differential equation** is simply an equation that has a derivative in it. For example, the equations

$$y' = 2x$$

$$y'' + 2y' - 3y = 0$$

and

$$f'(x) = -f(x)$$

are all differential equations. Differential equations arise in applications because sometimes it is possible to determine an equation involving the rate of change of a function without knowing what the function is to begin with. For example, we have seen how a production manager can know what the marginal cost function is and then use that knowledge to build the total cost function. For another example, you could know the rate at which you are paying off a car loan or a tuition loan without knowing exactly how much it is that you still owe.

A **solution** to a differential equation is a function that satisfies the equation. For example, $y = e^{3x}$ is a solution to the differential equation $y' - 3y = 0$. To see this, first compute each term: $y' = 3e^{3x}$ and $3y = 3e^{3x}$. Next, substitute each expression into the equation. We have $3e^{3x} - 3e^{3x} = 0$, which is true. A given differential equation may have more than one solution. We have seen that $y = e^{3x}$ is one solution to $y' = 3y$, but $y = 5e^{3x}$ is also a solution. If $y = 5e^{3x}$, then $y' = 5e^{3x} \cdot 3 = 15e^{3x}$ and $3y = 3(5e^{3x}) = 15e^{3x}$, so $y' = 3y$ and $y' - 3y = 0$ is again true.

In the previous section, we saw that $\int f(x)\,dx$ represents a collection of functions whose graphs are parallel. The problem of finding $\int f(x)\,dx$ can be viewed as solving a special form of differential equation—one of the form $y' = f(x)$.

Example 1 (*Compare Exercise 1*)
Find all solutions to the differential equation

$$y' = 5x^2 - 3$$

Solution This problem is the same as finding $\int (5x^2 - 3)\,dx$. Any solution is an antiderivative of $5x^2 - 3$; so all solutions are of the form $y = \frac{5}{3}x^3 - 3x + C$ for some constant C.

Particular Solutions to Differential Equations

An application may demand finding a **particular** solution to a differential equation. There may be an additional condition the solution must satisfy, and this additional condition may determined a specific value for C.

Example 2 (*Compare Exercise 41*)
The marketing research department tells us that if we increase the price of our soft drink, we will sell 50 fewer bottles per day for every penny we increase the price. As managers of the soft-drink company we want to know the precise price-demand equation. But, before we can determine this equation we need to know more information—we need to know how many bottles we can sell at a particular price. If the sales department tells us that we can sell 3500 bottles per day at a price of 30¢ per bottle, what is the price-demand equation for our soft drink?

Solution If y is the number of bottles we sell per day, and x is the price in pennies, then what the marketing research department has told us is that $dy/dx = -50$. Thus, $y = -50x + C$. Now we use the information from the sales department. We substitute the values $y = 3500$ and $x = 30$ into the equation $y = -50x + C$ and obtain $3500 = (-50)(30) + C$, so $C = 5000$. Our particular price-demand equation is now fully determined; it is $y = -50x + 5000$.

Example 3 (*Compare Exercise 17*)
Find the particular solution to $y' = x^2 + (18/x^3)$ that satisfies the added condition $y = 7$ when $x = 3$.

Solution We know y must be an antiderivative of $x^2 + (18/x^3)$. First, we find the collection of all antiderivatives;

$$\begin{aligned}\int\left(x^2 + \frac{18}{x^3}\right)dx &= \int (x^2 + 18x^{-3})\,dx \\ &= \frac{1}{3}x^3 + 18\left(-\frac{1}{2}\right)x^{-2} + C \\ &= \frac{1}{3}x^3 - \frac{9}{x^2} + C\end{aligned}$$

Thus, y is of the form

$$y = \frac{1}{3}x^3 - \frac{9}{x^2} + C$$

Now, we want to pick out a particular antiderivative from this collection. Let $x = 3$ and $y = 7$;

$$7 = \left(\frac{1}{3}\right)3^3 - \frac{9}{3^2} + C$$

$$7 = 9 - 1 + C$$

$$C = -1$$

ANSWER $y = \frac{1}{3}x^3 - \frac{9}{x^2} - 1$

Example 4 (*Compare Exercise 15*)
Find $y = f(x)$ if $f'(x) = x - 9\sqrt{x}$, and the point (4, 10) is on the graph of f.

Solution First, find all antiderivatives of f.

$$\int (x - 9\sqrt{x})\,dx = \int (x - 9x^{1/2})\,dx$$

$$= \frac{1}{2}x^2 - 9\left(\frac{2}{3}\right)x^{3/2} + C$$

$$= \frac{1}{2}x^2 - 6x^{3/2} + C$$

Thus, the solution is of the form $f(x) = \frac{1}{2}x^2 - 6x^{3/2} + C$. The additional condition states that (4, 10) is on the graph of f; that is, $f(4) = 10$. Substituting 10 for $f(x)$ and 4 for x in the equation

$$f(x) = \frac{1}{2}x^2 - 6x^{3/2} + C$$

we get

$$10 = \left(\frac{1}{2}\right)4^2 - 6(4^{3/2}) + C$$

Now, $4^2 = 16$ and $4^{3/2} = (4^{1/2})^3 = 2^3 = 8$, so we have

$$10 = 8 - 48 + C;$$

$$C = 50$$

ANSWER $y = \frac{1}{2}x^2 - 6x^{3/2} + 50.$

One of the most famous differential equations is one that revolutionized the way mankind viewed the physical universe. The equation is due to Isaac Newton and is $F = m \cdot a$, where F is force, m is mass, and a is acceleration. Don't be put off by the introduction of physics; we will only give a very specific use of this equation. The force of the earth's gravity on an object becomes weaker the farther the object is from the earth. But if the distance from the earth doesn't change much, then we can assume that the force is constant. (You weigh only slightly less in an airplane at 30,000 feet than you do at ground level.) Since we can assume that the force F is constant, then the acceleration of an object falling to earth is a constant. Physicists are able to measure the acceleration due to gravity as 32 ft/sec^2. Acceleration is the rate of change of velocity. If we use $v(t)$ to denote the velocity at time t, then the differential equation can be written

$$a = \frac{dv}{dt} = -32$$

(The negative sign indicates that when you drop a rock, it falls; its height decreases.)

Example 5 (*Compare Exercise 37*)
A baseball is thrown straight up at an initial speed of 96 feet per second. What is the ball's velocity function?

Solution The differential equation is $dv/dt = -32$. Thus, $v(t)$ has the form $v(t) = -32t + C$. The additional condition says that $v(0) = 96$. Let $v = 96$ and $t = 0$.

$$96 = -32 \cdot 0 + C$$

$$C = 96$$

The velocity function is given by $v(t) = -32t + 96$.

The solution to Example 5 can itself be interpreted as a differential equation. If $h(t)$ represents the height of the ball at time t, then the velocity is the rate of change of $h(t)$.

$$v(t) = \frac{dh}{dt}$$

Example 6 (*Compare Exercise 37*)
If a baseball is thrown straight up at a speed of 96 feet per second from a height of 6 feet, what is its height function?

Solution From Example 5, we know that the baseball's velocity at time t is given by $v(t) = -32t + 96$.

$$\frac{dh}{dt} = v(t) = -32t + 96$$

so the general form for $h(t)$ is given by

$$h(t) = \int v(t)\,dt = \int (-32t + 96)\,dt$$

$$= -16t^2 + 96t + C$$

Here if $t = 0$, then $h(0) = 6$

$$6 = -16 \cdot 0^2 + 96 \cdot 0 + C, \qquad \text{and} \qquad C = 6$$

The height function is given by $h(t) = -16t^2 + 96t + 6$.

Newton's equation, $F = m \cdot a$, can be used to tell where the baseball is at every moment of its flight and also how fast the ball is going. What makes Newton's equation so important, however, is that the same principle applies to other bodies as well. Using this equation, if one knows the position and velocity of a particular body at a given moment and knows the forces acting on this body, then one can derive an equation that gives the position and velocity of the body for all time. Thus, this differential equation can be used to derive the laws of planetary motion, predict solar and lunar eclipses, and guide our rockets through space.

The acceleration function is the derivative of the velocity function, which is the derivative of the position function. Examples 5 and 6 show how to compute $h(t)$ for a thrown rock when $h(0)$ and $h'(0)$ are given. We give another example that shows how to solve for a particular solution of a differential equation that involves the second derivative of the function. Such a differential equation is called a **second-order** differential equation.

Example 7 (*Compare Exercise 33*)
Find the particular function that satisfies

$$f''(x) = 6x - \frac{4}{x^3}$$

and the additional conditions $f'(1) = 9$ and $f(2) = 12$.

Solution The general form for $f'(x)$ is given by

$$\begin{aligned} f'(x) &= \int f''(x)\,dx \\ &= \int \left(6x - \frac{4}{x^3}\right) dx \\ &= \int (6x - 4x^{-3})\,dx \\ &= 3x^2 + 2x^{-2} + C_1 \end{aligned}$$

To solve for C_1, let $x = 1$; $f'(1) = 3 + 2 + C_1 = 5 + C_1$. We want $f'(1) = 9$; set $9 = 5 + C_1$ so $C_1 = 4$. Thus,

$$f'(x) = 3x^2 + 2x^{-2} + 4$$

Next, the general form for $f(x)$ is given by

$$f(x) = \int f'(x)\,dx$$
$$= \int (3x^2 + 2x^{-2} + 4)\,dx$$
$$= x^3 - 2x^{-1} + 4x + C_2$$
$$\text{Thus } f(2) = 8 - 1 + 8 + C_2 = 15 + C_2$$

We want $f(2) = 12 = 15 + C_2$ so $C_2 = -3$. The particular solution we want is

$$f(x) = x^3 - \frac{2}{x} + 4x - 3$$

Solution of $y' = \frac{1}{x}$

Up to now, we have been using the specific formula

$$\int x^p\,dx = \frac{1}{p+1}x^{p+1} + C,\ p \neq -1$$

The case $p = -1$ is special indeed. In Chapter 5, we showed that if $f(x) = \ln x$ then $f'(x) = 1/x$. The fact that $1/x$ is the derivative of $\ln x$ means that $\ln x$ is an antiderivative of $1/x$. Thus, you might expect the formula to be

$$\int \frac{1}{x}\,dx = \ln x + C$$

But this formula is only "half-right." Some special difficulties are caused by the domains of $\ln x$ and $1/x$; they are not the same; $\ln x$ is only defined when $x > 0$, while $1/x$ is defined for all $x \neq 0$.

$$\int \frac{1}{x}\,dx = \ln x + C \text{ on the interval } (0, \infty)$$

But now we must find an antiderivative for $1/x$ on the interval $(-\infty, 0)$. If x is negative, then $-x$ is positive (although it doesn't *look* like it! Remember that x itself is negative), so $\ln(-x)$ is defined. We compute the derivative of $\ln(-x)$ using the chain rule. Let $u = -x$ so that $y = \ln u$.

$$\frac{dy}{dx} = \frac{dy}{du}\cdot\frac{du}{dx}$$
$$= \left(\frac{1}{u}\right)(-1)$$
$$= \left(\frac{1}{-x}\right)(-1) = \frac{1}{x}$$

If $x < 0$ and $y = \ln(-x)$, then $\dfrac{dy}{dx} = \dfrac{1}{x}$. Thus,

$$\int \frac{1}{x}\,dx = \ln(-x) + C \text{ on the interval } (-\infty, 0)$$

The definition of absolute value allows us to combine the two ca
or $x > 0$) into one formula. Remember,

if x is positive, $|x| = x$
if x is negative, $|x| = -x$

The two formulas can now be combined; the derivative of $\ln|x|$ is $1/x$.
this information in the opposite order, we have the formula for $\int 1/x\,dx$.

$$\int \frac{1}{x}\,dx = \ln|x| + C$$

Example 8 (*Compare Exercise 19*)
If $\dfrac{dy}{dx} = \dfrac{1}{3x}$, find y.

Solution

$$y = \int \frac{1}{3x}\,dx = \frac{1}{3}\int \frac{1}{x}\,dx$$
$$= \frac{1}{3}\ln|x| + C$$

Warning! Sometimes, the formula $\int (1/x)\,dx = \ln|x| + C$ is misused. In trying to compute $\int \dfrac{1}{3x}\,dx$ for instance, some people will write $\int \dfrac{1}{3x}\,dx = \ln|3x| + C$. But, a careful use of the chain rule shows that this is not correct. If we let $g(x) = \ln|3x| + C$, then $g'(x) = \dfrac{1}{3x} \cdot 3$ (chain rule) $= \dfrac{1}{x}$, not $\dfrac{1}{3x}$. As you saw in the previous section, remember to be careful with numbers in the denominator of the integrand.

The formula for $\int \ln x\,dx$ requires a certain technique for evaluating indefinite integrals that we will not encounter until Chapter 7. We can, however, proceed with a discussion of finding antiderivatives of exponential functions.

Antiderivatives of Exponential Functions

We will obtain the antiderivative of e^{ax} the same way we found previous antiderivatives, by using the rule for derivatives in reverse. To find the derivative of e^{ax} you multiply e^{ax} by a; to find the antiderivative of e^{ax}, you divide e^{ax} by a.

$$\int e^{ax}\,dx = \frac{1}{a}e^{ax} + C$$

Check this formula by taking the derivative.

Example 9 (*Compare Exercise 21*)
Find $\int e^{2x}\,dx$.

Solution Here $a = 2$, so $\displaystyle\int e^{2x}\,dx = \frac{1}{2}e^{2x} + C.$

Example 10 (*Compare Exercise 23*)
Find $\displaystyle\int \frac{1}{e^x}\,dx$

Solution We again remind you, do not be tempted to use logarithms just because the integral is a quotient. Instead, rewrite $1/e^x$ as e^{-x}, and apply the formula with $a = -1$.

$$\int \frac{1}{e^x}\,dx = \int e^{-x}\,dx = -e^{-x} + C = -\frac{1}{e^x} + C$$

We conclude with more examples using these two formulas.

Example 11 (*Compare Exercise 27*)
Find $\displaystyle\int \left(x + \frac{1}{x^2}\right)^2 dx.$

Solution Do the algebra first.

$$\left(x + \frac{1}{x^2}\right)^2 = x^2 + \frac{2}{x} + \frac{1}{x^4}$$

$$\int \left(x + \frac{1}{x^2}\right)^2 dx = \int \left(x^2 + \frac{2}{x} + x^{-4}\right) dx$$

$$= \frac{1}{3}x^3 + 2\ln|x| - \frac{1}{3}x^{-3} + C$$

Example 12 (*Compare Exercise 29*)
Find $\displaystyle\int \frac{u^2 - 2u - 1}{u}\,du.$

Solution Again, perform the algebra first.

$$\frac{u^2 - 2u - 1}{u} = \frac{u^2}{u} - \frac{2u}{u} - \frac{1}{u}$$

$$= u - 2 - \frac{1}{u}$$

$$\int \frac{u^2 - 2u - 1}{u}\,du = \int \left(u - 2 - \frac{1}{u}\right) du$$
$$= \frac{1}{2}u^2 - 2u - \ln|u| + C$$

Example 13 Find $\int 1/x^3\,dx$.

Solution This example is to emphasize that you shouldn't use ln as the antiderivative of every quotient. The answer to this example is **not** $\ln|x^3| + C$. Rewrite $\int (1/x^3)\,dx$ as $\int x^{-3}\,dx$. The exponent is -3, not -1, and you should use the power rule:

$$\int \frac{1}{x^3}\,dx = \int x^{-3}\,dx = -\frac{1}{2}x^{-2} + C$$
$$= -\frac{1}{2x^2} + C$$

Remember: use $\ln|x|$ only as an antiderivative for x^{-1}.

Example 14 *(Compare Exercise 31)*
Find $\int (e^x + e^{-x})^2\,dx$

Solution

$$(e^x + e^{-x})^2 = (e^x)^2 + 2e^x e^{-x} + (e^{-x})^2$$
$$= e^{2x} + 2 + e^{-2x}$$

Thus,

$$\int (e^x + e^{-x})^2\,dx = \int (e^{2x} + 2 + e^{-2x})\,dx$$
$$= \frac{1}{2}e^{2x} + 2x - \frac{1}{2}e^{-2x} + C$$

6–2 Exercises

I.

Find the general solution for the differential equations given in Exercises 1 through 6.

1. *(See Example 1)* $y' = 21x^2 - 14x + 8$
2. $y' = x^4 - 5x^2 - 9$
3. $\dfrac{dy}{dx} = 8x - \dfrac{1}{x}$
4. $\dfrac{dy}{dx} = \sqrt{x} + x$
5. $f'(x) = \dfrac{1}{3x^2} + \dfrac{8}{x}$
6. $f'(x) = x^3 + 4x$

In Exercises 7 through 14, determine whether the statement is true or false.

7. $y = \ln 7x$ is a solution to $y' = 1/x$.
8. $y = \ln x^3$ is a solution to $y' = 3/x$.
9. $y = (2x + 1)^3 + 5$ is a solution to $y' = 6(2x + 1)^2$.
10. $y = x^2 - 3$ is a solution to $y' = (1/3)x^3 - 3x + C$.
11. $y = 4\sqrt{x}$ is a solution to $y' = 1/(2\sqrt{x})$.
12. $y = x^3$ is a solution to $y' = 3x^2 + C$.
13. $y = 4e^{-x}$ is a solution to $y'' = y$.
14. $y = 5e^{2x}$ is a solution to $y'' = 20y$.

II.

In Exercises 15 through 24, find the particular solution to each differential equation that satisfies the additional condition.

15. (*See Example 4*)
    $f'(x) = 3x^2 + 6$ and $f(0) = 4$.
16. $f'(x) = 2x - 1$ and $f(1) = 3$.
17. (*See Example 3*)
    $y' = x - (16/x^2)$ and $y = 2$ when $x = 8$.
18. $y' = 3\sqrt{x}$ and $y = 6$ when $x = 4$.
19. (*See Example 8*)
    $\dfrac{dy}{dx} = 5 - \dfrac{1}{x}$ and $y = 8$ when $x = -1$.
20. $y' = 4 + (1/x)$ and $y = 9$ when $x = -1$.
21. (*See Example 9*)
    $\dfrac{dy}{dx} = 8e^{2x}$ and $y = 14e^2$ when $x = 1$.
22. $y' = 4e^{-x}$ and $y = 6$ when $x = 0$.
23. (*See Example 10*)
    $y' = \dfrac{1}{e^x}$ and $y = 3$ when $x = 0$.
24. $y' = \dfrac{4}{e^{2x}}$ and $y = 10$ when $x = 1$.

III.

In Exercises 25 through 36, find the solution to each of the given differential equations and conditions.

25. Find the curve that satisfies $y' = 4x^2 + (1/x)$ and that goes through the point $(1, 8)$.
26. Find the curve that satisfies $y' = \sqrt{x}(x + 1)$ and that goes through the point $(4, 5)$.

27. (*See Example 11*)
What is $f(x)$ if $f'(x) = \left(x + \frac{1}{x}\right)^2$ and $f(1) = 7$?

28. Find $f(x)$ if $f'(x) = \left(2x + \frac{1}{x}\right)^2$ and $f(1) = 4$.

29. (*See Example 12*)
What is $f(x)$ if $f'(x) = \sqrt{x} + (3/x)$ and $f(1) = 5$?

30. Find $f(x)$ if $f'(x) = \dfrac{3x^2 - 4x + 1}{x^2}$ and $f(1) = 8$.

31. (*See Example 14*)
Find all solutions to $y' = (e^x + 1)^2$.

32. Find all solutions to $y' = \dfrac{e^{3x} + 1}{e^x}$.

33. (*See Example 7*) Find $f(x)$ if $f''(x) = 42x + 6$, $f'(0) = 4$, and $f(0) = 7$.

34. Find $f(x)$ if $f''(x) = 1/x^3$, $f'(1) = 8$, and $f(1) = -2$.

35. Find y if $y'' = 20$, $y' = 4$ when $x = 2$, and $y = 8$ when $x = 2$.

36. Find y if $y'' = 12$, $y' = 5$ when $x = 1$, and $y = 3$ when $x = 0$.

37. (*See Examples 5 and 6*)
A baseball is thrown straight up at an initial speed of 60 feet per second from a height of 8 feet. Express the height, $h(t)$, of the ball as a function of time.

38. A rock is thrown *down* from a bridge 84 feet above the river. The initial speed of the rock is 20 feet per second. If $h(t)$ is the height of the rock above the water t seconds after it is thrown, what is $h(t)$?

39. The marginal costs for a company have been estimated as being given by

$$MC(x) = \frac{x^2}{4} - 24x + 350$$

where x is the number of units manufactured per day. The present production level is $x = 6$, and daily costs at this level are 4000.

(a) What is the company's cost function?
(b) What is the company's fixed cost?

40. A company's marginal profit at a production level of x units per day is given by $MP(x) = 20 - \frac{1}{5}x$. If $x = 20$, the company breaks even.

(a) What is the company's profit function?
(b) What is the company's maximum possible profit?

41. (*See Example 2*)
If a company can sell 6200 candy bars per week at a price of 40¢ each, and if the demand decreases by 80 bars per week for every penny increase in price, what is the price-demand equation for this company's candy bars?

6–3 The Definite Integral

Total Change of a Function
The Definite Integral

Total Change of a Function

We introduced the derivative of a function as the instantaneous rate of change of that function. As a particular example, we saw that the velocity function is the derivative of the position function. Thus, if we know the height function of a falling rock, we can determine the velocity function of the rock; $v(t) = h'(t)$.

In this chapter we are posing a different, although related, problem; the problem now is to find an **antiderivative** for a given function. The height function is an antiderivative for the velocity function. However, we have just seen that knowing $v(t)$ is not enough to completely determine $h(t)$. The height function depends on the initial height of the rock, which is completely independent of the rock's velocity. But knowing the velocity function over a period of time is enough to compute **the distance travelled** during this period.

Example 1 *(Compare Exercise 49)*
How far does a rock travel in the third second after it is dropped?

Solution We know from the previous section that $v(t) = -32t$ for a dropped rock (dropped means $v(0) = 0$). Now,

$$h(t) = \int v(t)\,dt = \int -32t\,dt = -16t^2 + C$$

Thus, $h(0) = -16 \cdot 0^2 + C = C$, so C is the height of the rock at time $t = 0$. This height is frequently written h_0 and is called the initial height; $h(t) = -16t^2 + h_0$. The height of the rock after two seconds is

$$h(2) = -64 + h_0$$

The height of the rock after three seconds is

$$h(3) = -144 + h_0$$

The total distance travelled during the time interval $2 \leq t \leq 3$ is the difference in the heights.

$$\begin{aligned} h(3) - h(2) &= (-144 + h_0) - (-64 + h_0) \\ &= -80 \text{ feet} \end{aligned}$$

The h_0 terms subtracted out. The minus sign indicates that the height is less after three seconds than after two seconds; the rock has dropped 80 feet during the third second of its fall.

The velocity function allows us to compute the total distance travelled—the total change in height—without knowing the specific height function.

We now look at another example where we need to compute the total change in a function over an interval.

Example 2 (*Compare Exercise 47*)
A company that manufactures air conditioners has determined that its marginal cost function at the daily production level of x units is given by

$$MC(x) = \frac{x^2}{40} - 8x + 700$$

What will be the increase in the company's cost if production is raised from 120 units to 150 units per day?

Solution We are being asked to find the change in production cost: $C(150) - C(120)$. The change in cost is independent of the fixed costs. Notice how this independence again shows up in the computations. The marginal cost function MC is the derivative of the cost function C. Thus, the cost function C is an antiderivative of the marginal cost.

$$\begin{aligned} C(x) &= \int MC(x)\,dx \\ &= \int \left(\frac{x^2}{40} - 8x + 700\right) dx \\ &= \frac{1}{40}\cdot\frac{1}{3}x^3 - 8\cdot\frac{1}{2}x^2 + 700x + K \\ &= \frac{x^3}{120} - 4x^2 + 700x + K \end{aligned}$$

In this example, K represents the fixed costs. Thus,

$$\begin{aligned} C(150) &= \frac{150^3}{120} - 4(150)^2 + 700(150) + K \\ &= 43125 + K \end{aligned}$$

and

$$\begin{aligned} C(120) &= \frac{120^3}{120} - 4(120)^2 + 700(120) + K \\ &= 40800 + K \end{aligned}$$

The increase in costs is

$$C(150) - C(120) = (43125 + K) - (40800 + K) = 2325$$

Notice that again the constant term is subtracted out. Costs will increase by \$2325.

The Definite Integral

Now we analyze what was common to these two examples.

1. The problem was to find the total change in a function over an interval $[a, b]$. (The change in h over the interval $[2, 3]$ and the change in C over the interval $[120, 150]$.)
2. The function that we wanted the change in was not given, but the derivative of the function was given (velocity $= h'$, and marginal cost $= C'$).
3. To compute the desired answer, we needed an antiderivative of the given derivative. But because the total change is a difference of two values of the antiderivative $[h(3) - h(2)$ and $C(150) - C(120)]$, the total change is independent of the constant term of the antiderivative.

The first two steps are what you must recognize to solve problems like these; the actual solution of the problem occurs in step 3:

given f, find a function g that is an antiderivative of f, and then evaluate $g(b) - g(a)$.

The number obtained by the process of starting with a function f, finding an antiderivative g, and then evaluating $g(b) - g(a)$ has so many applications that it has its own name—the definite integral. The formal definition follows. We include the condition that f be continuous on $[a, b]$ because this condition guarantees that f does have an antiderivative g.

Definition

Let f be a continuous function on the interval $[a, b]$, and let g be an antiderivative of f on this interval. Then, the **definite integral of f from a to b**, written $\int_a^b f(x)\,dx$, is the number $g(b) - g(a)$.

$$\int_a^b f(x)\,dx = g(b) - g(a)$$

The number b is called the **upper limit of the integral**, and a is called the **lower limit**.

Notice the similarity in notation between the indefinite integral and the definite integral. But remember,

$\int f(x)\,dx$ is the collection of all antiderivatives of f;

$\int_a^b f(x)\,dx$ is a number.

In the definition of the definite integral, g is an antiderivative of f; thus, $g'(x) = f(x)$. We can now write the following:

The total change of a function g on an interval $[a, b]$ is given by

$$g(b) - g(a) = \int_a^b g'(x)\,dx$$

Example 3 (*Compare Exercise 5*)
Find $\int_1^3 (x^2 + 4)\,dx$.

Solution First, find an antiderivative of $x^2 + 4$. The simplest is the one whose additive constant of integration in 0. Let $g(x) = \frac{1}{3}x^3 + 4x$. Now, compute $g(3) - g(1)$.

$$g(3) = \frac{1}{3} \cdot 27 + 12 = 21$$

$$g(1) = \frac{1}{3} + 4 = 4\frac{1}{3}$$

$$g(3) - g(1) = 16\frac{2}{3}$$

$$\int_1^3 (x^2 + 4)\,dx = 16\frac{2}{3}$$

We now introduce a notation that allows us to compress the steps in the evaluation of $g(3) - g(1)$.

Notation

$$g(x)\big|_a^b = g(b) - g(a)$$

Example 4 (*Compare Exercise 1*)
Evaluate $(3x^2 - 5x)\big|_1^4$.

Solution Here, $g(x) = 3x^2 - 5x$, and we wish to compute $g(4) - g(1)$.

$$\begin{aligned}(3x^2 - 5x)\big|_1^4 &= (3 \cdot 16 - 5 \cdot 4) - (3 \cdot 1 - 5 \cdot 1)\\ &= 28 - (-2)\\ &= 30\end{aligned}$$

Warning! There is no calculus involved in this notation; its purpose is to shorten the computations after you have found an antiderivative of f. We repeat Example 3 to show you how the computation is shortened.

Example 3 (Redone)
Find $\int_1^3 (x^2 + 4)\,dx$.

Solution

$$\begin{aligned}\int_1^3 (x^2 + 4)\,dx &= \left[\tfrac{1}{3}x^3 + 4x\right]\Big|_1^3\\ &= (9 + 12) - (\tfrac{1}{3} + 4)\\ &= 16\tfrac{2}{3}\end{aligned}$$

The hard part is finding an antiderivative; the remaining steps are simply notation and evaluation.

We continue this section with more examples to let you become familiar with both the new notation and the definition of the definite integral. We will give more applications of the definite integral in the following sections.

Example 5 (*Compare Exercise 23*)
Compute $\int_1^4 (x - \sqrt{x})\,dx$.

Solution First, we rewrite $\sqrt{x}$ as $x^{1/2}$ so we can use the integral power rule.

$$\int_1^4 (x - \sqrt{x})\,dx = \int_1^4 (x - x^{1/2})\,dx$$

Second, find an antiderivative of $x - x^{1/2}$

$$\int_1^4 (x - x^{1/2})\,dx = \left(\frac{1}{2}x^2 - \frac{2}{3}x^{3/2}\right)\Bigg|_1^4$$

Now, do the evaluations

$$= \left[\left(\frac{1}{2}\right)16 - \left(\frac{2}{3}\right)(4)^{3/2}\right] - \left(\frac{1}{2} - \frac{2}{3}\right)$$

$$= \left[8 - \left(\frac{2}{3}\right)8\right] - \left(\frac{3}{6} - \frac{4}{6}\right)$$

$$= \frac{17}{6}$$

Example 6 (*Compare Exercise 19*)
Compute $\int_1^5 1/x\,dx$.

Solution

$\displaystyle\int_1^5 \frac{1}{x}\,dx = \ln\lvert x\rvert\Big|_1^5$	Found antiderivative
$= \ln 5 - \ln 1$	Computed $g(b) - g(a)$
$= \ln 5$	Used $\ln 1 = 0$

Warning!

$$\int_{-1}^5 \frac{1}{x}\,dx \neq \ln\lvert x\rvert\Big|_{-1}^5$$

because $\ln|x|$ is not an antiderivative of $1/x$ on the whole interval $[-1, 5]$. There is trouble at $x = 0$. The computation of $\int_a^b f(x)\,dx$ requires that f be continuous on $[a, b]$ and that you find g so that $g'(x) = f(x)$ for all x in $[a, b]$. Here, $f(x) = 1/x$ is not even defined for all x in $[-1, 5]$. In the exercises that follow, the integrands will be continuous on the interval $[a, b]$.

Example 7 *(Compare Exercise 21)*
Compute $\int_0^3 e^{2x}\,dx$.

Solution

$$\int_0^3 e^{2x}\,dx = \frac{1}{2}e^{2x}\Big|_0^3$$
$$= \frac{1}{2}e^6 - \frac{1}{2}e^0$$
$$= \frac{1}{2}e^6 - \frac{1}{2} \qquad \text{(Because } e^0 = 1\text{)}$$

The constant-times rule and the addition and subtraction rules for indefinite integrals carry over to definite integrals. We list these properties for emphasis.

Properties of the Definite Integral

Constant-times

1. $$\int_a^b c \cdot f(x)\,dx = c \cdot \int_a^b f(x)\,dx$$

Sum

2. $$\int_a^b [f(x) + g(x)]\,dx = \int_a^b f(x)\,dx + \int_a^b g(x)\,dx$$

Difference

3. $$\int_a^b [f(x) - g(x)]\,dx = \int_a^b f(x)\,dx - \int_a^b g(x)\,dx$$

Example 8 *(Compare Exercise 39)*
Compute $\int_2^5 4e^x\,dx$.

Solution

$$\int_2^5 4e^x\,dx = 4\int_2^5 e^x\,dx = 4\left(e^x\Big|_2^5\right)$$
$$= 4(e^5 - e^2)$$

Example 9 *(Compare Exercise 37)*
Compute $\int_1^3 6x^2\,dx + \int_1^3 5\,dx$.

Solution

$$\int_1^3 6x^2\,dx + \int_1^3 5\,dx = \int_1^3 (6x^2 + 5)\,dx$$
$$= (2x^3 + 5x)\Big|_1^3$$
$$= [2(27) + 5(3)] - [2(1) + 5(1)]$$
$$= 69 - 7 = 62$$

There is an important property of definite integrals that does not come directly from any property of indefinite integrals. This property comes from the observation that if c is between a and b, then the total change of a function over the interval $[a, b]$ is the sum of its change over the two intervals $[a, c]$ and $[c, b]$.

$$g(b) - g(a) = [g(b) - g(c)] + [g(c) - g(a)]$$

There is not a commonly agreed upon name for this property. We hope that the name we use helps you to remember what this property is.

The Gluing of Intervals Property

4. If $a < c < b$, then

$$\int_a^c f(x)\,dx + \int_c^b f(x)\,dx = \int_a^b f(x)\,dx$$

If two intervals have an endpoint in common, you can "glue" them together and evaluate the definite integral over the whole interval.

The discussion of the definite integral $\int_a^b f(x)\,dx$ so far has only involved situations where $a < b$. However, we could just as easily ask for the change in total costs if the production level goes from $x = 150$ to $x = 120$. Some applications may require the computation of $\int_a^b f(x)\,dx$ in cases other than $a < b$. There are exactly two such cases, $a = b$ and $a > b$. Our definition of definite integral leads us to the following two properties.

5. $\int_a^a f(x)\,dx = 0$
6. $\int_a^b f(x)\,dx = -\int_b^a f(x)\,dx$

In terms of the total change of f, these two properties say what you expect. The first says that the total change of f on $[a, a]$ is 0. The second says that the change in f as x goes from a to b is the opposite of the change in f as x goes from b to a. Think about the change in costs as production goes from 120 to 150 compared to the change in costs as production goes from 150 to 120.

Using these properties, it can now be shown that the requirement $a < c < b$ in the gluing property (4) can be generalized.

7. Independent of the ordering of a, b, and c,

$$\int_a^c f(x)\,dx + \int_c^b f(x)\,dx = \int_a^b f(x)\,dx$$

Example 10 (*Compare Exercise 29*)
Verify Equation 7 for $f(x) = 2x$, $a = -1$, $b = 2$, and $c = 3$.

Solution We want to show that

$$\int_{-1}^{2} 2x\,dx = \int_{-1}^{3} 2x\,dx + \int_{3}^{2} 2x\,dx$$

First, $\int_{-1}^{2} 2x\,dx = x^2|_{-1}^{2} = 4 - (-1)^2 = 4 - 1 = 3$
Second, $\int_{-1}^{3} 2x\,dx = x^2|_{-1}^{3} = 9 - 1 = 8.$
Third, $\int_{3}^{2} 2x\,dx = x^2|_{3}^{2} = 4 - 9 = -5.$
Finally, $3 = 8 + (-5)$ is true.

Example 11 (*Compare Exercise 41*)
If $\int_1^3 f(x)\,dx = 4$ and $\int_1^5 f(x)\,dx = 11$, what is $\int_3^5 f(x)\,dx$?

To use the "gluing" property, we must have the same number as the upper bound of one definite integral and the lower bound of the other definite integral. Here, the number that occurs in both integrals is 1, so we want 1 as the upper limit of one of the integrals. To accomplish this, we rewrite $\int_1^3 f(x)\,dx$ as $-\int_3^1 f(x)\,dx$. Thus,

$$\int_3^1 f(x)\,dx = -\int_1^3 f(x)\,dx = -4$$

Now, we can write

$$\int_3^1 f(x)\,dx + \int_1^5 f(x)\,dx = \int_3^5 f(x)\,dx$$

$$-4 \quad + \quad 11 \quad = \quad 7$$

ANSWER $\int_3^5 f(x)\,dx = 7$

6–3 Exercises

I.

Compute the following.

1. (*See Example 4*) $(2x^2 + x)|_{-1}^{3}$
2. $(x^3 - 3x^2 + x)|_{-1}^{4}$
3. $(\sqrt{x} - 4x)|_1^9$
4. $(x^{3/2} - x^{-1/2})|_4^{25}$

Compute the following definite integrals.

5. (*See Example 3*) $\int_1^3 (3x^2 - 4x + 1)\,dx$
6. $\int_2^5 (6x - 4)\,dx$
7. $\int_1^4 \sqrt{x}\,dx$
8. $\int_0^8 x^{1/3}\,dx$
9. $\int_2^5 \frac{1}{x^2}\,dx$
10. $\int_{-4}^{-1} \left(3x^2 - \frac{3}{x^2}\right) dx$
11. $\int_{-1}^{4} 3\,dx$
12. $\int_{-1}^{4} -3\,dx$
13. Verify that $\int_1^3 x\,dx + \int_3^6 x\,dx = \int_1^6 x\,dx$ by computing all three definite integrals.
14. Verify that $\int_{-1}^{2} 4x\,dx + \int_{-1}^{2} 7\,dx = \int_{-1}^{2} (4x + 7)\,dx$ by computing all three definite integrals.

II.

Compute the following.

15. $\ln x\Big|_1^e$

16. $e^x\Big|_0^{\ln 7}$

17. $\ln|x|\Big|_{-4}^{-1}$

18. $e^{x/2}\Big|_0^{\ln 16}$

Compute the following definite integrals.

19. *(See Example 6)* $\int_{-4}^{-1} \frac{1}{x}\,dx$

20. $\int_1^7 \frac{1}{x}\,dx$

21. *(See Example 7)* $\int_{-2}^{3} 4e^{-x}\,dx$

22. $\int_0^4 e^{3x}\,dx$

23. *(See Example 5)* $\int_1^{25} \left(\sqrt{x} - \frac{4}{\sqrt{x}}\right) dx$

24. $\int_0^{64} (x^{1/3} - 2x^{1/2})\,dx$

25. $\int_0^5 (x^4 - 9x^2 + 2)\,dx$

26. $\int_2^8 \frac{4}{x}\,dx$

27. $\int_2^8 \frac{1}{4x}\,dx$

28. $\int_1^3 (5e^{-x/2})\,dx$

Verify the equation $\int_a^b f(x)\,dx = \int_a^c f(x)\,dx + \int_c^b f(x)\,dx$ in the following two exercises by computing all three definite integrals.

29. *(See Example 10)* $f(x) = x^2$; $a = 2$, $b = 2$, $c = 4$

30. $f(x) = x + 5$; $a = 5$, $b = 1$, $c = -2$

III.

Evaluate the definite integrals in Exercises 31 through 36.

31. $\int_0^{\ln 8} e^{x/3}\,dx$

32. $\int_1^4 \left(x^2 + \frac{3}{x}\right)^2 dx$

33. $\int_{-1}^{3} x(2x + 3)\,dx$

34. $\int_1^5 \frac{6x^3 - 7x}{x}\,dx$

35. $\int_1^3 \frac{6x^2 + 4x - 8}{x^2}\,dx$

36. $\int_{-3}^{1} (x + 2)(x - 4)\,dx$

In Exercises 37 through 40, use $\int_1^3 f(x)\,dx = 7$ and $\int_1^3 g(x)\,dx = -2$ to compute the given definite integral.

37. *(See Example 9)* $\int_1^3 [f(x) + g(x)]\,dx$

38. $\int_1^3 -f(x)\,dx$

39. *(See Example 8)* $\int_1^3 2f(x)\,dx$

40. $\int_1^3 [2f(x) - 4g(x)]\,dx$

41. (*See Example 11*) If $\int_1^4 f(x)\,dx = 10$ and $\int_{-2}^4 f(x)\,dx = 11$, what is $\int_{-2}^1 f(x)\,dx$?

42. If $\int_2^6 f(x)\,dx = 11$ and $\int_2^8 f(x)\,dx = -1$, what is $\int_6^8 f(x)\,dx$?

In Exercises 43 and 44, use $\int_1^5 f(x)\,dx = 8$ and $\int_{-1}^5 f(x)\,dx = 10$ to compute the given definite integral.

43. $\int_5^1 f(x)\,dx$

44. $\int_{-1}^1 f(x)\,dx$

45. If $\int_2^5 f(x)\,dx = 6$, what is $\int_2^5 (2f(x) + 5)\,dx$?

46. If $\int_1^4 f(x)\,dx = 7$, what is $\int_4^1 (2f(x) + 3)\,dx$?

47. (*See Example 2*) A company has determined that its marginal cost is given by

$$MC(x) = \frac{x^2}{30} - 6x + 500$$

where x is the number of units manufactured per week.

(a) How will the total cost change if the company increases weekly production from 120 units to 150 units per week?

(b) How will the total cost change if the company decreases weekly production from 120 units to 90 units per week?

48. The marginal cost of a company is given by

$$MC(x) = \frac{x^2}{40} - 10x + 52{,}000$$

where x is the number of units produced monthly. How will the total monthly cost of the company be affected if the company changes production level from 300 units per month to

(a) 360 units per month?

(b) 270 units per month?

49. (*See Example 1*) A rock is thrown so that its velocity is given by $v(t) = -32t + 128$, where t is in seconds and $v(t)$ is in ft/sec.

(a) How far does the rock travel in the first 2 seconds?

(b) What is its change in height from $t = 3$ seconds to $t = 5$ seconds?

50. The price of a certain microcomputer has been steadily declining. Let $P(t)$ be its price t months after it was introduced. Then, $P'(t) = \frac{1}{4}t^2 - 5t - 24$. How much did the price drop during the first year?

6–4 The Definite Integral and Area

Introduction

One of the most important uses of the definite integral is to compute the area of certain figures. Thus, the definite integral is important in solving some geometric questions. Furthermore, the area between two curves often turns out to represent some other physical quantity in applications. We will see, for instance, that the area between the marginal revenue and the marginal cost curves can be interpreted as profit. Before we state the connection between area and the definite integral, we present two examples that may let you guess what the connection is.

Example 1 *(Compare Exercise 1)*
Find the area of the shaded rectangle in Figure 6–2.

Solution You know of course that the area of a rectangle is the base times the height. Here, the base is $4 - 1 = 3$ and the height is 2, so the area is $3 \cdot 2 = 6$. Now, notice that the top of the rectangle is the graph of $f(x) = 2$. If we compute

$$\int_1^4 f(x)\,dx = \int_1^4 2\,dx$$

we get

$$\int_1^4 2\,dx = 2x\Big|_1^4 = 8 - 2 = 6$$

The area!

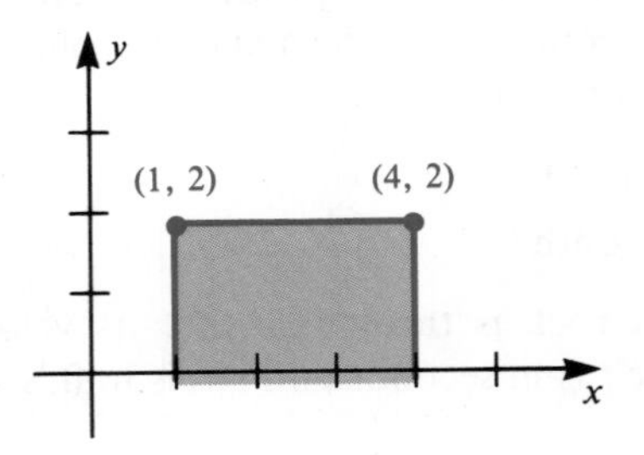

Figure 6–2

Example 2 *(Compare Exercise 3)*
Find the area of the triangle shown in Figure 6–3.

Solution The area of this triangle is $(1/2)b \cdot h = (1/2) \cdot 2 \cdot 4 = 4$. Now, think of the triangle as having the graph of $f(x) = 2x - 2$ as its "top." Then, if we compute

$$\int_1^3 f(x)\,dx = \int_1^3 (2x-2)\,dx$$

we get

$$\begin{aligned}\int_1^3 (2x-2)\,dx &= (x^2-2x)\Big|_1^3 \\ &= (9-6)-(1-2) \\ &= 4\end{aligned}$$

The area *again*!

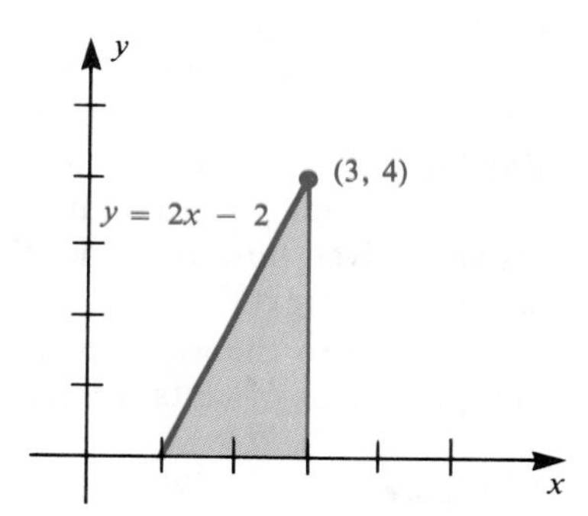

Figure 6–3

The Fundamental Theorem of Calculus

While we can't expect you to guess the exact relationship between the area and the definite integral from only two examples, perhaps your guess is close to the following theorem.

The Fundamental Theorem of Calculus

> If f is continuous on $[a, b]$ with $f(x) \geq 0$, and if R denotes the region bounded above by the graph of $y = f(x)$, bounded below by the x-axis, bounded on the left by the line $x = a$, and bounded on the right by the line $x = b$, then
>
> $$\text{the area of } R = \int_a^b f(x)\,dx$$

(Pretty formal—but your guess was probably closer than you realize.)

Before we go on with more examples and further applications, we would like to say a word about this theorem. Given our approach to the definite integral, you may find it strange that a particular application (computing areas) bears such an impressive title—the **fundamental** theorem of calculus. Because we (the authors) are trying to present calculus in terms of its applications while downplaying the theory, our approach has hidden the depth and significance of this theorem. By just stating the result, we are relying on your intuitive notion of what "the area of

R" means; we are just using $\int_a^b f(x)\,dx$ as a way to compute that area. But consider, for example, the area enclosed within a circle. You know the formula $A = \pi r^2$ for this area, because that's what has been in your textbooks and your teachers have told you that it is true, but how was this formula discovered in the first place? How can you talk about area inside a circle in terms of **square** inches? The problem of finding areas of plane figures is thousands of years old; the idea of antiderivative is only 300 years old. How are areas and antiderivatives connected?

Human beings find pleasure in seeing connections between topics that seem at first unrelated. In particular, mathematicians find this pleasure in the Fundamental Theorem of Calculus. Derivatives were invented as the language for rate of change and are used to solve the geometric problem of finding tangent lines. The Fundamental Theorem of Calculus says that there is a connection between two classic geometric problems, tangents and areas, that seem at first unrelated. It is a profound and beautiful theorem, and in a sense we regret that all we can do is tell you about it rather than letting you experience that appreciation for yourselves; but, such a development of the theory of area is surely beyond the scope of this text.

We return then to explaining exactly what this theorem does say, how to compute areas of certain figures, and how area can be used in applications.

Example 3 (*Compare Exercise 19*)
Find the area of the shaded region R indicated in Figure 6–4.

Solution R is bounded above by the graph of $y = x^2$, below by the x-axis, on the left by the line $x = -1$, and on the right by the line $x = 5$. Thus,

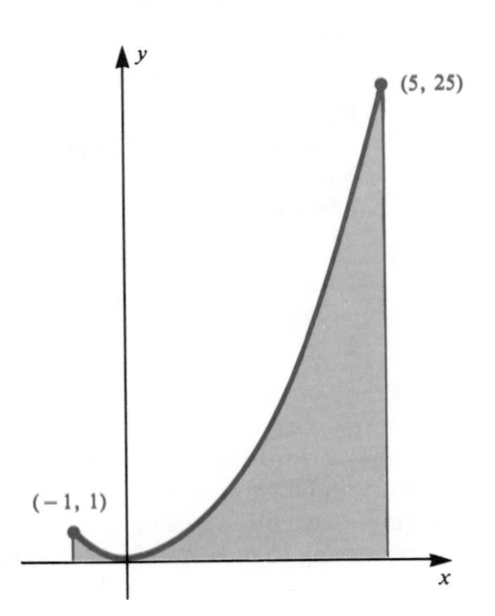

Figure 6–4

$$\text{the area of } R = \int_{-1}^{5} x^2\,dx = \frac{x^3}{3}\Bigg|_{-1}^{5}$$
$$= \frac{125}{3} - \frac{(-1)^3}{3}$$
$$= \frac{125}{3} + \frac{1}{3} = \frac{126}{3}$$

In this example, the graph of $y = f(x)$ actually comes down and touches the x-axis, but $\int_{-1}^{5} x^2\,dx$ gives the area of this region because $x^2 \geq 0$ on $[-1, 5]$.

Example 4 (*Compare Exercise 21*)
Find the area of the region bounded above by the graph of $y = 1/x$ and below by the interval $[1, 3]$.

Solution We have shortened the formal terminology. The interval $[1, 3]$ is on the x-axis; its leftmost point is 1 and its rightmost point is 3. The region is shown in Figure 6–5.

The area of the region is

$$\int_1^3 \frac{1}{x}\,dx = \ln|x|\Bigg|_1^3$$
$$= \ln 3 - \ln 1$$
$$= \ln 3 - 0 \qquad \text{(Remember that } \ln 1 = 0\text{)}$$
$$= \ln 3$$

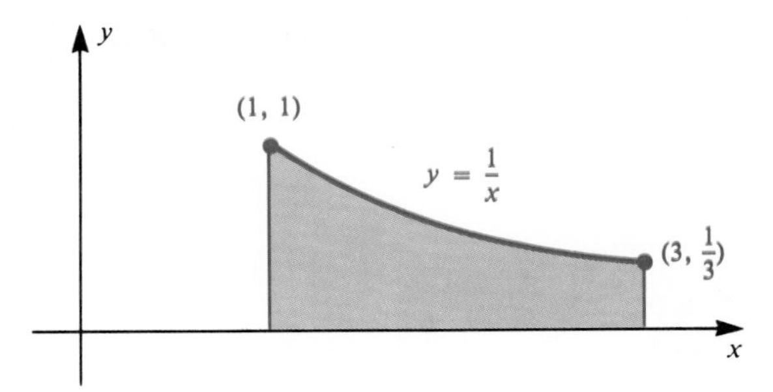

Figure 6–5

This example now gives us a geometric interpretation of logarithms; $\ln 3$ is the area between the curve $y = 1/x$ and the x-axis from $x = 1$ to $x = 3$. Thus, if we can find a numerical method of computing this area that doesn't involve antiderivatives, then we have a way to get a numerical approximation to $\ln 3$, such as is found in tables; $\ln 3 \approx 1.099$.

In fact, there is nothing special about the number 3. If b is any number bigger than 1, the region shown in Figure 6–6 has area $\ln b$.

For $b > 1$, we have an important formula for $\ln b$ that does not use exponents.

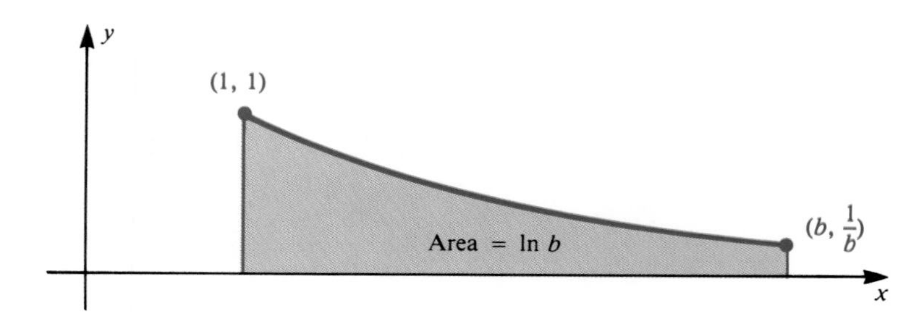

Figure 6–6

$$\ln b = \int_1^b \frac{1}{x}\,dx$$

In fact, this formula is valid for all $b > 0$. Using numerical techniques to approximate the area of this region would allow you to find the approximate value of $\ln b$. Doing this for various values of b would then allow you to construct a table of values for the natural logarithm. (Someone has to construct those tables somehow!)

Let's look at another problem concerning area, again using the curve $y = 1/x$.

Example 5 (*Compare Exercise 21*)
Find the area under the curve $y = 1/x$ and above the interval [2, 6]. See Figure 6–7.

Solution

$$\text{The area of } R = \int_2^6 \frac{1}{x}\,dx = \ln|x|\Big|_2^6$$

$$= \ln|6| - \ln|2|$$

$$= \ln 6 - \ln 2 = \ln\frac{6}{2} = \ln 3$$

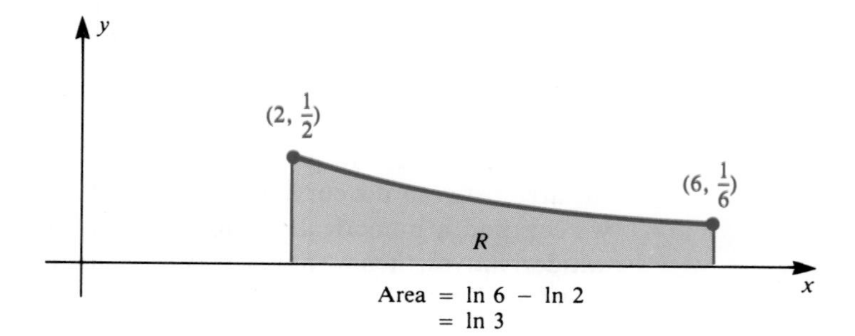

Figure 6–7

Example 5 provides an interesting aspect of the area under the curve $y = 1/x$. The area under this curve and above the interval $[1, p]$ is given by

$$\int_1^p \frac{1}{x}\,dx = \ln|x|\Big|_1^p = \ln p - \ln 1 = \ln p$$

The area under the curve and above the interval $[q, pq]$ is

$$\int_q^{pq} \frac{1}{x}\,dx = \ln|x|\Big|_q^{pq} = \ln(pq) - \ln q = \ln \frac{p \cdot q}{q} = \ln p$$

The areas are the same! In Examples 4 and 5, $p = 3$ and $q = 2$.

Integrands with Negative Values

We now look at what happens if the values of f are negative. How does this change the relationship between area and the definite integral? We do some examples to find out.

Example 6 (*Compare Exercise 9*)
Find the area of the rectangle bounded *above* by the x-axis, below by the line $y = -3$, on the left by the line $x = 2$, and on the right by the line $x = 6$. Compare this area with $\int_2^6 -3\,dx$. See Figure 6–8.

Solution This rectangle measures 3 by 4, so its area is 12. If we compute $\int_a^b f(x)\,dx$, however, with $a = 2$, $b = 6$, and $f(x) = -3$, we get

$$\begin{aligned}\int_2^6 -3\,dx &= -3x\Big|_2^6 \\ &= -18 - (-6) \\ &= -18 + 6 = -12\end{aligned}$$

the negative of the area.

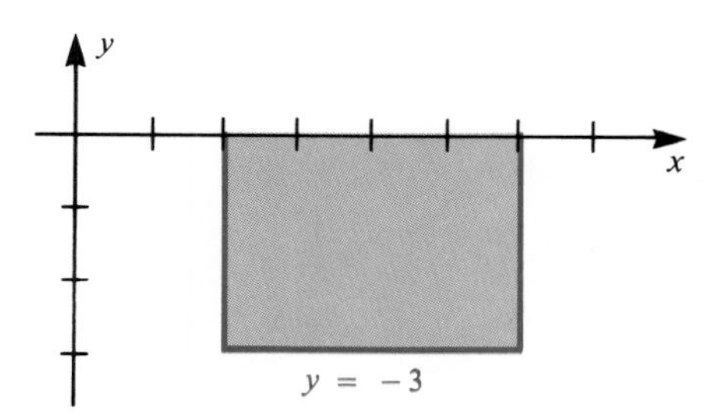

Figure 6–8

In fact, the general statement is as follows:

Theorem

If R is the region bounded *above* by the x-axis, below by the graph of $y = f(x)$ [so that $f(x) \le 0$], on the left by the line $x = a$, and on the right by the line $x = b$, then

$$\text{the area of } R = -\int_a^b f(x)\,dx$$

Put another way,

$$\text{if } f(x) \leq 0 \text{ on } [a, b], \text{ then } \int_a^b f(x)\,dx \text{ is the}$$

negative of the area of R.

But, there *is* a definite integral that gives us the area of R because

$$-\int_a^b f(x)\,dx = -1 \int_a^b f(x)\,dx = \int_a^b -1 \cdot f(x)\,dx$$
$$= \int_a^b - f(x)\,dx$$

So put yet another way,

$$\text{if } f(x) \leq 0 \text{ on } [a, b], \text{ then the area of } R = \int_a^b -f(x)\,dx$$

Figures 6–9 and 6–10 tell the geometry of this story. If $f(x) \leq 0$, define $g(x) = -f(x)$ so that $g(x) \geq 0$.

Figure II is obtained by flipping Figure I around the x-axis. The areas of the two figures are the same.

$$\text{area I} = \text{area II} = \int_a^b -f(x)\,dx$$

Next, we consider what happens when f has both positive and negative values on $[a, b]$.

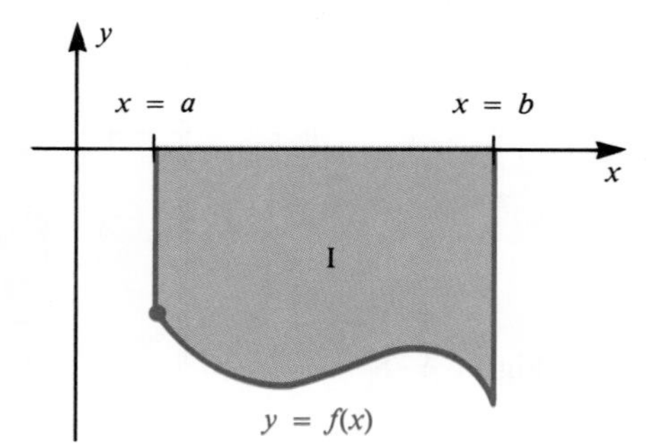

Figure 6–9

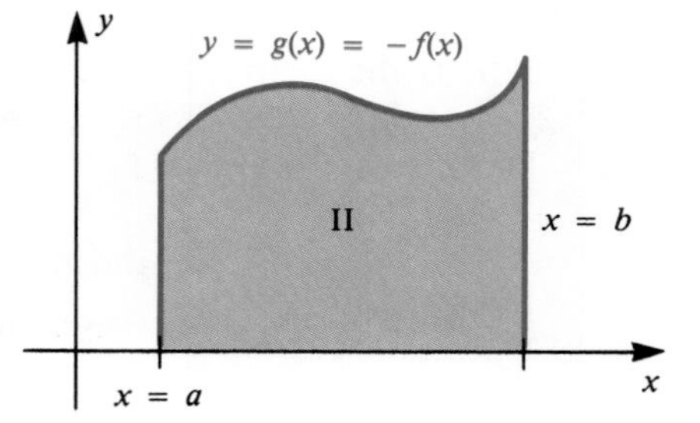

Figure 6–10

Example 7 (*Compare Exercise 27*)
Find the area of the shaded region in Figure 6–11. Compare the area with $\int_0^4 (2x - 2)\,dx$.

Solution Using geometry, we can split the region into two triangles.

$$\text{Area I} = \left(\frac{1}{2}\right)b \cdot h = \left(\frac{1}{2}\right)1 \cdot 2 = 1$$

$$\text{Area II} = \left(\frac{1}{2}\right)b \cdot h = \left(\frac{1}{2}\right)3 \cdot 6 = 9$$

$$\text{Total Area} = 9 + 1 = 10$$

Next,

$$\int_0^4 (2x - 2)\,dx = (x^2 - 2x)\Big|_0^4$$
$$= (16 - 8) - (0 - 0) = 8$$

This is neither the area of the region nor the negative of the area.

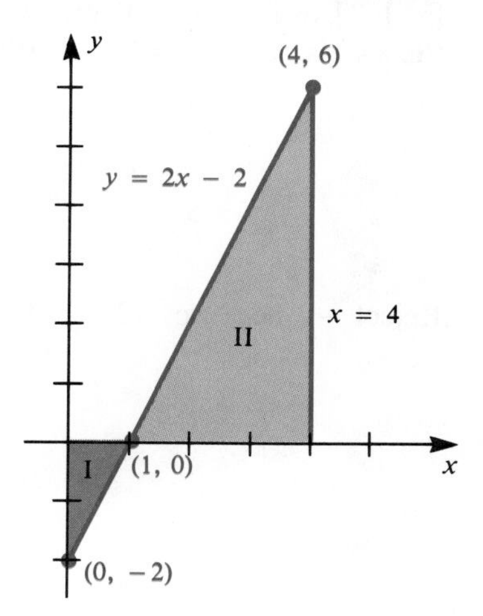

Figure 6–11

What happens in Example 7 is that the graph of $f(x) = 2x - 2$ crosses the x-axis between 0 and 4, so $f(x)$ is positive over some of the interval and negative over some of the interval. When we computed the total area, we broke the region into two pieces. Notice what happens when we break the definite integral into two corresponding integrals.

$$\int_0^4 (2x - 2)\,dx = \int_0^1 (2x - 2)\,dx + \int_1^4 (2x - 2)\,dx$$
$$= (x^2 - 2x)\Big|_0^1 + (x^2 - 2x)\Big|_1^4$$
$$= [(1 - 2) - (0 - 0)] + [16 - 8 - (1 - 2)]$$
$$= -1 + 9$$

$\int_0^4 (2x - 2)\,dx = 8$ because the area of region I was subtracted from the area of region II. The area of the whole region is

$$1 + 9 = -(-1) + 9 = -\int_0^1 (2x - 2)\,dx + \int_1^4 (2x - 2)\,dx$$

The next example again shows how to compute the area of a region between the x-axis and the graph of a function $y = f(x)$ over the interval $[a, b]$ when the sign of the function changes between a and b.

Warning! To compute areas, you must find out where the graph crosses the x-axis, and then compute the definite integrals over intervals on which f has a constant sign.

Example 8 (*Compare Exercise 31*)
Find the area of the region bounded by the lines $x = -1$, $x = 3$, the x-axis, and the curve $y = x^2 - 4$.

Solution The region is shaded in Figure 6–12.

Because the graph of $f(x) = x^2 - 4$ crosses the x-axis inside the interval $[-1, 3]$, one definite integral will not give the area. To see where the graph crosses the x-axis, set $f(x) = 0$ and solve.

$$\begin{aligned} f(x) &= 0 \\ x^2 - 4 &= 0 \\ x^2 &= 4 \\ x &= \pm 2 \end{aligned}$$

Reject -2 because -2 is not in $[-1, 3]$. The curve crosses the x-axis at $x = 2$.

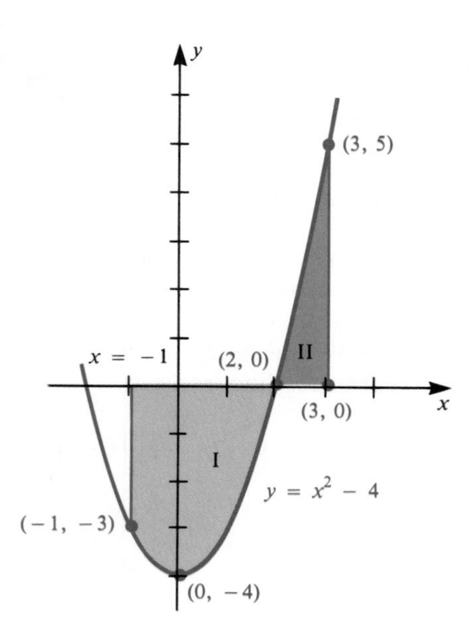

Figure 6–12

The total area is area I plus area II. Because $y = x^2 - 4$ lies below the x-axis from $x = -1$ to $x = 2$,

$$\begin{aligned}\text{Area I} &= -\int_{-1}^{2} (x^2 - 4)\,dx \\ &= -\left(\frac{x^3}{3} - 4x\right)\Bigg|_{-1}^{2} \\ &= -\left[\left(\frac{8}{3} - 8\right) - \left(\frac{-1}{3} - (-4)\right)\right] \\ &= 9\end{aligned}$$

From $x = 2$ to $x = 3$, however, the curve lies above the x-axis, so

$$\begin{aligned}\text{Area II} &= \int_{2}^{3} (x^2 - 4)\,dx \\ &= \left(\frac{x^3}{3} - 4x\right)\Bigg|_{2}^{3} \\ &= \left(\frac{27}{3} - 12\right) - \left(\frac{8}{3} - 8\right) \\ &= \frac{7}{3}\end{aligned}$$

The total area is $9 + \frac{7}{3} = \frac{34}{3}$.

Remember that in Example 7 we computed $\int_0^4 (2x - 2)\,dx = 8$, and 8 is not the area of any associated region.

Remember: To associate $\int_a^b f(x)\,dx$ with either the area of the region or the negative of this area, the function f cannot change sign between the endpoints a and b.

Area Between Curves

Sometimes, applications force you to look at two curves simultaneously (like marginal revenue and marginal cost). There can be a significance to the area of a region bounded by these two curves and two vertical lines. We have just seen how to compute the area of a region bounded by a curve $y = f(x)$, the x-axis, and the two vertical lines $x = a$ and $x = b$. Now we look at the more general problem of finding the area of a region between two curves. Again, we proceed by example.

Example 9 (*Compare Exercise 25*)
Find the area of the region bounded by the curves $y = 6 - x$ and $y = x^2$, and by the lines $x = -2$ and $x = 1$.

Solution These curves meet when $x = -3$ and $x = 2$, so they do not cross in the region we are interested in, which is shown in Figure 6–13.

We can represent the area we want as one area minus another area. See Figure 6–14.

$$\text{Area of I} = \text{area of II} - \text{area III}$$

$$= \int_{-2}^{1} (6 - x)\,dx - \int_{-2}^{1} x^2\,dx$$

We combine these two integrals into one since their limits of integration are the same.

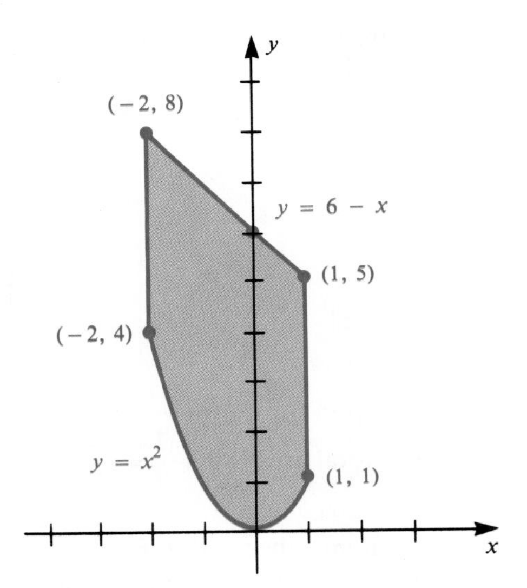

Figure 6–13

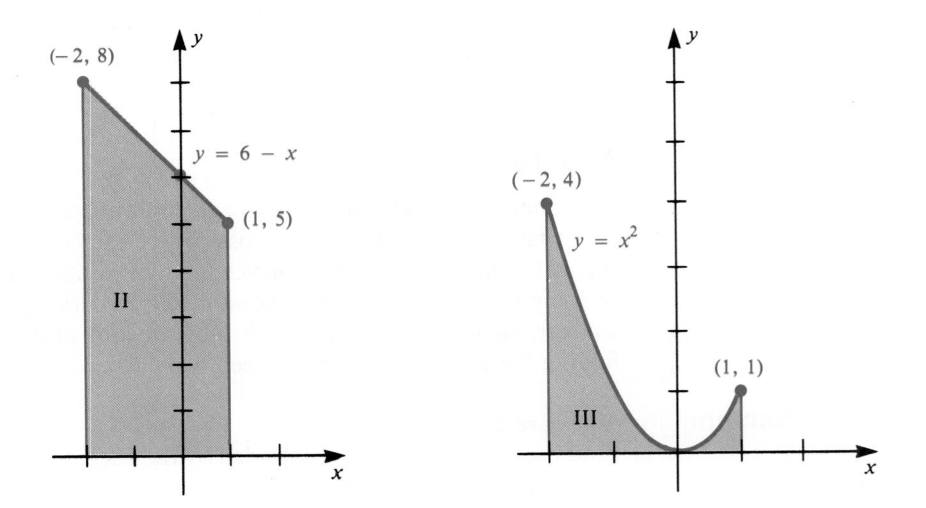

Figure 6–14

$$\begin{aligned}\text{Area of I} &= \int_{-2}^{1} (6 - x - x^2)\,dx \\ &= \left(6x - \frac{1}{2}x^2 - \frac{1}{3}x^3\right)\Bigg|_{-2}^{1} \\ &= \left(6 - \frac{1}{2} - \frac{1}{3}\right) - \left(-12 - \frac{4}{2} - \left(-\frac{8}{3}\right)\right) \\ &= \frac{33}{2}\end{aligned}$$

The general formula follows from computations similar to the one we gave in this specific example. See Figure 6–15.

Area Between Two Curves

If $f(x) \geq g(x)$ on the interval $[a, b]$, and if R is the region bounded above by $y = f(x)$, below by $y = g(x)$, on the left by the line $x = a$, and on the right by the line $x = b$, then

$$\text{The area of } R = \int_a^b [f(x) - g(x)]\,dx$$

Finally, what if the curves cross so that $f(x) \geq g(x)$ on some subinterval and $f(x) \leq g(x)$ on another subinterval? In this case, you must proceed in a manner similar to what we did in Example 8, where the graph of the function crossed the x-axis. This time you find where the graphs of f and g cross by solving the equation $f(x) = g(x)$. This will allow you to compute the area with two or more definite integrals, again as we did in Example 8. Remember that the integrand should be of the form (larger function − smaller function). Example 10 demonstrates how to do this.

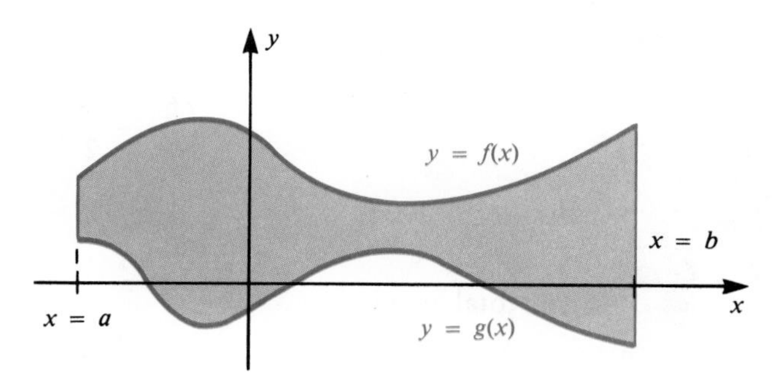

Figure 6–15

Example 10 (*Compare Exercise 33*)
Find the area of the region bounded by the curves $y = 5 - x$ and $y = x^2 - 4x + 5$, and by the lines $x = 1$ and $x = 4$.

Solution The curves cross when

$$5 - x = x^2 - 4x + 5$$

Solving, we have

$$0 = x^2 - 3x$$
$$0 = x(x - 3)$$
$$x = 0, x = 3$$

The solution $x = 0$ does not concern us since 0 is not in the interval [1, 4]. A quick sketch yields the shaded region of Figure 6–16. The curves meet when $x = 3$. From $x = 1$ to $x = 3$, $y = 5 - x$ is the top boundary of the region, so

$$\begin{aligned}
\text{Area I} &= \int_1^3 [(5 - x) - (x^2 - 4x + 5)]\,dx \\
&= \int_1^3 [-x^2 + 3x]\,dx \\
&= \left(-\frac{1}{3}x^3 + \frac{3}{2}x^2\right)\Big|_1^3 \\
&= \left(-\frac{27}{3} + \frac{27}{2}\right) - \left(-\frac{1}{3} + \frac{3}{2}\right) \\
&= \frac{10}{3}
\end{aligned}$$

From $x = 3$ to $x = 4$, $y = x^2 - 4x + 5$ is on top, so

$$\begin{aligned}
\text{Area II} &= \int_3^4 [(x^2 - 4x + 5) - (5 - x)]\,dx \\
&= \int_3^4 [x^2 - 3x]\,dx \\
&= \left(\frac{1}{3}x^3 - \frac{3}{2}x^2\right)\Big|_3^4 \\
&= \left(\frac{64}{3} - \frac{48}{2}\right) - \left(\frac{27}{3} - \frac{27}{2}\right) \\
&= \frac{11}{6}
\end{aligned}$$

The total area is

$$\frac{10}{3} + \frac{11}{6} = \frac{31}{6}$$

We conclude with one application of area. For more applications, see Section 6–5.

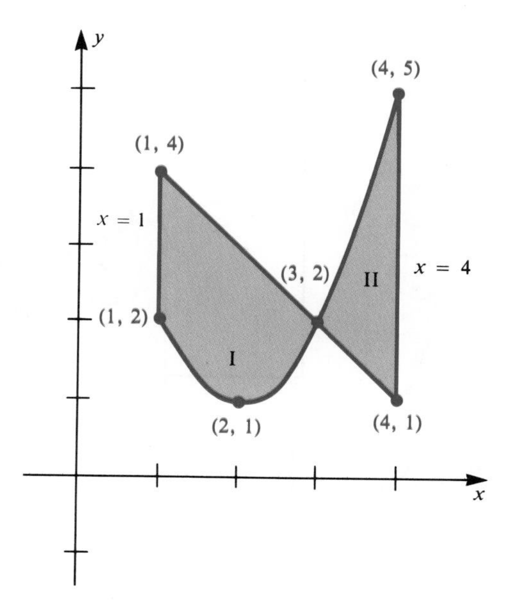

Figure 6–16

Application

Let C be the cost function of a company, let R be its revenue function, and let x be the number of units it produces daily. Then, its profit function P is given by

$$P(x) = R(x) - C(x)$$

and so,

$$P'(x) = R'(x) - C'(x)$$

If production is increased from a units per day to b units per day, the change in profit is

$$\begin{aligned} P(b) - P(a) &= \int_a^b P'(x)\,dx \\ &= \int_a^b (R'(x) - C'(x))\,dx \end{aligned}$$

If $R'(x) \geq C'(x)$, this last integral gives the area between the two curves over the interval $[a, b]$. See Figure 6–17.

The shaded area in Figure 6–17 gives the increase in profit when production is increased from a units to b units. Let q be the largest production level where $R'(x) = C'(x)$, that is, where marginal revenue = marginal cost.

As b moves to the right, the area of the shaded region increases until $b = q$. If the production level becomes bigger than q, then the curves cross on $[a, b]$. See Figure 6–18.

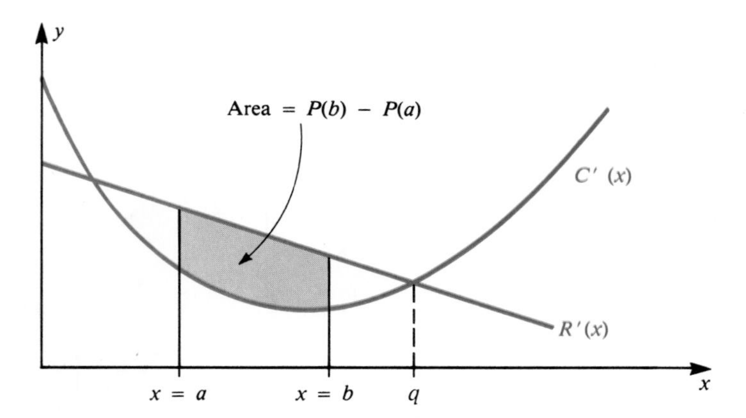

Figure 6–17

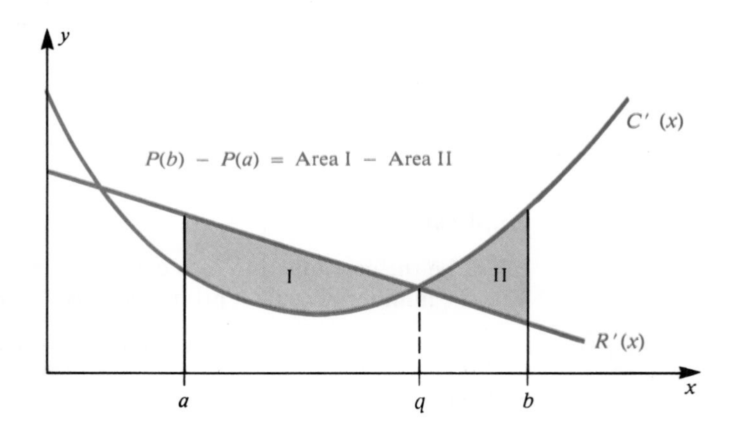

Figure 6–18

We have a geometric interpretation in terms of areas. The change in profit remains

$$P(b) - P(a) = \int_a^b P'(x)\,dx = \int_a^b (R'(x) - C'(x))\,dx$$

$$= \text{area I} - \text{area II} < \text{area I}$$

If b is larger than q, the total profit is less than when $b = q$.

We see again that with revenue and cost functions like those given in the above figures, the maximum profit will occur when the production level is the largest number x such that $R'(x) = C'(x)$. (*Compare Exercise 37*)

Often in applications, it is the marginal functions that are known first. If $R'(x)$ and $C'(x)$ are so complicated that we can't compute $\int_a^b (R'(x) - C'(x))dx$ by finding antiderivatives, then this geometric interpretation gives us another method of computing profit. We can compute the profit by computing the area of a certain region. We will explore this method and other applications dealing with area in the next section.

6–4 Exercises

I.

Compute each of the following areas by using formulas for the area of rectangles and triangles. Then write a definite integral that gives the area, evaluate the definite integral, and compare your answers.

1. (*See Example 1*)

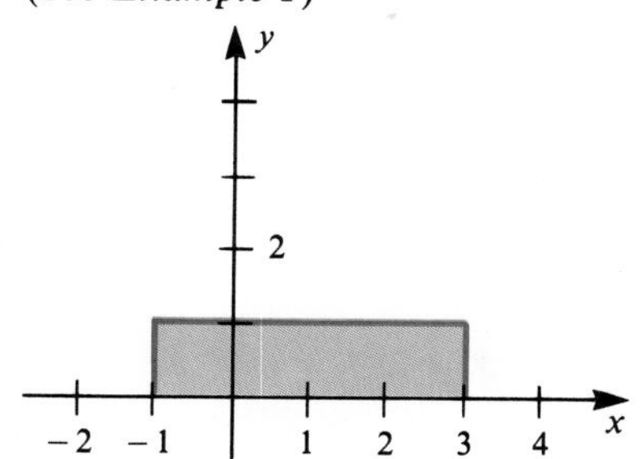

2.

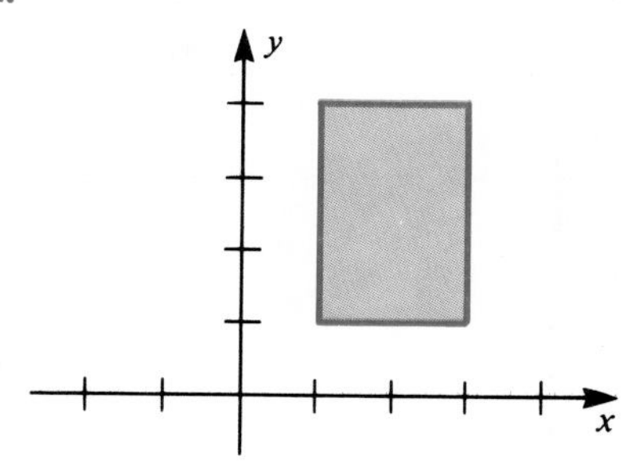

3. (*See Example 2*)

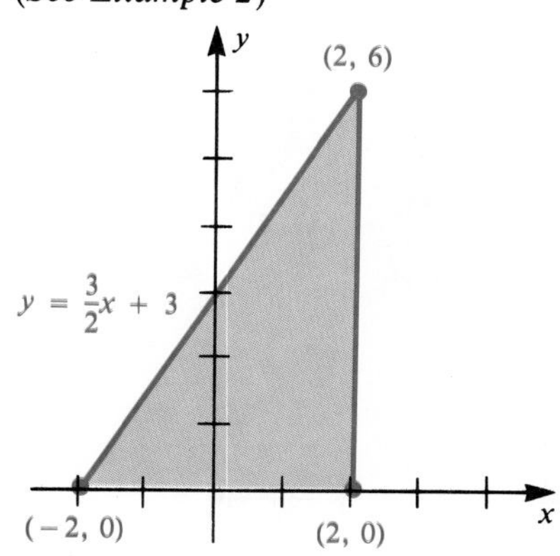

4.

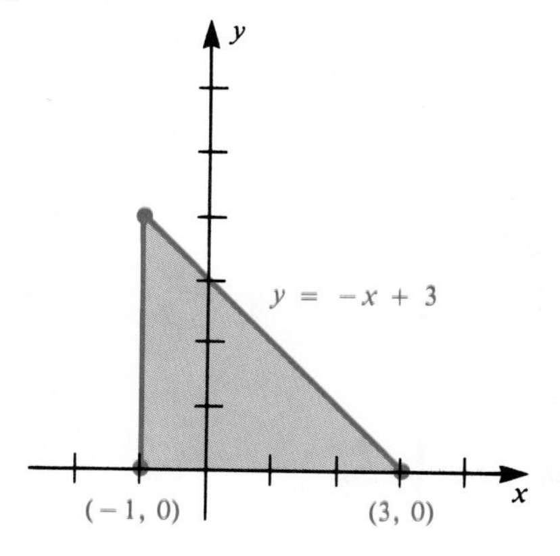

5. Find the area of the region between the graph of $f(x) = 4$, the x-axis, and the vertical lines $x = -3$ and $x = 2$.
6. Find the area of the region between the graph of $f(x) = 2x - 2$, the x-axis, and the vertical lines $x = 1$ and $x = 6$.

II.

Compute each of the following areas by geometry, using formulas for the area of rectangles and triangles. Then write a definite integral that gives the area, evaluate the definite integral, and compare your answers.

7.

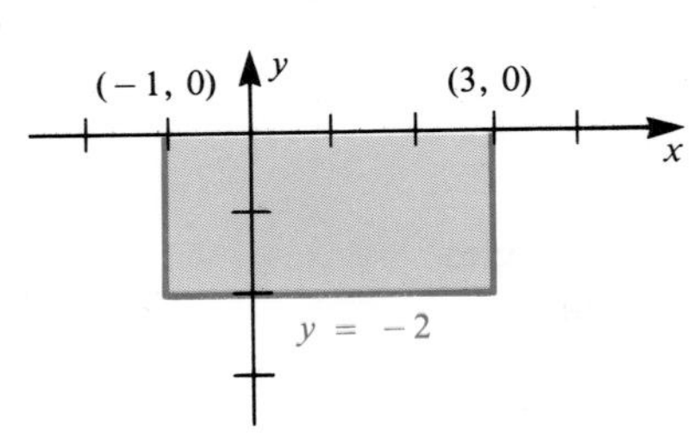

8.

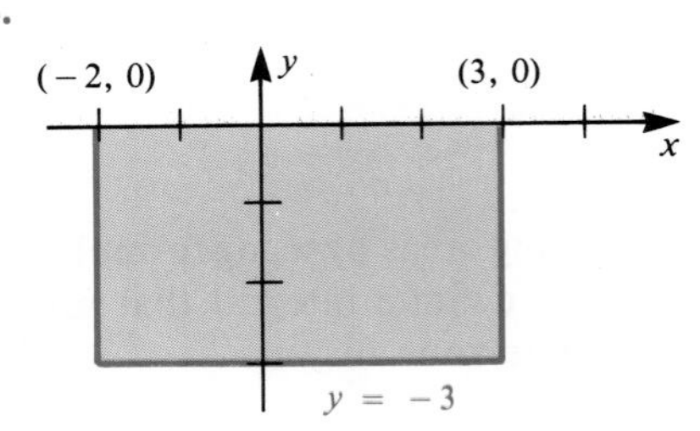

9. (*See Example 6*)

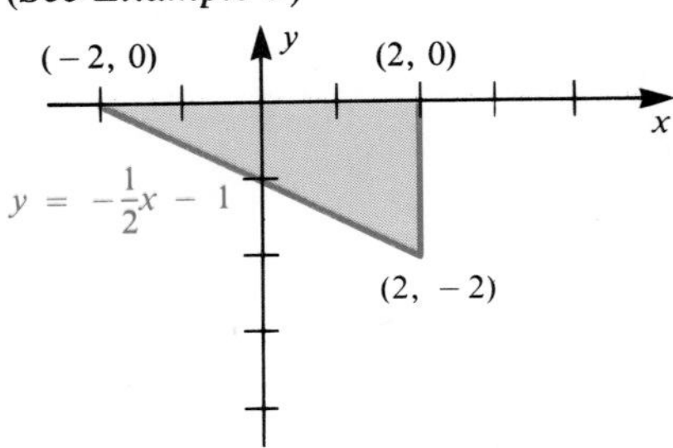

10.

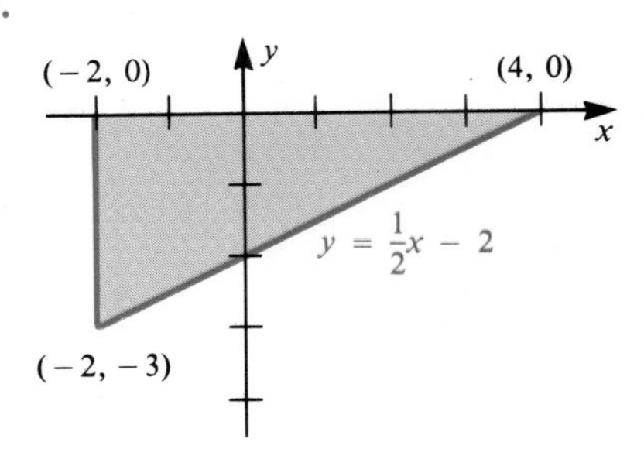

11.

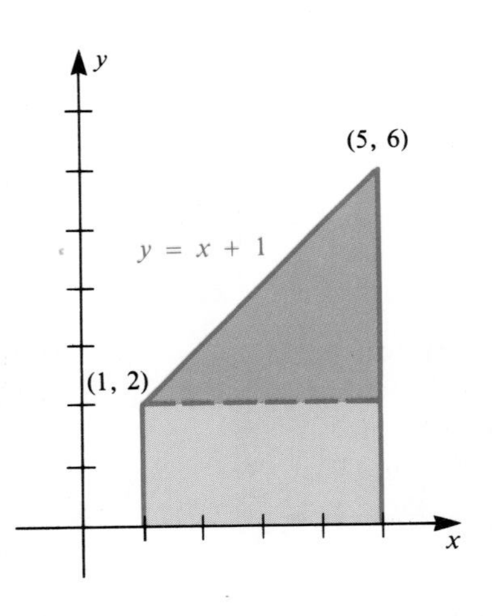

12.

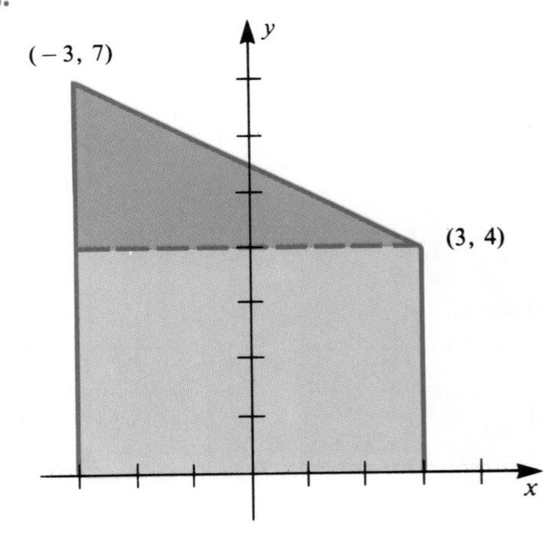

In Exercises 13 through 18, find the area of the indicated region.

13. The region between the graph of $f(x) = 2x - 1$, the x-axis, and the vertical lines $x = 2$ and $x = 7$.

14. The region between the graph of $f(x) = 5 - x$, the x-axis, and the vertical lines $x = -3$ and $x = 1$.

15. The region between the graph of $f(x) = -4$, the x-axis, and the vertical lines $x = 0$ and $x = 6$.

16. The region between the graph of $f(x) = 2x + 1$, the x-axis, and the vertical lines $x = -5$ and $x = -1$.

17. The region between the graph of $f(x) = 5 - 3x$, the x-axis, and the vertical lines $x = 2$ and $x = 6$.

18. The region between the graph of $f(x) = x - 9$, the x-axis, and the vertical lines $x = 3$ and $x = 5$.

Sketch each of the following regions and use a definite integral to compute the area of the region.

19. (*See Example 3*) The region between the graph of $f(x) = x^2 - 2x + 1$, the x-axis, and the vertical lines $x = 0$ and $x = 3$.

20. The region between the graph of $f(x) = 4 - x^2$, the x-axis, and the vertical lines $x = -1$ and $x = 2$.

21. (*See Examples 4 and 5*) The region between the graph of $f(x) = 2/x$, the x-axis, and the vertical lines $x = 2$ and $x = 10$.

22. The region between the graph of $y = -3/x$, the x-axis, and the vertical lines $x = 1$ and $x = 3$.

23. The region bounded above by the graph of $y = x^2 + 1$, below by the graph of $y = x - 4$, and between the vertical lines $x = 1$ and $x = 3$.

24. The region bounded above by the graph of $y = 2x + 5$, below by the graph of $y = 6 - x^2$, and between the vertical lines $x = 2$ and $x = 5$.

25. (*See Example 9*) The region bounded above by the graph of $y = x^3 + 2x^2 + 5x + 3$, below by the line $y = x + 2$, on the left by the y-axis, and on the right by the line $x = 2$.

26. The region bounded above by the graph of $y = 6 - x^2$, below by the graph of $y = x^2 - 4$, and between the vertical lines $x = -2$ and $x = 1$.

III.

Use geometry to compute the area of the shaded regions. Then compute the area using definite integrals.

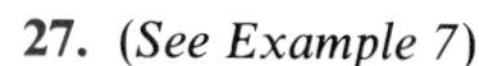

27. (*See Example 7*)

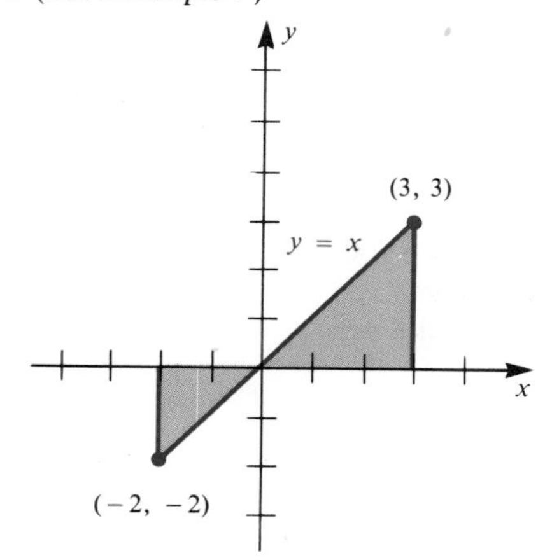

28.

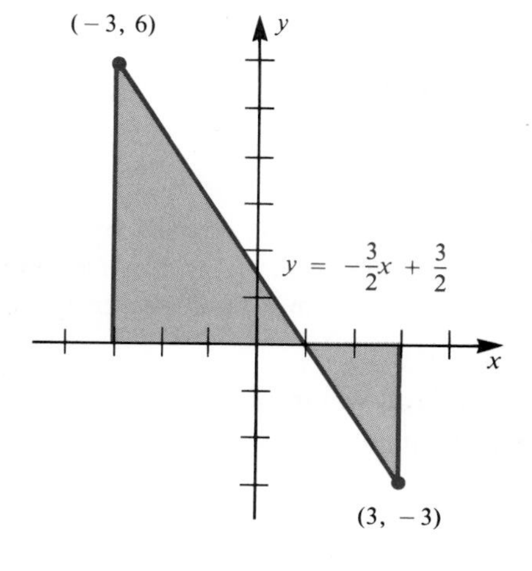

29.

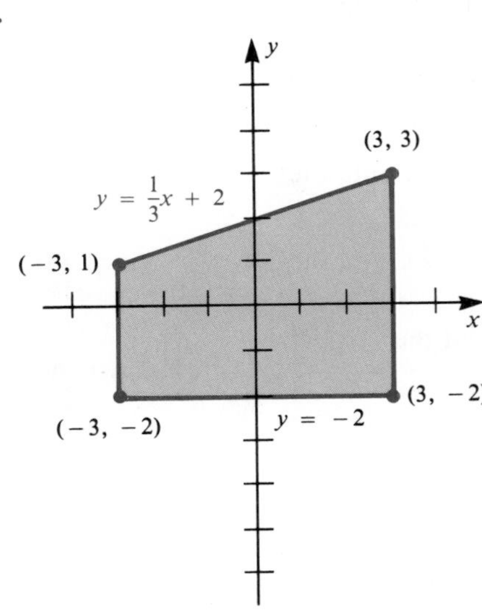

30.

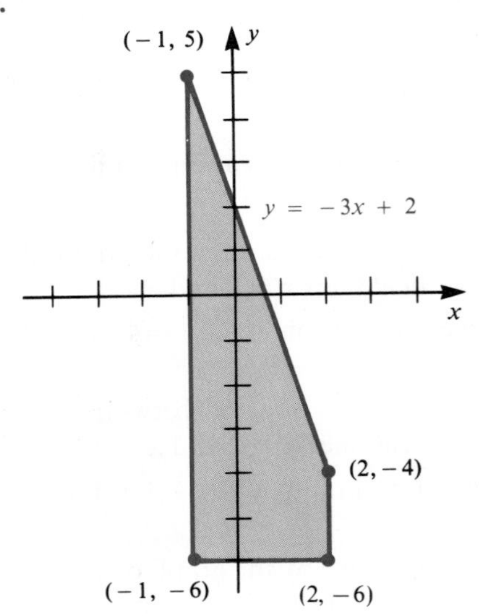

31. (*See Example 8*) The region between the x-axis, the graph of $y = 3x - 6$, and the vertical lines $x = 1$ and $x = 4$.
32. The region between the graph of $y = x^2 - 9$, the x-axis, and the vertical lines $x = 2$ and $x = 5$.
33. (*See Example 10*) The region between the graph of $y = x^2 + x + 2$, the line $y = 3x + 5$, and the vertical lines $x = -3$ and $x = 0$.
34. The region between the graph of $y = x^2 + x - 5$ and the line $y = x + 4$.
35. The region between the graph of $y = x^3 + 3$ and the line $y = x + 3$.
36. The region between the graph of $y = x^3 + x + 1$ and the line $y = 5x + 1$.
37. (*See Application*) A company's marginal cost is $MC(x) = x^2 - 20x + 300$, and its marginal revenue is $MR(x) = 276 - 6x$, where x is the number of units produced per day. The company is presently producing 10 units per day.

    (a) What production level will maximize profit?
    (b) How much will the company's profit increase if the production level is changed to maximize profit?

38. A company's marginal cost function is given by $MC(x) = x^2 - 30x + 500$, and its marginal revenue function is given by $MR(x) = 480 - 9x$, where x is the weekly production. The company is presently producing 24 units per week.

    (a) What production level will maximize the company's profits?
    (b) How much will profit increase if production level is changed to maximize profit?

6–5 Applications of Area

Average Value of a Function
Consumers' Surplus
Flows and Accumulations
The Gini Index

In the last section, we saw that the definite integral can be used to compute the area of a region that is bounded on the left and right by the vertical lines $x = a$ and $x = b$ and that is bounded above and below by the graphs of functions. Further, we saw that the change in profit can be interpreted as a certain area. We continue in this geometric vein, trying to obtain new insights into applications by viewing the problems geometrically.

Average Value of a Function

When introducing the derivative, we used the concept of the average speed of a car over shorter and shorter intervals to provide us with some feeling for what "instantaneous rate of change" means. Integration is the opposite of differentiation. We now suppose that the odometer in our car is broken but that we can keep track of the velocity by looking at the speedometer. Let $t = 0$ correspond to the time we started our trip, and let $v(t)$ be our velocity at time t hours after we start. Suppose that during the fourth and fifth hours of our trip we were going through periods of construction and heavy traffic, so our speed on the highway varied considerably. The graph of $v(t)$ is drawn in Figure 6–19.

If we wanted to complain later about the difficulty of this trip, we would probably want to complain about our average speed during these two hours.

$$\text{Average speed} = \frac{\text{distance travelled}}{\text{elapsed time}}$$

We have seen earlier in this chapter that the distance travelled is $\int_3^5 v(t)\,dt$; the total time is $5 - 3 = 2$. (Notice the limits of integration; for the first hour of our trip, $0 \le t \le 1$. The second hour corresponds to the interval $[1, 2]$; the third hour corresponds to the interval $[2, 3]$; the fourth hour corresponds to the interval

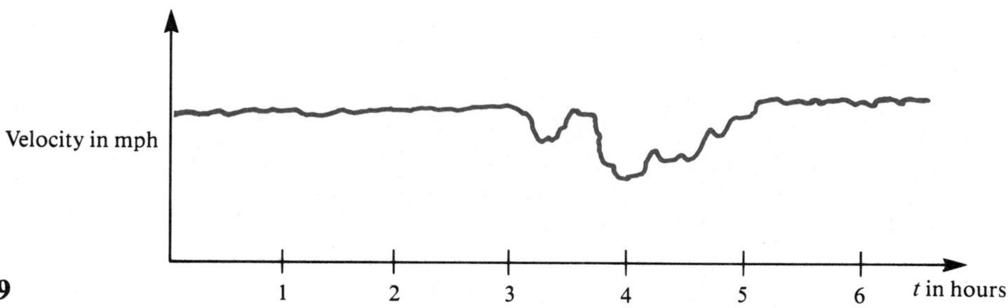

Figure 6–19

[3, 4]; and the fifth hour corresponds to the interval [4, 5]. So, the fourth and fifth hours correspond to $3 \leq t \leq 5$.) The average speed during this time is

$$\frac{\int_3^5 v(t)\,dt}{5-3}$$

We generalize this concept to any continous function over an interval $[a, b]$.

Definition

> The **average value of the function f over the interval $[a, b]$** is
>
> $$\frac{1}{b-a}\int_a^b f(x)\,dx$$

We now interpret the average value of a function over $[a, b]$ geometrically, and use area to see what this definition gives us in contexts other than average velocity.

Let R be the region bounded above by the graph of $y = f(x)$, below by the x-axis, on the left by the line $x = a$, and on the right by the line $x = b$. If we let AVE stand for the average value of f on the interval $[a, b]$, we have

$$AVE = \frac{1}{b-a}\int_a^b f(x)\,dx$$

so,

$$(AVE)(b-a) = \int_a^b f(x)\,dx = \text{area of } R$$

The average value of f is the average height of the graph of $y = f(x)$. The average height of the graph is that number which, when multiplied by the length of the interval, gives the area of R.

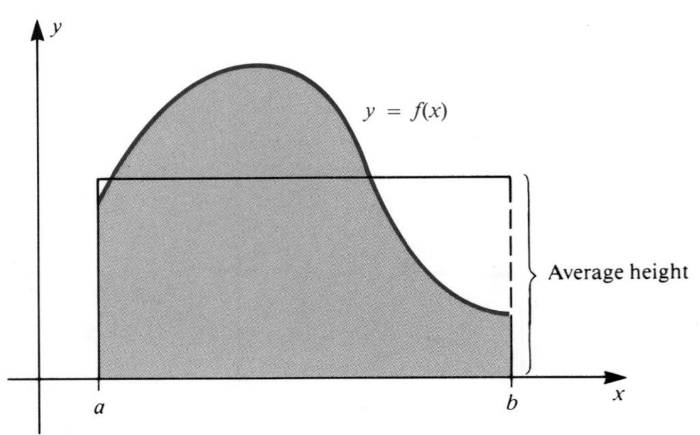

Figure 6–20 Area of rectangle = Area of region under the curve

This idea is shown in Figure 6–20 and gives a geometric motivation for the definition of the average value of f on $[a, b]$ as

$$\frac{1}{b-a}\int_a^b f(x)\,dx$$

Example 1 (*Compare Exercise 1*)
What is the average value of $f(x) = x^2$ on the interval $[-3, 3]$?

Solution First,

$$\int_{-3}^{3} x^2\,dx = \left.\frac{x^3}{3}\right|_{-3}^{3}$$
$$= \frac{27}{3} - \left(-\frac{27}{3}\right)$$
$$= 9 + 9$$
$$= 18$$

Next, the length of the interval, $b - a$, is $3 - (-3) = 6$. The average value of x^2 on $[-3, 3]$ is $\frac{18}{6} = 3$.

Example 2 (*Compare Exercises 17 and 19*)
If the population (in millions) of a country is given by $P(t) = (43.2)e^{.02t}$, where $t = 0$ corresponds to 1980 and t is in years, what is the average population of the country during the decade of the 1980's? [Notice that $P'(t) = .02P(t)$, so we are modeling a 2% growth rate.]

Solution The interval is $[0, 10]$, so the average population is

$$\frac{1}{10-0}\int_0^{10} (43.2)e^{.02t}\,dt = \frac{1}{10}(43.2)\int_0^{10} e^{.02t}\,dt$$
$$= 4.32\left(\left.\frac{1}{.02}e^{.02t}\right|_0^{10}\right)$$
$$= 216\left(\left.e^{.02t}\right|_0^{10}\right)$$
$$= 216(e^{.2} - e^0)$$
$$= 216(e^{.2} - 1)$$
$$\approx 216(1.221 - 1)$$
$$\approx 47.8$$

The average population is about 47.8 million people.

Consumers' Surplus

How do you quantify the individual satisfaction that comes when you've gotten a "good buy"? It's hard to do, but economists have a method for quantifying society's group satisfaction with certain price levels, and that method is called

consumers' surplus. The explanation of how consumers' surplus is computed and what it means relies heavily on the interpretation of integrals as areas.

Suppose that the price-demand curve of a certain commodity is given by

$$p = 250 - \frac{x^2}{10^5}$$

Remember that x is the quantity of goods that can be sold at price p. It seems backwards to think of p as a function of x; as consumers, we usually think of demand as a function of price. The reason for viewing the relationship this way rests on the geometric interpretation of consumers' surplus that this viewpoint provides.

If the market price of this commodity is \$160, then the corresponding value of x is 3000. Thus, consumers will buy 3000 units at \$160 each for a total cost to the consumers of (3000)(\$160) = \$480,000. This number is represented by the shaded area in Figure 6–21.

Presumably, the consumers who are willing to pay \$220 for this good are even happier with the market price of \$160 than are those consumers who are willing to pay \$180. How can we measure this aggregate happiness? Economists call $p(x)$ the **value** of xth unit purchased and define the **total value** of the 3000 units to the consumer by

$$\text{Total value} = \int_0^{3000} p(x)\,dx$$

With $p(x) = 250 - \dfrac{x^2}{10^5}$ we have

$$\begin{aligned}\text{Total value} &= \int_0^{3000} \left(250 - \frac{x^2}{10^5}\right) dx \\ &= 250x - \frac{x^3}{3 \cdot 10^5}\Bigg|_0^{3000} \\ &= 660{,}000\end{aligned}$$

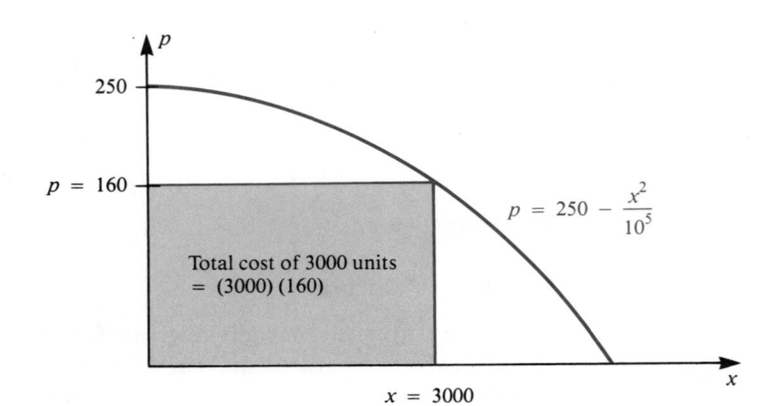

Figure 6–21

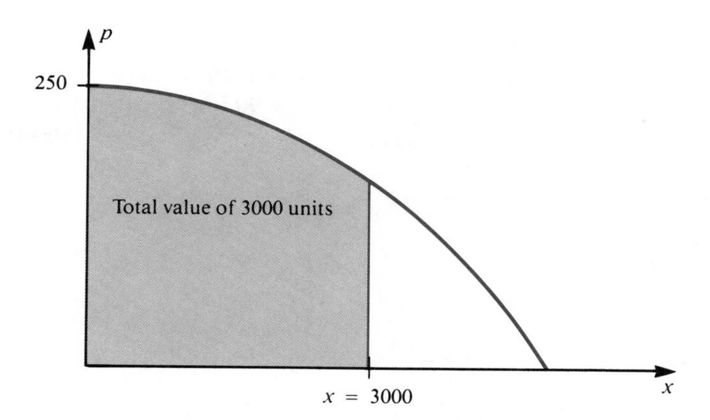

Figure 6–22

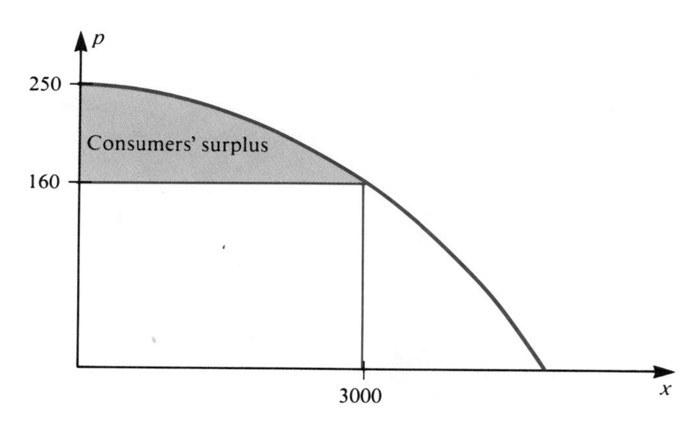

Figure 6–23

See Figure 6–22 for a geometric interpretation of this number.

Next, the **consumers' surplus** is defined to be the total value minus the total cost. Here, consumers' surplus is 660,000 − 480,000 = 180,000. See Figure 6–23 for a geometric interpretation of this quantity.

Notice that the consumers' surplus is

$$\int_0^{3000} p(x)\,dx - \int_0^{3000} 160\,dx = \int_0^{3000} (p(x) - 160)\,dx$$

This now gives us a general formula for consumers' surplus.

Definition

> If the market price is p_1, and the demand at that price is x_1, then the **consumers' surplus** $= \int_0^{x_1} (p(x) - p_1)\,dx$

Example 3 (*Compare Exercise 7*)

If $p = 250 - \dfrac{x^2}{10^5}$, what is the consumers' surplus when the market price is 90?

Solution Before we begin the solution, we remark that consumers should be happier with a market price of 90 than with 160, so our answer should be bigger than 180,000. To find x_1, substitute $p = 90$ into the demand-price equation and solve for x.

$$90 = 250 - \frac{x^2}{10^5}$$

$$-160 = -\frac{x^2}{10^5}$$

$$16{,}000{,}000 = x^2$$

$$x = 4000 \qquad \text{(Reject } x = -4000 \text{ because demand} \geq 0)$$

Thus, $x_1 = 4000$ and $p_1 = 90$, so the

$$\begin{aligned} \text{Consumers' surplus} &= \int_0^{4000} \left(250 - \frac{x^2}{10^5} - 90\right) dx \\ &= \int_0^{4000} \left(160 - \frac{x^2}{10^5}\right) dx \\ &= \left(160x - \frac{x^3}{3 \cdot 10^5}\right)\Bigg|_0^{4000} \\ &= \left[(160)(4000) - \frac{4000^3}{3 \cdot 10^5}\right] - (0 - 0) \\ &= 426{,}666.67 \end{aligned}$$

Consumers bought more items (4000 vs. 3000) at a lower price (90 vs. 160) and were more than twice as happy as measured by the consumers' surplus (426,666 vs. 180,000).

Flows and Accumulations

Economists define investment as a flow of resources into the production of new capital. Capital here refers to inventory, equipment, and other resources of production of the firm. We use $K(t)$ to represent the total capital of a firm at time t (capital starts with c, but c is reserved for **c**ost functions), and $I(t)$ to represent the amount of **in**vestment being made at time t. By definition, investment is the rate of change of capital;

$$I = \frac{dK}{dt}$$

This is analogous to any flow into (or out of) a container. For example, we can think of measuring the flow of gas into a balloon. The gas in the balloon corresponds to the total capital of the firm, and the flow of gas into or out of the balloon at time t corresponds to $I(t)$.

Firms may be able to project their investment flow over a period of time and then ask what the total change in capital will be. If the period of time is given by $a \le t \le b$, then the total change of capital over this interval is $K(b) - K(a)$. This total change of capital is called the **capital formation** for this period and is given by the definite integral of the flow function.

$$K(b) - K(a) = \int_a^b I(t)\,dt$$

Example 4 (*Compare Exercise 11*)
Suppose that a company plans to increase its investment flow and has determined that the model that best suits its capabilities for the next 12 months is given by

$$I(t) = -\frac{1}{30}t^2 + t + 1$$

Here, the units of capital are millions of dollars, and t is measured in months. What will be the company's capital formation during the second quarter of this period?

Solution The capital formation for the entire twelve month period is the area under the curve

$$I(t) = -\frac{1}{30}t^2 + t + 1$$

over the interval [0, 12]. We have to be careful to choose the limits of integration that represent the second quarter.

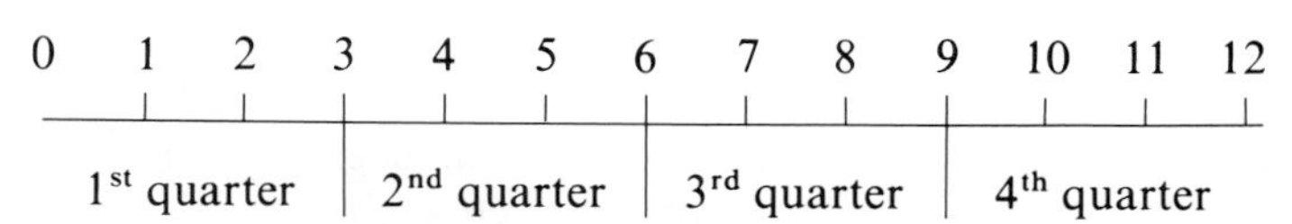

The capital formation for the second quarter is represented by the shaded region in Figure 6–24.

We can compute this area using the fundamental theorem of calculus.

$$\begin{aligned}
\text{Capital formation} &= \text{area} \\
&= \int_3^6 \left(-\frac{1}{30}t^2 + t + 1\right) dt \\
&= \left(-\frac{1}{90}t^3 + \frac{1}{2}t^2 + t\right)\Bigg|_3^6 \\
&= \left(-\frac{216}{90} + \frac{36}{2} + 6\right) - \left(-\frac{27}{90} + \frac{9}{2} + 3\right) \\
&= 14.4
\end{aligned}$$

The company's capital will increase by \$14.4 million.

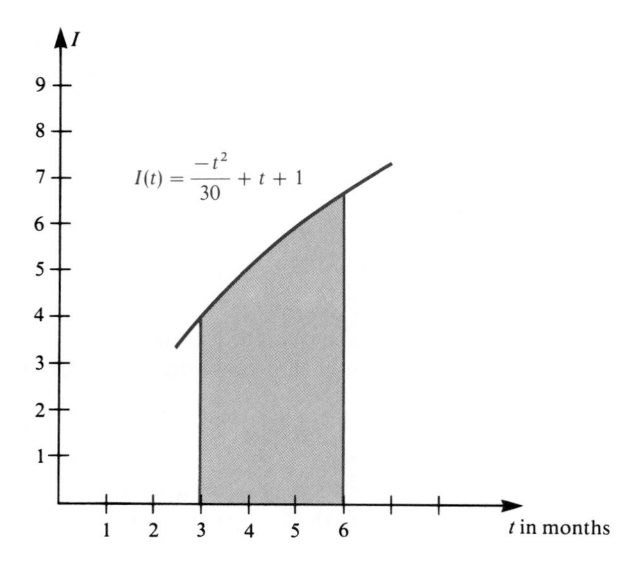

Figure 6–24

The importance of realizing that the capital formation can be viewed as area under a curve becomes even clearer when we cannot find an antiderivative for the function describing $I(t)$. For example, suppose the company decides that $I(t) = \sqrt{t^3 + 1}$ is a better model for their investment flow. This $I(t)$ does not have an antiderivative that can be expressed using simple functions. Recognizing certain quantities as areas will allow you to use area-geometric computations to calculate an approximate value of

$$\int_a^b I(t)\,dt$$

As mentioned at the beginning, this flow-accumulation model works with any flow problem, such as the population of a given political entity in terms of flow rate of people [(births + immigration) − (deaths + emigration)]; the trade balance of a country [exports − imports]; or the amount of water in a reservoir. Frequently, the flow function will be determined from data and will not have an antiderivative that can be found easily. Numerical techniques must be used in these cases.

The Gini Index

In Chapter 1, we introduced the notion of a Lorenz curve for measuring the income distribution of a given country. The Lorenz curve for a given country is the graph of

$$F(x) = y, \qquad 0 \le x \le 1$$

where $F(x) = y$ means that the lower $100x\%$ of the families (ranked by income) earned $100y\%$ of the total income. Thus, $F(.4) = .25$ would mean that the lower 40% of families earned 25% of the total income. The line of perfect equality is the

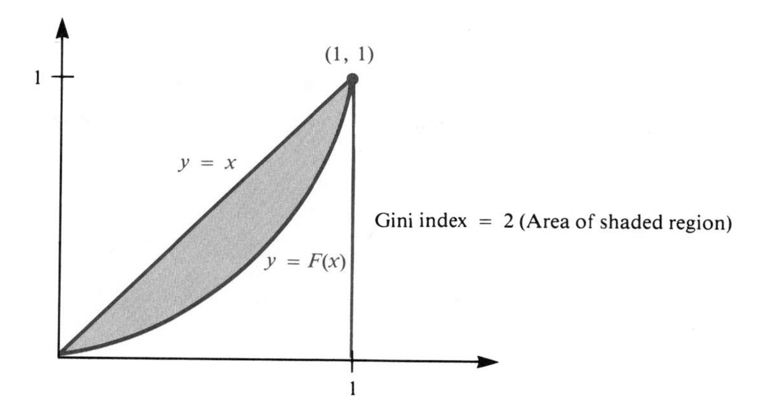

Figure 6–25

line $y = x$. The area between this line and the Lorenz curve gives a measure of the income inequality of the country—the larger this area, the more the income distribution of the country deviates from perfect equality. The problem of comparing one country's income distribution to another country's is a difficult one. The difficulties come from gathering reliable data that are comparable. The measure that economists use for comparing the income distribution between various countries is called the **Gini index**. Referring to Figure 6–25, the Gini index for a given country is defined to be the ratio

$$\frac{\text{Area of shaded region}}{\text{Total area of lower triangle}}$$

Because the area of the lower triangle is $\frac{1}{2}$, the Gini index is twice the shaded area. If F is the income distribution for a given country (F is defined above), then

Definition

The Gini index $= 2\int_0^1 [x - F(x)]\,dx$

The function F is arrived at by finding the curve of a given type that best approximates the data collected. The type of curve economists use arises in advanced statistics and is called a beta distribution. To keep our computations simple, we will use polynomial approximations to the data. We require that

$$F(0) = 0 \qquad \text{(0\% of population must earn 0\% of total income)}$$

and

$$F(1) = 1 \qquad \text{(100\% of population earns 100\% of total income)}$$

The data for the United States in 1971 is given in the table below. (It is a historical fact that these data varied *very* little in the U.S. for the period from 1950–83.)

Percentile of Population	Cumulative Percent of Total Income
0	0
20	5
40	15.8
60	32.3
80	55.4
100	100.0

The cubic $F(x) = .91x^3 - .25x^2 + .34x$ closely approximates this data (for example, $F(.4) = .154$). Using this cubic, we get an approximation to the Gini index for the United States in 1971.

$$\begin{aligned}
\text{The Gini index} &= 2\int_0^1 [x - F(x)]\,dx \\
&= 2\int_0^1 [x - (.91x^3 - .25x^2 + .34x)]\,dx \\
&= 2\int_0^1 [-.91x^3 + .25x^2 + .66x]\,dx \\
&= 2\left(\left[-\frac{.91}{4}x^4 + \frac{.25}{3}x^3 + \frac{.66}{2}x^2\right]\Bigg|_0^1\right) \\
&= 2\left(-\frac{.91}{4} + \frac{.25}{3} + \frac{.66}{2}\right) \\
&= .37
\end{aligned}$$

The actual Gini index computed by economists using a beta distribution is .39. Those of Sweden and Brazil are given in Exercises 15 and 16.

Example 5 (*Compare Exercise 15*)
Use the function $F(x) = 1.95x^4 - 2.95x^3 + 2.08x^2 - .08x$ to approximate the Gini index for the United States in 1971. (This function is a fourth-degree polynomial that gives a good approximation to the data.)

Solution

$$\begin{aligned}
\text{The Gini index} &= 2\int_0^1 [x - F(x)]\,dx \\
&= 2\int_0^1 [x - (1.95x^4 - 2.95x^3 + 2.08x^2 - .08x)]\,dx \\
&= 2\int_0^1 [-1.95x^4 + 2.95x^3 - 2.08x^2 + 1.08x]\,dx
\end{aligned}$$

$$= 2\left(-\frac{1.95}{5}x^5 + \frac{2.95}{4}x^4 - \frac{2.08}{3}x^3 + \frac{1.08}{2}x^2\Big|_0^1\right)$$

$$= 2\left(-\frac{1.95}{5} + \frac{2.95}{4} - \frac{2.08}{3} + \frac{1.08}{2}\right)$$

$$= .39$$

6–5 Exercises

Compute the average value of each function on the given interval.

1. (*See Example 1*) $f(x) = x^2 - 2$ on $[-4, 4]$
2. $f(x) = 2x + 3$ on $[1, 6]$
3. $f(x) = e^{-x/2}$ on $[0, 4]$
4. $f(x) = \sqrt{x}$ on $[4, 25]$
5. $f(x) = \dfrac{1}{x^3}$ on $[2, 6]$
6. $f(x) = \dfrac{3}{x}$ on $[1, 6]$
7. (*See Example 3*) The consumer demand for a certain product is given by $p = 400 - 6x$. The market price is 190.

   (a) Determine how many units will be sold.
   (b) What is the total cost to the consumer of these units?
   (c) What is the total value to the consumer of these units?
   (d) What is the consumers' surplus?

8. For a certain product, the consumer demand is given by $p = 510 - 8x$, and the market price is 110.

   (a) How many units will be sold at this price?
   (b) What is the total cost of these units to the consumer?
   (c) What is the total value of these units to the consumer?
   (d) What is the consumers' surplus?

9. For a certain product, the consumer demand is given by $p = 300 - (x^2/10^5)$. The market price is 210.

   (a) How many units will be sold?
   (b) What is the total cost of these units to the consumer?
   (c) What is the total value of these units to the consumer?
   (d) What is the consumers' surplus?

10. If the consumer demand function is $p = 360 - (x^2/10^5)$, and the market price is 200, what is the consumers' surplus?
11. (*See Example 4*) The investment flow of a company during a given year is $I(t) = 30e^{t/20}$; capital is measured in thousands of dollars, and t is measured in months. What was the company's capital formation

    (a) during the second quarter of the year?
    (b) during the third quarter?

12. The investment flow of a company during a given year is

$$I(t) = 6 + \frac{1}{2}t + \frac{1}{12}t^2$$

where capital is measured in millions of dollars, and t is measured in months. What was the company's capital formation

(a) during the third quarter?

(b) during the second half?

13. Oil is flowing out of a well at a rate of $12e^{-t/10}$ thousand gallons per year t years after it starts. How much oil flowed out of the well during the first 20 years of production?

14. Propane gas is flowing into a furnace at the rate of

$$10 + 2t - \frac{1}{6}t^2 \text{ cubic feet per hour}$$

t hours after it was started.

(a) How much gas is used during the first four hours?

(b) How much gas is used during the fifth hour?

15. (*See Example 5*) The income distribution for Brazil in 1972 is given by the following table.*

Percentile of Population	Cumulative Percentage of Total Income
0	0
20	2.0
40	7.0
60	16.4
80	33.4
100	100

The Gini index for Brazil in 1972 is .61.

(a) The curve $F(x) = .7x - 2.4x^2 + 2.7x^3$ is a good cubic approximation to the data. What is the approximation to the Gini index using this cubic?

(b) The curve $F(x) = 5.3x^4 - 7.7x^3 + 3.9x^2 - .5x$ is a good fourth-degree polynomial approximation to the data. What is the approximation to the Gini index obtained by using this fourth-degree polynomial?

---

* These data were taken from W. van Ginneken and J. Park, eds., *Generating Internationally Comparable Income Distribution Estimates* (Geneva: International Labour Office, 1984).

16. The income distribution for Sweden in 1979 is given by the following table.*

Percentile of Population	Cumulative Percentage of Total Income
0	0
20	7.2
40	20
60	37.4
80	62.8
100	100

Sweden's Gini index for this data is .30.

(a) A good cubic fit to this data is given by $F(x) = .33x + .20x^2 + .47x^3$. What is the approximation to the Gini index given by this cubic?

(b) A good quartic fit to this data is given by $F(x) = .20x + .95x^2 - .78x^3 + .63x^4$. What is the approximation to the Gini index given by this quartic?

17. (*See Example 2*) One estimate of the world's population is given by $P(t) = 4.9e^{.02t}$, with $t = 0$ corresponding to January 1, 1985. The units for $P(t)$ are billions of people, and t is in years.

(a) Estimate the average population of the earth from January 1, 1985 to January 1, 1990.

(b) Estimate the average population of the earth from January 1, 1990 to January 1, 1995.

18. The price, in dollars, of a typical three-bedroom house in a certain area from 1980 to 1985 is estimated by $P(t) = 300t^2 + 500t + 42{,}000$; t is years since 1980. What was the average price of such a house during this period?

19. (*See Example 2*) The velocity of a decelerating train is given by

$$V(t) = 88 - 4t - \frac{1}{50}t^2$$

so that it comes to rest 20 seconds after the brakes are applied. V is measured in feet per second, and time, t, is in seconds.

(a) What was the train's average speed during these 20 seconds?

(b) What was its average speed during the first 10 seconds?

---

* These data were taken from W. van Ginneken and J. Park, eds., *Generating Internationally Comparable Income Distribution Estimates* (Geneva: International Labour Office, 1984).

6–6 Summation and Definite Integrals

Approximation of Area
Summations Using Integrals
Volumes of Revolution (Optional)

We have seen several applications of the definite integral, and that the definite integral and the notion of area are closely related. In this section, we look at this relationship again in three settings:

1. the concept of area can be used to approximate definite integrals by summations;
2. certain summations, in turn, can be approximated by definite integrals; and
3. the summation formulas themselves show how the definite integral can be used in the computation of other quantities, such as volume.

Approximation of Area

Although we have pointed to this section as a method of numerically approximating numbers, like $\int_0^1 \sqrt{x^3 + 1}\,dx$, that cannot be computed using the fundamental theorem of calculus, we start with a more familiar integrand so that you can compare results.

Example 1 (*Compare Exercise 5*)
Find the area of the region bounded above by the curve $y = x^2 + 2$, below by the x-axis, on the left by the line $x = 1$, and on the right by the line $x = 4$. See Figure 6–26.

Solution We want the area of the shaded region in Figure 6–26. How would you have approached this problem if this had been the first problem presented in this course, before we had any notion of derivatives or antiderivatives? One way might be to carefully draw the region on lined graph paper and then to count the squares inside the region. This number would be too small, for there are squares that would be partially inside and partially outside the region. You might then come up with some sort of estimation technique for the area of the region inside these squares.

This would be a good approach and is based on using the area of squares to estimate the area of a more complicated figure. The approximation technique we will use is much like that outlined above. We will use rectangles instead of squares. Furthermore, we will not be dependent on lines already drawn on the paper but will draw our own rectangles. The bases of these rectangles will all lie on the x-axis, and for simplicity, the bases will all be the same length. We want the bases to cover the interval [1, 4] with no overlap. We start three rectangles, so the length of each base is

$$\frac{4 - 1}{3} = 1$$

The bases are drawn in Figure 6–27.

Now, we want to choose a height for each rectangle. For the first rectangle, if we use the height $f(1) = 3$, the top of the rectangle will be entirely inside the

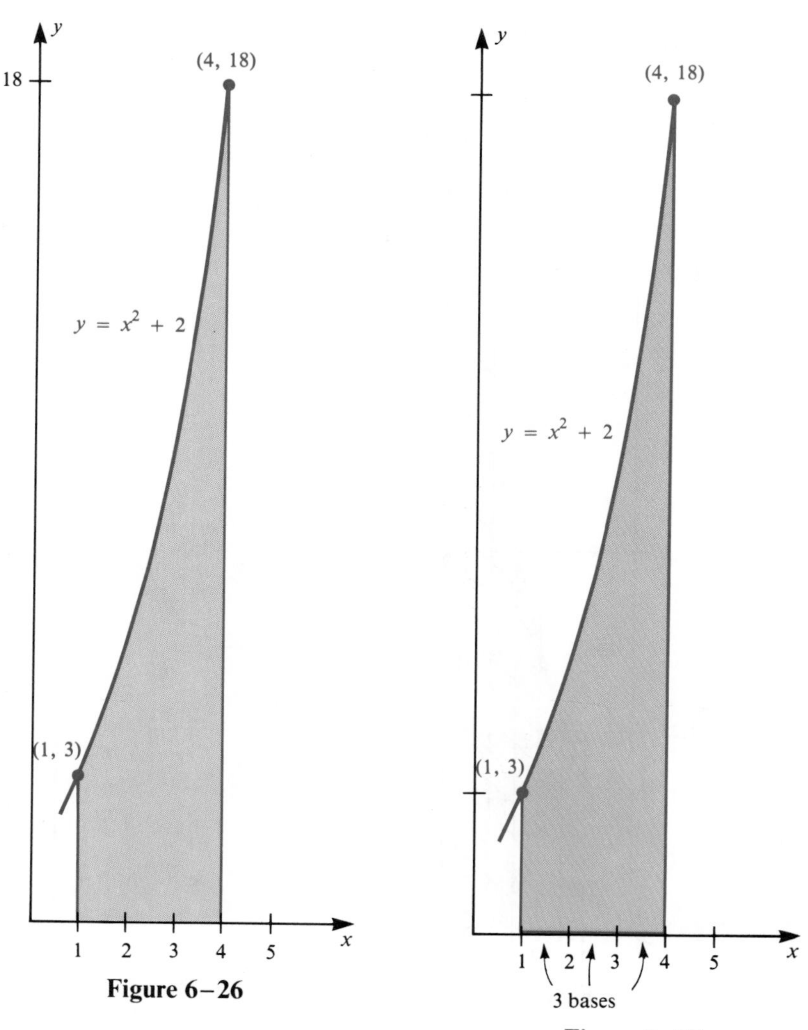

Figure 6–26

Figure 6–27

region; if we use the height $f(2) = 6$, the top of the rectangle will be entirely outside the region. We split the difference and use the height $f(1.5) = (1.5)^2 + 2 = 4.25$. Similarly, for the next rectangle we use the height $f(2.5) = 8.25$; and for the last rectangle, we use the height $f(3.5) = 14.25$. We let c_k stand for the midpoint of the base of kth rectangle.

We now have a region shaped like the one in Figure 6–28.

The area of this figure is the sum of the areas of the rectangles:

$$\text{sum of (height)(base)} = (4.25)(1) + (8.25)(1) + (14.25)(1) = 26.75$$

If we used more rectangles with smaller bases, and again used the value of f at the midpoint of each base as the height of the rectangle, then we would obviously get a better approximation. Doubling the number of rectangles to six, we would get a region like that shown in Figure 6–29.

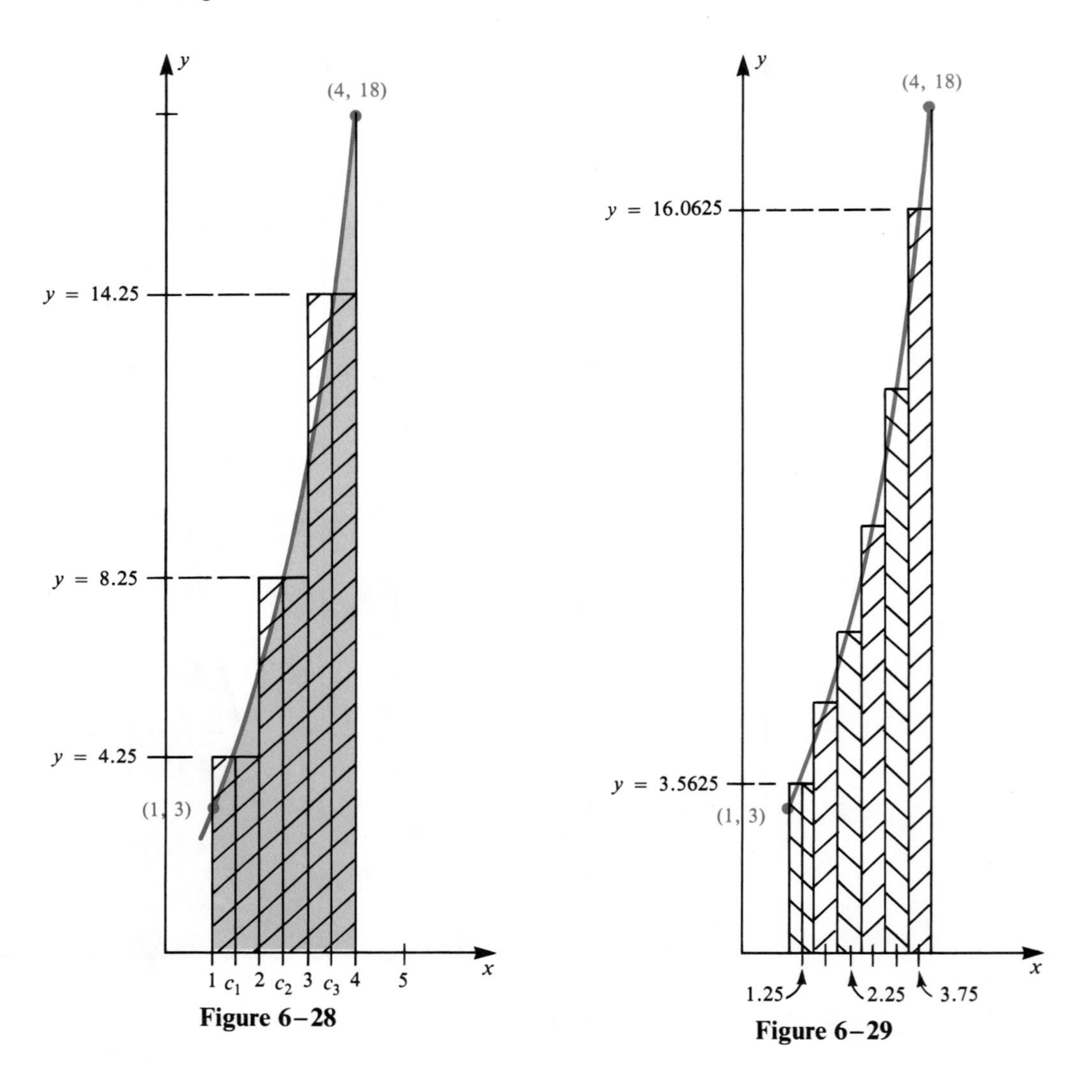

Figure 6–28

Figure 6–29

We have labeled only two of the heights to avoid making the picture cluttered. The following table computes the height of each rectangle.

Base of Rectangle	Midpoint of Base = c_k	Height of Graph at $x = c_k$
[1, 1.5]	1.25	$f(1.25) = (1.25)^2 + 2 = 3.5625$
[1.5, 2]	1.75	$f(1.75) = \quad = 5.0625$
[2, 2.5]	2.25	$f(2.25) = \quad = 7.0625$
[2.5, 3]	2.75	$f(2.75) = \quad = 9.5625$
[3, 3.5]	3.25	$f(3.25) = \quad = 12.5625$
[3.5, 4]	3.75	$f(3.75) = \quad = 16.0625$

To find the area of the region in Figure 6.32 we compute, as before, the sum of the areas of the rectangles. Note that the length of each base is (the total length of the interval divided by the number of rectangles) $= \frac{3}{6} = .5$

$$\begin{aligned}\text{The sum of (height)(length of base)} &= [(3.5625)(.5) + (5.0625)(.5)\\ &\quad + (7.0625)(.5) + (9.5625)(.5)\\ &\quad + (12.5625)(.5) + (16.0625)(.5)]\\ &= 26.9375\end{aligned}$$

We have computed two approximations to the area of R; the first approximation was 26.75, and the second was 26.9375. To compare these approximations to the actual area, we compute that area by using the Fundamental Theorem of Calculus.

$$\begin{aligned}\text{Area } R = \int_1^4 (x^2 + 2)\,dx &= \left(\frac{x^3}{3} + 2x\right)\Bigg|_1^4\\ &= \left(\frac{64}{3} + 8\right) - \left(\frac{1}{3} + 2\right)\\ &= 27\end{aligned}$$

The actual area is 27, so our approximations were pretty good.

There are many methods of numerically approximating a given definite integral. This particular method is commonly called the **midpoint approximation**, due to the method of selecting the height of each rectangle. The bases of the rectangles are all taken to have the same length; the interval $[a, b]$ is divided into n equal intervals, so the length of each is

$$\frac{b - a}{n}$$

The Midpoint Approximation Formula

> If the interval $[a, b]$ is divided into n equal subintervals, each of length $(b - a)/n$, and if c_k is the midpoint of the kth interval, then
>
> $$\int_a^b f(x)\,dx \approx f(c_1)\frac{b-a}{n} + f(c_2)\frac{b-a}{n} + \cdots + f(c_n)\frac{b-a}{n}$$

We use the midpoint approximation method again in Example 2 to find a numerical approximation to ln 3.

Example 2 (*Compare Exercise 17*)
Use the midpoint approximation method with $n = 4$ to approximate

$$\ln 3 = \int_1^3 \frac{1}{x}\,dx$$

Solution Recall that

$$\int_1^3 \frac{1}{x}\,dx = \ln|x|\Big|_1^3 = \ln 3 - \ln 1 = \ln 3$$

We are indeed computing a numerical approximation to ln 3. The interval $[a, b]$ is $[1, 3]$. With $n = 4$, each base has length

$$\frac{3-1}{4} = \frac{1}{2}$$

Thus, the interval $[1, 3]$ is partitioned into four subintervals as follows:

c_1 c_2 c_3 c_4

1 1.5 2 2.5 3

c_k is the midpoint of the kth interval.

$$c_1 = 1.25 \qquad c_2 = 1.75 \qquad c_3 = 2.25 \qquad c_4 = 2.75$$

$f(x) = \dfrac{1}{x}$ so,

$$f(c_1) = \frac{1}{1.25} \qquad f(c_2) = \frac{1}{1.75} \qquad f(c_3) = \frac{1}{2.25} \qquad f(c_4) = \frac{1}{2.75}$$

The sum of the areas of the rectangles is the

sum of (height)(length of base)

$$= \left(\frac{1}{1.25}\right)\left(\frac{1}{2}\right) + \left(\frac{1}{1.75}\right)\left(\frac{1}{2}\right) + \left(\frac{1}{2.25}\right)\left(\frac{1}{2}\right) + \left(\frac{1}{2.75}\right)\left(\frac{1}{2}\right)$$

$$= \frac{1}{2.5} + \frac{1}{3.5} + \frac{1}{4.5} + \frac{1}{5.5}$$

$$= .4000 + .2857 + .2222 + .1818$$

$$\approx 1.0897 \qquad \text{(To 4 places)}$$

If the intermediate steps in a calculation are rounded to four places, it is generally a good idea to round off the final answer to three places. This gives us 1.090 as our approximation to ln 3. [In fact, the decimal expansion of ln 3 begins 1.0986....]

ANSWER $\ln 3 \approx 1.090$

You can find formulas that give bounds on the size of the error in approximating the definite integral by this method in texts that go more deeply into the theory of approximation methods.

Summations Using Integrals

The relation between summations and integrals works both ways.

Example 3 (*Compare Exercise 33*)
The number of people using a city's rapid transit system since it was modernized in 1985 is growing. Officials estimated that k years after the modernization was completed, $100(6k^2 + 500k + 6000)$ customers would use the system that year.

This estimate means that in the first year, roughly

$$100(6 \cdot 1^2 + 500 \cdot 1 + 6000) = 650{,}600$$

customers used the system. In the second year, about

$$100(6 \cdot 2^2 + 500 \cdot 2 + 6000) = 702{,}400$$

customers used the system. If the official estimates are good, approximately how many customers will have used the modernized system by 1995?

Solution The answer can be computed by letting $N(k)$ be the number of customers in the kth year and then computing the sum

$$N(1) + N(2) + \cdots + N(10)$$

Notice that if we do something peculiar at this stage, we can start to get a quick method of approximating this number. We multiply each $N(k)$ by 1. Now the sum looks like

$$N(1) \cdot 1 + N(2) \cdot 1 + N(3) \cdot 1 + \cdots + N(10) \cdot 1$$

Multiplying each number by 1 makes this sum look like a summation that approximates an integral over the interval [0, 10] when the interval [0, 10] has been divided into 10 subintervals each of length 1. This is not an approximation using the midpoint of each interval to evaluate the function, but here the right-hand endpoint of each interval is used. Thus, this summation is approximately

$$\int_0^{10} N(t)\, dt$$

We have replaced the variable k in the formula for N by the variable t.

$$\begin{aligned}\int_0^{10} N(t)\, dt &= \int_0^{10} 100(6t^2 + 500t + 6000)\, dt \\ &= 100\left[(2t^3 + 250t^2 + 6000t)\Big|_0^{10}\right] \\ &= 100[2000 + 25{,}000 + 60{,}000] \\ &= 8{,}700{,}000\end{aligned}$$

The system will be used by about 8.7 million customers.

Example 4 (*Compare Exercise 35*)
The annual payroll of a certain company has been growing at the continuous rate of 12%. At the beginning of 1984, the payroll was $450,000. What will be the total payroll expenditures for the company from the beginning of 1990 to the beginning of 1995?

Solution Let $P(t)$ be the payroll (in thousands of dollars) t years after January 1, 1984. We are given that $P'(t) = .12P(t)$ and that $P(0) = 450$. Thus, the function we should use for $P(t)$ is $P(t) = 450e^{.12t}$. The beginning of 1990 corresponds to $t = 6$, and the beginning of 1995 corresponds to $t = 11$. Thus, this period corresponds to $6 \le t \le 11$.

$$\int_6^{11} 450e^{.12t}\,dt = \left[450\left(\frac{1}{.12}\right)e^{.12t}\right]\Bigg|_6^{11}$$
$$= 3750e^{.12t}\Big|_6^{11}$$
$$= 3750(e^{1.32} - e^{.72})$$
$$\approx 3750(3.743 - 2.054)$$
$$\approx 6334 \text{ thousand dollars}$$

The company's payroll will total about \$6.3 million during this period.

Example 5 (*Compare Exercise 29*)
Suppose that you save \$2000 per year in a retirement fund that pays 6% continuously-compounded interest. You start saving at age 25. How much will your retirement fund be worth when you retire at age 65? (We suppose, for simplicity, that your contributions are made on January 1 of each year, and that you retire on December 31.)

Solution Your first contribution to the fund earns interest for 40 years. The first \$2000 will grow to $2000e^{(.06)(40)}$ [using $A(t) = Pe^{Rt}$ with $P = 2000$, $R = .06$, and $t = 40$]. The second \$2000 is only in the account for 39 years; it will grow to $2000e^{(.06)(39)}$. Your contribution of \$2000 in the kth year will be in the fund for $(40 - k)$ years and will grow to $2000e^{(.06)(40-k)}$.

Instead of summing these 40 numbers, we proceed as we did in Example 3. Replace k by t and integrate. Total worth of the fund will be approximately

$$\int_0^{40} 2000e^{(.06)(40-t)}\,dt = 2000\int_0^{40} e^{(.06)(40-t)}\,dt$$

Bring 2000 outside the integral; multiply out in the exponent

$$= 2000\int_0^{40} e^{2.4-.06t}\,dt$$

Use $e^{a-b} = e^a e^{-b}$

$$= 2000\int_0^{40} e^{2.4}e^{-.06t}\,dt$$

Bring $e^{2.4}$ outside

$$= 2000e^{2.4}\int_0^{40} e^{-.06t}\,dt$$

Integrate

$$= 2000e^{2.4}\left(-\frac{1}{.06}\right)e^{-.06t}\Big|_0^{40}$$
$$= 2000e^{2.4}\left(-\frac{100}{6}\right)(e^{-2.4} - e^0)$$

$$= 2000e^{2.4}\left(-\frac{100}{6}\right)(e^{-2.4} - 1)$$

$$= 2000e^{2.4}\left(\frac{100}{6}\right)(1 - e^{-2.4})$$

$$\approx 334{,}105.88$$

ANSWER Your retirement fund will be worth about $334,000, over a third of a million dollars. You personally will have contributed $80,000, and you will have earned over $254,000 in interest.

Volumes of Revolution

The approximation

$$\int_a^b f(x)\,dx \approx f(c_1)\frac{b-a}{n} + \cdots + f(c_n)\frac{b-a}{n}$$

allows us to go from summations to definite integrals in all types of settings. One further geometric application is presented here. We derive the volume of a sphere whose radius is four. The particular value four plays no special role, so you will be able to see how to derive the general formula for the volume of a sphere. The key to this approach is to think of a sphere as being the solid swept out by rotating a semicircle about the x-axis. See Figures 6–30 and 6–31.

The area of the semicircle can be approximated with rectangles (see Figure 6–32); and by rotating these rectangles, we can arrive at an approximation to the area of the sphere. See Figure 6–33.

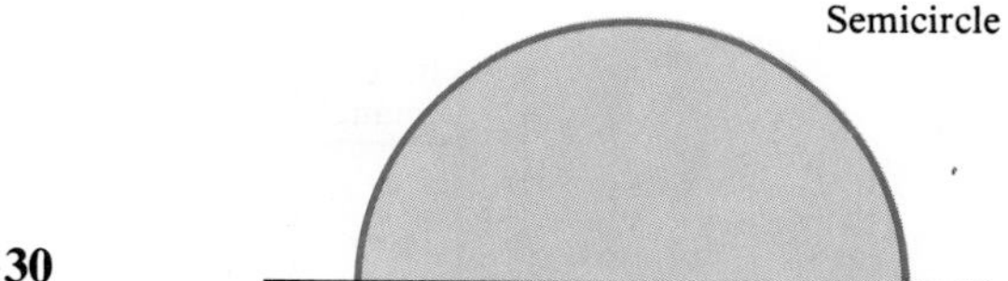

Figure 6–30

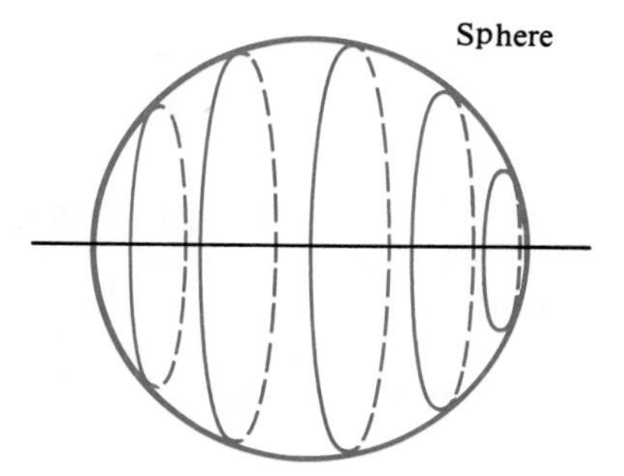

Figure 6–31

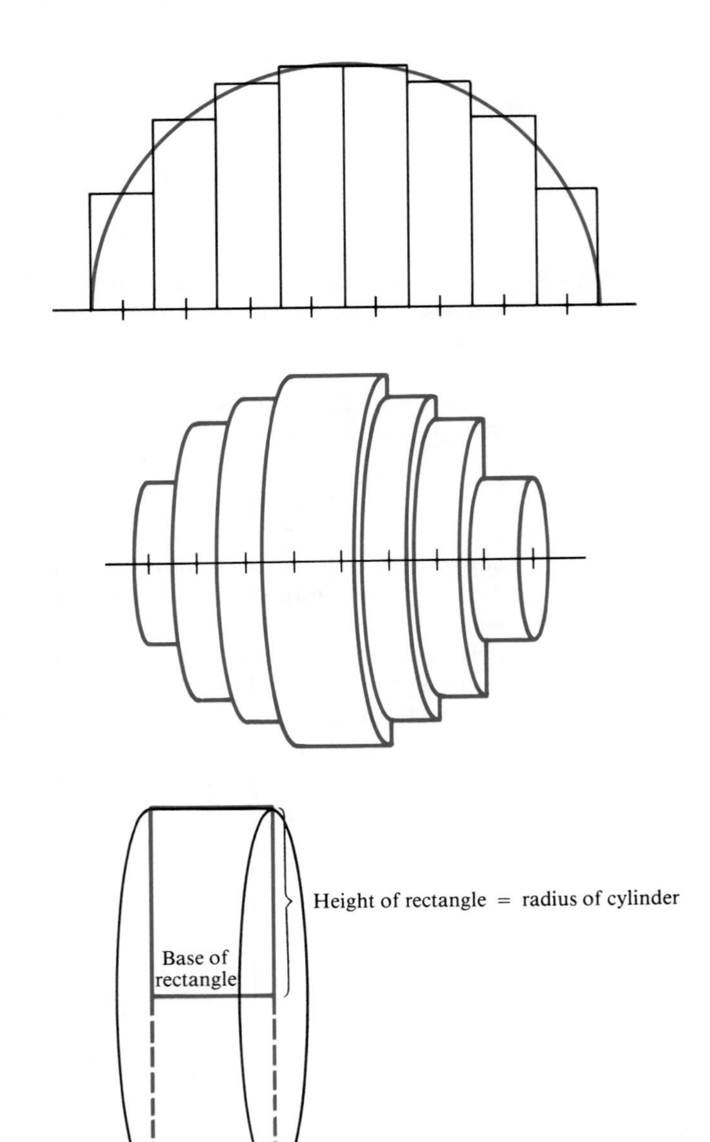

Figure 6–32

Figure 6–33

Figure 6–34

When you rotate a rectangle 360° about its base, you get a cylinder. The thickness of the cylinder is the length of the base of the rectangle, and the radius of the cylinder is the height of the rectangle. See Figure 6–34.

If $L =$ (length of base of rectangle) and $H =$ (height of rectangle), then the volume of the cylinder is $\pi H^2 \cdot L$. Now, the original semicircle is the region under the curve $y = \sqrt{16 - x^2}$ over the interval $-4 \leq x \leq 4$. If we divide the interval into n equal subintervals, each will have length

$$L = \frac{4 - (-4)}{n} = \frac{8}{n}$$

If c_k is the midpoint of the kth interval, the height H of the kth rectangle will be

$$H = f(c_k) = \sqrt{16 - c_k{}^2}$$

Therefore, the volume of the kth cylinder will be

$$\pi H^2 \cdot L = \pi(\sqrt{16 - c_k{}^2})^2\left(\frac{8}{n}\right)$$

$$= \pi(16 - c_k^2)\left(\frac{8}{n}\right)$$

Summing all these volumes we have

$$\pi(16 - c_1^2)\left(\frac{8}{n}\right) + \pi(16 - c_2^2)\left(\frac{8}{n}\right) + \cdots + \pi(16 - c_n^2)\left(\frac{8}{n}\right)$$

Remember that we got the 8 as $4 - (-4) = b - a$. Thus, this sum is approximately

$$\int_{-4}^{4} \pi(16 - x^2)\,dx = \pi \int_{-4}^{4} (16 - x^2)\,dx$$

We now compute this definite integral.

$$\pi \int_{-4}^{4} (16 - x^2)\,dx = \pi\left(16x - \frac{x^3}{3}\right)\Big|_{-4}^{4}$$

$$= \pi\left[\left(64 - \frac{64}{3}\right) - \left(-64 + \frac{64}{3}\right)\right]$$

$$= \pi\,\frac{256}{3}$$

The volume of a sphere whose radius is 4 inches is $\frac{256}{3}\pi$ cubic inches.

The general formula for a solid obtained by rotation of a region is obtained by similar considerations, replacing -4 by a, 4 by b, and $\sqrt{16 - x^2}$ by $f(x)$.

Volume Formula

> If the region bounded above by the curve $y = f(x)$, below by the x-axis, on the left by the line $x = a$, and on the right by the line $x = b$ is rotated about the x-axis, then the volume of the solid of revolution is given by
>
> $$\text{volume} = \pi \int_a^b [f(x)]^2\,dx$$

Example 6 (*Compare Exercise 27*)
Find the volume of the sphere of radius five.

Solution Rotate the semicircle bounded above by $f(x) = \sqrt{25 - x^2}$. See Figure 6–35.

$$\begin{aligned}
\text{Volume} &= \pi \int_{-5}^{5} (\sqrt{25 - x^2})^2\,dx \\
&= \pi \int_{-5}^{5} (25 - x^2)\,dx \\
&= \pi\left(25x - \frac{x^3}{3}\right)\Bigg|_{-5}^{5} \\
&= \pi\left[\left(125 - \frac{125}{3}\right) - \left(-125 + \frac{125}{3}\right)\right] \\
&= \pi\frac{500}{3}
\end{aligned}$$

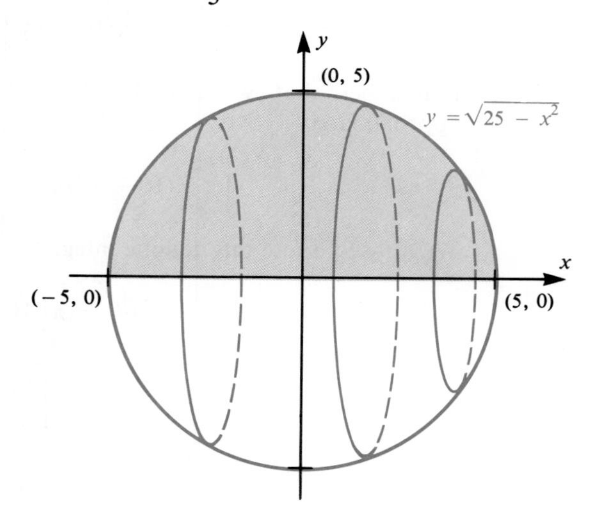

Figure 6–35

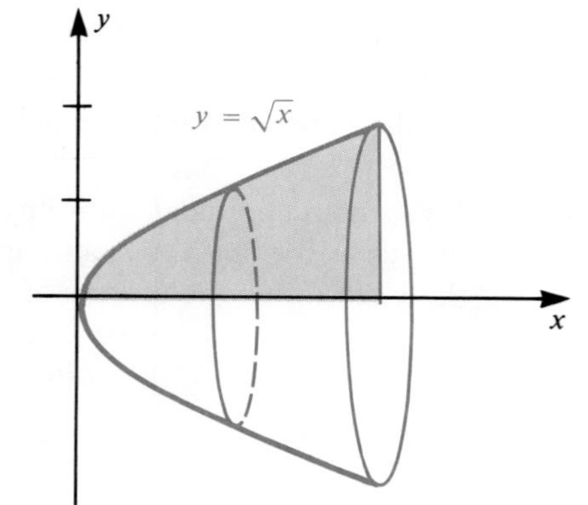

Figure 6–36

Example 7 (*Compare Exercise 23*)
The inside of a soup bowl has a shape like that formed by rotating the curve $y = \sqrt{x}$, $0 \le x \le 3$, about the x-axis. What is the volume of soup the bowl can hold? See Figure 6–36.

Solution

$$\text{Volume} = \pi \int_0^3 (\sqrt{x})^2\,dx$$
$$= \pi \int_0^3 x\,dx$$
$$= \frac{\pi}{2} x^2 \Big|_0^3$$
$$= \frac{9\pi}{2}$$

ANSWER The volume is $9\pi/2$.

6–6 Exercises

I.

Use the midpoint approximation formula to approximate $\int_a^b f(x)\,dx$ with n giving the number of subintervals. Then evaluate each integral and compare your answers.

1. $\int_1^3 4\,dx$ **(a)** $n = 2$ **(b)** $n = 4$
2. $\int_{-1}^2 2\,dx$ **(a)** $n = 3$ **(b)** $n = 6$
3. $\int_0^3 2x\,dx$ **(a)** $n = 3$ **(b)** $n = 6$
4. $\int_0^4 x\,dx$ **(a)** $n = 4$ **(b)** $n = 8$
5. (*See Example 1*) $\int_1^3 x^2\,dx$ **(a)** $n = 2$ **(b)** $n = 4$
6. $\int_{-1}^2 x^2\,dx$ **(a)** $n = 3$ **(b)** $n = 6$

II.

Use the midpoint formula to approximate $\int_a^b f(x)\,dx$ with n giving the number of subintervals to use in the approximation. Evaluate the integral and compare your answers.

7. $\int_{1}^{3} (2x^2 - 3x)\,dx$ **(a)** $n = 2$ **(b)** $n = 4$

8. $\int_{-1}^{5} (4 - 3x^2)\,dx$ **(a)** $n = 3$ **(b)** $n = 6$

9. $\int_{1}^{5} \frac{1}{x^2}\,dx$ **(a)** $n = 2$ **(b)** $n = 4$

10. $\int_{0}^{4} \sqrt{x}\,dx$ **(a)** $n = 4$ **(b)** $n = 8$

11. $\int_{1}^{3} \frac{1}{x^3}\,dx$ **(a)** $n = 2$ **(b)** $n = 4$

Approximate the following summation by using the given integral.

12. $1 + 2 + 3 + 4 + \cdots + 100$; $\int_{0}^{100} x\,dx$

13. $1^2 + 2^2 + 3^2 + \cdots + 60^2$; $\int_{0}^{60} x^2\,dx$

14. $1^3 + 2^3 + 3^3 + \cdots + 20^3$; $\int_{0}^{20} x^3\,dx$

Compute the volumes given by the following integrals.

15. $\pi \int_{-6}^{6} (\sqrt{36 - x^2})^2\,dx$

16. $\pi \int_{0}^{4} (x^2)^2\,dx$

III.

17. (*See Example 2*) Estimate ln 3 by approximating $\int_{2}^{6} \frac{1}{x}\,dx$ with

(a) $n = 4$.
(b) $n = 8$.

18. Estimate ln 3 by approximating $\int_{3}^{9} \frac{1}{x}\,dx$ with

(a) $n = 6$.
(b) $n = 12$. (*Compare results with Example 2.*)

19. **(a)** Estimate ln 4 by approximating $\int_{1}^{4} \frac{1}{x}\,dx$ with

(i) $n = 3$. **(ii)** $n = 6$.

(b) Estimate ln 4 by approximating $\int_{2}^{8} \frac{1}{x}\,dx$ with

(i) $n = 6$. **(ii)** $n = 12$.

20. **(a)** Estimate ln 9 by approximating $\int_1^9 \frac{1}{x}\,dx$ with $n = 8$.

(b) $\ln 9 = \ln 3^2 = 2\ln 3 = 2\int_1^3 \frac{1}{x}\,dx$. *(Compare your results in a with twice the approximations to ln 3 obtained in Example 2.)*

21. Approximate $\int_0^2 \sqrt{x^2+1}$ with $n = 4$.

22. Approximate $\int_0^3 \frac{1}{\sqrt{x^2+1}}$ with $n = 6$.

In Exercises 23 through 28, find the volume of the solid obtained by rotating about the x-axis the region bounded above by $y = f(x)$, below by the x-axis, and between the vertical lines $x = a$ and $x = b$.

23. (*See Example 7*) $f(x) = x + 1$, $a = -1$, $b = 3$

24. $f(x) = \frac{1}{x}$, $a = 2$, $b = 5$

25. $f(x) = x^2 - 3$, $a = 2$, $b = 4$

26. $f(x) = \frac{1}{x^3}$, $a = \frac{1}{2}$, $b = 3$

27. (*See Example 6*) $f(x) = \sqrt{R^2 - x^2}$, $a = -R$, $b = R$. (This gives the formula for the volume of a sphere with radius R.)

28. $f(x) = \frac{R}{H}x$, $a = 0$, $b = H$. (This gives the formula for the volume of a cone with radius R and height H.)

29. (*See Example 5*) If \$1000 is deposited into an account annually for 20 years, and the account pays 7% interest compounded continuously, what will the account be worth at the end of the 20 years?

30. If \$1000 is deposited into an account annually for 10 years, and the account pays 7% interest compounded continuously, what will the account be worth at the end of the 10 years?

31. If \$1000 is deposited into an account annually for 10 years, and the account pays 8% interest compounded continuously, what will the account be worth at the end of 10 years?

32. If \$1000 is deposited into an account annually for 30 years, and the account pays 6% interest compounded continuously, what will the account be worth at the end of 30 years?

33. (*See Example 3*) The birth pattern of ducks in a certain park has been estimated to be approximated by the following formula: $N(t) = 3t^3 - 36t^2 + 108t + 750$, where t is time in years with $t = 0$ corresponding to 1980, and $N(t)$ is the number of births that took place at time t. Estimate the number of ducks born between 1982 and 1986.

34. A mining company estimates that the amount of phosphate it mines can be described by the following function: $W(t) = 4t^3 - 72t^2 + 420t + 7000$. Here, t is time measured in years from 1970, and the units of W are thousands

of tons. The quantity mined at time t is $W(t)$. Estimate the quantity of phosphate mined between 1975 and 1985.

35. (*See Example 4*) A manufacturing company estimates that its costs at time t since 1975 are described by $C(t) = 3t^2 + 60t + 7000$. Time t is measured in years from 1975, and the units of C are thousands of dollars. Estimate the total costs incurred from 1977 to 1987.

36. The number of new automobiles sold at time t, on a world-wide basis since 1965, has been estimated to be described by $N(t) = 6t^2 + 72t + 9000$, where t is the time in years measured from 1965, and $N(t)$ is the number of cars in units of one hundred. Estimate the total number of new cars that were sold between 1969 and 1976.

IMPORTANT TERMS

6–1	**Antiderivative**	**Indefinite integral**	**Integration**
	Antidifferentiation	$\int f(x)\,dx$	**Additive constant of integration**
	Integrand		
6–2	**Differential equation**	**Solution to a differential equation**	**Particular solution**
	Second-order differential equation	$\int e^{ax}\,dx$	$\int \frac{1}{x}\,dx$
6–3	**Total change of a function on an interval**	**Definite integral of f from a to b**	$\int_a^b f(x)\,dx$
	Upper limit of integration	**Lower limit of integration**	$f(x)\|_a^b$
	"Gluing of intervals" property for definite integrals		
6–4	**Fundamental theorem of calculus**	**Area between curves**	$\ln b = \int_1^b \frac{1}{x}\,dx$
6–5	**Average value of a function on an interval**	**Consumers' surplus**	**Investment flow and capital formation**
	Flows and accumulations	**Gini index**	
6–6	**Midpoint approximation formula**	**Approximating an integral by a summation**	**Approximating a summation by an integral**
	Volume of revolution		

REVIEW EXERCISES

Find the indicated indefinite integral in Exercises 1 through 10.

1. $\int \left(6x^2 + \frac{6}{x^2}\right) dx$

2. $\int \left(\sqrt{x} - \frac{1}{\sqrt{x}}\right) dx$

3. $\int \left(x^3 - \frac{1}{x}\right)^2 dx$

4. $\int \left(x + \frac{1}{x^2}\right)^2 dx$

5. $\int \frac{4x^2 + 8}{x} dx$

6. $\int \frac{6x - 4}{x^2} dx$

7. $\int (e^{6x} + e^{-4x})\, dx$

8. $\int \frac{1}{e^{3x}} dx$

9. $\int (e^x + 4)^2\, dx$

10. $\int (e^x + e^{-x})\, dx$

Compute each of the definite integrals given in Exercises 11 through 16.

11. $\int_1^9 (3\sqrt{x} + 4x - 1)\, dx$

12. $\int_{-2}^1 (9x^2 - 8x + 4)\, dx$

13. $\int_0^4 (e^{2x} - x)\, dx$

14. $\int_1^e \frac{1}{x} dx$

15. $\int_{\ln 3}^{\ln 8} e^x\, dx$

16. $\int_1^{e^5} \frac{1}{x} dx$

Verify the equations in Exercises 17 and 18 by evaluating each definite integral.

17. $\int_{-1}^1 (6x^2 - 8x)\, dx + \int_1^2 (6x^2 - 8x)\, dx = \int_{-1}^2 (6x^2 - 8x)\, dx$

18. $\int_2^1 (12x^2 - 4x)\, dx + \int_1^3 (12x^2 - 4x)\, dx = \int_2^3 (12x^2 - 4x)\, dx$

In Exercises 19 through 22, find the particular solution that satisfies the given differential equation and the additional condition or conditions.

19. $y' = 16x^3 - 12x + 9$, with $y = 4$ when $x = 1$

20. $y' = 6\sqrt{x} - 5$, with $y = 13$ when $x = 4$

21. $y'' = 12x - 2$, with $y' = 4$ and $y = 9$ when $x = 0$

22. $y'' = 18$, with $y' = 6$ and $y = 10$ when $x = 2$

23. From a height of 2 meters, a ball is thrown straight up into the air with a velocity of 30 meters per second; the deceleration due to gravity is -9.8 m/sec^2. Find a formula for $h(t)$, the ball's height in meters t seconds after it is released.

24. The marginal cost function for a company producing x hundred units per day is given by

$$MC(x) = \frac{1}{100} x^2 - 6x + 7000$$

How much will the company's daily costs increase if production is increased from 900 units per day to 1000 units?

25. Find the area of the region bounded above by the curve $y = 3x^2 + 1$, below by the x-axis, on the left by the line $x = -1$, and on the right by the line $x = 2$.

26. Find the area of the region bounded on the left by the line $x = -2$ and on the right by the line $x = 3$, and that lies between the two curves $y = x^2 + 4x + 2$ and $y = x^2 - 2x + 8$.

27. Find the area of the region between the curve $y = x^2 - x + 1$ and the line $y = 2x + 5$.

28. If the price-demand equation for a certain product is $p = 680 - 6x$, and the market price is 80, then what is the consumers' surplus?

29. What is the average value of $f(x) = \sqrt{x}$ on the interval $[0, 9]$?

30. What is the Gini index for a country whose income distribution function is given by $F(x) = x^4 - .6x^3 + .3x^2 + .3x$?

31. Use the midpoint approximation formula to approximate $\int_1^5 \frac{1}{x}\,dx$

(a) with $n = 4$.

(b) with $n = 8$.

32. If \$1000 is deposited into an account annually for 30 years, and the account pays 6% interest, continuously compounded, what will the account be worth at the end of the 30 years?

7

Topics in this chapter can be applied to:
Total Cost Functions ● Production Volumes ● Marginal Revenue and Total Revenue Changes ● Cost and Production Management ● Determining Municipal Income from Local Industry ● Evaluation of Purchasing Rates

Techniques of Integration

- Integration by Substitution
- Integration by Parts
- Improper Integrals
- Integration Using Tables

7–1 Integration by Substitution

Introduction
Method of Substitution
Substitution and the Definite Integral
Substitutions Involving Algebra

Introduction

The process of finding an antiderivative of a given function is generally more difficult than the process of finding its derivative. Given a particular function, f, if we can recognize it as the sum, product, quotient, and composition of simpler functions, we can apply the corresponding rules of differentiation to compute the derivative of f. Applying these rules correctly requires care and practice, but at least there are rules to follow. So far we have seen some rules of integration that parallel similar differentiation rules. We have three general rules regarding indefinite integrals:

constant-times	**1.** $\int c \cdot f(x)\,dx = c \cdot \int f(x)\,dx$;
sum	**2.** $\int [f(x) + g(x)]\,dx = \int f(x)\,dx + \int g(x)\,dx$;
difference	**3.** $\int [f(x) - g(x)]\,dx = \int f(x)\,dx - \int g(x)\,dx$.

And we have three rules for specific functions:

power rule

4. $\int x^n\,dx = \frac{1}{n+1}x^{n+1} + C, n \neq -1;$

5. $\int \frac{1}{x}\,dx = \ln|x| + C;$

6. $\int e^{ax}\,dx = \frac{1}{a}e^{ax} + C.$

Unfortunately, there is no antidifferentiation rule that corresponds to the quotient rule, and the rules that correspond to the chain rule and product rule are only of limited use. Indeed there are functions, like $f(x) = \sqrt{x^3 + 1}$, whose antiderivative cannot be written in simple terms. In fact, while mathematicians will talk about **rules** of differentiation, they use terms for antidifferentiation that don't carry the same guarantee of success—terms like "methods" or "techniques" of integration.

The method we look at in this section is called "substitution" and corresponds to the chain rule. Because so many functions involve composition of simpler functions, this method must be used frequently. We start with two examples showing how substitution leads to a generalized power rule for integration, just as we had a generalized power rule for differentiation.

Example 1 Find $\int \sqrt{5x + 1}\,dx$.

Solution Rewriting the integrand, we have

$$\int \sqrt{5x + 1}\,dx = \int (5x + 1)^{1/2}\,dx$$

Now we try to find an antiderivative by using the power rule. Let $f(x) = \frac{2}{3}(5x + 1)^{3/2}$. Next, we check to see if $f(x)$ is an antiderivative of $(5x + 1)^{1/2}$ by computing $f'(x)$.

$$f'(x) = \frac{2}{3}\cdot\frac{3}{2}(5x + 1)^{1/2}\cdot 5 \qquad \text{(Chain rule)}$$

$$= 1(5x + 1)^{1/2}\cdot 5 = 5\sqrt{5x + 1}$$

$$f'(x) = 5\sqrt{5x + 1} \neq \sqrt{5x + 1}$$

We are off by the factor of 5 that was forced into the computation by the chain rule. We were led to the generalized power rule for differentiation by substituting a single letter for the quantity to be raised to a power. We proceed here in the same manner. We have $5x + 1$ raised to a power, so we let $u = 5x + 1$. Then, $\int (5x + 1)^{1/2}\,dx$ becomes $\int u^{1/2}\,dx$. But this last expression doesn't make sense. The variable in the integrand (here u) must match the variable following the d (here x). We must continue the substitution process and substitute an appropriate term for the dx.

To replace the dx, we use the formalism of differentials.

$$\text{If } u = f(x), \qquad \text{then} \qquad du = f'(x)\,dx$$

Thus, if $u = 5x + 1$, then $du = 5\,dx$. Solving for dx, we have $\frac{1}{5}\,du = dx$. Next, we make the complete substitution:

$$\int (5x + 1)^{1/2}\,dx = \int u^{1/2}\left(\frac{1}{5}\,du\right) = \frac{1}{5}\int u^{1/2}\,du$$

Now we can use the power rule on $\frac{1}{5}\int u^{1/2}\,du$:

$$= \frac{1}{5}\left[\frac{2}{3}u^{3/2} + C\right]$$

$$= \frac{2}{15}u^{3/2} + C$$

As when using the chain rule, the last step is to rewrite the answer in terms of x. Since $u = 5x + 1$, we have

$$\frac{2}{15}u^{3/2} + C = \frac{2}{15}(5x + 1)^{3/2} + C$$

Finally, we have our answer:

$$\int (5x + 1)^{1/2}\,dx = \frac{2}{15}(5x + 1)^{3/2} + C$$

We check this answer:

If $f(x) = \frac{2}{15}(5x + 1)^{3/2}$, then

$$f'(x) = \frac{2}{15}\cdot\frac{3}{2}(5x + 1)^{1/2}\cdot 5 = (5x + 1)^{1/2}$$

Notice particularly how the substitution for the dx in the integration relates to the use of the chain rule in the differentiation. This is one of the important reasons that the expression dx occurs in the integral; it is a valuable tool in the method of substitution.

We go through the same procedure more succinctly in the next example.

Example 2 (*Compare Exercise 1*)
Find $\int (6x - 8)^3\,dx$.

Solution We replace $6x - 8$ by a single variable. Let $u = 6x - 8$; then $du = 6\,dx$, so $\frac{1}{6}\,du = dx$. Substituting for both the integrand and the differential gives us

$$\int (6x - 8)^3\,dx = \int u^3\left(\frac{1}{6}\,du\right)$$

$$= \int \frac{1}{6}u^3\,du$$

$$= \frac{1}{24}u^4 + C$$

Resubstitute $6x - 4$ for u.

$$\int (6x - 8)^3\, dx = \frac{1}{24}(6x - 8)^4 + C$$

As we did in Example 1, you should check this answer by taking the derivative of $\frac{1}{24}(6x - 8)^4$.

We say that integration by substitution is the integration technique that corresponds to the chain rule because just as we did with the chain rule, we are substituting a new variable u for a function of x. With this substitution, we are writing

$$\int \left(\frac{dy}{dx}\right) dx = \int \left(\frac{dy}{du} \cdot \frac{du}{dx}\right) dx$$

$$= \int \left(\frac{dy}{du}\right)\left(\frac{du}{dx}\, dx\right) = \int \left(\frac{dy}{du}\right) du$$

We find this last integral in terms of u and then complete the process by substituting for u to end up with an expression in terms of x.

Method of Substitution

Example 3 (*Compare Exercise 3*)
Find $\int 10x\sqrt{5x^2 + 1}\, dx$.

Solution As with the chain rule, we look for a composition of functions, especially for a function raised to a power. Here, we have $\sqrt{5x^2 + 1} = (5x^2 + 1)^{1/2}$. So we start by letting $u = 5x^2 + 1$. Remember that $u = f(x)$ means $du = f'(x)\,dx$; here, $u = 5x^2 + 1$ means $du = 10x\,dx$. If we rewrite the original integral, then the substitution becomes clearer.

$$\int 10x\sqrt{5x^2 + 1}\, dx = \int \sqrt{5x^2 + 1}(10x\, dx)$$

$$= \int \sqrt{u}\, du$$

Now we integrate.

$$\int \sqrt{u}\, du = \int u^{1/2}\, du = \frac{2}{3}u^{3/2} + C$$

Finally, resubstitute $5x^2 + 1$ for u to end up with an expression in x.

$$\frac{2}{3}u^{2/3} + C = \frac{2}{3}(5x^2 + 1)^{3/2} + C$$

ANSWER $\int 10x\sqrt{5x^2 + 1}\, dx = \frac{2}{3}(5x^2 + 1)^{3/2} + C$

[Check by computing the derivative of $\frac{2}{3}(5x^2 + 1)^{3/2}$.]

Example 4 (*Compare Exercise 7*)
Find $\int (x^3 + 2)^4 x^2\,dx$.

Solution Here, $x^3 + 2$ is raised to a power, so we let $u = x^3 + 2$; $(x^3 + 2)^4$ will become u^4. Next, compute du.

$$du = 3x^2\,dx$$

We divide by 3 to isolate the $x^2\,dx$:

$$\frac{1}{3}\,du = x^2\,dx$$

Now we make those substitutions.

$$\int (x^3 + 2)^4 x^2\,dx = \int u^4 \frac{1}{3}\,du$$

$$= \int \frac{1}{3} u^4\,du \qquad \text{Integrate } u^4 \text{ to get}$$

$$= \frac{1}{15} u^5 + C \qquad \text{Finally, resubstitute } u = x^3 + 2$$

$$= \frac{1}{15}(x^3 + 2)^5 + C$$

ANSWER $\int (x^3 + 2)^4 x^2\,dx = \frac{1}{15}(x^3 + 2)^5 + C$. Check this answer by differentiating $\frac{1}{15}(x^3 + 2)^5$.

Example 5 (*Compare Exercise 11*)
Find $\int xe^{x^2}\,dx$.

Solution The generalized power rule was not the only form of the chain rule that we used. We also had to use the chain rule when computing the derivative of $e^{f(x)}$. This integrand involves an expression of this form, so as with the chain rule, we replace the exponent by a single letter:

$$\text{let } u = x^2$$

$$du = 2x\,dx$$

so,

$$\frac{1}{2}\,du = x\,dx$$

We have

$$\int xe^{x^2}\,dx = \int e^{x^2}(x\,dx)$$

Now we make the substitutions

$$= \int e^u \left(\frac{1}{2}\,du\right)$$

$$= \int \frac{1}{2}\,e^u\,du$$

Now integrate:

$$= \frac{1}{2}\,e^u + C$$

resubstituting,

$$= \frac{1}{2}\,e^{x^2} + C$$

ANSWER $\int xe^{x^2}\,dx = \frac{1}{2}e^{x^2} + C$. Check this answer.

Example 6 (*Compare Exercise 13*)
Find $\displaystyle\int \frac{2x+3}{(x^2+3x+6)^4}\,dx$.

Solution The parentheses here again indicate composition. Let $u = x^2 + 3x + 6$, so $(x^2 + 3x + 6)^4$ becomes u^4, and $du = (2x + 3)\,dx$.

$$\int \frac{2x+3}{(x^2+3x+6)^4}\,dx = \int \frac{1}{(x^2+3x+6)^4}\,(2x+3)\,dx$$

$$= \int \frac{1}{u^4}\,du$$

(Remember that there is no quotient rule for antiderivatives; we must rewrite $1/u^4 = u^{-4}$.) Rewriting, we have

$$= \int u^{-4}\,du$$ Now integrate

$$= \frac{-1}{3}\,u^{-3} + C$$

$$= \frac{-1}{3u^3} + C$$ Finally, resubstitute $u = x^2 + 3x + 6$

$$= \frac{-1}{3(x^2+3x+6)^3} + C$$

ANSWER $\displaystyle\int \frac{2x+3}{(x^2+3x+6)^4}\,dx = \frac{-1}{3(x^2+3x+6)^3} + C$. Check the answer.

Example 7 (*Compare Exercise 9*)

What is $\int \frac{4x}{x^2 + 6} dx$?

Solution This substitution is not immediately pointed out by parentheses or a radical or an exponent. But remembering that we must rewrite quotients and writing

$$\int \frac{4x}{x^2 + 6} dx = \int 4x(x^2 + 6)^{-1} dx$$

introduces the parentheses that can serve as an indicator of the substitution.

$$\text{Let } u = x^2 + 6$$

Then,

$$du = 2x\,dx$$

We need to substitute for $4x\,dx$, so multiply by 2 and get

$$2\,du = 4x\,dx$$

$$\int 4x(x^2 + 6)^{-1} dx = \int u^{-1} 2\,du = 2 \int u^{-1} du$$

We cannot use the power rule with the exponent -1; the indefinite integral of u^{-1} is $\ln|u| + C$.

$$2 \int u^{-1} du = 2 \int \frac{1}{u} du$$

$$= 2 \ln|u| + C$$

$$= 2 \ln|x^2 + 6| + C$$

ANSWER $\int \frac{4x}{x^2 + 6} dx = 2 \ln|x^2 + 6| + C$. Check the answer.

In the last example, because $x^2 + 6 > 0$ for all x, we could rewrite the expression $\ln|x^2 + 6|$ without using the absolute value bars as $\ln(x^2 + 6)$. But remember, the absolute value is necessary when integrating; in general,

$$\int \frac{1}{u} du = \ln|u| + C$$

A general pattern to look for is

$$\int \frac{f'(x)}{f(x)} dx$$

Example 8 shows how this pattern can involve an exponential function.

Example 8 (*Compare Exercise 15*)

Find $\int \frac{e^{2x}}{e^{2x} - 5} dx$.

Solution Again, we look to the denominator of the quotient.

$$\text{Let } u = e^{2x} - 5$$

Then, $$du = 2e^{2x}\,dx$$

Solving for $e^{2x}\,dx$, we have

$$\frac{1}{2}du = e^{2x}\,dx$$

$$\int \frac{1}{e^{2x} - 5} e^{2x}\,dx = \int \frac{1}{u}\frac{1}{2}du$$

$$= \frac{1}{2}\ln|u| + C$$

$$= \frac{1}{2}\ln|e^{2x} - 5| + C$$

ANSWER $\int \frac{e^{2x}}{e^{2x} - 5}dx = \frac{1}{2}\ln|e^{2x} - 5| + C$. This time, the absolute value signs cannot be omitted. (Check the answer.)

Example 9 Find $\int \frac{(8 + \ln x)^2}{x}dx$.

Solution There is a quotient here, and after the last three examples you may anticipate the substitution $u = x$ because the denominator is x. One of the points in this example is that the substitution $u = x$ never does any good; all it does is change letters without simplifying the integral. In this case, the substitution $u = x$ leads to $\int \frac{(8 + \ln u)^2}{u}du$, the same indefinite integral.

Here, $(8 + \ln x)$ is raised to a power, so we try

$$u = 8 + \ln x$$

Then, $$du = \frac{1}{x}dx$$

$$\int \frac{(8 + \ln x)^2}{x}dx = \int (8 + \ln x)^2 \frac{1}{x}dx$$

$$= \int u^2\,du$$

$$= \frac{1}{3}u^3 + C$$

$$= \frac{1}{3}(8 + \ln x)^3 + C$$

ANSWER $\int \frac{(8 + \ln x)^2}{x} dx = \frac{1}{3}(8 + \ln x)^3 + C$. (Check the answer.)

These examples have shown some general patterns to look for:

$$1. \int [f(x)]^p f'(x)\,dx = \frac{1}{p+1}[f(x)]^{p+1} + C, p \neq -1$$

$$2. \int \frac{f'(x)}{f(x)} dx = \ln|f(x)| + C$$

and

$$3. \int e^{f(x)} \cdot f'(x)\,dx = e^{f(x)} + C$$

In each of these integrals, the substitution $u = f(x)$ leads to the following corresponding integrals:

$$1'. \int u^p\,du = \frac{1}{p+1} u^{p+1} + C, p \neq -1$$

$$2'. \int \frac{du}{u} = \ln|u| + C$$

$$3'. \int e^u\,du = e^u + C$$

Substitution and the Definite Integral

You may have noticed that so far all our examples have been indefinite integrals. There's a reason for this; definite integrals are harder. When substituting into a definite integral, not only do you have to substitute for the integrand and the differential, you also have to substitute for the limits of integration.

Example 10 (*Compare Exercise 21*)
Compute $\int_0^4 2x\sqrt{x^2 + 9}\,dx$.

Solution First, we rewrite the integrand in preparation for the power rule.

$$\int_0^4 2x\sqrt{x^2 + 9}\,dx = \int_0^4 2x(x^2 + 9)^{1/2}\,dx$$

We let $u = x^2 + 9$, so $du = 2x\,dx$. Now the limits of integration must be taken care of. The 0 and the 4 are values of x. We must substitute the corresponding values of u, with $u = x^2 + 9$.

When $x = 0$, $u = 0^2 + 9 = 9$; replace the 0 with 9.
When $x = 4$, $u = 4^2 + 9 = 25$; replace the 4 with 25.

Substituting, we have $\int_0^4 \sqrt{x^2+9}(2x\,dx) = \int_9^{25} u^{1/2}\,du$. Now,

$$\begin{aligned}\int_9^{25} u^{1/2}\,du &= \frac{2}{3}u^{3/2}\Big|_9^{25}\\ &= \frac{2}{3}(25)^{3/2} - \frac{2}{3}(9)^{3/2}\\ &= \frac{2}{3}(5)^3 - \frac{2}{3}(3)^3\\ &= \frac{250}{3} - \frac{54}{3} = \frac{196}{3}\end{aligned}$$

ANSWER $\int_0^4 2x\sqrt{x^2+9}\,dx = \frac{196}{3}$.

Remember that $\int_0^4 2x\sqrt{x^2+9}\,dx$ is a number, and to keep equality throughout the computations, the limits of integration must be changed. Notice that there was no final resubstitution; the answer does not involve a variable. We were not asked to find the family of antiderivatives, but to compute a number.

Substitutions Involving Algebra

Sometimes, the substitution leaves a variable left over even after taking care of the differential. In the above examples, we have seen how to handle the cases that involve the arithmetic of multiplying or dividing by constants. If there is still some of the original variable left over, go back to the substitution equation and see if you can use that equation again.

Example 11 (*Compare Exercise 27*)
Find $\int x\sqrt{x+5}\,dx$.

Solution

$$\text{We let } u = x+5$$
$$du = dx$$

Then, $x\sqrt{x+5}\,dx$ becomes $x\sqrt{u}\,du$, but there is an x left over. Go back to the original substitution (here, $u = x+5$), and try to express this extra x in terms of u.

$$u = x+5 \qquad \text{means} \qquad x = u-5$$

$$\text{Now, } \int x\sqrt{x+5}\,dx = \int (u-5)\sqrt{u}\,du$$
$$= \int (u-5)u^{1/2}\,du$$

Multiply through by $u^{1/2}$ to obtain

$$= \int u^{3/2} - 5u^{1/2}\,du$$

Now integrate.

$$= \frac{2}{5}u^{5/2} - \frac{10}{3}u^{3/2} + C$$

$$= \frac{2}{5}(x+5)^{5/2} - \frac{10}{3}(x+5)^{3/2} + C$$

ANSWER $\int x\sqrt{x+5}\,dx = \frac{2}{5}(x+5)^{5/2} - \frac{10}{3}(x+5)^{3/2} + C.$

Example 12 (*Compare Exercise 29*)

$$\int \frac{x^3}{x^2+9}\,dx$$

Solution As before, the denominator attracts our attention.

$$\text{Let } u = x^2 + 9$$

$$du = 2x\,dx$$

$$\frac{1}{2}\,du = x\,dx$$

$$\int \frac{x^3}{x^2+9}\,dx = \int \frac{1}{x^2+9}\,x^2(x\,dx)$$

$$= \int \frac{1}{u}\,x^2\left(\frac{1}{2}\,du\right)$$

This time, there is an x^2 left over. We go back to $u = x^2 + 9$, and solve for x^2.

$$x^2 = u - 9$$

Now our integral is

$$\int \frac{1}{u}(u-9)\frac{1}{2}\,du = \frac{1}{2}\int \frac{u-9}{u}\,du \qquad \text{Divide } (u-9) \text{ by } u$$

$$= \frac{1}{2}\int \left(1 - \frac{9}{u}\right)du \qquad \text{Now integrate}$$

$$= \frac{1}{2}(u - 9\ln|u|) + C \qquad \text{Resubstitute}$$

$$= \frac{1}{2}(x^2 + 9 - 9\ln|x^2+9|) + C$$

ANSWER $\int \frac{x^3}{x^2+9}\,dx = \frac{1}{2}(x^2 + 9 - 9\ln|x^2+9|) + C.$

7–1 Exercises

I.

Evaluate the following indefinite integrals (use the indicated substitution when given).

1. (*See Example 2*) $\int (5x+1)^3\,dx$; let $u = 5x + 1$
2. $\int \sqrt{9x-4}\,dx$; let $u = 9x - 4$
3. (*See Example 3*) $\int (2x^2+3x)^4(4x+3)\,dx$; let $u = 2x^2 + 3x$
4. $\int (5x^3+4x)^5(15x^2+4)\,dx$; let $u = 5x^3 + 4x$
5. $\int 3(7u^2+3u-2)^2(14u+3)\,du$
6. $\int (x^3-3x^2+2)^3(3x^2-6x)\,dx$
7. (*See Example 4*) $\int (4x^3-6x)^3(2x^2-1)\,dx$
8. $\int \frac{4x}{x^2-3}\,dx$; let $u = x^2 - 3$
9. (*See Example 7*) $\int \frac{4}{4x-3}\,dx$
10. $\int \frac{2x}{x^2+1}\,dx$
11. (*See Example 5*) $\int 2x^2e^{x^3}\,dx$
12. $\int e^{6x}\,dx$
13. (*See Example 6*) $\int \frac{4x-2}{\sqrt{x^2-x+5}}\,dx$
14. $\int \frac{12x-2}{3x^2-x+4}\,dx$
15. (*See Example 8*) $\int \frac{e^x}{e^x-7}\,dx$
16. $\int \frac{e^{4x}}{e^{4x}+2}\,dx$

II.

17. (*See Example 9*) $\int \frac{1}{x(1+\ln x)^2}\,dx$
18. $\int \frac{\sqrt{\ln x}}{x}\,dx$
19. $\int \frac{e^{1/x^2}}{x^3}\,dx$
20. $\int \frac{(\sqrt{x}+6)^5}{\sqrt{x}}\,dx$

Fill in the missing limits of integration, then compute the definite integral.

21. (*See Example 10*) $\int_3^7 \frac{x}{x^2-4}\,dx = \int_{-}^{-} \frac{1}{2u}\,du$, with $u = x^2 - 4$

22. $\int_{-1}^{5} x^2\sqrt{x^3+9}\,dx = \int_{-}^{-} \frac{1}{3}\sqrt{u}\,du$, $u = x^3 + 9$

23. $\int_{-1}^{3} xe^{x^2}\,dx = \int_{-}^{-} \frac{1}{2}e^u\,du$

24. $\int_0^1 (2x-1)(3x^2-3x+4)^3\,dx = \int_{-}^{-} \frac{1}{3}u^3\,du$

III.

Find the following indefinite integrals.

25. $\int \frac{x}{e^{x^2}}\,dx$

26. $\int \frac{1}{x\ln x}\,dx$

27. (*See Example 11*) $\int x\sqrt{3x-4}\,dx$

28. $\int x\sqrt{5-x}\,dx$

29. (*See Example 12*) $\int x^3(x^2+5)^3\,dx$

30. $\int \frac{4x^5}{x^3+10}\,dx$

Compute the following definite integrals.

31. $\int_0^5 \frac{2x^3}{x^2+1}\,dx$

32. $\int_4^{11} x\sqrt{x+5}\,dx$

33. A company with fixed costs of \$150 has modeled its marginal cost function by

$$MC(x) = x\sqrt{x^2+9}$$

What is the company's total cost function?

34. What is the area under the curve $y = xe^{x^2}$, above the x-axis, and bounded on the right by the line $x = 2$?

35. If a company's marginal production t hours after opening in the morning is

$$\frac{t}{\sqrt{t^2+1}}, \qquad 0 \le t \le 8$$

what is its total volume of production over the eight-hour period?

36. A particle is moving with a velocity given by

$$V(t) = 6t\sqrt{t^2+16}$$

where V is given in feet per second, and t is in seconds. How far does the particle travel during the interval $0 \le t \le 3$?

37. A company's marginal revenue is given by

$$MR(x) = \frac{20x}{x^2 + 1}$$

What would be the total change in revenue if x increased from 3 to 7?

7–2 Integration by Parts

The method of integration discussed in this section is called **integration by parts**. This technique is based on the product rule for differentiation. The formula to use looks almost silly in theory. It is in practice that its usefulness becomes apparent.

The product rule for differentiation states that

$$(f(x) \cdot g(x))' = g(x) \cdot f'(x) + f(x) \cdot g'(x)$$

We integrate both sides:

$$\int (f(x) \cdot g(x))' \, dx = \int [g(x) \cdot f'(x) + f(x) \cdot g'(x)] \, dx$$

$$= \int g(x) \cdot f'(x) \, dx + \int f(x) \cdot g'(x) \, dx$$

Now, $\int (f(x) \cdot g(x))' \, dx = f(x) \cdot g(x) + C$ because $f(x) \cdot g(x)$ is an antiderivative of its derivative, $(f(x) \cdot g(x))'$. Using this to evaluate the left-hand side, we have

$$f(x) \cdot g(x) + C = \int g(x) \cdot f'(x) \, dx + \int f(x) \cdot g'(x) \, dx$$

Now subtract $\int g(x) \cdot f'(x) \, dx$ from both sides.

$$f(x) \cdot g(x) + C - \int g(x) \cdot f'(x) \, dx = \int f(x) \cdot g'(x) \, dx$$

We don't have to write $+ C$ anymore; the indefinite integral $\int g(x) \cdot f'(x) \, dx$ by itself indicates a family of functions that we know only up to an additive constant. If we change the order of the equation, we get the formula we want.

The Integration by Parts Formula

$$\int f(x) \cdot g'(x) \, dx = f(x) \cdot g(x) - \int g(x) \cdot f'(x) \, dx$$

As we indicated above, this formula doesn't *look* very useful. The idea is this: Try to recognize the integrand you are given as the product of two functions, one of which you can integrate fairly easily; you will differentiate the other one. Call the one you can integrate $g'(x)$, so its integral is $g(x)$. Apply this formula and hope that $\int g(x) \cdot f'(x) \, dx$ is an easier integral than the one you started with. Examples show how useful this method can be.

Example 1 (*Compare Exercise 1*)
Find $\int x^2 \ln x \, dx$.

Solution We factor the integral into two parts: $\ln x$ and x^2. We can integrate x^2; if $g'(x) = x^2$, then $g(x) = \frac{1}{3}x^3$. (We do not worry about $+C$ until the *very* end.) The other part of the integrand in the formula is called $f(x)$; here, $f(x) = \ln x$, so $f'(x) = \frac{1}{x}$.

The parts formula

$$\int f(x) \cdot g'(x) \, dx = f(x) \cdot g(x) - \int g(x) \cdot f'(x) \, dx$$

now gives us

$$\int (\ln x)x^2 \, dx = (\ln x)\left(\frac{x^3}{3}\right) - \int \frac{x^3}{3} \cdot \frac{1}{x} \, dx$$

We simplify $\frac{x^3}{x}$ to x^2 and continue.

$$= \frac{x^3 \ln x}{3} - \frac{1}{3} \int x^2 \, dx$$

It worked! We can find $\int x^2 \, dx$. Thus,

$$\int (\ln x)x^2 \, dx = \frac{x^3 \ln x}{3} - \frac{1}{3}\left(\frac{1}{3}x^3\right) + C$$

(This is the last step, so now include the $+C$.)

ANSWER $\int x^2 \ln x \, dx = \frac{x^3 \ln x}{3} - \frac{x^3}{9} + C$

There is a change of notation that makes the integration by parts formula easier to remember and easier to use.

Denote $f(x)$ by u, so $f'(x) \, dx$ becomes du.

Denote $g'(x) \, dx$ *by* dv, so $g(x)$ is replaced by v.

The formula now is

$$\int u \, dv = uv - \int v \, du$$

We redo Example 1 so that you can see how this notation works.

Example 1 (Redone) (*Compare Exercise 1*)
Find $\int x^2 \ln x \, dx$.

Solution We integrate x^2 and differentiate $\ln x$, so

$$u = \ln x \qquad dv = x^2 \, dx$$

$$du = \frac{1}{x} \, dx \qquad v = \frac{x^3}{3}$$

Using

$$\int u\,dv = uv - \int v\,du$$

we have

$$\int (\ln x)(x^2\,dx) = (\ln x)\left(\frac{x^3}{3}\right) - \int \frac{x^3}{3}\cdot\frac{1}{x}\,dx$$

$$= \frac{x^3 \ln x}{3} - \frac{1}{9}x^3 + C, \text{ as before.}$$

Example 2 (*Compare Exercise 3*)
Find $\int xe^{3x}\,dx$.

Solution We factor xe^{3x} into parts, x and e^{3x}. We choose to differentiate the x and integrate the e^{3x}. (The other choice will be discussed after this solution.)

We let $\quad u = x \quad$ and $\quad dv = e^{3x}\,dx$

Thus, $\quad du = dx \quad$ and $\quad v = \frac{1}{3}e^{3x}$

Then,

$$\int u\,dv = uv - \int v\,du$$

becomes

$$\int xe^{3x}\,dx = x\left[\frac{1}{3}e^{3x}\right] - \int \frac{1}{3}e^{3x}\,dx$$

$$= \frac{xe^{3x}}{3} - \frac{1}{3}\int e^{3x}\,dx$$

Since $\int e^{3x}\,dx = \frac{1}{3}e^{3x} + C$, we conclude with

$$= \frac{xe^{3x}}{3} - \frac{e^{3x}}{9} + C$$

ANSWER $\quad \int xe^{3x}\,dx = \frac{xe^{3x}}{3} - \frac{e^{3x}}{9} + C.$

In discussing integration by substitution, we were able to give you some things to look for that would suggest what the substitution might be: parentheses, radicals, and denominators. We can also give you some specific forms to watch for to use integration by parts. Knowledge of these forms just comes from experience. For instance, we look again at the integral of Example 2. How did we choose to treat xe^{3x} the way we did? What would happen if we switched the roles of e^{3x} and x? After all, we can integrate x very easily, too.

Example 2 Revisited Find $\int xe^{3x}\,dx$.

Attempt at Solution This time, we let $u = e^{3x}$ and $dv = x\,dx$

Then, $du = 3e^{3x}$ $v = \dfrac{x^2}{2}$

and

$$\int u\,dv = uv - \int v\,du$$

becomes

$$\int xe^{3x}\,dx = e^{3x}\frac{x^2}{2} - \int \frac{3}{2}x^2e^{3x}\,dx$$

The new integral $\frac{3}{2}\int x^2e^{3x}\,dx$ is not any simpler. In fact, it is more complicated than the one we started with, $\int xe^{3x}\,dx$. This choice of u and dv didn't work to get us a simpler integral. This choice of parts doesn't work. If in some problem, your choice of u and dv leads to a more complicated integral than the one you started with, go back and try a different choice of parts.

Experience tells us that when confronted with an integral of the form $\int x^ne^{ax}\,dx$, where n is a positive integer, let

$$u = x^n \qquad \text{and} \qquad dv = e^{ax}\,dx$$

Example 3 shows that this choice will help you find the integral even if you have to be persistent.

Example 3 (*Compare Exercise 5*)
Find $\int x^2e^{4x}\,dx$.

Solution The hint says let $u = x^2$ and $dv = e^{4x}\,dx$

Then $du = 2x\,dx$ $v = \dfrac{1}{4}e^{4x}$

and

$$\int u\,dv = uv - \int v\,du$$

becomes

$$\begin{aligned}\int x^2e^{4x}\,dx &= x^2\frac{1}{4}e^{4x} - \int \frac{1}{4}e^{4x}2x\,dx \\ &= \frac{x^2e^{4x}}{4} - \frac{1}{2}\int xe^{4x}\,dx\end{aligned}$$

We can't evaluate this last integral immediately, but it is simpler than the one we started with, and the hint does say to be persistent. Remembering the successful solution to Example 2, we use parts again to find $\int xe^{4x}\,dx$.

Following the hint again,

$$u = x \qquad \text{and} \qquad dv = e^{4x}\,dx$$

$$du = dx \qquad\qquad v = \frac{1}{4}e^{4x}$$

$$\int u\,dv = uv - \int v\,du$$

becomes

$$\int xe^{4x}\,dx = x\frac{1}{4}e^{4x} - \int \frac{1}{4}e^{4x}\,dx$$

$$= \frac{xe^{4x}}{4} - \frac{1}{16}e^{4x} + C$$

Now we go back to our original problem. We had

$$\int x^2e^{4x}\,dx = \frac{x^2e^{4x}}{4} - \frac{1}{2}\left[\int xe^{4x}\,dx\right]$$

$$= \frac{x^2e^{4x}}{4} - \frac{1}{2}\left[\frac{xe^{4x}}{4} - \frac{e^{4x}}{16}\right] + C$$

See Comment (2) below regarding the $+ C$ term.

ANSWER $\displaystyle\int x^2e^{4x}\,dx = \frac{x^2e^{4x}}{4} - \frac{xe^{4x}}{8} + \frac{e^{4x}}{32} + C.$

Two Comments:

1. Watch out for an arithmetic error when using parts twice. Notice that $-\frac{1}{2}$ must be multiplied through the whole expression for $\int xe^{4x}dx$.
2. We can write $+C$ after completing the antidifferentiation. To find the integral, you find *one* antiderivative and then add the arbitrary constant term.

If k is a positive integer, you can also use integration by parts to compute integrals of the form

$$\int x^n(\ln x)^k\,dx$$

$$\text{Let } u = (\ln x)^k \qquad \text{and} \qquad dv = x^n\,dx$$

This is what we did in Example 1 with $n = 2$ and $k = 1$. In the next example, we show how this technique allows us to find the integral of $\ln x$.

Example 4 (*Compare Exercise 7*)
Find $\int \ln x\,dx$.

Solution Following the hint with $n = 0$ and $k = 1$, we let

$$u = \ln x \qquad \text{and} \qquad dv = dx$$

Then, $du = \frac{1}{x}\,dx$ and $v = x$

Applying $\int u\,dv = uv - \int v\,du$,

we have
$$\int \ln x\,dx = (\ln x)x - \int x\frac{1}{x}\,dx$$
$$= x\ln x - \int 1\,dx$$
$$= x\ln x - x + C$$

ANSWER $\int \ln x\,dx = x\ln x - x + C$.

Example 5 (*Compare Exercise 9*)
Find $\int x(\ln x)^2\,dx$.

Solution Following the hint with $n = 1$ and $k = 2$, we let

$$u = (\ln x)^2 \qquad \text{and} \qquad dv = x\,dx$$
$$du = 2(\ln x)\frac{1}{x}\,dx \qquad\qquad v = \frac{x^2}{2}$$
$$\int u\,dv = uv - \int v\,du$$

becomes

$$\int (\ln x)^2 x\,dx = (\ln x)^2\frac{x^2}{2} - \int \frac{x^2}{2}2(\ln x)\frac{1}{x}\,dx$$
$$= \frac{x^2(\ln x)^2}{2} - \int x\ln x\,dx$$

We use parts on $\int x\ln x\,dx$, following the same hint.

$$u = \ln x \qquad \text{and} \qquad dv = x\,dx$$
$$du = \frac{1}{x}\,dx \qquad\qquad v = \frac{x^2}{2}$$
$$\int x\ln x\,dx = \frac{(\ln x)x^2}{2} - \int \frac{x^2}{2}\frac{1}{x}\,dx$$
$$= \frac{(\ln x)x^2}{2} - \int \frac{x}{2}\,dx$$
$$= \frac{(\ln x)x^2}{2} - \frac{x^2}{4} + C$$

Returning to our original equation, we have

$$\int x(\ln x)^2\,dx = \frac{x^2(\ln x)^2}{2} - \left(\frac{x^2\ln x}{2} - \frac{x^2}{4}\right) + C$$

$$= \frac{x^2(\ln x)^2}{2} - \frac{x^2\ln x}{2} + \frac{x^2}{4} + C$$

Example 6 Find $\int (x^2 - e^{-x})\,dx$.

Solution Don't get "parts happy" and think all integrals are now to be attacked by choosing u and dv.

$$\int (x^2 - e^{-x})\,dx = \frac{1}{3}x^3 - \frac{1}{-1}e^{-x} + C$$

$$= \frac{x^3}{3} + e^{-x} + C$$

7–2 Exercises

I.

Use integration by parts to find the following integrals.

1. (*See Example 1*) $\int x \ln x\,dx$
2. $\int 4x^2 \ln x\,dx$
3. (*See Example 2*) $\int xe^{5x}\,dx$
4. $\int 3xe^x\,dx$

II.

Use parts (repeat if necessary).

5. (*See Example 3*) $\int x^2e^{2x}\,dx$
6. $\int x^3e^{-x}\,dx$
7. (*See Example 4*) $\int x^3\ln x\,dx$
8. $\int x^5\ln x\,dx$
9. (*See Example 5*) $\int x^2(\ln x)^2\,dx$
10. $\int (\ln x)^2\,dx$

III.

Find the following indefinite integrals; you may or may not have to use parts. You may have to use a substitution together with parts.

11. $\int x^3e^{x^2}\,dx$ (*Hint*: Let $u = x^2$ and $dv = xe^{x^2}\,dx$)

12. $\int \frac{x}{e^{4x}}\,dx$ 13. $\int 4xe^{x^2}\,dx$ 14. $\int x^2(\ln x^2)\,dx$

15. $\int \frac{\ln x}{x^2}\,dx$ 16. $\int \frac{\ln x}{x}\,dx$

17. What is the area under the curve te^{-t} and above the interval $1 \le t \le 2$?
18. A company's marginal cost for producing x units is $MC(x) = x\ln[x + 1]$. What is the total change in its cost if production goes from eight units to ten units?

7–3 Improper Integrals

Unbounded Intervals
Improper Integrals
Applications

Unbounded Intervals

Suppose that a certain service business finds that its customers arrive at the average rate of 50 per hour, and that the usual transaction takes 4 minutes. The business could be a bank, a cosmetics department, or a supermarket. The business must decide how many service personnel to hire so that there are enough to handle busy times. At the same time, the business doesn't want to hire so many people that most of them are often idle. For example, the firm needs to know how likely it is that 10 customers will need service all at once. The firm may ask questions like, "What is the probability there will be more than 30 customers in 15 minutes?" or "If we hire 6 sales personnel, what is the probability that one of them will be idle for more than 10 minutes at a time?"

These questions theoretically involve no upper bound on the number of customers or the length of time an employee is idle. To handle questions like these, the applied science of statistics frequently gets involved in integrals over unbounded regions—integrals that at first glance seem about as far removed from applications as possible. We approach the concept of integrating over unbounded regions by looking at some simple functions. Figure 7–1 gives the graph of $y = 1/x^2$ for $x \ge 1$.

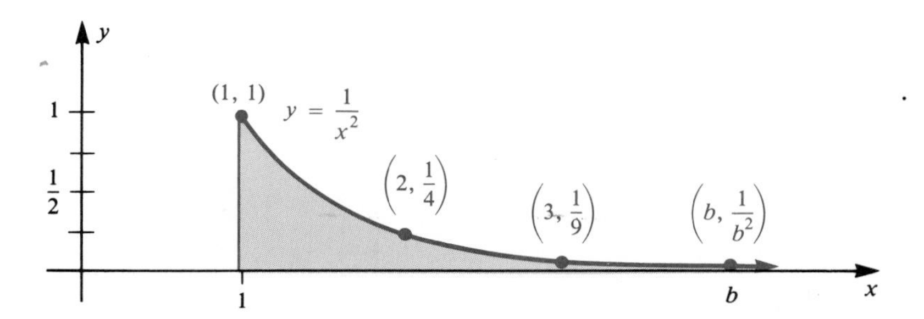

Figure 7–1

If we can, we want to make sense of the question, "What is the area of the shaded region?" We start by just considering the area of that part of the region between $x = 1$ and $x = b$. That area is given by

$$\int_1^b \frac{1}{x^2}\,dx = -\frac{1}{x}\Big|_1^b = -\frac{1}{b} - \left(-\frac{1}{1}\right)$$
$$= 1 - \frac{1}{b}$$

No matter how far to the right you place b, the area of the region over $[1, b]$ is less than 1. For large values of b, $1 - (1/b)$ is very close to 1. As $b \to \infty$, the quantity

$$\left(1 - \frac{1}{b}\right) \to 1$$

$$\lim_{b\to\infty}\left(\int_1^b \frac{1}{x^2}\,dx\right) = \lim_{b\to\infty}\left(1 - \frac{1}{b}\right) = 1$$

These calculations lead us to say that the area of the shaded region in Figure 7–1 is 1. (*Compare Exercise 1*)

Now we look at another simple function, $y = 1/x$, and ask the same question as we did with $y = 1/x^2$. See Figure 7–2.

What is the area of the region bounded above by $y = 1/x$, below by the x-axis, on the left by the line $x = 1$, but unbounded on the right? Using the same approach as above, we first fix a number b and compute the area of that part of the region which is bounded on the right by the line $x = b$. This area is given by

$$\int_1^b \frac{1}{x}\,dx = \ln|x|\Big|_1^b$$
$$= \ln|b| - \ln|1| \qquad (b > 0 \text{ and } 1 > 0)$$
$$= \ln b - \ln 1 \qquad (\ln 1 = 0)$$
$$= \ln b$$

Again, the next step is to look at what happens as b grows arbitrarily large. But this time, $\lim_{b\to\infty} \ln b$ doesn't exist; as b grows arbitrarily large, so does $\ln b$. We conclude that the shaded region in Figure 7–2 does not have a finite area.

These two examples may play havoc with your intuition. First, it may seem "obvious" to many that an unbounded region cannot have a finite area. Second, it

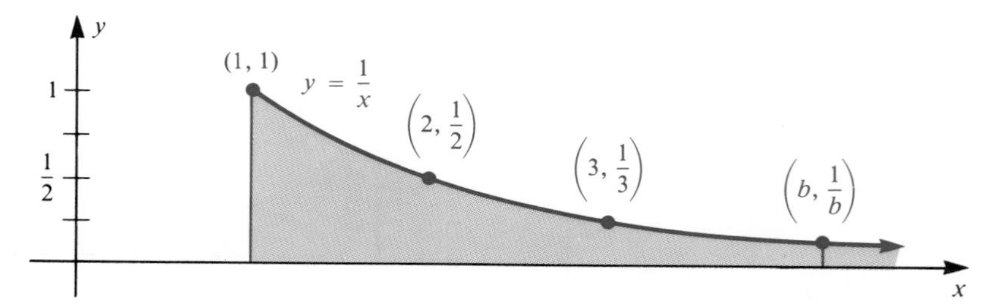

Figure 7–2

is hard to believe that there is that much difference in behavior between the region under $y = 1/x^2$ and the region under $y = 1/x$. These examples teach us that we must be very careful about jumping to "obvious" conclusions about unbounded regions.

Example 1 (*Compare Exercise 1*)
Does the region bounded above by the graph $y = 1/x^4$, below by the x-axis, and on the left by the line $x = 2$ have finite area or not?

Solution The integral

$$\int_2^b \frac{1}{x^4}\,dx$$

gives the area of the part of this region that lies to the left of the line $x = b$. We compute the area of this subregion and then see what happens as $b \to \infty$.

$$\int_2^b \frac{1}{x^4}\,dx = -\frac{1}{3x^3}\bigg|_2^b = -\frac{1}{3b^3} + \frac{1}{24}$$

The area of this subregion is $\frac{1}{24} - \frac{1}{3b^3}$, and as $b \to \infty$, $\left(\frac{1}{24} - \frac{1}{3b^3}\right) \to \frac{1}{24}$.

ANSWER The area of this unbounded region is $\frac{1}{24}$.

Improper Integrals

Based on these examples, we now make the following definitions.

Definition

The integral $\int_a^\infty f(x)\,dx$ is called an **improper integral** (the upper limit of the integral is not a number), and we define

$$\int_a^\infty f(x)\,dx = \lim_{b \to \infty} \int_a^b f(x)\,dx$$

if this limit exists.

If $\lim_{b \to \infty} \int_a^b f(x)\,dx$ exists, then we say that $\int_a^\infty f(x)\,dx$ is **convergent.**

If $\lim_{b \to \infty} \int_a^b f(x)\,dx$ does not exist, then we say that $\int_a^\infty f(x)\,dx$ is **divergent.**

Thus, from our discussion above we have

1. $\displaystyle\int_1^\infty \frac{1}{x^2}\,dx$ is convergent, while
2. $\displaystyle\int_1^\infty \frac{1}{x}\,dx$ is divergent.

Example 2 (*Compare Exercise 9*)
Evaluate $\int_0^\infty e^{-2x}\,dx$ or determine that it is divergent.

Solution By definition,

$$\int_0^\infty e^{-2x}\,dx = \lim_{b\to\infty} \int_0^b e^{-2x}\,dx$$

Next, we compute the definite integral $\int_0^b e^{-2x}\,dx$:

$$\begin{aligned}\int_0^\infty e^{-2x}\,dx &= \lim_{b\to\infty}\left[-\frac{1}{2}e^{-2x}\Big|_0^b\right]\\ &= \lim_{b\to\infty}\left[-\frac{1}{2}(e^{-2b}-e^0)\right]\\ &= \lim_{b\to\infty}\left[-\frac{1}{2}\left(\frac{1}{e^{2b}}-1\right)\right]\end{aligned}$$

Now compute the limit.

As $b \to \infty$, $e^{2b} \to \infty$, so $\dfrac{1}{e^{2b}} \to 0$.

Thus, $$\lim_{b\to\infty}\left[-\frac{1}{2}\left(\frac{1}{e^{2b}}-1\right)\right] = -\frac{1}{2}(0-1) = \frac{1}{2}$$

ANSWER $\int_0^\infty e^{-2x}\,dx$ is convergent, and $\int_0^\infty e^{-2x}\,dx = \frac{1}{2}$.

Since $e^{-2x} \geq 0$ on $[0, \infty)$, we can interpret this integral geometrically; the region under the curve $y = e^{-2x}$, above the x-axis, and to the right of the y-axis has area $= \frac{1}{2}$.

Example 3 (*Compare Exercise 15*)
Evaluate $\displaystyle\int_1^\infty \frac{1}{\sqrt{x}}\,dx$, or show that it is divergent.

Solution Using the definition, we write

$$\begin{aligned}\int_1^\infty \frac{1}{\sqrt{x}}\,dx &= \lim_{b\to\infty}\int_1^b \frac{1}{\sqrt{x}}\,dx\\ &= \lim_{b\to\infty}\int_1^b x^{-1/2}\,dx\\ &= \lim_{b\to\infty}\left[2x^{1/2}\Big|_1^b\right]\\ &= \lim_{b\to\infty}[2\sqrt{b}-2]\end{aligned}$$

But as $b \to \infty$, $\sqrt{b} \to \infty$ also, so this limit does not exist.

ANSWER $\int_1^\infty \frac{1}{\sqrt{x}}\,dx$ diverges. The region bounded above by the curve $y = \frac{1}{\sqrt{x}}$, below by the x-axis, and on the left by the line $x = 1$ does not have finite area.

Next, we look at integrals over intervals of the form $(-\infty, b]$.

Definition

$$\int_{-\infty}^{b} f(x)\,dx = \lim_{a\to-\infty} \int_a^b f(x)\,dx, \text{ if this limit exists.}$$

As before, the improper integral is called **convergent** if it has a finite limit; otherwise, the integral is called **divergent**.

Example 4 (*Compare Exercise 13*)

Evaluate $\int_{-\infty}^{-2} \frac{4}{x^5}\,dx$ or determine that the integral is divergent. Interpret the answer geometrically.

Solution

$$\begin{aligned}\int_{-\infty}^{-2} \frac{4}{x^5}\,dx &= \lim_{a\to-\infty} \int_a^{-2} \frac{4}{x^5}\,dx \\ &= \lim_{a\to-\infty} \left(-\frac{1}{x^4}\Big|_a^{-2}\right) \\ &= \lim_{a\to-\infty} \left(-\frac{1}{16} + \frac{1}{a^4}\right) \\ &= -\frac{1}{16}\end{aligned}$$

We have found that the integral is convergent. For the geometric interpretation, we sketch the graph of $y = \frac{4}{x^5}$ for $-\infty < x \le -2$. See Figure 7–3.

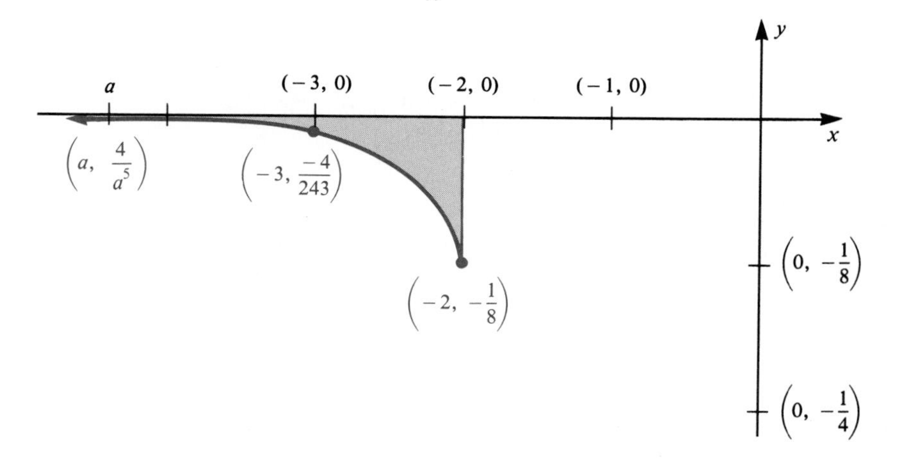

Figure 7–3

If R is the region bounded above by the x-axis, below by the curve $y = \frac{4}{x^5}$, and on the right by the line $x = -2$, then the area of $R = \frac{1}{16}$. Note that the integral gives the negative of the area because $\frac{4}{x^5} < 0$ on the interval $(-\infty, -2]$.

One of the most important curves in statistical applications is the normal, or bell-shaped, curve. See Figure 7–4 where the standard normal distribution curve is drawn.

The curve is the graph of the function

$$f(x) = \frac{1}{\sqrt{2\pi}} e^{-x^2/2}, \qquad -\infty < x < \infty$$

Areas of regions under this curve correspond to probabilities. To talk about the area under this curve, we have to extend our class of improper integrals. Notice that this function is defined for all real numbers; we will have to talk about integrals of functions defined over intervals that are unbounded in both directions.

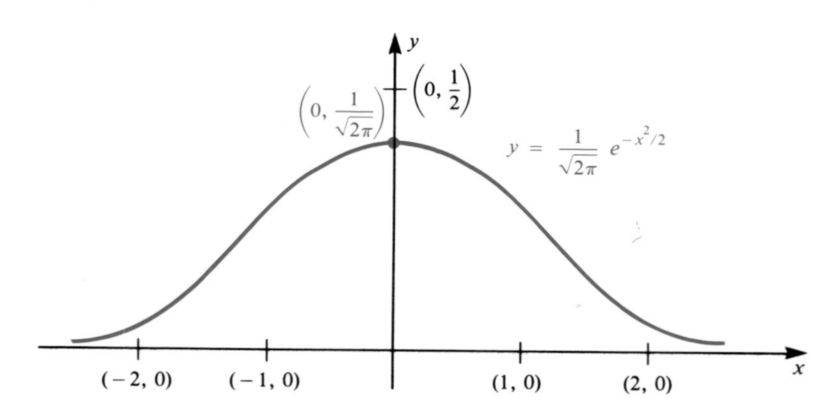

Figure 7–4

Definition

> $\int_{-\infty}^{\infty} f(x)\,dx = \int_{-\infty}^{0} f(x)\,dx + \int_{0}^{\infty} f(x)\,dx$, if both the improper integrals on the right exist.

The terms convergent and divergent are used as before. Note that to discuss the convergence of $\int_{-\infty}^{\infty} f(x)\,dx$ you have to handle two different improper integrals separately.

Example 5 (*Compare Exercise 7*)
Evaluate $\int_{-\infty}^{\infty} e^{-2x}\,dx$ or show that it is divergent.

Solution We consider $\int_{-\infty}^{0} e^{-2x}\,dx$ and $\int_{0}^{\infty} e^{-2x}\,dx$ separately. Example 2 tells us that $\int_{0}^{\infty} e^{-2x}\,dx = \frac{1}{2}$, so now we turn our attention to $\int_{-\infty}^{0} e^{-2x}\,dx$.

$$\int_{-\infty}^{0} e^{-2x}\,dx = \lim_{a\to-\infty} \int_{a}^{0} e^{-2x}\,dx$$
$$= \lim_{a\to-\infty} \left[-\frac{1}{2} e^{-2x} \Big|_{a}^{0} \right]$$
$$= \lim_{a\to-\infty} \left[-\frac{1}{2}(e^{0} - e^{-2a}) \right]$$
$$= \lim_{a\to-\infty} \left[-\frac{1}{2}(1 - e^{-2a}) \right]$$
$$= \lim_{a\to-\infty} \left[-\frac{1}{2} + \frac{1}{2} e^{-2a} \right]$$

To compute this limit, note that as $a \to -\infty$, $-2a \to +\infty$ (we use the + for emphasis), so $e^{-2a} \to \infty$. Thus, $\int_{-\infty}^{0} e^{-2x}\,dx$ diverges. Therefore, $\int_{-\infty}^{\infty} e^{-2x}\,dx$ diverges also.

ANSWER $\int_{-\infty}^{\infty} e^{-2x}\,dx$ diverges.

Applications

Example 6 (*Compare Exercise 23*)
Customers arrive at a certain service facility at the rate of 25 per hour. The probability that there will be a gap of six minutes or more between successive customers is given by the improper integral (6 minutes = $\frac{1}{10}$ hour)

$$\int_{1/10}^{\infty} 25e^{-25t}\,dt$$

What is this probability?

Solution

$$\int_{1/10}^{\infty} 25e^{-25t}\,dt = \lim_{b\to\infty} \int_{1/10}^{b} 25e^{-25t}\,dt$$
$$= \lim_{b\to\infty} \left[-e^{-25t} \Big|_{1/10}^{b} \right]$$
$$= \lim_{b\to\infty} [-e^{-25b} + e^{-2.5}]$$
$$= \lim_{b\to\infty} \left[\frac{1}{e^{2.5}} - \frac{1}{e^{25b}} \right]$$
$$= \frac{1}{e^{2.5}} \approx 0.082$$

The probability of waiting at least six minutes between customers is about 0.082.

Recall that in Chapter 6, we used definite integrals to approximate certain summations. We conclude this section with an application showing how improper integrals can be used to approximate "infinite summations."

A consulting firm in a certain city has estimated that 70¢ of every dollar earned in that city is spent within the city. This 70¢ then is earned within the city, so 70% of this amount will be spent within the city. This progression goes on indefinitely.

Example 7 (*Compare Exercise 27*)
A company has a monthly payroll of $1,000,000. What is the total amount of income this payroll generates in the local community if 70¢ of every dollar earned in this community is spent in the community?

Solution First, there is the $1,000,000 generated directly by the company. But now we start to take the multiplier effect into account; 70% of this $1,000,000 is spent within the local community, thus generating another $700,000 of income. This $700,000 generates another 70% of $700,000 (that is $490,000) of income, and so on. To find the total of all these amounts, we want to sum

$$\begin{aligned}&1{,}000{,}000 + (0.7)(1{,}000{,}000) + (0.7)[(0.7)(1{,}000{,}000)] + \cdots \\ &\quad = 1{,}000{,}000[1 + 0.7 + (0.7)^2 + (0.7)^3 + \cdots]\end{aligned}$$

There is a formula for this sum (which is called a geometric series), but we are interested here in approximating this sum by an integral. The variable in the summation is in the exponent, and so our integral is $1{,}000{,}000 \int_0^\infty (0.7)^t\,dt$. We compute

$$\begin{aligned}\int_0^\infty (0.7)^t\,dt &= \lim_{b\to\infty} \int_0^b (0.7)^t\,dt \\ &= \lim_{b\to\infty} \int_0^b e^{(\ln 0.7)t}\,dt \\ &= \lim_{b\to\infty} \left[\frac{1}{\ln 0.7} e^{(\ln 0.7)t}\Big|_0^b\right] \\ &= \lim_{b\to\infty} \left[\frac{1}{\ln 0.7}\left(e^{(\ln 0.7)b} - e^0\right)\right]\end{aligned}$$

Now $e^0 = 1$, and because $\ln 0.7 < 0$, $(\ln 0.7 \approx -0.356675)$, as $b \to \infty$, $e^{(\ln 0.7)b} \to 0$

$$\begin{aligned}&= \left(\frac{1}{\ln 0.7}\right)(0 - 1) = -\frac{1}{\ln 0.7} \\ &\approx \frac{-1}{-0.356675} \\ &\approx 2.8037\end{aligned}$$

Remember to multiply by $1,000,000 to get the final answer. The total income generated in the local community by this company's payroll is about $2.8 million.

7–3 Exercises

I.

Find the area of the given region or state "no finite area."

1. (*See Example 1*) The region under the curve $y = \frac{1}{x^3}$, above the x-axis, and to the right of the line $x = 1$.
2. The region under the curve $y = \frac{1}{\sqrt{x}}$, above the x-axis, and to the right of the line $x = 4$.
3. The region under the curve $y = e^{-x}$, above the x-axis, and to the right of the line $x = -2$.
4. The region under the curve $y = \frac{1}{x}$, above the x-axis, and to the right of the line $x = 100$.
5. The region under the curve $y = e^{2x}$, above the x-axis, and to the *left* of the line $x = 0$.
6. The region under the curve $y = \frac{1}{x^4}$, above the x-axis, and to the left of the line $x = -2$.
7. (*See Example 5*) The region under the curve e^{-x} and above the x-axis. (The region is unbounded on the left and on the right.)
8. The region under the curve $y = \frac{|x|}{x^2 + 1}$ and above the x-axis.

II.

Compute the following integrals or state that they are divergent. Give a geometric interpretation if possible.

9. (*See Example 2*) $\int_4^{\infty} e^{-x}\,dx$
10. $\int_1^{\infty} e^x\,dx$
11. $\int_1^{\infty} \frac{1}{x^4}\,dx$
12. $\int_0^{\infty} \frac{1}{(x+2)^2}\,dx$
13. (*See Example 4*) $\int_{-\infty}^{0} e^{4x}\,dx$
14. $\int_{-\infty}^{-1} \frac{1}{x^2}\,dx$
15. (*See Example 3*) $\int_4^{\infty} \frac{1}{\sqrt{x}}\,dx$
16. $\int_{-\infty}^{-4} \frac{1}{\sqrt{-x}}\,dx$

III.

17. $\int_{-\infty}^{-1} \frac{1}{x}\,dx$
18. $\int_2^{\infty} \frac{2x}{(1+x^2)^2}\,dx$
19. $\int_0^{\infty} x^2 e^{-x^3}\,dx$
20. $\int_4^{\infty} x^{-3/2}\,dx$

21. $\int_{-\infty}^{\infty} x^{-4/3}\,dx$

22. $\int_{-\infty}^{\infty} \frac{2x}{(1 + x^2)^3}\,dx$

23. (*See Example 6*) If customers arrive at a service facility at the rate of ten per hour, then the probability that there will be more than five minutes between successive customers is given by the improper integral

$$\int_{1/12}^{\infty} 10e^{-10t}\,dt$$

What is this probability?

24. A certain brand of refrigerator claims that its product lasts an average of ten years. Based on this claim, the probability that one of their refrigerators will last at least five years is given by

$$\frac{1}{10}\int_{5}^{\infty} e^{-t/10}\,dt$$

What is this probability?

25. The production of an oil well t years after it began production is given by $P(t) = 3e^{-t/6}$. What is the total production of this well?

26. The revenue produced from sales of a certain product t months after it was introduced is $R(t) = 10e^{-2t}$. What is total revenue produced by this product?

27. (*See Example 7*) Suppose that a utility company gives its customers in a certain state rebates totaling $2 million, and that 90% of any money earned in the state is spent in the state. How much money will be spent in the state as a result of these rebates?

7–4 Integration Using Tables

Certain integrals occur frequently because they involve combinations of linear factors, quadratic factors, and square roots. These functions are commonly used to model real world situations, and because integrals of this type occur so often, there are tables that contain their antiderivatives. We have included a rather limited table of such integrals to give you some practice in manipulating formulas. This table is printed inside the back cover. There are tables that are quite extensive; for instance, the 12th edition of the C.R.C. Standard Mathematical Tables contains over 29 pages of integral formulas. The purpose here is to give you some practice in recognizing the correct formula to use and to demonstrate how to modify your particular integral so that you can use one of the integrals in the table. To use the table, search through the integrals on the left until you find an integrand that has the same **form** as yours; then evaluate the integral using the formula on the right.

Warning! Notice that the table, as is customary in such tables, does not include the additive constant term; you must remember to include it in your answer.

Example 1 (*Compare Exercise 5*)

Evaluate the integral $\int \frac{x}{(3x+2)^2}\,dx$, using the table inside the back cover.

Solution Looking through the integrals in the table, we find that integral 16 is the one we want. That integral is $\int \frac{x}{(ax+b)^2}\,dx$. Setting $a = 3$ and $b = 2$ gives our integral. The formula in the table is

$$\int \frac{x}{(ax+b)^2}\,dx = \frac{b}{a^2(ax+b)} + \frac{1}{a^2}\ln|ax+b|$$

Substituting $a = 3$ and $b = 2$, and remembering that the table does not include the additive constant, we get

$$\int \frac{x}{(3x+2)^2}\,dx = \frac{2}{9(3x+2)} + \frac{1}{9}\ln|3x+2| + C$$

Sometimes, you must modify your integral somewhat to get it in the form that appears in the table.

Example 2 (*Compare Exercise 3*)

Find $\int \frac{6x}{(3x+2)^2}\,dx$.

Solution You will not find an integral of the form $\int \frac{cx}{(ax+b)^2}\,dx$. You must remember that

$$\int \frac{6x}{(3x+2)^2}\,dx = 6\int \frac{x}{(3x+2)^2}\,dx$$

Now you can apply formula 16, as you did in Example 1.

$$\int \frac{6x}{(3x+2)^2}\,dx = 6\int \frac{x}{(3x+2)^2}\,dx = 6\left[\frac{2}{9(3x+2)} + \frac{1}{9}\ln|3x+2|\right] + C$$

Sometimes, the table combines two formulas into one, especially when the only difference is a sign difference.

Example 3 (*Compare Exercise 1*)

Find $\int \sqrt{x^2-9}\,dx$.

Solution This integral is a special case of integral 1, which is $\int \sqrt{x^2 \pm a^2}\,dx$. Here, we will let $a^2 = 9$ so that $a = 3$ and use the $-$ sign in the formula: The table gives

$$\int \sqrt{x^2 \pm a^2}\,dx = \frac{x}{2}\sqrt{x^2 \pm a^2} \pm \frac{a^2}{2}\ln|x + \sqrt{x^2 \pm a^2}|$$

With $a = 3$ and using $-$ in the four places the formula has $\pm$, we have

$$\int \sqrt{x^2-9}\,dx = \frac{x}{2}\sqrt{x^2-9} - \frac{9}{2}\ln|x + \sqrt{x^2-9}| + C$$

Again, we show how you must sometimes do some arithmetic to modify your integral before you can apply the formula.

Example 4 (*Compare Exercise 7*)

Find $\displaystyle\int \frac{1}{\sqrt{4x^2 + 25}}\,dx$.

Solution The closest integral in the table is number 2

$$\int \frac{1}{\sqrt{x^2 \pm a^2}}\,dx$$

and we can choose the + case.

The arithmetic here is a bit more complicated than it was in Example 2, but the idea is to write the integral so that the coefficient of x^2 is 1.

$$\sqrt{4x^2 + 25} = \sqrt{4\left(x^2 + \frac{25}{4}\right)} = \sqrt{4}\,\sqrt{x^2 + \frac{25}{4}}$$

$$= 2\sqrt{x^2 + \frac{25}{4}}$$

Thus,

$$\int \frac{1}{\sqrt{4x^2 + 25}}\,dx = \int \frac{1}{2\sqrt{x^2 + \frac{25}{4}}}\,dx = \frac{1}{2}\int \frac{1}{\sqrt{x^2 + \frac{25}{4}}}\,dx$$

We have now manipulated our integral so that formula 2 can be used with $a^2 = \frac{25}{4}$. The table gives, choosing the + sign,

$$\int \frac{1}{\sqrt{x^2 + a^2}}\,dx = \ln|x + \sqrt{x^2 + a^2}|$$

so

$$\int \frac{1}{\sqrt{4x^2 + 25}}\,dx = \frac{1}{2}\int \frac{1}{\sqrt{x^2 + \frac{25}{4}}}\,dx$$

$$= \frac{1}{2}\ln\left|x + \sqrt{x^2 + \frac{25}{4}}\right| + C$$

You should also notice the use of a^2 in the table. This is to indicate to you that the number appearing in that place in your particular integral must be positive. A constant that is not squared in the table may be positive or negative.

Example 5 (*Compare Exercise 9*)

Find $\displaystyle\int \frac{1}{x(2x - 5)}\,dx$.

Solution This integral exactly matches the form of integral 17

$$\int \frac{1}{x(ax+b)}\,dx$$

with $a = 2$ and $b = -5$. Using $a = 2$ and $b = -5$ in

$$\int \frac{1}{x(ax+b)}\,dx = \frac{1}{b}\ln\left|\frac{x}{ax+b}\right| + C$$

gives

$$\int \frac{1}{x(2x-5)}\,dx = \frac{-1}{5}\ln\left|\frac{x}{2x-5}\right| + C$$

The following example illustrates that you may have to use algebra to rearrange the given integral into a form that is found in the table.

Example 6 (*Compare Exercise 17*)

Evaluate $\displaystyle\int \frac{1}{4u^3 + 4u^2 + u}\,du.$

Solution The table does not contain any integral of this type. However, let us factor the denominator—all the denominators of the integrals in the table are in factored form.

$$\int \frac{1}{4u^3 + 4u^2 + u}\,du = \int \frac{1}{u(4u^2 + 4u + 1)}\,du$$
$$= \int \frac{1}{u(2u+1)^2}\,du$$

We can now recognize this as the same form as integral 18,

$$\int \frac{1}{x(ax+b)^2}\,dx$$

Letting $a = 2$ and $b = 1$ and using the variable u in place of x, we can use formula 18

$$\int \frac{1}{x(ax+b)^2}\,dx = \frac{1}{b(ax+b)} + \frac{1}{b^2}\ln\left|\frac{x}{ax+b}\right| + C$$

to get

$$\int \frac{1}{4u^3 + 4u^2 + u}\,du = \int \frac{1}{u(2u+1)^2}\,du$$
$$= \frac{1}{2u+1} + \ln\left|\frac{u}{2u+1}\right| + C$$

As our final example, we point out that sometimes you must do a substitution to manipulate your integral into a form you can find in the table.

Example 7 *(Compare Exercise 20)*

Find $\int \frac{e^x}{e^{2x} - 9}\,dx$.

Solution $e^{2x} = (e^x)^2$, so $e^{2x} - 9 = (e^x)^2 - 9$. This denominator is of the form "something-squared-minus-something-squared," so we hope to use integral 10,

$$\int \frac{1}{x^2 - a^2}\,dx$$

Substitute $u = e^x$. Then, $e^{2x} - 9 = (e^x)^2 - 9$ becomes $u^2 - 9$; Also, $u = e^x$ means $du = e^x\,dx$.

Thus,
$$\int \frac{e^x}{e^{2x} - 9}\,dx = \int \frac{1}{u^2 - 9}\,du$$

Formula 10,

$$\int \frac{1}{x^2 - a^2}\,dx = \frac{1}{2a} \ln\left|\frac{x - a}{x + a}\right| + C$$

becomes, with $a^2 = 9$ so that $a = 3$ and u playing the role of x,

$$\int \frac{1}{u^2 - 9}\,du = \frac{1}{6} \ln\left|\frac{u - 3}{u + 3}\right| + C$$

Remember to back-substitute $u = e^x$ for the final answer.

$$\int \frac{e^x}{e^{2x} - 9}\,dx = \frac{1}{6} \ln\left|\frac{e^x - 3}{e^x + 3}\right| + C$$

7–4 Exercises

I.

1. *(See Example 3)* $\int \frac{1}{x^2 - 25}\,dx$

2. $\int \frac{x}{(x - 4)^2}\,dx$

3. *(See Example 2)* $\int \frac{5}{36 - x^2}\,dx$

4. $\int \sqrt{x^2 + 100}\,dx$

5. *(See Example 1)* $\int \frac{x}{(3x - 2)^2}\,dx$

6. $\int \frac{3}{x(x - 6)}\,dx$

7. *(See Example 4)* $\int \sqrt{4x^2 + 100}\,dx$

8. $\int \frac{1}{3x^2 - 48}\,dx$

9. *(See Example 5)* $\int \frac{6}{x(8 - 2x)}\,dx$

10. $\int \frac{5x}{12 - 3x}\,dx$

II.

11. $\int \frac{6}{\sqrt{100 + 25x^2}}\,dx$

12. $\int \frac{\sqrt{3 + x^2}}{x}\,dx$

13. $\int \sqrt{(x + 2)^2 + 25}\,dx$

14. $\int \frac{1}{(x + 2)^2 - 9}\,dx$

15. $\int \sqrt{(2x + 1)^2 + 100}\,dx$

16. $\int \frac{e^{2x}}{4e^x + 5}\,dx$

17. *(See Example 6)*
$\int \frac{1}{x^2 + 3x}\,dx$

18. $\int \frac{x}{x^2 + 6x + 9}\,dx$

19. $\int \frac{3x}{x^2 - 4x + 4}\,dx$

20. *(See Example 7)*
$\int \frac{e^x}{e^x\sqrt{e^{2x} + 9}}\,dx$

III.

21. $\int_0^4 \sqrt{x^2 + 9}\,dx$

22. $\int_3^6 \frac{dx}{\sqrt{4x^2 - 25}}$

23. What is the area under the curve

$$y = \frac{1}{x^2\sqrt{x^2 - 1}}$$

from $x = 2$ to $x = 4$?

24. What is the area under the curve

$$y = \frac{\sqrt{25 + 9x^2}}{4x}$$

from $x = 1$ to $x = 4$?

25. The marginal cost of producing x car stereos is given by

$$MC(x) = \frac{6x}{2x + 5}$$

How much do costs increase if the production level goes from $x = 10$ to $x = 20$?

26. A company purchased raw materials at the rate of $4\sqrt{x^2 + 25}$ tons on the xth day of the business month. What is the total they purchased during the first ten days of the business month?

IMPORTANT TERMS

7–1	**Integration by substitution**	**Relation to the chain rule**	**Change of the limits of integration for definite integrals**
7–2	**Integration by parts**	**Relation to the product rule for differentiation**	
7–3	**Unbounded interval**	**Improper integral**	**Convergent integral**
	Divergent integral	**Normal curve**	
7–4	**Table of integrals**		

REVIEW EXERCISES

Compute the integrals in Exercises 1 through 12 by using substitution, integration by parts, or both.

1. $\int x^2 e^{x^3}\,dx$
2. $\int \sqrt{3x+4}\,dx$
3. $\int \frac{1}{5x+8}\,dx$
4. $\int \frac{x}{e^x}\,dx$
5. $\int \frac{1}{x\ln x}\,dx$
6. $\int 2x^3 e^{x^2}\,dx$
7. $\int x\ln(3x)\,dx$
8. $\int xe^{(x+1)}\,dx$
9. $\int xe^{(x^2+1)}\,dx$
10. $\int \sqrt{x}\ln x\,dx$
11. $\int x\ln x^2\,dx$
12. $\int x^2\sqrt{x+1}\,dx$

Evaluate each of the integrals in Exercises 13 through 16 or show that the integral is divergent.

13. $\int_1^{\infty} \frac{x}{1+x^2}\,dx$
14. $\int_{-\infty}^{-1} \frac{1}{x^2}\,dx$
15. $\int_1^{\infty} \frac{1+e^x}{e^{2x}}\,dx$
16. $\int_{-\infty}^{\infty} \frac{x^3}{(1+x^4)^2}\,dx$

You may use the table of integrals to help you find the integrals in Exercises 17 through 21. You may also need to use substitution.

17. $\int \frac{1}{4x^2-100}\,dx$
18. $\int \frac{e^x}{e^{2x}-4}\,dx$
19. $\int \frac{1}{\sqrt{x^2+2x+10}}\,dx$ *Hint:* $x^2+2x+10=(x+1)^2+9$
20. $\int \frac{e^{2x}}{e^x+5}\,dx$
21. $\int \frac{1}{x^2+6x}\,dx$

8

Topics in this chapter can be applied to:
Advertising and Public Awareness ● Manufacturing Output and Cost Projections ● Cobb-Douglas Production Function ● Management of Quality Control ● Results of Advertising ● Production Management ● Marginal Productivity of Labor and Capital ● Maximizing Revenue ● Maximizing Demand ● Maximizing Profit ● Minimizing Costs

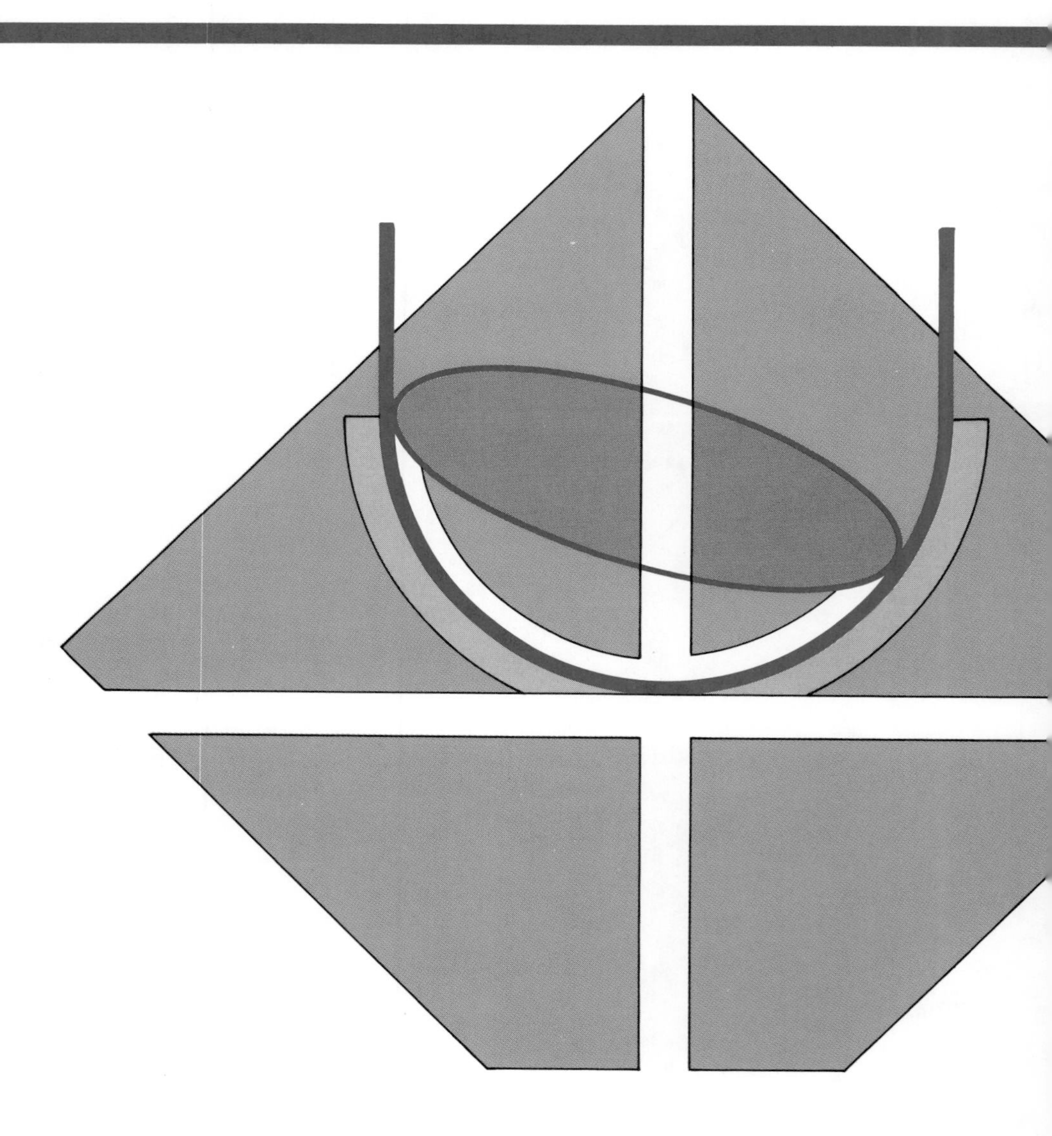

Calculus of Functions of Several Variables

- Functions of Several Variables
- Partial Derivatives
- Maxima and Minima of Functions of Two Variables
- Lagrange Multipliers
- Multiple Integrals

8–1 Functions of Several Variables

Introduction
Applications
Graphs of Functions of Two Variables

Introduction

To this point, the functions we have studied have all been functions of one variable. The function defined by $f(x) = 3x^2 + 2x - 4$, for example, is a function of the single variable x. We did see several applications, however, where the function we were analyzing really depended on more than one variable. For example,

$$\text{Revenue} = (\text{number of units sold}) \cdot (\text{price per unit})$$

$$R = x \cdot p$$

R depends on both x and p.

When discussing revenue, however, we generally assumed that there was a cost-demand curve that allowed us to solve for one of x or p in terms of the other. We could then express R as a function of one variable. In a perfectly competitive market, however, one firm's output will not affect the market price. Nevertheless, this market price may vary for other reasons. For that firm, x and p are independent variables.

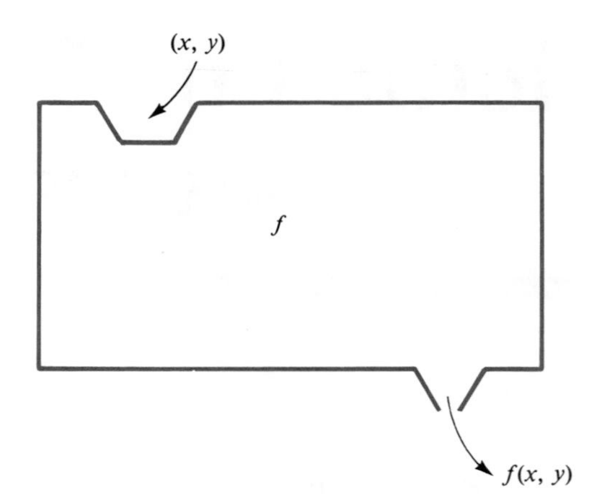

Figure 8–1

Similarly, a company may start an advertising campaign to raise the public's awareness of a new product. The company may want to measure the public's awareness, not just as a function of total amount spent on advertising, but rather as a function of the expenditures on newspaper ads, billboards, radio, and television commercials. Consumer awareness would then be a function of several variables. To handle problems like these, we need to extend our mathematical vocabulary to include functions of more than one variable.

The general idea is the same; the only thing we're changing is the allowable input. For a function of two variables, the input will be ordered pairs. See Figure 8–1.

We write the rule for evaluating $f(x, y)$ in a manner similar to that with one variable. For example, we can write

$$f(x, y) = 3x^2y + 4xy + y^3$$

Corresponding to each value of x and y, there will be a unique value of $f(x, y)$. For example, when $x = 1$ and $y = 2$, we get

$$f(1, 2) = 3(1)^2(2) + 4(1)(2) + (2)^3 = 22$$

Just as we wrote $y = f(x)$, we write $z = f(x, y)$. Thus, we say that when $(x, y) = (1, 2)$, the value of z is 22. Here, x and y are two independent variables, and z is a function of these two variables.

Example 1 (*Compare Exercise 1*)

$$z = f(x, y) = 3x^2y + 4xy + y^3$$

(a) What is $f(-1, 3)$?

(b) What is z when $x = 2$ and $y = -1$?

Solution

(a) $f(-1, 3) = 3(-1)^2(3) + 4(-1)(3) + (3)^3 = 24$.

(b) $f(2, -1) = 3(2)^2(-1) + 4(2)(-1) + (-1)^3 = -21$, so the value of z is -21 when $x = 2$ and $y = -1$.

Functions of three or more variables are handled in a similar manner. For example,

$$f(x, y, z) = 4xy - 2x^2z$$

is a function of the three variables, x, y, and z. If $x = 1$, $y = -2$, and $z = 3$, then

$$f(1, -2, 3) = 4(1)(-2) - 2(1)^2(3) = -8 - 6 = -14$$

Example 2 (*Compare Exercise 3*)
If $w = f(x, y, z) = 2y - xz$,

(a) What is $f(-1, 3, 5)$?
(b) What is the value of w when $x = -2$, $y = 6$, and $z = 4$?

Solution **(a)** $f(-1, 3, 5) = 2 \cdot 3 - (-1)(5) = 6 + 5 = 11$.
(b) $f(-2, 6, 4) = 2 \cdot 6 - (-2)(4) = 12 + 8 = 20$. So, $w = 20$ if $x = -2$, $y = 6$, and $z = 4$.

Applications

Different products may use some of the same resources in their manufacture. For instance, both bread and donuts require flour. If a company manufactures both bread and donuts and decides to increase its production of bread, that increase in the demand for flour may allow the company's supplier of flour to raise the price of the flour, thereby raising the cost of manufacturing donuts. The costs of manufacturing these two products are interrelated.

Example 3 (*Compare Exercise 13*)
The cost, in hundreds of dollars, to the company for manufacturing x thousand loaves of bread and y thousand dozen donuts is given by $C(x, y) = 6x + 7y + xy$. How much does it cost the company to manufacture 40,000 loaves of bread and 2000 dozen donuts?

Solution Here, $x = 40$ and $y = 2$, so

$$C(40, 2) = 6 \cdot 40 + 7 \cdot 2 + 40 \cdot 2$$
$$= 334$$

The company's manufacturing costs are \$33,400.

The amount of a product that a company can manufacture depends upon how much labor and how much capital are devoted to the manufacture of this particular product. (Capital in this context includes buildings, machinery, and so on.) Economists can measure these expenditures in meaningful units, but for our purposes we will ignore the units and let x be the amount of labor used and y be the amount of capital. Often, the number of units produced is described by a function $f(x, y)$ of the type

$$f(x, y) = Cx^a y^{1-a}$$

where C and a are appropriate constants. This frequently used function is called a **Cobb-Douglas production function**.

Example 4 (*Compare Exercise 17*)
Suppose that the manufacturing process of a company is described by the function

$$f(x, y) = 9x^{2/3}y^{1/3}$$

What is the number of units produced when 64 units of labor are used, and 27 units of capital are invested?

Solution We get

$$\begin{aligned} f(64, 27) &= 9(64)^{2/3}(27)^{1/3} \\ &= 9(4)^2(3) \\ &= 432 \end{aligned}$$

There will be 432 units produced.

Geometric formulas can also be given in terms of several variables. For instance, the volume of a right circular cylinder is given by the area of the base times the height. See Figure 8–2.

If r is the radius of the base, then the area of the base is πr^2. If we let h be the height of the cylinder, then the volume V can be interpreted as a function of the two variables r and h.

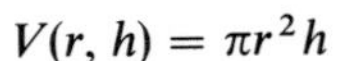

$$V(r, h) = \pi r^2 h$$

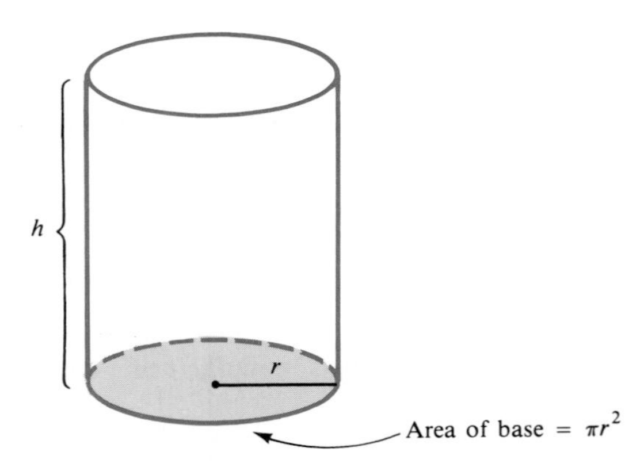

Figure 8–2

Example 5 Find the volume of a cylinder whose base has radius 2 inches and whose height is 3 inches.

Solution When $r = 2$ and $h = 3$, the value of V is

$$V(2, 3) = \pi(2)^2 3 = 12\pi$$

The volume is 12π cubic inches.

Graphs of Functions of Two Variables

We obtained insight into the behavior of a function of a single variable by looking at its graph, the set of points $(x, f(x))$. We called this graph the curve given by $y = f(x)$. A function of two variables has a graph in three-dimensional space; this graph is usually called a **surface**. We shall not go into a detailed discussion of graphs of functions of two variables. These graphs involve three dimensions, and so are difficult to draw on a piece of two-dimensional paper. Furthermore, the graphs of functions of more than two variables involve more than three dimensions, and so our spatial intuition is of little use in this case. Nevertheless, we would like to have a few pictures available to help give you some insight into the geometric significance of the topics we will be discussing, such as partial rates of change and relative extreme values.

With this aim in mind, we introduce the concept of a three-dimensional coordinate system and illustrate how the graph of a function like $f(x, y) = x^2 + y^2$ is a surface in three-dimensional space.

A rectangular Cartesian coordinate system of three dimensions can be obtained by starting with the familiar Cartesian plane and putting a third axis through the origin perpendicular to the plane. The three axes, called the x-, y- and z-axes, are at right angles to one another. Taken in pairs, these axes define planes called the coordinate planes. These planes are the xy-plane, the xz-plane, and the yz-plane as is shown in Figure 8–3.

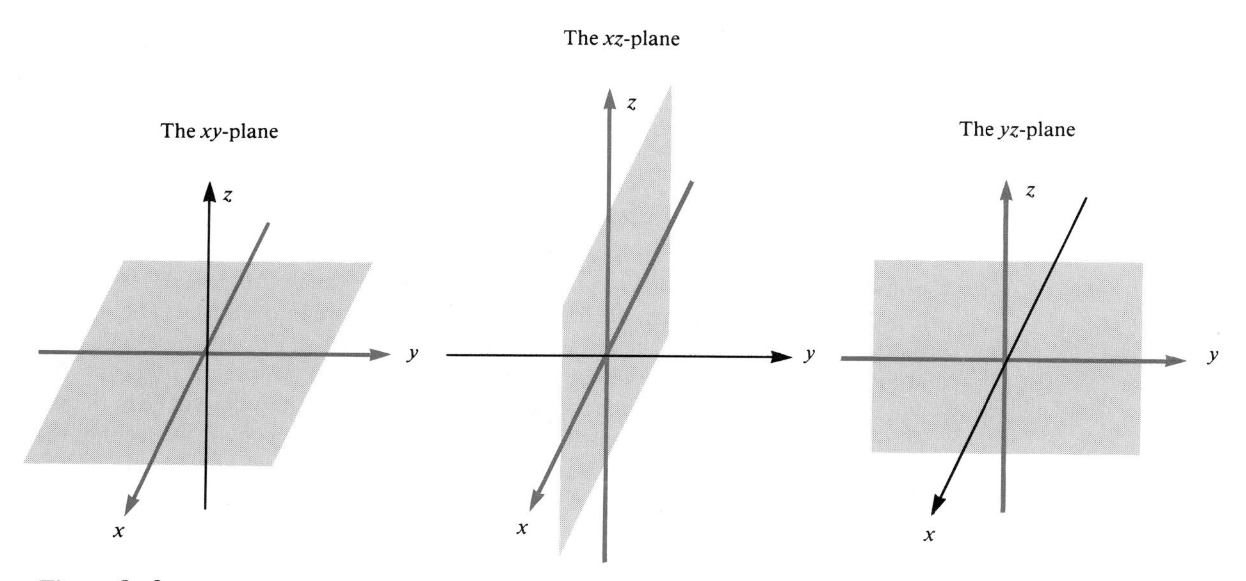

Figure 8–3

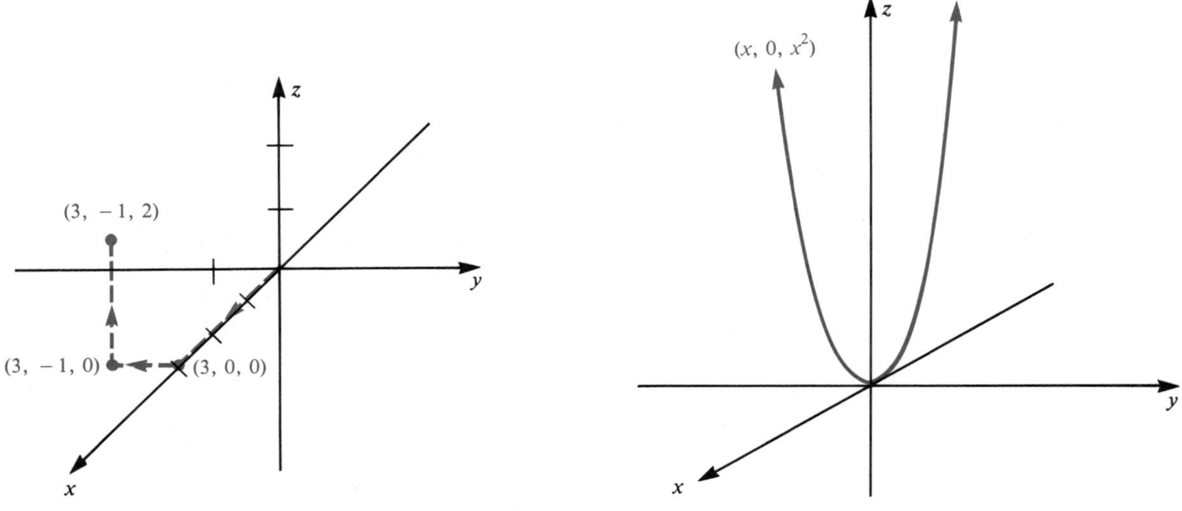

Figure 8–4

As in the Cartesian plane, we introduce units of measurement along the axes. We can then locate a point, P, by knowing three coordinates that give the directed distance from the respective axes. In Figure 8–4, P would be described as the point with coordinates $(3, -1, 2)$.

Try to visualize the yz-plane being the same as the plane of this page, and the x-axis is coming straight out at you. To locate $(3, -1, 2)$, you start at the origin $(0, 0, 0)$ and come straight out the x-axis three units ($x = 3$), go left for one unit ($y = -1$), and then go up two units ($z = 2$).

Now consider the function $z = f(x, y)$, where

$$z = x^2 + y^2$$

The graph of this function is in three-dimensional space and consists of all those points (x, y, z), such that $z = x^2 + y^2$. For example, the point $(3, 4, 25)$ is on this graph. If we hold one variable fixed, we can use what we know about functions of one variable to get some idea of what the graph looks like. For instance, let's look at all points on the graph whose second coordinate is 0. The set of *all* points for which $y = 0$ is the xz-plane, so the points on the graph with $y = 0$ will lie in the xz-plane. Turning now to the equation $z = x^2 + y^2$ and setting $y = 0$, we see that the equation becomes $z = x^2$, the equation of a parabola. The set of points $(x, 0, x^2)$ forms a parabola in the xz-plane. See Figure 8–5.

Next, we consider the points on the graph for which $x = 0$. These points will satisfy the equation $z = y^2$ and will form a parabola in the yz-plane. This is the set of points $(0, y, y^2)$. Figure 8–6 includes both parabolas. Sometimes, it is hard to visualize these graphs.

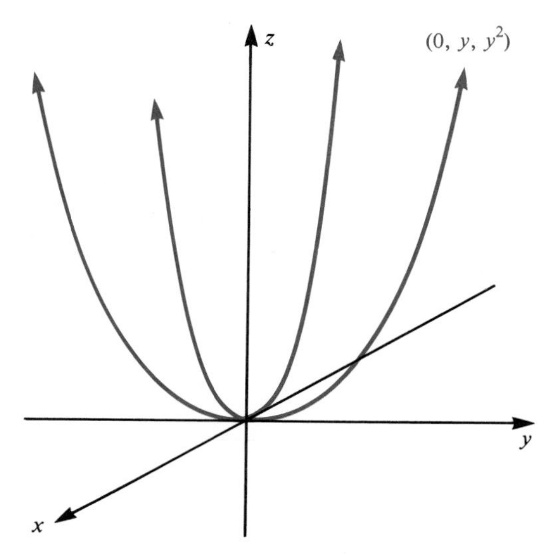

Figure 8–6

Next, we let z be constant; let us consider all the points on the graph for which $z = 4$. These points will satisfy the equation $4 = x^2 + y^2$, which is an equation of a circle. The set of points on the graph $z = x^2 + y^2$ with $z = 4$ is a circle of radius 2; the circle is four units above the xy-plane. This is the set of points $(x, y, 4)$ with $x^2 + y^2 = 4$. All three curves are drawn in Figure 8–7.

If we look at all points on this surface that are nine units high, then $z = 9$, and we would get a circle of radius 3. All cross sections of the surface parallel to the xy-plane will be circles.

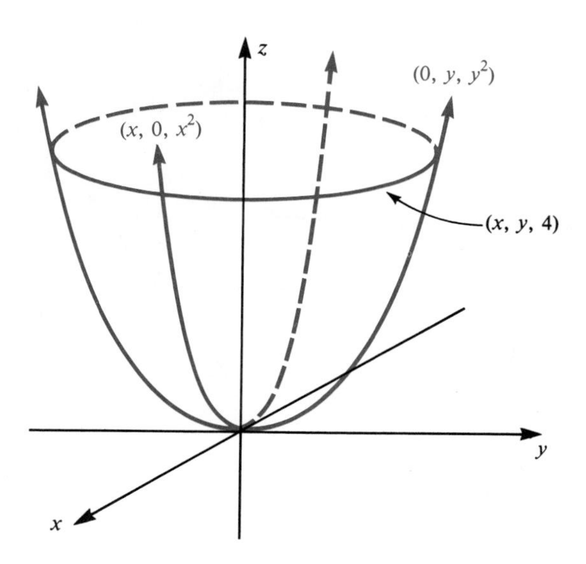

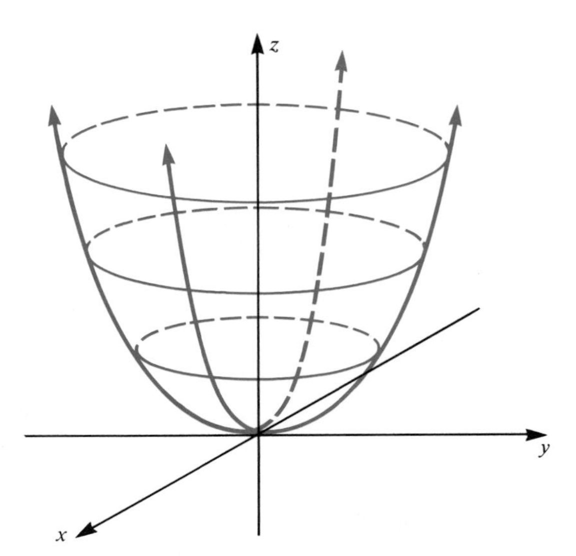

Figure 8–8

Putting all this together, we find that the graph of $z = x^2 + y^2$ is a **paraboloid**, shown in Figure 8–8. The graph is like a bowl resting on the origin.

All functions of two variables, $z = f(x, y)$, that we shall discuss will have graphs consisting of smooth surfaces in three-dimensional space. In general, we can visualize the graph of such a function in the following manner. Let $(a, b, 0)$ be a point in the xy-plane, and let $f(a, b) = c$. Then, $P = (a, b, c)$ is on the graph of $z = f(x, y)$. If $c > 0$, then P is c units above the point $(a, b, 0)$. If $c < 0$, then P is $|c|$ units below the point $(a, b, 0)$. If we plot all such points, we get a surface S, the graph of $f(x, y)$, as in Figure 8–9.

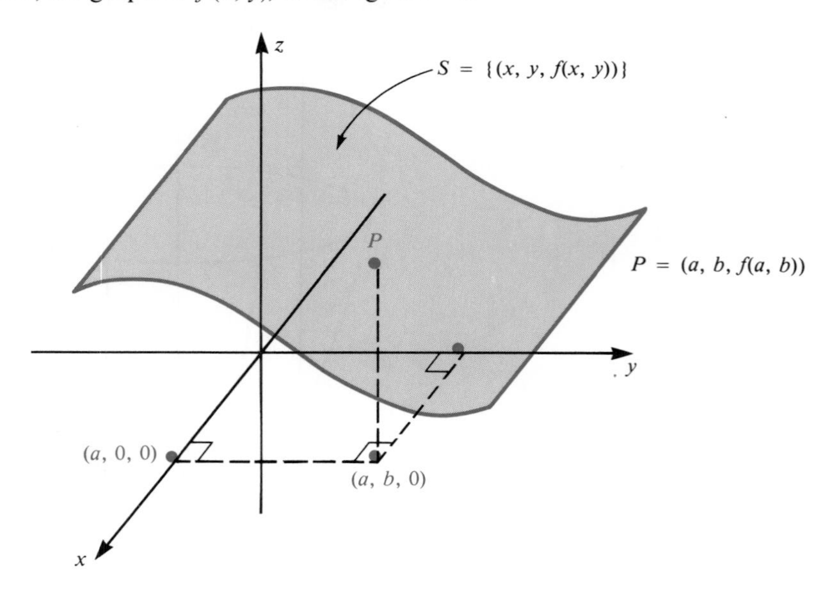

Figure 8–9

Example 6 (*Compare Exercise 7*)
Sketch the graph of $z = x^2 + y^2 + 3$.

Solution If we let $g(x, y) = x^2 + y^2 + 3$, then $g(x, y) = f(x, y) + 3$, where $f(x, y) = x^2 + y^2$. The surface $z = g(x, y)$ lies three units above the surface $z = f(x, y)$; we obtain the graph of $z = x^2 + y^2 + 3$ by sliding the graph of $z = x^2 + y^2$ up a distance of three units. See Figure 8–10.

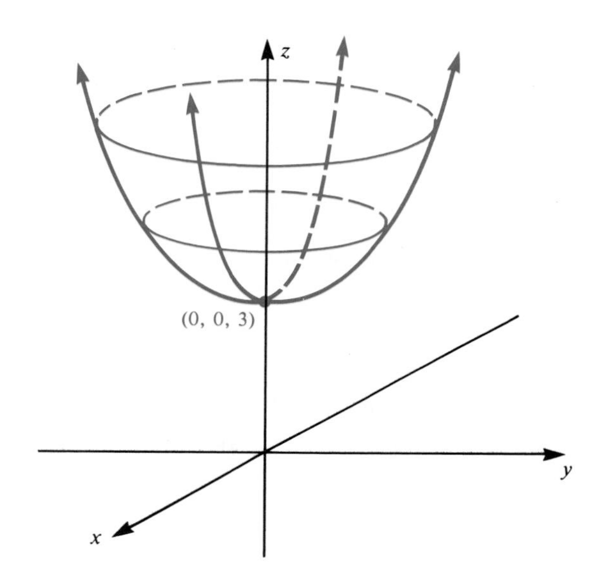

Figure 8–10

8–1 Exercises

I.

In Exercises 1 through 6, determine the values of the functions at the given points.

1. (*See Example 1*) $f(x, y) = 4xy + 3$ at the point $(1, 2)$
2. $g(x, y) = 6x + 2y + x^3 + 4$ at $(3, 1)$
3. (*See Example 2*) $g(x, y, z) = x^2y + 4xz + y^3$ at $(-1, 2, 1)$
4. $f(x, y, z) = 2xyz - x^3 + 3y^2 + 4$ at $(0, 1, 2)$
5. $f(x, y) = 4ye^{x+2}$ at $(1, 2)$
6. $g(x, y) = x^2 \ln y - 4$ at $(3, 1)$

II.

7. (*See Example 6*) Sketch the graph of $z = x^2 + y^2 - 2$
8. Sketch the graph of $z = -x^2 - y^2 + 1$
9. Sketch the locations of the following points in a coordinate system in three-dimensional space.

$$L = (1, 0, 0),$$
$$Q = (0, 3, 1),$$
$$M = (0, 0, 4),$$
$$R = (1, 2, 4),$$
$$N = (0, 2, 0),$$
$$S = (2, -1, 3)$$
$$P = (1, 2, 0),$$

10. Give an equation describing the set of all points in three-dimensional space that are

    (a) in the yz-plane.
    (b) three units above the xy-plane.
    (c) a distance of five units from the origin.

11. Sketch the graph of the function $z = 3$ in three-dimensional space.
12. Sketch the graph of $z = y^2$ in three-dimensional space

III.

13. (*See Example 3*) A company manufactures two models of vans, A and B. The total cost of manufacturing x units of model A and y units of model B is

$$C(x, y) = \frac{x^3}{20} + \frac{y^3}{10} + 100x + 200y + 20xy$$

Determine the cost of manufacturing 40 units of model A and 100 units of model B.

14. A company sells two products, X and Y. The total revenue obtained on selling x units of X and y units of Y is given by

$$R(x, y) = -2x^2 + 500x - 4y^2 + 600y + 50xy$$

Compute the total revenue on selling 50 units of X and 40 units of Y.

15. The cost of quality control in a manufacturing line is a function of the numbers of inspections x and y made per week at two inspection points P and Q, respectively.

$$C(x, y) = 4x^2 + 2y^2 - 4y$$

Determine the cost if 20 inspections per week are made at P, and 10 inspections per week are made at Q.

16. There is a relationship between the number of units N of an item sold by a company and the amounts spent on television advertisements, x, and on newspaper advertisements, y. This relationship is

$$N(x, y) = 1600x + 1200y - 4x^2 - y^2$$

If $x = 500$ and $y = 800$, compute the anticipated sales.

In the next two exercises, f is the firm's Cobb-Douglas production, x is the number of units of labor, and y is the number of units of capital.

17. (*See Example 4*) What is the production when $x = 81$ and $y = 16$ if $f(x, y) = 10x^{1/4}y^{3/4}$?

18. What is the production when $x = 27$ and $y = 64$ if $f(x, y) = 7x^{1/3}y^{2/3}$?

8–2 Partial Derivatives

Definition of Partial Derivatives
Higher Order Partial Derivatives
Geometry of Partial Derivatives
Marginal Analysis in Several Variables

Definition of Partial Derivatives

We used the derivative of a function of a single variable to talk about the rate of change of the function and to find relative maximum and minimum values of the function. We would like to carry over as much of the machinery of derivatives as possible to this new setting of several variables. Let us look at a specific example and see what we can do.

Suppose that a company produces two products whose costs of production are related, say milk and cheese. Let x represent the amount of milk produced, and let y be the amount of cheese. The company is using

$$C(x, y) = \frac{x^3}{300} + \frac{y^3}{50} + 10x + 30y + 2xy$$

as its cost function and is presently manufacturing 30 units of milk and 10 units of cheese. The company's present costs are

$$C(30, 10) = 90 + 20 + 300 + 300 + 600 = 1310$$

What will happen if the amount of cheese produced is held fixed at $y = 10$, and the company varies the amount of milk produced? The function that gives the cost when $y = 10$ and x varies is

$$C(x, 10) = \frac{x^3}{300} + \frac{10^3}{50} + 10x + (30)(10) + 2 \cdot x \cdot 10$$

$$= \frac{x^3}{300} + 30x + 320$$

This is a function of one variable, and so we can take its derivative just as we did before:

$$\frac{d}{dx} C(x, 10) = \frac{x^2}{100} + 30$$

We now have a marginal cost function for x when $y = 10$. When $x = 30$,

$$\frac{x^2}{100} + 30 = 39$$

Thus, if the company were to increase x from 30 to 31, **holding y constant at $y = 10$**, the company's costs would increase approximately 39 units, from 1310 to approximately 1349.

We could now ask the same kind of question, holding x fixed and letting y vary. When we hold one variable fixed and let the other one vary, all the rules of differentiation of a function of one variable still apply. So, our strategy for dealing with rates of change of functions of several variables is to hold all but one variable fixed and use all the rules for derivatives of a function of one variable.

One thing we must modify is our method of evaluating these rates of change. We don't want to have to plug in specific values of y first. We found formulas for $f'(x)$ so that we could find rates of change at any value of x. Similarly, we will want to know rates of change of functions of two variables at any production level (x, y). First, we must change notation. We will not be computing the derivative of $C(x, y)$. $C(x, y)$ is after all a function of two independent variables. The $\frac{dC}{dx}$ notation will be reserved for functions that depend on x alone. Instead, the notation used is $\frac{\partial C}{\partial x}$, and this is called the **partial derivative of C with respect to x**. To see what this implies, we return to

$$C(x, y) = \frac{x^3}{300} + \frac{y^3}{50} + 10x + 30y + 2xy$$

Now, think of y as some constant number whose specific value is unknown. To find $\frac{\partial C}{\partial x}(x, y)$, we compute the derivative of $C(x, y)$ with respect to x while holding y constant.

$$\frac{\partial C}{\partial x}(x, y) = \frac{x^2}{100} + 0 + 10 + 0 + 2y$$

Thus,

$$\frac{\partial C}{\partial x}(30, 10) = 9 + 10 + 20 = 39$$

as before.

For contrast, now we find $\frac{\partial C}{\partial y}(x, y)$.

$$\frac{\partial C}{\partial y}(x, y) = 0 + \frac{3y^2}{50} + 0 + 30 + 2x$$

$$\frac{\partial C}{\partial y}(30, 10) = 6 + 30 + 60 = 96$$

If x is held fixed at 30, and y is increased from 10 to 11, the total costs to the company will increase approximately 96 units, from 1310 to approximately 1406.

Although we do not have a derivative of $C(x, y)$, we do have two partial derivatives. The notation $C_x(x, y)$ is also widely used to mean the same as $\frac{\partial C}{\partial x}(x, y)$. Similarly, $C_y(x, y)$ means the same as $\frac{\partial C}{\partial y}(x, y)$. Since both notations are commonly used, you will need to be familiar with both, and we will use both.

In summary,

> Let f be a function of two independent variables x and y. The **partial derivative of f with respect to x** is found by keeping y constant and differentiating with respect to x, and is denoted by
>
> $$\frac{\partial f}{\partial x} \quad \text{or} \quad f_x$$
>
> The **partial derivative of f with respect to y** is computed by keeping x constant and differentiating with respect to y, and is denoted by
>
> $$\frac{\partial f}{\partial y} \quad \text{or} \quad f_y$$

Example 1 *(Compare Exercise 1)*

If $f(x, y) = 4x^2y^3$, compute $\frac{\partial f}{\partial x}$ and $\frac{\partial f}{\partial y}$.

Solution Keeping y constant and differentiating with respect to x,

$$\frac{\partial f}{\partial x}(x, y) = 4(2x)y^3$$
$$= 8xy^3$$

Keeping x constant and differentiating with respect to y,

$$\frac{\partial f}{\partial y}(x, y) = 4x^2(3y^2)$$
$$= 12x^2y^2$$

Example 2 (*Compare Exercise 5*)
Compute f_x and f_y if $f(x, y) = 4x^2y^3 + e^{3xy^2}$

Solution Holding y fixed and differentiating with respect to x,

$$f_x(x, y) = 8xy^3 + 3y^2e^{3xy^2}$$

Next, hold x fixed and differentiate with respect to y:

$$f_y(x, y) = 12x^2y^2 + 6xye^{3xy^2}$$

Notice that the chain rule carries over to several variables the same way the other differentiation rules do. In particular,

$$\text{if } f(x, y) = e^{g(x,y)}, \qquad \text{then} \qquad f_x(x, y) = e^{g(x,y)} \cdot g_x(x, y)$$

Example 3 shows another instance when the chain rule must be used.

Example 3 Compute f_x and f_y if $f(x, y) = \ln(x^2 + 4y^3)$.

Solution

$$f_x(x, y) = \frac{1}{x^2 + 4y^3}\,2x = \frac{2x}{x^2 + 4y^3}$$

$$f_y(x, y) = \frac{1}{x^2 + 4y^3}\,12y^2 = \frac{12y^2}{x^2 + 4y^3}$$

Another commonly used notation parallels the $\dfrac{dy}{dx}$ notation used in the one variable case. If we let $z = f(x, y)$, then $\dfrac{\partial z}{\partial x} = \dfrac{\partial f}{\partial x}$ and $\dfrac{\partial z}{\partial y} = \dfrac{\partial f}{\partial y}$.

Example 4 (*Compare Exercise 3*)
Compute $\dfrac{\partial z}{\partial x}$ and $\dfrac{\partial z}{\partial y}$ for the function

$$z = 2x^2y^2 - 4xy + y^2$$

at the point (1, 2).

Solution Holding y constant and differentiating with respect to x,

$$\frac{\partial z}{\partial x} = 4xy^2 - 4y$$

At the point (1, 2), $\dfrac{\partial z}{\partial x}(1, 2) = 4(1)(2)^2 - 4(2) = 8$. Keeping x constant and differentiating with respect to y,

$$\frac{\partial z}{\partial y} = 4x^2y - 4x + 2y$$

At the point (1, 2), $\dfrac{\partial z}{\partial y}(1, 2) = 4(1)^2(2) - 4(1) + 2(2) = 8$.

These concepts can be extended to functions of more than two variables. If a function f depends on the three variables x, y, and z, then $\frac{\partial f}{\partial x}$ is obtained by differentiating f with respect to x, keeping y and z constant; $\frac{\partial f}{\partial y}$ and $\frac{\partial f}{\partial z}$ are defined analogously.

Example 5 (*Compare Exercise 7*)
Find $\frac{\partial f}{\partial x}$, $\frac{\partial f}{\partial y}$, and $\frac{\partial f}{\partial z}$ for

$$f(x, y, z) = 2x^2yz - x^4z^3 + 2y^4$$

Solution Keeping y and z constant and differentiating with respect to x,

$$\frac{\partial f}{\partial x}(x, y, z) = 4xyz - 4x^3z^3$$

Keeping x and z constant,

$$\frac{\partial f}{\partial y}(x, y, z) = 2x^2z + 8y^3$$

Keeping x and y constant,

$$\frac{\partial f}{\partial z}(x, y, z) = 2x^2y - 3x^4z^2$$

Example 6 (*Compare Exercise 9*)
If $w = \sqrt{x^2 + yz}$, find $\frac{\partial w}{\partial x}$, $\frac{\partial w}{\partial y}$, and $\frac{\partial w}{\partial z}$.

Solution Rewrite $w = (x^2 + yz)^{1/2}$. Now treat y and z as constants and differentiate with respect to x.

$$\frac{\partial w}{\partial x} = \frac{1}{2}(x^2 + yz)^{-1/2} \cdot (2x + 0) = \frac{x}{\sqrt{x^2 + yz}}$$

In a similar fashion,

$$\frac{\partial w}{\partial y} = \frac{1}{2}(x^2 + yz)^{-1/2}(0 + z) = \frac{z}{2\sqrt{x^2 + yz}}$$

and

$$\frac{\partial w}{\partial z} = \frac{1}{2}(x^2 + yz)^{-1/2}(0 + y) = \frac{y}{2\sqrt{x^2 + yz}}$$

Higher Order Partial Derivatives

Remember that the second derivative was important in finding the relative extreme values of a given function of one variable. What is the analogous tool for functions of two or more variables? Notice that if f is a function of two variables, so is $\frac{\partial f}{\partial x}$. Thus, we can take partial derivatives of the function $\frac{\partial f}{\partial x}$, and keeping the terminology as familar as possible, we say that a partial derivative of $\frac{\partial f}{\partial x}$ is a second partial derivative of f. Just as there are two partial derivatives of f, there are two partial derivatives of $\frac{\partial f}{\partial x}$. We have

$$\frac{\partial}{\partial x}\left(\frac{\partial f}{\partial x}\right) \quad \text{and} \quad \frac{\partial}{\partial y}\left(\frac{\partial f}{\partial x}\right)$$

Again, we introduce some less cumbersome notation

$$\frac{\partial}{\partial x}\left(\frac{\partial f}{\partial x}\right) \text{ is written as } \frac{\partial^2 f}{\partial x^2} \text{ or as } f_{xx}$$

$$\frac{\partial}{\partial y}\left(\frac{\partial f}{\partial x}\right) \text{ is written as } \frac{\partial^2 f}{\partial y \partial x} \text{ or as } f_{xy}$$

Similarly,

$$\frac{\partial}{\partial x}\left(\frac{\partial f}{\partial y}\right) = \frac{\partial^2 f}{\partial x \partial y} = f_{yx}$$

and

$$\frac{\partial}{\partial y}\left(\frac{\partial f}{\partial y}\right) = \frac{\partial^2 f}{\partial y^2} = f_{yy}$$

Example 7 (*Compare Exercise 11*)

Let $f(x, y) = 3x^2 + 4xy^3 + 5y$. Find $\frac{\partial^2 f}{\partial x^2}, \frac{\partial^2 f}{\partial y \partial x}, \frac{\partial^2 f}{\partial x \partial y}$, and $\frac{\partial^2 f}{\partial y^2}$.

Solution First, we have to find $\frac{\partial f}{\partial x}$ and $\frac{\partial f}{\partial y}$.

$$\frac{\partial f}{\partial x}(x, y) = 6x + 4y^3$$

$$\frac{\partial f}{\partial y}(x, y) = 12xy^2 + 5$$

To get $\frac{\partial^2 f}{\partial x^2}$, we differentiate $\frac{\partial f}{\partial x}$ with respect to x.

$$\frac{\partial^2 f}{\partial x^2}(x, y) = \frac{\partial}{\partial x}(6x + 4y^3) = 6$$

To find $\frac{\partial^2 f}{\partial y \partial x}$, differentiate $\frac{\partial f}{\partial x}$ with respect to y.

$$\frac{\partial^2 f}{\partial y \partial x}(x, y) = \frac{\partial}{\partial y}(6x + 4y^3) = 12y^2$$

Next, find $\frac{\partial^2 f}{\partial x \partial y}$ by differentiating $\frac{\partial f}{\partial y}$ with respect to x.

$$\frac{\partial^2 f}{\partial x \partial y}(x, y) = \frac{\partial}{\partial x}(12xy^2 + 5) = 12y^2$$

Finally,

$$\frac{\partial^2 f}{\partial y^2}(x, y) = \frac{\partial}{\partial y}(12xy^2 + 5) = 24xy$$

Observe that in Example 8, $\frac{\partial^2 f}{\partial y \partial x}$ and $\frac{\partial^2 f}{\partial x \partial y}$ are the same. This is no accident; there is a theorem guaranteeing that if either $\frac{\partial^2 f}{\partial x \partial y}$ or $\frac{\partial^2 f}{\partial y \partial x}$ is continuous, then they both are continuous, and further, that $\frac{\partial^2 f}{\partial x \partial y} = \frac{\partial^2 f}{\partial y \partial x}$. Because of this theorem, we didn't stress the ordering of $\partial x \partial y$ or $\partial y \partial x$ when writing the second partials. The ordering makes a theoretical difference, but not a practical difference for all the functions that we will encounter. $\left(\text{Although there are functions where } \frac{\partial^2 f}{\partial x \partial y} \neq \frac{\partial^2 f}{\partial y \partial x}.\right)$

One can continue taking partial derivatives of higher order than two, just as we did in the case of a single variable. We will not do so here however.

Geometry of Partial Derivatives

The derivative of a function of one variable has a geometrical interpretation; $f'(x)$ is the slope of the line tangent to the graph at the point $(x, f(x))$. A partial derivative of a function of two variables also has a geometrical interpretation as the slope of a tangent line. We illustrate this geometrical interpretation in terms of the function $f(x, y) = x^2 + y^2$. From the previous section, we know that the graph of this function looks like Figure 8–11, on page 474.

The partial derivatives of this function are

$$f_x(x, y) = 2x \qquad \text{and} \qquad f_y(x, y) = 2y$$

Consider the point P on the graph above $(3, 4, 0)$. The third coordinate of P is

$$f(3, 4) = z = 3^2 + 4^2 = 25$$
$$P = (3, 4, 25)$$

The corresponding partial derivatives are

$$f_x(3, 4) = 2 \cdot 3 = 6 \qquad \text{and} \qquad f_y(3, 4) = 2 \cdot 4 = 8$$

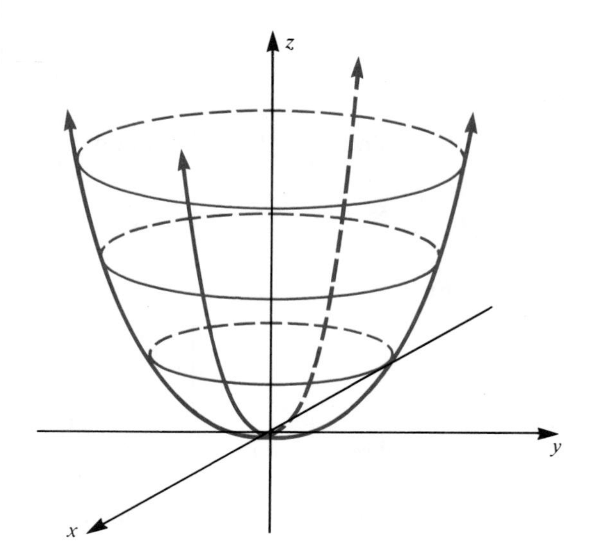

Figure 8–11

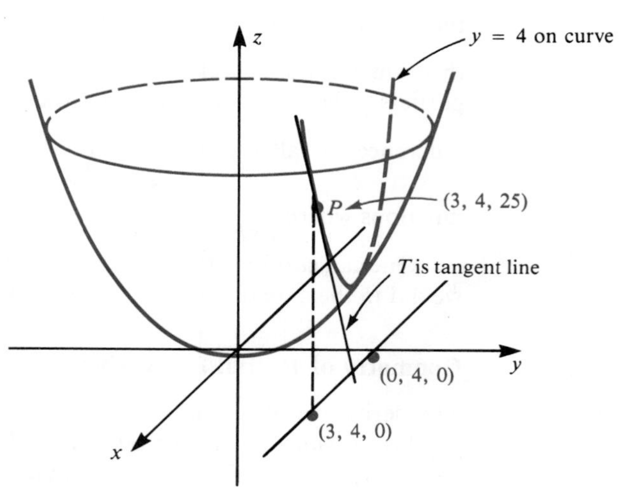

Figure 8–12

We now give geometrical significance to the two numbers 6 and 8. See Figure 8–12.

Consider the curve on the surface, passing through P, that lies in a plane parallel to the xz-plane. On such a curve, all the points have the same second coordinate, namely $y = 4$. Points on this curve are of the form $(x, 4, z)$, and the change in z is with respect to x only. $f_x(x, 4)$ is the slope in the plane $y = 4$ of the line tangent to this curve at an arbitrary point $(x, 4, z)$. Thus, $f_x(3, 4) = 6$ tells us that the slope of the line tangent to the curve at the point P is 6. We label this tangent line T in Figure 8–12.

Similarly, $f_y(3, 4)$ is the slope of the line that is tangent to the surface at P and that lies in the plane through P parallel to the yz-plane. See Figure 8–13.

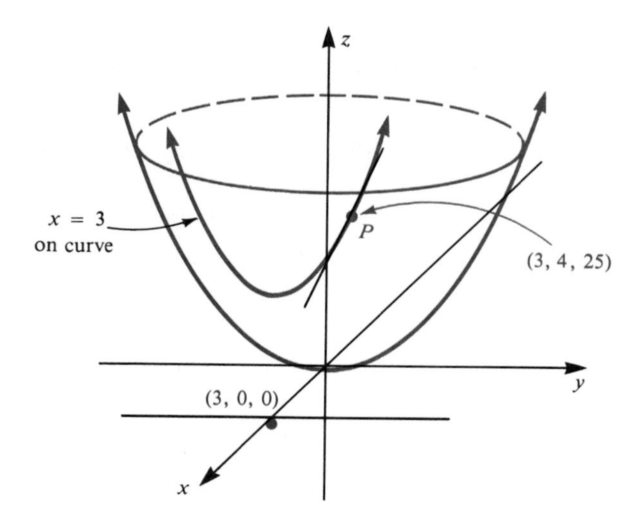

Figure 8–13

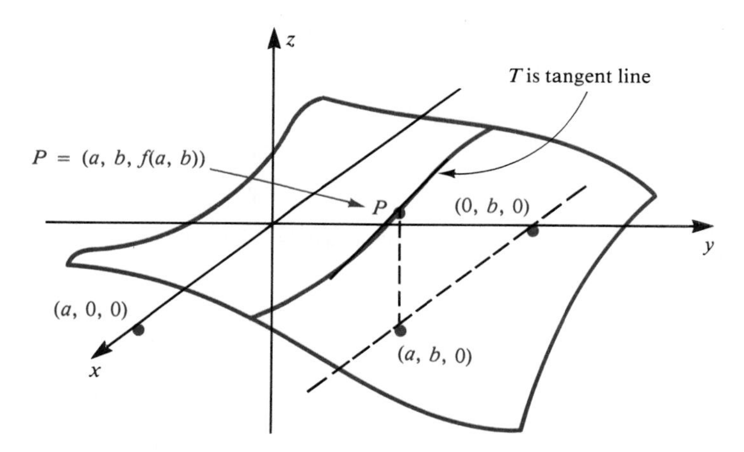

Figure 8–14

A general picture is given in Figure 8–14.

Let C be the curve on the surface through P that lies in the plane $y = b$ parallel to the xz-plane. The slope of the tangent T to this curve at P is $f_x(a, b)$.

Marginal Analysis in Several Variables

In the case of a single variable, the slope of the tangent line is the rate of change of the function. We have just seen that we can carry over to several variables this geometric relation between slope and rate of change. Now we see how marginal analysis can also be carried over to functions of more than one variable. We introduced the class of **Cobb-Douglas production functions** in the last section; they are functions of the type

$$f(x, y) = Cx^a y^{1-a}$$

where C and a are appropriate constants.

Generally, x is the amount of labor, and y is the amount of capital. The functions $\frac{\partial f}{\partial x}$ and $\frac{\partial f}{\partial y}$ are called the **marginal productivity of labor** and the **marginal productivity of capital**, respectively. $\frac{\partial f}{\partial x}$ and $\frac{\partial f}{\partial y}$ evaluated at a point (x, y) are interpreted as follows: $\frac{\partial f}{\partial x}(x, y)$ is approximately the additional output obtained on raising labor from x to $(x + 1)$ units, while capital is held constant at y units. $\frac{\partial f}{\partial y}(x, y)$ is approximately the additional output obtained on raising capital from y to $(y + 1)$ units, while labor is held constant at x units.

Example 8 (*Compare Exercise 25*)
If a company's production function is given by

$$f(x, y) = 9x^{2/3}y^{1/3}$$

find the marginal productivities of labor and capital when $x = 64$ and $y = 27$.

Solution

$$\frac{\partial f}{\partial x}(x, y) = 9\left(\frac{2}{3}\right)x^{-1/3}y^{1/3} = \frac{6y^{1/3}}{x^{1/3}}$$

$$\frac{\partial f}{\partial y}(x, y) = 9\left(\frac{1}{3}\right)x^{2/3}y^{-2/3} = \frac{3x^{2/3}}{y^{2/3}}$$

When $x = 64$ and $y = 27$, we get

$$\frac{\partial f}{\partial x}(64, 27) = \frac{6(27)^{1/3}}{(64)^{1/3}} = \frac{6(3)}{4} = \frac{9}{2}$$

$$\frac{\partial f}{\partial y}(64, 27) = \frac{3(64)^{2/3}}{(27)^{2/3}} = \frac{3(16)}{9} = \frac{16}{3}$$

Thus, raising labor by one unit from 64 to 65 while maintaining capital at 27 units would increase output by about $4\frac{1}{2}$ units. On the other hand, raising capital by one unit, from 27 to 28, while maintaining the level of labor at 64 units would increase output by approximately $5\frac{1}{3}$ units.

8–2 Exercises

I.

1. (*See Example 1*) If $f(x, y) = x^2 + 3xy$, compute $\frac{\partial f}{\partial y}$ and $\frac{\partial f}{\partial x}$.
2. If $f(x, y) = x^2y - y^3$, compute $\frac{\partial f}{\partial y}$ and $\frac{\partial f}{\partial x}$.

3. (*See Example 4*) If $z = 5x^2y^2 + 2y$, compute $\dfrac{\partial z}{\partial y}$ and $\dfrac{\partial z}{\partial x}$.

4. Compute $\dfrac{\partial z}{\partial y}$ and $\dfrac{\partial z}{\partial x}$ for $z = e^{4x} + yx^3$.

5. (*See Example 2*) If $f(x, y) = \dfrac{x}{x^2 + y^2}$, compute f_x and f_y.

6. If $f(x, y) = x \ln|y|$, compute f_x and f_y.

7. (*See Example 5*) If $f(x, y, z) = \dfrac{x^2y}{z^3}$, compute $\dfrac{\partial f}{\partial x}$, $\dfrac{\partial f}{\partial y}$, and $\dfrac{\partial f}{\partial z}$.

8. If $f(x, y, z) = e^x + y^2$, compute $\dfrac{\partial f}{\partial x}$, $\dfrac{\partial f}{\partial y}$, and $\dfrac{\partial f}{\partial z}$.

9. (*See Example 6*) For $w = \ln|xy + yz^2|$, compute $\dfrac{\partial w}{\partial x}$, $\dfrac{\partial w}{\partial y}$, and $\dfrac{\partial w}{\partial z}$.

10. Compute $\dfrac{\partial w}{\partial x}$, $\dfrac{\partial w}{\partial y}$, and $\dfrac{\partial w}{\partial z}$ if $w = xy^2z^3$.

II.

Compute $\dfrac{\partial f}{\partial y}$, $\dfrac{\partial f}{\partial x}$, $\dfrac{\partial^2 f}{\partial x^2}$, $\dfrac{\partial^2 f}{\partial y \partial x}$, $\dfrac{\partial^2 f}{\partial x \partial y}$, and $\dfrac{\partial^2 f}{\partial y^2}$ for each of the functions in Exercises 11 through 18. Observe that $\dfrac{\partial^2 f}{\partial y \partial x} = \dfrac{\partial^2 f}{\partial x \partial y}$ for these functions.

11. (*See Example 7*) $f(x, y) = 2x^3y$

12. $f(x, y) = x^2 + 2y^2$

13. $f(x, y) = 2x^2y - 3xy^2$

14. $f(x, y) = 4x^2y - 3x^3 - y^4 + 2$

15. $f(x, y) = 2e^{x^2+y}$

16. $f(x, y) = x^2 \ln(xy)$

17. $f(x, y) = 2x + e^{x^2y}$

18. $f(x, y) = y \ln(x^2 + 3y)$

Calculate f_x, f_y, f_{xx}, f_{yy}, and f_{xy} at the point p for each of the functions in Exercises 19 through 24.

19. $f(x, y) = x^2y$; $p = (2, 3)$

20. $f(x, y) = 3x^2 + 2xy$; $p = (1, 1)$

21. $f(x, y) = xy - 4xy^3$; $p = (-1, 0)$

22. $f(x, y) = 4xy^2 - xy + 2$; $p = (1, 2)$

23. $f(x, y) = x \ln(xy)$; $p = (l, 1)$

24. $f(x, y) = x^2e^{(x^2-y^2)}$; $p = (2, 3)$

III.

25. (*See Example 8*) The manufacturing process of a company is described by the Cobb-Douglas production function $f(x, y) = 10x^{1/4}y^{3/4}$.

    (a) Determine the number of units manufactured when 16 units of labor are used and 625 units of capital are invested.

(b) Find the marginal productivity of labor and the marginal productivity of capital.

(c) What are the values of these marginal functions when 16 units of labor and 625 units of capital are being used?

(d) What is the approximate change in the production level if the use of labor is increased from 16 units to 17 units while the level of capital used is held constant at 625 units?

(e) What is the approximate change in the production level if the number of units of labor is held constant at 16, but the amount of capital used is *decreased* from 625 units to 624 units?

26. A company manufactures two products, X and Y. The total cost incurred on manufacturing x units of X and y units of Y is

$$C(x, y) = \frac{x^3}{900} + \frac{y^3}{600} + 60x + 40y + 20xy$$

(a) What is the total cost of manufacturing 90 units of X and 120 units of Y?

(b) Compute $\frac{\partial C}{\partial x}$ (the marginal cost of X) and $\frac{\partial C}{\partial y}$ (the marginal cost of Y).

(c) Evaluate $\frac{\partial C}{\partial x}(90, 120)$ and $\frac{\partial C}{\partial y}(90, 120)$.

(d) What will be the approximate change in costs if the company increases the production of X from 90 units to 91 while holding the production of Y fixed at 120?

(e) What is the approximate change in production costs if the company *decreases* the production of Y from 120 units to 118 units while holding the production of X fixed at 90 units?

27. The total revenue function of a company that sells x units of trash compactors and y units of garbage disposals is

$$R(x, y) = -3x^2 + 400x - 2y^2 + 500y$$

(a) What is the total revenue if $x = 10$ and $y = 15$?

(b) Compute $\frac{\partial R}{\partial x}$ and $\frac{\partial R}{\partial y}$.

(c) Evaluate $\frac{\partial R}{\partial x}(10, 15)$ and $\frac{\partial R}{\partial y}(10, 15)$.

(d) How will the company's revenue change if x increases from 10 to 11 while y remains 15?

(e) How will the company's revenue change if y decreases from 15 to 14 while x remains 10?

8–3 Maxima and Minima of Functions of Two Variables

Extreme Values and Critical Points
The Second Derivative Test for Functions of Two Variables
An Application

Derivatives helped us find relative extreme values of a function of one variable. In this section, we will see how partial derivatives help us find relative extreme values for a function of two variables. We will restrict our discussion here to nice functions of two variables. "Nice" means that all the second order partial derivatives exist and are continuous. We stay with two variables for simplicity. There are analogous methods for functions of more than two variables, but these methods are more complicated.

Extreme Values and Critical Points

A function f has a **relative maximum** at the point (a, b) if $f(a, b) \geq f(x, y)$ for all points (x, y) "near" (a, b). A **relative minimum** is defined analogously. We can visualize maxima and minima of functions of two variables geometrically, as shown in Figure 8–15. This time the graph gives you more than the outline of a mountain; now you can see the whole mountain.

Again, by letting only one coordinate vary at a time, we try to see how much of our knowledge of a single variable we can apply to several variables. We begin by holding the second variable fixed and supposing that $f(a, b)$ is a relative maximum. Then, $f(a, b) \geq f(x, y)$ for all (x, y) near (a, b); this in turn implies that $f(a, b) \geq f(x, b)$ for all x near a. Thus, $f(x, b)$ has a relative maximum when $x = a$. We know from the calculus of a single variable that this means that $\frac{\partial f}{\partial x}(a, b) = 0$. Similarly, $f(a, y)$ will have a relative maximum when $y = b$, so $\frac{\partial f}{\partial y}(a, b) = 0$. Geometrically, the highest point on a north-south trail across the

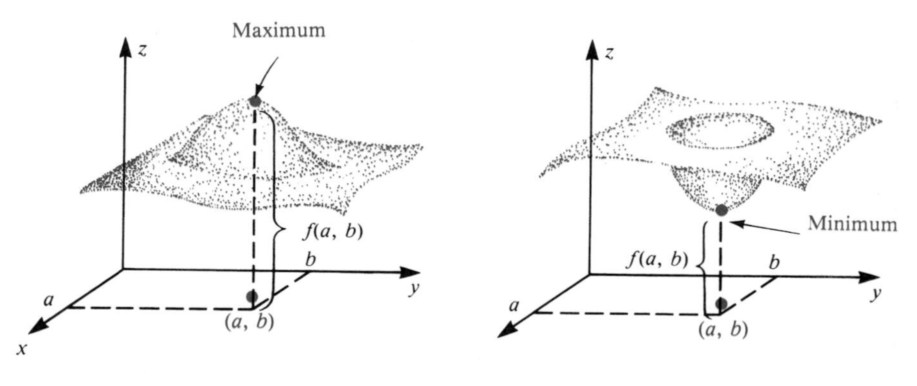

Figure 8–15 (i) (ii)

top of a mountain will be at the top of the mountain, and the same is true of an east-west trail. Since relative extrema occur when both partial derivatives equal 0, we make a definition similar to the one we made in the one variable case.

Definition

The point (a, b) is called a **critical point** of the function f if *both* $\dfrac{\partial f}{\partial x}(a, b) = 0$ and $\dfrac{\partial f}{\partial y}(a, b) = 0$.

Example 1 (*Compare Exercise 1*)

Find the critical points of the function defined by

$$f(x, y) = x^2 + 2y^2 - 2x + 8y + 12$$

Solution

$$f_x(x, y) = 2x - 2$$
$$f_y(x, y) = 4y + 8$$

We must have both $2x - 2 = 0$ and $4y + 8 = 0$. The first equation tells us that $x = 1$, and the second says that $y = -2$. There is only one critical point, $(1, -2)$.

Example 2 (*Compare Exercise 3*)

Find the critical points of

$$f(x, y) = 2x^3 - 3x^2 + 12y^2 - 12xy - 18x + 10$$

Solution

$$f_x(x, y) = 6x^2 - 6x - 12y - 18$$
$$f_y(x, y) = 24y - 12x$$

At the critical point, the two equations $f_x = 0$ and $f_y = 0$ must be true simultaneously. We have to solve the system of equations

$$6x^2 - 6x - 12y - 18 = 0 \qquad (1)$$

and

$$24y - 12x = 0 \qquad (2)$$

From equation (2), we have $24y = 12x$ or $y = \frac{1}{2}x$. We substitute $\frac{1}{2}x$ for y in equation (1), getting

$$6x^2 - 6x - 12\left(\frac{1}{2}x\right) - 18 = 0$$
$$6x^2 - 12x - 18 = 0$$
$$x^2 - 2x - 3 = 0$$
$$(x - 3)(x + 1) = 0$$

The solutions are $x = 3$ and $x = -1$. Since $y = \frac{1}{2}x$, when $x = 3$, $y = \frac{3}{2}$ and when $x = -1$, $y = -\frac{1}{2}$.

The two critical points are $(3, \frac{3}{2})$ and $(-1, -\frac{1}{2})$.

The relation between the critical points of f and the points where f has a relative extreme value is the same for functions of several variables as it is for functions of one variable. The critical points are the candidates for finding relative extrema, but (a, b) may be a critical point for f even though $f(a, b)$ is not a relative extreme value.

Example 3 Show that $f(x, y) = 2 - (x - 1)^2 + (y - 3)^2$ does not have a relative extreme value at its critical point.

Solution

$$f_x(x, y) = -2(x - 1) \qquad \text{and} \qquad f_y(x, y) = 2(y - 3)$$

$f_x = 0$ when $x = 1$, and $f_y = 0$ when $y = 3$. The point $(1, 3)$ is the only critical point, and we now show that f does not have a relative extreme value at $(1, 3)$. First, $f(1, 3) = 2$. Next, if $(x, 3)$ is close to $(1, 3)$ but not equal to $(1, 3)$, then $f(x, 3) = 2 - (x - 1)^2 < 2$; so, $f(1, 3) = 2$ cannot be a relative minimum. Furthermore, if $(1, y)$ is close to $(1, 3)$ but not equal to $(1, 3)$, then $f(1, y) = 2 + (y - 3)^2 > 2$; so, $f(1, 3) = 2$ cannot be a relative maximum.

The only critical point of f is $(1, 3)$, and f does not have a relative extreme value at $(1, 3)$.

Figure 8–16 shows what the graph of f looks like. As we saw in Example 3, $f(x, 3)$ has a relative maximum when $x = 1$, but $f(1, y)$ has a relative minimum when $y = 3$. The graph of f bends up in one direction and down in the other. The graph of f is shaped somewhat like a saddle above $(1, 3)$, so the point $(1, 3)$ is called a **saddle point** of the function.

The task of classifying critical points is more complicated in the several variable case than with just one variable, as is seen in the second derivative test for two variables.

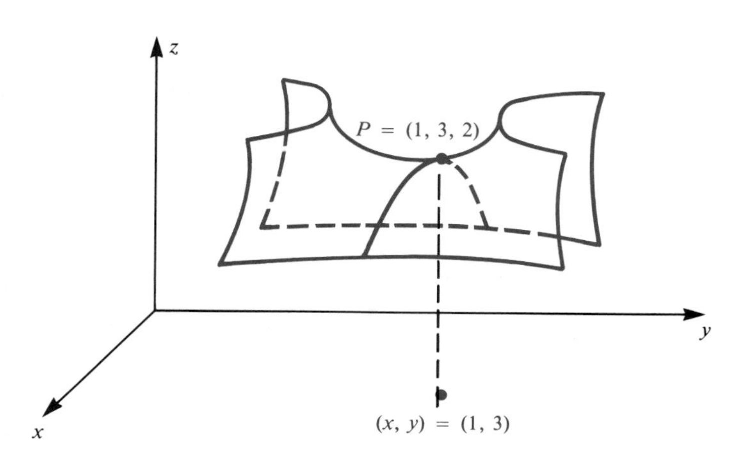

Figure 8–16

The Second Derivative Test for Functions of Two Variables

Let (a, b) be a critical point.

Define

$$M(x, y) = f_{xx}(x, y)f_{yy}(x, y) - [f_{xy}(x, y)]^2$$

1. If $M(a, b) > 0$, then $f(a, b)$ is a relative extreme value.
   i) if $f_{xx}(a, b) < 0$, then $f(a, b)$ is a relative maximum.
   ii) if $f_{xx}(a, b) > 0$, then $f(a, b)$ is a relative minimum.
2. If $M(a, b) < 0$, then $f(a, b)$ is neither a relative maximum nor a relative minimum.
3. If $M(a, b) = 0$, this test gives no information.

Case 1 indicates (i) a relative maximum or (ii) a relative minimum as illustrated in Figure 8–15.

Case 2 indicates a saddle-shaped graph as in Figure 8–16.

In case 3, that is, if $M(a, b) = 0$, we have to resort to other techniques to find out if $f(a, b)$ is a maximum or a minimum. This text will not go further into case 3.

The following examples show how this test works.

Example 4 (*Compare Exercise 7*)
Find the relative maxima and minima of

$$f(x, y) = x^2 + 2y^2 - 2x + 8y + 12$$

Solution This is the same function as in Example 1. The first partial derivatives are

$$f_x(x, y) = 2x - 2 \qquad \text{and} \qquad f_y(x, y) = 4y + 8$$

We have already seen that there is only one critical point, namely $(1, -2)$. Now, we use the second derivatives to determine whether this critical point leads to a relative maximum or minimum value of f. The second derivatives are

$$f_{xx}(x, y) = 2, \qquad f_{yy}(x, y) = 4, \qquad f_{xy}(x, y) = 0$$

These functions are all constant, so

$$M(1, -2) = (2)(4) - (0)^2 = 8$$

$M(1, -2) > 0$, so part 1 of the test says that $f(1, -2)$ *is* a relative extremum. Now we find out which kind.

$$f_{xx}(1, -2) = 2$$

$f_{xx}(1, -2) > 0$, so (ii) tells us that f has a relative minimum at the point $(1, -2)$. This minimum value is

$$f(1, -2) = (1)^2 + 2(-2)^2 - 2(1) + 8(-2) + 12 = 3$$

Example 5 (*Compare Exercise 13*)
Find the relative maxima and minima of

$$f(x, y) = 2x^3 - 3x^2 + 12y^2 - 12xy - 18x + 10$$

Solution This is the function from Example 2. We already have

$$f_x(x, y) = 6x^2 - 6x - 12y - 18$$
$$f_y(x, y) = 24y - 12x$$

and two critical points, $(3, \frac{3}{2})$ and $(-1, -\frac{1}{2})$.

Next,
$$f_{xx}(x, y) = 12x - 6$$
$$f_{xy}(x, y) = -12$$

and
$$f_{yy}(x, y) = 24$$

1. Test $(3, \frac{3}{2})$

$$f_{xx}(3, \tfrac{3}{2}) = 30 \qquad f_{xy}(3, \tfrac{3}{2}) = -12 \qquad f_{yy}(3, \tfrac{3}{2}) = 24$$
$$M(3, \tfrac{3}{2}) = (30)(24) - (-12)^2$$
$$= 720 - 144 = 576 > 0$$

Thus $f(3, \frac{3}{2})$ *is* a relative extremum.

Next $f_{xx}(3, \frac{3}{2}) = 30 > 0$

so $f(3, \frac{3}{2})$ is a relative minimum

2. Test $(-1, -\frac{1}{2})$

$$f_{xx}(-1, -\tfrac{1}{2}) = -18;\ f_{xy}(-1, -\tfrac{1}{2}) = -12;\ f_{yy}(-1, -\tfrac{1}{2}) = 24$$
$$M(-1, -\tfrac{1}{2}) = (-18)(24) - (-12)^2$$
$$= -432 - 144 = -576 < 0$$

We are in case 2, and there is no need to do any more calculation. $f(-1, -\frac{1}{2})$ *is not* a relative extreme value.

An Application

Generally speaking, the demand for a certain commodity and the price of that commodity are related. Moreover, that relationship often involves other factors. For example, the demand for a certain company's computers will depend not only on the price of the computer, but also on the price that the company charges for supporting services and software.

Example 6 (*Compare Exercise 15*)
Suppose that a firm charges x for its computer and y for a certain package of programs, where x and y are in thousands of dollars. The firm estimates its

revenue function as

$$R(x, y) = -2x^2 + 40x - 4y^2 + 48y$$

Find the values of x and y which lead to the maximum revenue for the company.

Solution The partial derivatives of R are

$$R_x(x, y) = -4x + 40 \qquad \text{and} \qquad R_y(x, y) = -8y + 48$$

The critical points must simultaneously satisfy the equations

$$-4x + 40 = 0 \qquad \text{and} \qquad -8y + 48 = 0$$

There is a single critical point, $x = 10$, $y = 6$. Now, we use the second derivative test at this critical point. The second derivatives are

$$R_{xx}(x, y) = -4 \qquad R_{yy}(x, y) = -8 \qquad R_{xy}(x, y) = 0$$

Next, evaluate $M(10, 6)$

$$M(10, 6) = (-4)(-8) - (0)^2 = 32$$

Thus, $M(10, 6) > 0$ and R does have a relative extreme value at $(10, 6)$. Furthermore, $R_{xx}(10, 6) = -4 < 0$, so R has a relative maximum at $x = 10$, $y = 6$.

There is no "single critical point test" for several variables as there was in the one variable situation. However, it is true that if a function has the form $f(x, y) = ax^2 + bx + cy^2 + dy + e$ with both a and c negative, then the relative maximum is in fact the maximum value of the function. So in this case, we can conclude that the company will maximize its revenue by charging \$10,000 for the computer and \$6000 for its software.

8–3 Exercises

I.

Find the critical points for the functions defined in Exercises 1 through 6.

1. (*See Example 1*) $f(x, y) = x^2 + y^2 + 3$
2. $f(x, y) = 2x^2 + y^2 + 2x$
3. (*See Example 2*) $f(x, y) = x^2 + xy + y^2$
4. $f(x, y) = 3xy - x^2 - y^2$
5. $f(x, y) = 4x^3 + y^2 - 12x^2 - 36x$
6. $f(x, y) = x^2 - 2x + xy$

II.

Find the critical points for each function defined in Exercises 7 through 14. Then, use the second derivative test to determine if the function has a relative extreme value at the critical point. If f does have a relative extremum, determine whether it is a relative maximum or minimum.

7. (*See Example 4*) $f(x, y) = x^2 + 2y^2 - 4x + 4y - 3$
8. $f(x, y) = -2x^2 - 2xy - y^2 + 4x + 2y$
9. $f(x, y) = x^2 - xy + y^2 + 2x + 2y + 3$
10. $f(x, y) = x^2 + xy + y^2 - 4x - 5y$
11. $f(x, y) = y^3 + x^2 - 6xy + 3x + 6y - 7$
12. $f(x, y) = x^3 + 3x^2 + y^2 - 4y + 3$
13. (*See Example 5*) $f(x, y) = 2x^3 + 2x^2 - 6xy + 3y^2 - 4x + 5$
14. $f(x, y) = 2x^4 + y^4 - x^2 - 2y^2$

III.

15. (*See Example 6*) A company manufactures two products, X and Y, each affecting the demand and supply of the other. Let x be the price of X, and let y be the price of Y. The demand D for product X depends on both and x and y and is given by

$$D(x, y) = -x^2 + 20x - 6y^2 + 60y + 550$$

Determine the prices x and y that maximize this demand D.

16. A firm produces two products, X and Y, each affecting the supply of the other. The supply S for product X is the following function of the prices, x of X and y of Y.

$$S(x, y) = -2x^2 + 400x - y^2 + 480y + 700$$

Find the prices x and y that maximize the supply of X.

17. A manufacturing company produces two items, X and Y. When its resources are divided between producing x units of X and y units of Y, then its daily profit function is given by

$$P(x, y) = -x^2 + 300x - 2y^2 + 400y + 8000$$

Determine the daily production of items X and Y that leads to maximum profit.

18. The U.S. postal service puts a limit to the size of a package that can be sent by mail. The sum of the length and girth of the package must be less than or equal to 100 inches. Determine the dimensions of the box with the largest volume that can be sent by mail.

19. The number of shutdowns N (either intentional or due to malfunctions) of a certain machine in an industrial plant is a function of the number of replacements made of two components X and Y. If x replacements of X and y replacements of Y are made per month, the number of shutdowns is

$$N(x, y) = 2x^3 - 12x^2 + 2y^3 - 18y^2 + 290$$

How many replacements of each component should be made per month to minimize the number of shutdowns?

20. A company plans to manufacture large rectangular storage containers. These containers are to have a volume of 1250 cubic feet. The cost of the material to be used is \$4 per square foot for the base, \$2 per square foot for the sides, and \$1 per square foot for the top. Find the dimensions of the container that will minimize cost.

8–4 Lagrange Multipliers

Introduction to Constraints
The Method of Lagrange Multipliers for $f(x, y)$
The Method of Lagrange Multipliers for $f(x, y, z)$

Introduction to Constraints

In the previous section, we introduced the second derivative test for functions of two variables and used this test to determine the relative maxima and minima of such functions. Now we change the problem somewhat; in this section, we will show how to find the maximum or minimum of a function of x and y when these two variables are subject to a certain **constraint**. For example, we discussed earlier in this chapter the Cobb-Douglas production function, $f(x, y) = Cx^a y^{1-a}$, where C and a are constants. This function is used by economists to model the amount of a good produced when x units of labor and y units of capital go into the process.

Naturally, any manufacturer has only limited resources and must decide how to allocate these resources between labor and capital. This limitation on resources is an example of what is called a **constraint on x and y**.

Specifically, suppose that the manufacturing process of a company is described by

$$f(x, y) = 9x^{2/3}y^{1/3}$$

and that each unit of labor costs \$200 and each unit of capital costs \$300. Thus, x units of labor cost \$200x, and y units of capital cost \$300y. If \$90,000 is available for production, then the company must operate under the constraint $200x + 300y = 90{,}000$. The company will want to find the values of x and y that maximize $f(x, y) = 9x^{2/3}y^{1/3}$, subject to the constraint $200x + 300y - 90{,}000 = 0$.

We will solve this particular problem in Example 2, but will begin with an easier problem. The method used to solve this type of problem, the method of Lagrange multipliers, is named after a French mathematician, Joseph Louis Lagrange, and introduces a new variable, traditionally represented by λ (lambda), the Greek letter corresponding to our letter L. (Notice that Lagrange's last name begins with L.) We will demonstrate the method in Example 1 and then give the general procedure.

Example 1 (*Compare Exercise 1*)
Find the extreme values of $f(x, y) = x^2 + y^2$, subject to the constraint $x + 2y = 10$.

Solution First, rewrite the constraint as $x + 2y - 10 = 0$. Second, define a new function F by

$$\begin{aligned} F(x, y, \lambda) &= f(x, y) + \lambda(x + 2y - 10) \\ &= x^2 + y^2 + \lambda(x + 2y - 10) \end{aligned}$$

Third, compute the partial derivatives of this new function F. In this case, we have

$$\frac{\partial F}{\partial x}(x, y, \lambda) = 2x + \lambda$$

$$\frac{\partial F}{\partial y}(x, y, \lambda) = 2y + 2\lambda$$

$$\frac{\partial F}{\partial \lambda}(x, y, \lambda) = x + 2y - 10$$

Fourth, set each of these derivatives equal to zero and solve the corresponding system of equations. Here, we have the system

$$2x + \lambda = 0 \qquad (1)$$
$$2y + 2\lambda = 0 \qquad (2)$$
$$x + 2y - 10 = 0 \qquad (3)$$

The procedure we use to solve this particular system is a general procedure that you will be able to use in further problems.

Solve equations (1) and (2) for λ:

From (1), $\qquad \lambda = -2x$

and from (2), $\qquad \lambda = -y$

Set these expressions for λ equal to each other:

$$-y = -2x$$

and solve for either x or y. Here, we solve for y.

$$y = 2x$$

Now we have y in terms of x, so we substitute $2x$ for y in equation (3). This gives us an equation that only involves x, so we can solve for x.

$$\begin{aligned} x + 2(2x) - 10 &= 0 \\ 5x - 10 &= 0 \\ x &= 2 \end{aligned}$$

To find y, we go back to the expression for y in terms of x. We had $y = 2x$, so when $x = 2$, $y = 4$.

The solutions to this system of equations give us the candidates for the extreme values of f. Here we have only one candidate: $(x, y) = (2, 4)$.

Now, $f(2, 4) = 4 + 16 = 20$. But how do we know whether this is a minimum or maximum value of f? If we evaluate $f(x, y)$ at some other point that satisfies the constraint $x + 2y - 10 = 0$, the point $(10, 0)$ for example, we see that $f(10, 0) = 100$, so 20 cannot possibly be the maximum. But we cannot evaluate the function everywhere; we must try to analyze the situation. If (x, y) is such that $|x|$ or $|y|$ is large, then $f(x, y) = x^2 + y^2$ is also large. Thus, $f(x, y)$ does have a minimum on the line whose equation is $x + 2y - 10 = 0$, the constraint equation.

ANSWER Subject to the constraint $x + 2y - 10 = 0$, $f(x, y) = x^2 + y^2$ does not have a maximum value; the minimum value of $f(x, y)$ is $f(2, 4) = 20$.

We now outline the general method for finding the extreme values of f, subject to some constraining equation.

The Method of Lagrange Multipliers for $f(x, y)$

The Method of Lagrange Multipliers for $f(x, y)$

To find the relative extreme values of f under a constraint,

1. rewrite the constraint in the form $g(x, y) = 0$;
2. define a new function F by

$$F(x, y, \lambda) = f(x, y) + \lambda g(x, y)$$

3. compute $\frac{\partial F}{\partial x}$, $\frac{\partial F}{\partial y}$, and $\frac{\partial F}{\partial \lambda}$;
4. then, set these partial derivatives equal to 0 and solve that system of equations. The solutions to this system give you the points where f will have its extrema, subject to the constraint.
5. Finally, analyze the function and the constraint to determine whether the functional value is a maximum or minimum, or neither.

Thus, the points (x, y, λ) that satisfy the system $\frac{\partial F}{\partial x} = 0$, $\frac{\partial F}{\partial y} = 0$, and $\frac{\partial F}{\partial \lambda} = 0$ play the same role as critical points; they tell you where to look for extreme values.

Also, notice that the equation $\frac{\partial F}{\partial \lambda}(x, y, \lambda) = 0$ is the same equation as the constraint, $g(x, y) = 0$. Finally, we review the way we solved the system of equations $\frac{\partial F}{\partial x} = 0$, $\frac{\partial F}{\partial y} = 0$, and $\frac{\partial F}{\partial \lambda} = 0$. We solved for λ in equations (1) and (2). Equating these expressions for λ gave us an equation involving only x and y. We used that equation to express y in terms of x (sometimes of course, you may wish to express x in terms of y), and substituted that expression into equation (3), giving us an equation involving only x. We solved this equation for x and then found the corresponding values of y.

We return to the problem of maximizing production that led us into the discussion of Lagrange multipliers.

Example 2 (*Compare Exercise 11*)
Maximize $9x^{2/3}y^{1/3}$, subject to the constraint $200x + 300y = 90{,}000$.

Solution The function f is given by $f(x, y) = 9x^{2/3}y^{1/3}$, and we rewrite the constraint as $g(x, y) = 200x + 300y - 90{,}000 = 0$.

Next, define

$$F(x, y, \lambda) = 9x^{2/3}y^{1/3} + \lambda(200x + 300y - 90{,}000)$$

Now, we compute the partial derivatives of F.

$$\frac{\partial F}{\partial x}(x, y, \lambda) = 9\left(\frac{2}{3}\right)x^{-1/3}y^{1/3} + 200\lambda = 6x^{-1/3}y^{1/3} + 200\lambda$$

$$\frac{\partial F}{\partial y}(x, y, \lambda) = 9\left(\frac{1}{3}\right)x^{2/3}y^{-2/3} + 300\lambda = 3x^{2/3}y^{-2/3} + 300\lambda$$

$$\frac{\partial F}{\partial \lambda}(x, y, \lambda) = 200x + 300y - 90000$$

Setting these derivatives equal to zero gives the following system:

$$6x^{-1/3}y^{1/3} + 200\lambda = 0 \qquad (1)$$

$$3x^{2/3}y^{-2/3} + 300\lambda = 0 \qquad (2)$$

$$200x + 300y - 90000 = 0 \qquad (3)$$

Solving equation (1) for λ gives

$$200\,\lambda = -6x^{-1/3}y^{1/3}$$

$$\lambda = \frac{-3}{100}x^{-1/3}y^{1/3}$$

Solving equation (2) for λ gives

$$300\,\lambda = -3x^{2/3}y^{-2/3}$$

$$\lambda = \frac{-1}{100}x^{2/3}y^{-2/3}$$

Equating these expressions for λ, we have

$$\frac{-3}{100}x^{-1/3}y^{1/3} = \frac{-1}{100}x^{2/3}y^{-2/3}$$

Multiplying both sides by $-100x^{1/3}y^{2/3}$ gives

$$3y = x$$

This time, we have x in terms of y; substituting $3y$ for x in equation 3 gives

$$\begin{aligned} 200(3y) + 300y - 90{,}000 &= 0 \\ 900y &= 90{,}000 \\ y &= 100 \end{aligned}$$

Since $x = 3y$, the corresponding value of x is 300.

One candidate for maximizing production is $(x, y) = (300, 100)$. Furthermore, notice that other critical points are introduced into this example, because $\frac{\partial F}{\partial x}$ does not exist when $x = 0$, and $\frac{\partial F}{\partial y}$ does not exist when $y = 0$. Referring to the constraint $200x + 300y - 90{,}000 = 0$, $y = 300$ when $x = 0$, and $x = 450$ when $y = 0$.

$$\begin{aligned} f(0, 300) &= 0 \\ f(300, 100) &= 9(300)^{2/3}100^{1/3} \\ f(450, 0) &= 0 \end{aligned}$$

Moreover, the physical constraints from the application also require that both $x \geq 0$ and $y \geq 0$. Production will be maximized when the company uses 300 units of labor and 100 units of capital. At $200 per unit for labor and $300 per unit for capital, the resources of $90,000 should be distributed as follows: $60,000 for labor and $30,000 for capital.

The Method of Lagrange Multipliers for $f(x, y, z)$

The method of Lagrange multipliers can also be used with functions of more than two variables. We will give an example showing how it works for three variables. The procedure is essentially unchanged.

Let $f(x, y, z)$ be a function, subject to the constraint $g(x, y, z) = 0$. Define a new function $F(x, y, z, \lambda)$ by

$$F(x, y, z, \lambda) = f(x, y, z) + \lambda g(x, y, z)$$

Solve the system of equations

$$\frac{\partial F}{\partial x} = 0, \qquad \frac{\partial F}{\partial y} = 0, \qquad \frac{\partial F}{\partial z} = 0, \qquad \frac{\partial F}{\partial \lambda} = 0$$

The values of x, y, and z that give maxima or minima of f subject to the constraint are among the simultaneous solutions of these equations.

The technique for solving the system $\frac{\partial F}{\partial x} = 0$, $\frac{\partial F}{\partial y} = 0$, $\frac{\partial F}{\partial z} = 0$, $\frac{\partial F}{\partial \lambda} = 0$ is slightly more complicated than in the previous examples. We solve for λ in each of the three equations $\frac{\partial F}{\partial x} = 0$, $\frac{\partial F}{\partial y} = 0$, and $\frac{\partial F}{\partial z} = 0$. Next, these expressions for

λ are used *in pairs* to write first y and then z in terms of x. These expressions for y and z are then substituted into the equation $\dfrac{\partial F}{\partial \lambda} = 0$ to give an equation involving only x. We solve this equation for x and then go back and find the corresponding values of y and z.

Example 3 (*Compare Exercise 9*)
Find the minimum value of the function

$$f(x, y, z) = 10xy + 4xz + 6yz$$

subject to the constraint $xyz = 240$, where x, y, and z are all positive.

Solution Applications frequently require that the variables be positive, and we will show at the end of the solution that this requirement is necessary here in order that f have a minimum. We construct the function

$$F(x, y, z, \lambda) = 10xy + 4xz + 6yz + \lambda(xyz - 240)$$

The partial derivatives of F are

$$F_x(x, y, z, \lambda) = 10y + 4z + \lambda(yz)$$
$$F_y(x, y, z, \lambda) = 10x + 6z + \lambda(xz)$$
$$F_z(x, y, z, \lambda) = 4x + 6y + \lambda(xy)$$
$$F_\lambda(x, y, z, \lambda) = xyz - 240$$

We now must solve the following system:

$$10y + 4z + \lambda(yz) = 0 \qquad (1)$$
$$10x + 6z + \lambda(xz) = 0 \qquad (2)$$
$$4x + 6y + \lambda(xy) = 0 \qquad (3)$$
$$xyz - 240 = 0 \qquad (4)$$

Solving (1), (2), and (3) for λ, we get, respectively

$$\lambda = -\frac{10y + 4z}{yz}$$

$$\lambda = -\frac{10x + 6z}{xz}$$

$$\lambda = -\frac{4x + 6y}{xy}$$

The constraint, $xyz = 240$, means that x, y, and $z \neq 0$, so we have no problems with the denominators possibly being equal to 0. We now equate the first two expressions for λ.

$$-\frac{10y + 4z}{yz} = -\frac{10x + 6z}{xz}$$

Cross multiply:

$$\begin{aligned} -(10xyz + 4xz^2) &= -(10xyz + 6yz^2) \\ 4xz^2 &= 6yz^2 \\ 4x &= 6y \\ y &= \frac{2x}{3} \end{aligned}$$

We have solved for y in terms of x. Next, equate the first and third expressions for λ.

$$-\frac{10y + 4z}{yz} = -\frac{4x + 6y}{xy}$$

Again, cross multiply:

$$\begin{aligned} -(10xy^2 + 4xyz) &= -(4xyz + 6y^2z) \\ 10xy^2 &= 6y^2z \\ 10x &= 6z \\ z &= \frac{5x}{3} \end{aligned}$$

Having arrived at expressions for y and z in terms of x, substitute them into equation (4). We get

$$\begin{aligned} (x)\left(\frac{2x}{3}\right)\left(\frac{5x}{3}\right) - 240 &= 0 \\ \frac{10x^3}{9} &= 240 \\ 10x^3 &= 2160 \\ x^3 &= 216 \\ x &= 6 \end{aligned}$$

Since $y = \dfrac{2x}{3}$ and $z = \dfrac{5x}{3}$, we have $x = 6$, $y = 4$, and $z = 10$; and,

$$f(6, 4, 10) = 10 \cdot 6 \cdot 4 + 4 \cdot 6 \cdot 10 + 6 \cdot 4 \cdot 10 = 720$$

To see that 720 is the minimum of f for x, y, and z, all positive, we use the constraint $xyz = 240$ to write $z = \dfrac{240}{xy}$. Next, rewrite $10xy + 4xz + 6yz$ as

$$10xy + 4x\left(\frac{240}{xy}\right) + 6y\left(\frac{240}{xy}\right) = 10xy + \frac{960}{y} + \frac{1440}{x}$$

If xy is large, then the corresponding value of f will be large due to the $10xy$ term. If xy is small, then at least one of the terms, $\frac{960}{y}$ or $\frac{1440}{x}$, will be large. Thus, f must have a minimum at (6, 4, 10). This function behaves like the function $h(x) = x + \frac{1}{x}$ in the one variable case, which also indicates why we restricted the domain to x, y, and z positive. If we let $x > 0$ but both y and $z < 0$, f does not have a minimum value.

Exercise 19 shows how a function involving three variables, under constraint, can arise in practice, and how for the problem to be meaningful, all three variables must be positive.

We complete this section by mentioning that some applications may involve more than one constraint. For example, you may want to find the extreme values of a function f subject to two constraints, $g(x, y, z) = 0$ and $h(x, y, z) = 0$. To do this, construct a function F that involves two Lagrange multipliers, λ and μ, one for each constraint. Define

$$F(x, y, z, \lambda, \mu) = f(x, y, z) + \lambda g(x, y, z) + \mu h(x, y, z)$$

The values of x, y, and z which yield maxima or minima of f are among the solutions of the system of five equations,

$$\frac{\partial F}{\partial x} = 0, \quad \frac{\partial F}{\partial y} = 0, \quad \frac{\partial F}{\partial z} = 0, \quad \frac{\partial F}{\partial \lambda} = 0, \quad \text{and} \quad \frac{\partial F}{\partial \mu} = 0$$

The algebraic manipulation involved in solving such a system is more involved than the ones we have discussed, but the basic method is the same.

8–4 Exercises

In Exercises 1 through 12, find the maximum or minimum value of the given function under the given constraint.

1. (*See Example 1*) Minimum value of $x^2 + 4y^2 + 6$, subject to $x - 4y = 9$
2. Maximum value of $x^2 + xy - 3y^2$, subject to $x + 2y = 4$
3. Maximum value of $8x - x^2 + 4y - y^2$, subject to $x + y = 8$
4. Maximum value of $x^2 - y^2$, subject to $2x + y = 6$
5. Minimum value of $3x^2 + y^2 + 3xy - 60x - 32y + 504$, subject to $x + y = 10$
6. Maximum value of $12xy - 3x^2 - y^2$, subject to $x + y = 16$
7. Minimum value of $6x^2 + 5y^2 - xy$, subject to $2x + y = 24$
8. Maximum and minimum values of $4x + 3y$, subject to $x^2 + y^2 - 9 = 0$

9. (*See Example 3*) Minimum value of $x^2 + y^2 + z^2$, subject to $x + 2y + 4z - 21 = 0$
10. Maximum value of x^2yz subject to $x + y + z = 16, x > 0, y > 0, z > 0$
11. (*See Example 2*) A manufacturing process is described by the Cobb-Douglas function

$$f(x, y) = 10x^{1/4}y^{3/4}$$

where x is the units of labor, y is the units of capital, and \$60,000 is available for production. How many units of labor and how many units of capital should be used in order to maximize production if each unit of labor costs \$100 and each unit of capital costs \$200?

12. The profit that a company makes from employing workers x hours at regular time and y hours at overtime per day is estimated to be described by the function

$$p(x, y) = 6xy + \frac{x^2}{2} - 2y^2$$

The company has a rule that employees who work overtime cannot work more than a total of nine hours in any day. How should regular time and overtime be divided to ensure maximum profit?

13. A company manufactures refrigerators at two plants, X and Y, and sells the refrigerators at a town Z. The cost of manufacturing and transporting x refrigerators from X to Z and y refrigerators from Y to Z is

$$C(x, y) = 2x^2 + 4y^2 - xy$$

Town Z wants 126 refrigerators. How many should come from X and how many from Y?

14. A company manufactures a certain product on two production lines. The total daily profit on manufacturing x items on the one line and y items on the other is $20xy - 2x^2 - y^2$. The company wants to produce 92 items per day. How many items should it manufacture on each line?
15. The cost of quality control in a manufacturing line is a function of the numbers of inspections x and y made per week at two points, X and Y:

$$C = 4x^2 + 2y^2 - 4y$$

Safety regulations require that 22 inspections be made each week. How many inspections should be made at each location to minimize cost?

16. A company manufactures a certain product on two separate production lines, X and Y. The profit realized on producing x units on line X and y units on line Y is

$$P = 20x + 16y - 2x^2 - y^2$$

The company wants to manufacture 64 units per week. How many should be produced on each line to maximize profit?

17. The relationship between the number of units N of an item sold and the amounts x and y (in dollars) spent on two advertising media is given by

$$N = 1600x + 1200y - 4x^2 - y^2$$

If \$20,000 is available to spend on advertising, how should this be distributed between the media to maximize sales?

18. A company uses two types of communication equipment. The total cost of using x of one type and y of the other type is

$$C = 4x^2 + y^2 - 4xy$$

If $x + y = 6$, how should the distribution of the equipment be allocated to minimize cost?

19. To construct a free-standing garage, a construction company charges \$5.00 per square foot for walls, \$3.00 per square foot for the ceiling, and \$4.00 per square foot for the floor. These costs include both labor and material and allow for all the required doors and windows. What are the dimensions of the largest garage (in terms of volume) that can be built for \$9000?

8–5 Multiple Integrals

Iterated Integrals
The Double Integral
Volume
Average Value of $f(x, y)$

So far in this chapter, we have introduced functions of several variables, shown how to compute partial derivatives, and discussed the problem of finding extreme values, both with and without constraints. Remembering how we developed the calculus for functions of a single variable, you may have anticipated this section's topic. Just as we did for functions of one variable, we will avoid the *theory* of integration, treating only the "nice" cases and concentrating on the computational aspects.

Iterated Integrals

If f is a function of two variables, x and y, then you can compute $\frac{\partial f}{\partial x}$ by treating y as a constant and differentiating with respect to x. Similarly, you can compute $\int f(x, y)\,dx$ by treating y as a constant and antidifferentiating with respect to x.

Example 1 *(Compare Exercise 1)*
Find $\int (6x^2 + y - 2xy)\,dx$.

Solution

$$\int (6x^2 + y - 2xy)\,dx = 2x^3 + xy - x^2y + g(y)$$

Note that rather than $+C$, we have written $+g(y)$, where $g(y)$ denotes an expression that involves only constants and y, but not x. We mean by $g(y)$ the most general expression so that $\frac{\partial g}{\partial x} = 0$. Recall that $\int f(x)\,dx$ is the family of all functions whose derivative is $f(x)$. Similarly, $\int f(x, y)\,dx$ is the family of all functions whose partial derivative with respect to x is $f(x, y)$.

We can also treat y as the variable and x as a constant.

Example 2 *(Compare Exercise 3)*
Find $\int (6x^2y - e^{xy})\,dy$.

Solution

$$\int (6x^2y - e^{xy})\,dy = 3x^2y^2 - \frac{1}{x}e^{xy} + h(x)$$

Next, we introduce the definite integral, which is evaluated in the same manner as with functions of one variable. There is a difference in the answer however; $\int_a^b f(x)\,dx$ is a number, while $\int_a^b f(x, y)\,dx$ is, in general, an expression in y. The common aspect is that neither involves x.

Example 3 *(Compare Exercise 5)*
Compute $\int_1^3 (12xy^2 - 4x + 2y)\,dx$.

Solution

$$\begin{aligned}\int_1^3 (12xy^2 - 4x + 2y)\,dx &= (6x^2y^2 - 2x^2 + 2xy)\Big|_1^3 \\ &= (54y^2 - 18 + 6y) - (6y^2 - 2 + 2y) \\ &= 48y^2 + 4y - 16\end{aligned}$$

Notice that, as was the case when evaluating definite integrals involving a function of one variable, you only need to find one particular antiderivative to perform the evaluation.

Example 4 Compute $\int_{-1}^2 (x^2 - 6y + 5)\,dy$.

Solution

$$\begin{aligned}\int_{-1}^2 (x^2 - 6y + 5)\,dy &= (x^2y - 3y^2 + 5y)\Big|_{-1}^2 \\ &= (2x^2 - 12 + 10) - (-x^2 - 3 - 5) \\ &= 3x^2 + 6\end{aligned}$$

We are now in a position to evaluate an expression such as

$$\int_{-1}^2 \int_2^4 (3x^2 + 2xy)\,dx\,dy$$

which is called an **iterated integral**. This expression stands for two integrals, with the inner integral being computed first. The brackets are commonly omitted, but better notation might be

$$\int_1^2 \left(\int_2^4 (3x^2 + 2xy)\,dx \right) dy$$

Example 5 (*Compare Exercise 9*)
Compute $\int_1^2 \int_2^4 (3x^2 + 2xy)\,dx\,dy$.

Solution Performing the inside integration first, we have

$$\begin{aligned}\int_2^4 (3x^2 + 2xy)\,dx &= (x^3 + x^2y)\Big|_2^4 \\ &= (4^3 + 4^2y) - (2^3 + 2^2y) \\ &= 56 + 12y\end{aligned}$$

Thus, the inner evaluation gives us $56 + 12y$. Our next evaluation is

$$\begin{aligned}\int_1^2 (56 + 12y)\,dy &= (56y + 6y^2)\Big|_1^2 \\ &= (112 + 24) - (56 + 6) \\ &= 74\end{aligned}$$

Thus,

$$\int_1^2 \int_2^4 (3x^2 + 2xy)\,dx\,dy = 74$$

The iteration can be done in either order.

Example 6 (*Compare Exercise 13*)
Evaluate $\int_0^3 \int_{-1}^2 (4y + 6x^2y)\,dy\,dx$.

Solution First,

$$\begin{aligned}\int_{-1}^2 (4y + 6x^2y)\,dy &= (2y^2 + 3x^2y^2)\Big|_{-1}^2 \\ &= (8 + 12x^2) - (2 + 3x^2) \\ &= 6 + 9x^2\end{aligned}$$

Second,

$$\begin{aligned}\int_0^3 (6 + 9x^2)\,dx &= (6x + 3x^3)\Big|_0^3 \\ &= (18 + 81) - (0 + 0) \\ &= 99\end{aligned}$$

Thus,

$$\int_0^3 \int_{-1}^2 (4y + 6x^2y)\,dy\,dx = 99$$

The Double Integral

Example 6 asked you to evaluate

$$\int_0^3 \int_{-1}^2 (4y + 6x^2y)\,dy\,dx$$

Notice that Example 7 has the same function as the integrand, but that the order of integration has been reversed.

Example 7 (*Compare Exercise 17*)
Evaluate $\int_{-1}^2 \int_0^3 (4y + 6x^2y)\,dx\,dy$.

Solution This time, the first computation is

$$\int_0^3 (4y + 6x^2y)\,dx = (4yx + 2x^3y)\Big|_0^3$$
$$= (12y + 54y) - (0 + 0)$$
$$= 66y$$

Next,

$$\int_{-1}^2 66y\,dy = 33y^2\Big|_{-1}^2$$
$$= 132 - 33$$
$$= 99$$

The answers to Examples 6 and 7 are the same. This situation is not unique.

In general, if $f(x, y)$ is a nice function,

$$\int_a^b \int_c^d f(x, y)\,dy\,dx = \int_c^d \int_a^b f(x, y)\,dx\,dy$$

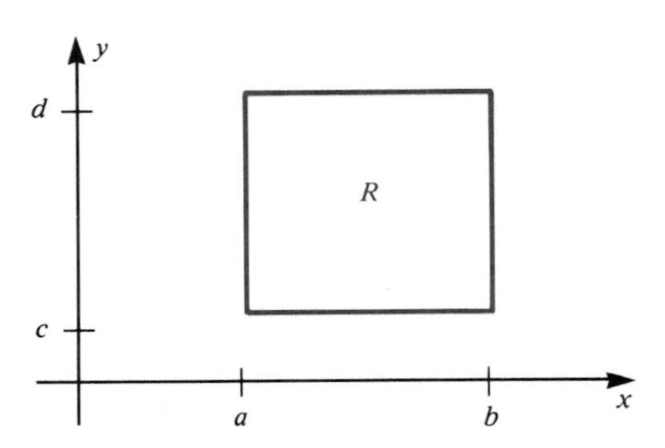

Figure 8–17

This independence of the order of integration allows us to make the following definition. We let R denote the rectangle $\{(x, y) \text{ with } a \le x \le b \text{ and } c \le y \le d\}$. (See Figure 8–17.) For the type of functions we are considering, we can define the double integral of f over the rectangle R to be the number obtained as either iterated integral. This common number is written $\iint_R f$.

Example 8 (*Compare Exercise 17*)
Evaluate $\iint_R f$ where $R = \{(x, y) \text{ with } 2 \le x \le 4 \text{ and } 1 \le y \le 2\}$ and $f(x, y) = 3x^2 + 2xy$.

Solution We can evaluate $\iint_R f$ as

$$\int_1^2 \int_2^4 (3x^2 + 2xy)\,dx\,dy$$

This iterated integral was found to be equal to 74 in Example 5. Thus, we have $\iint_R f = 74$.

Volume

We have shown how to compute a double integral by using an iterated integral, and now we discuss the geometric interpretation of double integrals. Suppose that $f(x, y) \ge 0$ for all points (x, y) in a rectangle R. Then the graph of f is a surface lying above the xy-plane, as shown in Figure 8–18.

There is a three-dimensional solid bounded above by the graph of f and below by the rectangle R. The volume of this solid is $\iint_R f$. We will not go into a rigorous discussion of the theory of volumes, and in fact we restrict our discussion to volumes of solids bounded below by a rectangle in the xy-plane and above by the graph of a function.

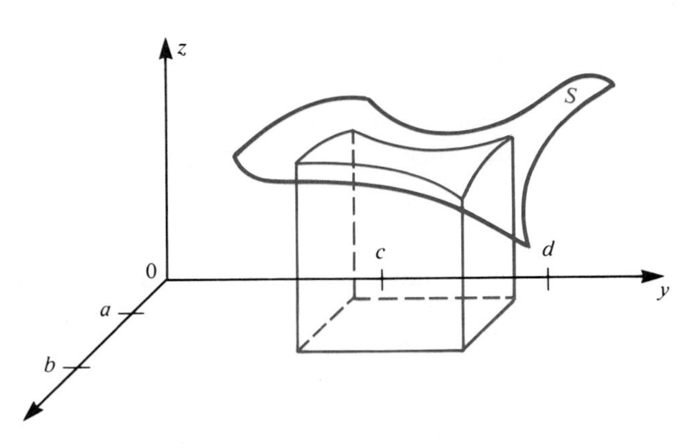

Figure 8–18

Example 9 (*Compare Exercise 21*)
Determine the volume of the solid under the graph of $f(x, y) = 3x^2 + 2y$ and having as base the rectangle R defined by $1 \leq x \leq 2, 2 \leq y \leq 4$.

Solution The volume will be given by the iterated integral

$$\begin{aligned}\int_2^4 \int_1^2 (3x^2 + 2y)\,dx\,dy &= \int_2^4 \left(\int_1^2 (3x^2 + 2y)\,dx \right) dy \\ &= \int_2^4 \left((x^3 + 2xy)\Big|_1^2 \right) dy \\ &= \int_2^4 [(8 + 4y) - (1 + 2y)]\,dy \\ &= \int_2^4 (7 + 2y)\,dy \\ &= (7y + y^2)\Big|_2^4 \\ &= (28 + 16) - (14 + 4) \\ &= 26\end{aligned}$$

The volume under the surface and above the given rectangular region is 26.

Volume plays the same role in discussing functions of two variables that area does in the case of one variable. For example, if $f(x, y)$ is the amount of money spent on leisure activities as a function of age (x) and income (y), volumes can be used to compare total amounts spent by different groups. Furthermore, statistics and decision-making theory use probability distributions that depend on more than one variable. In this case, the double integral over a rectangle R gives the probability that the two variables lie inside the rectangle. The double integral is also closely connected to double summations, and we saw in Chapter 6 that the connection between summations and the definite integral leads to various applications of the integral. However, we will not pursue these topics here. Our aim is simply to give you a brief introduction to double integrals and their evaluation. We conclude with an application that shows how applications of the double integral can closely parallel some of the applications we saw of the definite integral.

Average Value of $f(x, y)$

The double integral is used to define the average value of a function over a rectangular region in a natural extension of the definition of the average value of a function of a single variable. For purposes of our picture, we again assume that $f(x, y) \geq 0$ over R, so the graph of f will be as shown in Figure 8–19.

Geometrically, you can think of $f(x, y)$ as the height from the point (x, y) to the surface S. We are interested in the average value of all such heights. Denote

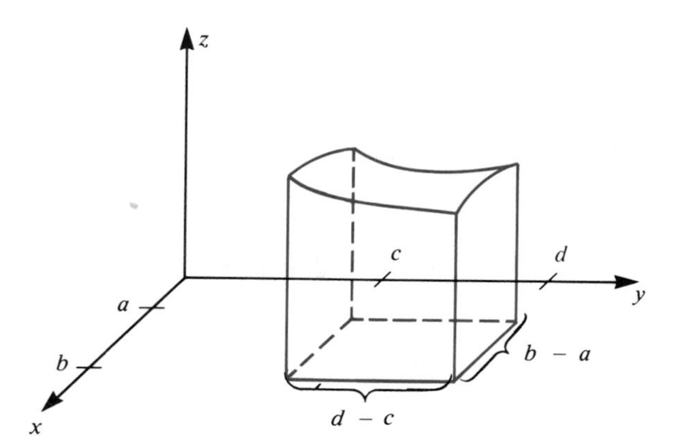

Figure 8–19

this average height by H. The area of R is $(b - a)(d - c)$. We call H the average height if the volume of the box having base R and height H is equal to the volume of the solid below the graph and over R. Thus,

$$(\text{Area of } R) \times H = \text{volume under the surface.}$$

$$(b - a)(d - c) \times H = \int_c^d \int_a^b f(x, y)\,dx\,dy$$

$$H = \frac{1}{(b - a)(d - c)} \int_c^d \int_a^b f(x, y)\,dx\,dy$$

Definition

The average value of $f(x, y)$, denoted H, over a region $a \le x \le b$, $c \le y \le d$, is given by

$$H = \frac{1}{(b - a)(d - c)} \int_c^d \int_a^b f(x, y)\,dx\,dy$$

Example 10 (*Compare Exercise 25*)
Compute the average value of the function $f(x, y) = 4x^3y$ over the region $0 \le x \le 1$, $2 \le y \le 4$.

Solution We get

$$H = \frac{1}{(1 - 0)(4 - 2)} \int_2^4 \int_0^1 4x^3y\,dx\,dy$$

$$= \frac{1}{2} \int_2^4 \left[\int_0^1 4x^3y\,dx \right] dy$$

$$= \frac{1}{2} \int_2^4 \left[x^4y \Big|_0^1 \right] dy$$

$$= \frac{1}{2}\int_{2}^{4} y\,dy = \frac{1}{2}\left[\frac{y^2}{2}\Big|_2^4\right]$$

$$= \frac{1}{2}(8 - 2)$$

$$= 3$$

The average value of the function over this region is 3.

8–5 Exercises

I.

Do the antidifferentiation indicated in Exercises 1 through 4.

1. *(See Example 1)* $\int (12x^3y - 6x)\,dy$

2. $\int (12x^3y - 6x)\,dx$

3. *(See Example 2)* $\int \frac{x}{y}\,dy$

4. $\int \frac{x}{y}\,dx$

Evaluate the given expression in Exercises 4 through 8.

5. *(See Example 3)* $\int_1^4 (10xy + 2x - 4y)\,dx$

6. $\int_1^4 (10xy + 2x - 4y)\,dy$

7. $\int_{-1}^{2} (6x^2y + 7 - 4x)\,dy$

8. $\int_{-1}^{2} (6x^2y + 7 - 4x)\,dx$

II.

In Exercises 9 through 16, evaluate the given integrals.

9. *(See Example 5)* $\int_0^1 \int_0^1 4xy\,dx\,dy$

10. $\int_1^2 \int_0^2 2x\,dx\,dy$

11. $\int_0^1 \int_2^3 (x + y - 4)\,dx\,dy$

12. $\int_1^3 \int_0^2 (8x + 4y)\,dx\,dy$

13. *(See Example 6)* $\int_0^4 \int_1^5 (6x^2y + 2x)\,dy\,dx$

14. $\int_{-1}^{2} \int_0^1 (12x^3y^2 - 6x^2y)\,dy\,dx$

15. $\int_{-1}^{1}\int_{2}^{4}(3y^2 + 4y^3)\,dx\,dy$

16. $\int_{0}^{1}\int_{0}^{2} e^{x+y}\,dx\,dy$

III.

Compute $\iint_R f$ two different ways for the function f and the region R given in Exercises 17 through 20.

17. (*See Examples 7* and *8*) $f(x, y) = x^2 - y^2$, $-1 \le x \le 3$, $1 \le y \le 2$
18. $f(x, y) = 4x - 8y$, $1 \le x \le 4$, $-1 \le y \le 1$
19. $f(x, y) = \dfrac{x}{y}$, $1 \le x \le 3$, $1 \le y \le e$
20. $f(x, y) = 4ye^{2x}$, $1 \le x \le 2$, $1 \le y \le 2$

In each of Exercises 21 through 24, compute the volume of the region that is under the graph of the given function and that has the given rectangle as its base.

21. (*See Example 9*) $f(x, y) = 2x + 4y$, $1 \le x \le 3$, $0 \le y \le 2$
22. $f(x, y) = 4xy + 6y^2$, $0 \le x \le 1$, $1 \le y \le 3$
23. $f(x, y) = 3x^2 + 6xy^2$, $1 \le x \le 2$, $-2 \le y \le 3$
24. $f(x, y) = 6x^2y + 2y$, $-2 \le x \le 1$, $0 \le y \le 4$

In each of Exercises 25 through 28, determine the average value of the given function over the given region.

25. (*See Example 10*) $f(x, y) = 4xy$, $0 \le x \le 2$, $0 \le y \le 2$
26. $f(x, y) = 4xy + 6$, $0 \le x \le 2$, $0 \le y \le 2$ (Discuss your answer in the light of your answer to Exercise 25)
27. $f(x, y) = 3x^2 + 6xy^2$, $0 \le x \le 3$, $-1 \le y \le 3$
28. $f(x, y) = 3y^2 - 4x^3$, $1 \le x \le 2$, $0 \le y \le 4$

IMPORTANT TERMS

8–1	**Function of several variables** **Coordinate planes**	**Cobb-Douglas production function** **Surface**	**Three-dimensional space**
8–2	**Partial derivative** $\dfrac{\partial^2 f}{\partial x^2}, \dfrac{\partial^2 f}{\partial y^2}, \dfrac{\partial^2 f}{\partial x \partial y}$, **and** $\dfrac{\partial^2 f}{\partial y \partial x}$; f_{xx}, f_{yy}, f_{yx}, **and** f_{xy} **Marginal productivity of capital**	f_x **and** $\dfrac{\partial f}{\partial x}$, f_y **and** $\dfrac{\partial f}{\partial y}$ **Equality of** $\dfrac{\partial^2 f}{\partial x \partial y}$ **and** $\dfrac{\partial^2 f}{\partial y \partial x}$	**Second partial derivatives** **Marginal productivity of labor**

8–3	**Relative extreme value for a function of two variables**	**Critical points**	**Saddle point**
	Second derivative test for a function of two variables		
8–4	**Lagrange multipliers**	**Constraint**	**Extreme value of f, subject to a constraint on x and y**
	Lagrange multiplier method for two variables	**Lagrange multiplier method for three variables**	
8–5	**Multiple integrals**	**Iterated integrals**	**Double integral**
	Reversing the order of integration	**Volume of solid**	**Average value of $f(x, y)$ over a rectangle**

REVIEW EXERCISES

For each of the functions given in Exercises 1 through 8, compute $f(x, y)$, $\frac{\partial f}{\partial x}$, $\frac{\partial f}{\partial y}$, $\frac{\partial^2 f}{\partial x^2}$, $\frac{\partial^2 f}{\partial y^2}$, and $\frac{\partial^2 f}{\partial x \partial y}$ at the indicated point p.

1. $f(x, y) = x^2 + y^3 - 4xy^2$; $p = (1, -2)$
2. $f(x, y) = \frac{y}{x^2 + 1}$; $p = (-1, 4)$
3. $f(x, y) = e^{xy^2}$; $p = (0, 2)$
4. $f(x, y) = \ln(x^2 + y^4)$; $p = (2, -1)$
5. $f(x, y) = \frac{x}{y}$; $p = (-4, 2)$
6. $f(x, y) = \sqrt{x^2 + y^3}$; $p = (1, 2)$
7. $f(x, y) = x\sqrt{y^2 + 9}$; $p = (2, 4)$
8. $f(x, y) = e^{(x-2y)}$; $p = (3, 1)$

Find the critical points for each of the functions given in Exercises 9 through 12. Then use the second derivative test to determine if f has a relative maximum, relative minimum, or neither at each critical point.

9. $f(x, y) = x^2 - y^2 - 6x - 4y + 20$
10. $f(x, y) = x^2 + y^3 - 6xy$
11. $f(x, y) = x^3 + y^3 - 6xy$
12. $f(x, y) = x^2 + y^2 + 8xy - 18x - 12y + 10$

Use the method of Lagrange multipliers in Exercises 13 through 16.

13. Find the maximum and the minimum of $f(x, y) = 6x + 8y$ if (x, y) must be on the circle $x^2 + y^2 = 25$.
14. Find the minimum $f(x, y) = x^2 + y^2$, subject to the constraint $xy = 4$.

15. Find the maximum of $f(x, y, z) = xyz$, subject to the constraint $2x + y + z = 6$ and x, y, and z are all positive.

16. Find the minimum of $f(x, y, z) = 3xy + xz + 2yz$, subject to the constraint $xyz = 6$ and x, y, and z must all be positive.

Compute the iterated integrals in Exercises 17 through 20.

17. $\int_{-1}^{2} \int_{0}^{3} (x^2 - yx)\, dy\, dx$

18. $\int_{0}^{9} \int_{1}^{4} \sqrt{xy}\, dx\, dy$

19. $\int_{-1}^{3} \int_{0}^{1} yxe^{x^2}\, dx\, dy$

20. $\int_{1}^{4} \int_{2}^{5} \frac{1}{x + y}\, dx\, dy$

21. What is the area under the surface $z = x^2 + y^2$ and above the rectangle $-1 \leq x \leq 2$, $0 \leq y \leq 3$?

Answers to Selected Exercises

CHAPTER 0

Section 0–1, page 4

1. -13 **3.** 23 **5.** -30 **7.** -35 **9.** -5 **11.** -7 **13.** -10 **15.** -8 **17.** 3 **19.** $\frac{9}{5}$
21. $\frac{17}{12}$ **23.** $-\frac{11}{12}$ **25.** $\frac{13}{20}$ **27.** $-\frac{1}{35}$ **29.** $\frac{2}{3}$ **31.** $\frac{5}{14}$ **33.** $\frac{1}{15}$ **35.** $\frac{8}{15}$ **37.** $\frac{12}{35}$ **39.** $\frac{20}{33}$
41. $\frac{4}{3}$ **43.** $\frac{15}{8}$ **45.** $\frac{171}{40}$ **47.** $-6a - 22b$ **49.** $-10a - 50b$

Section 0–2, page 7

1. -3 **3.** 4 **5.** 2 **7.** $-\frac{3}{4}$ **9.** $-\frac{7}{4}$ **11.** $\frac{20}{11}$ **13.** $\frac{15}{23}$ **15.** 6
17. (a) \$242 **(b)** \$412 **(c)** 950 miles **19. (a)** \$7.14 **(b)** \$22.05 **(c)** 61 lbs

Section 0–3, page 11

1.

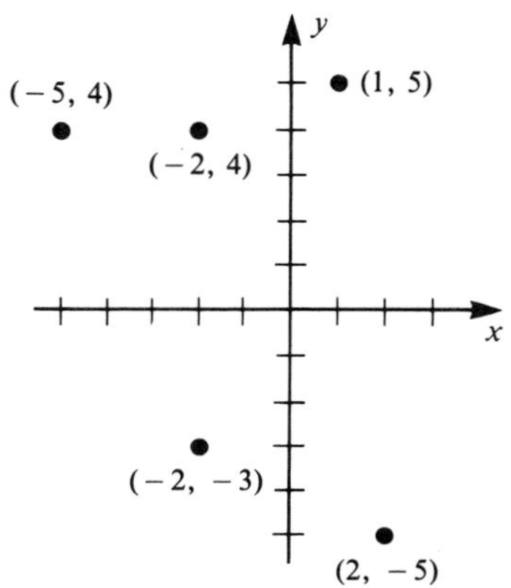

3.

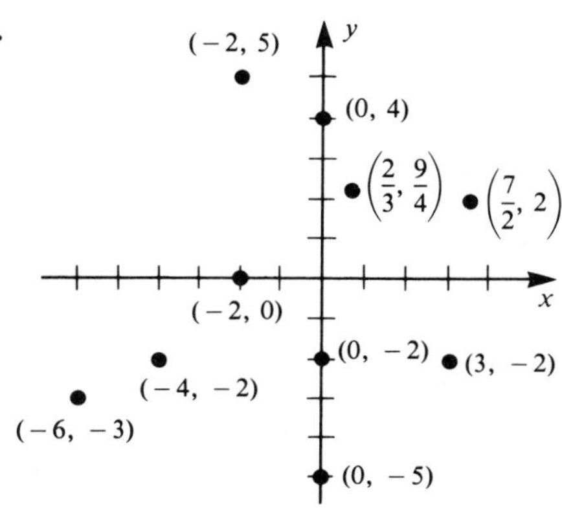

5. **(a)** x coordinate is negative, y coordinate is positive.
(b) x and y coordinates are both negative.
(c) x coordinate is positive, y coordinate is negative.

7.

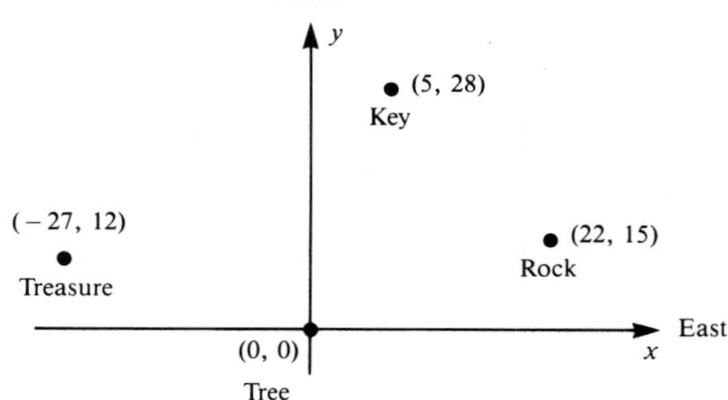

Section 0–4, page 18

1. **(a)** $9 - 3$ is a positive number 6, so true.
(b) $4 - 0$ is a positive number 4, so true.
(c) $-5 - 0$ is not a positive number, so false.
(d) $-3 - (-15)$ is a positive number 12, so true.
(e) $\frac{5}{6} - \frac{2}{3}$ is a positive number $\frac{1}{6}$, so true.

3. $x < \frac{9}{2}$ **5.** $x \geq -16$ **7.** $x < -4$ **9.** $0 \leq x < 12$

11. $x < -9$

13. $x < -24$

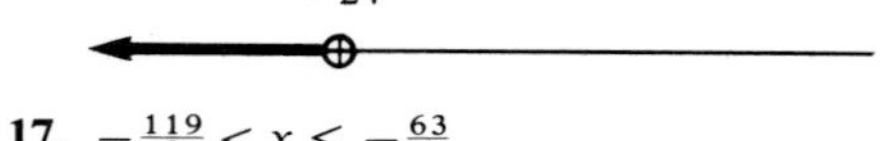

15. $x < \frac{7}{9}$

17. $-\frac{119}{2} < x \leq -\frac{63}{2}$

$-\frac{119}{2}$ $-\frac{63}{2}$

19. $(-\infty, -1]$ **21.** $\left(-\infty, \frac{1}{9}\right]$ **23.** $[7, 13)$ **25.** $x \leq \frac{1}{2}$ **27.** $\left[-11, -\frac{23}{3}\right)$ **29.** 8, 9, or 10

Section 0–5, page 25

1. **(a)** 4 **(c)** 2 **(e)** $-\frac{3}{4}$ **2.** **(a)** 0 **(c)** $\frac{1}{2}$ **(e)** $\frac{13}{8}$

3. **(a)** $7x^2 + 1$ **(c)** $4x^3 - 5x + 5$ **(e)** $x^3 - 7x^2 + 3x + 7$

4. **(a)** $x^2 - x - 6$ **(c)** $3x^3 + 11x^2 + 5x - 3$ **(e)** $4x^5 - 2x^4 + 6x^3 + x + 1$

5. **(a)** $x(x + 3)$ **(c)** $5x^2(x^2 - 5x + 2)$ **(e)** $(x + 1)(x + 5)$ **(g)** $(x + 2)(2x - 3)$ **(i)** $(2x - 1)(2x + 3)$ **(k)** $(x - 2)(4x - 3)$

6. **(a)** $x = -3$ or 1 **(c)** $x = -1$ or $\frac{3}{2}$ **(e)** $x = \frac{4}{3}$ or $-\frac{1}{2}$

7. **(a)** $x = \frac{3}{2}$ or -1 **(c)** $x = -\frac{1}{4}$ or 3 **(e)** $x = \frac{4}{3}$ or $-\frac{1}{3}$ **(g)** $x = -\frac{1}{2}$ or 1

8. **(a)** $\frac{x^2 + x - 1}{x - 1}$ **(c)** $\frac{-x}{x + 2}$ **(e)** $\frac{(5x - 1)(3x + 2)}{(2x + 1)(4x - 1)}$ **(g)** $\frac{5}{4x - 1}$ **(i)** $\frac{(4x + 3)(2x + 1)}{4x(5x - 1)}$ **(k)** $\frac{x(2x - 3)}{14}$

(m) $\frac{2(3x - 1)}{x(x - 1)}$ **(o)** $\frac{9x^2 + 2x - 3}{(x - 1)(x + 1)}$ **(q)** $\frac{-6x^2 + 1}{x(5x - 1)}$ **(s)** $\frac{5x + 7}{x^2}$

Section 0–6, page 30

1. 25 **3.** 8 **5.** 1 **7.** $\frac{1}{3}$ **9.** $\frac{1}{16}$ **11.** $\frac{1}{81}$ **13.** 3 **15.** 2 **17.** 4 **19.** -3 **21.** 32
23. 9 **25.** $\frac{1}{8}$ **27.** $-\frac{1}{3}$ **29.** y^{20} **31.** x^{42} **33.** does not exist **35.** $\frac{1}{3}$ **37.** $\frac{1}{64}$

CHAPTER 1

Section 1–1, page 39

1. **(a)** 13 **(b)** -5 **(c)** 8 **3.** **(a)** -1 **(b)** 13 **(c)** 2 **5.** all x **7.** all $x \neq 9$ **9.** all $x \geq 8$; $[8, \infty)$
11. all x **13.** $[-1, 5]$ **15.** $f(x) = x^2$ **17.** $f(x) = 3x + 2$ **19.** $f(x) = x^2 - 9$
21. **(a)** 139 **(b)** 49 **(c)** 7 **23.** **(a)** 4 **(b)** 2 **(c)** not defined **25.** all $x \neq \pm 3$ **27.** all $x \neq 3$ or -1
29. $(4, \infty)$ **31.** **(a)** 16 **(b)** -4 **(c)** not defined **33.** **(a)** 10 **(b)** not defined **(c)** not defined
35. **(a)** $3w + 7$ **(b)** $3x + 13$ **37.** **(a)** $-7x - 17$ **(b)** $-7x + 25$
39. **(a)** $h^2 - 4$ **(b)** $x^2 + (2h - 4)x + h^2 - 4h$ **41.** **(a)** $2h^2 + 8h + 11$ **(b)** $2x^2 + 4wx + 2w^2 - 4x - 4w + 5$
43. **(a)** $E(x) = 3 + .5x$ **(b)** 14 rides

Section 1–2, page 53

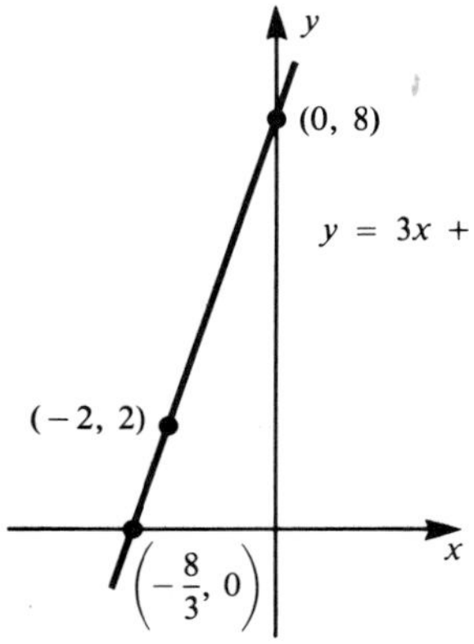

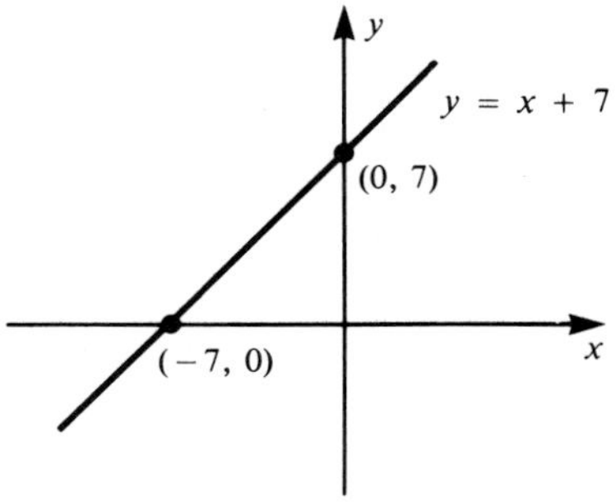

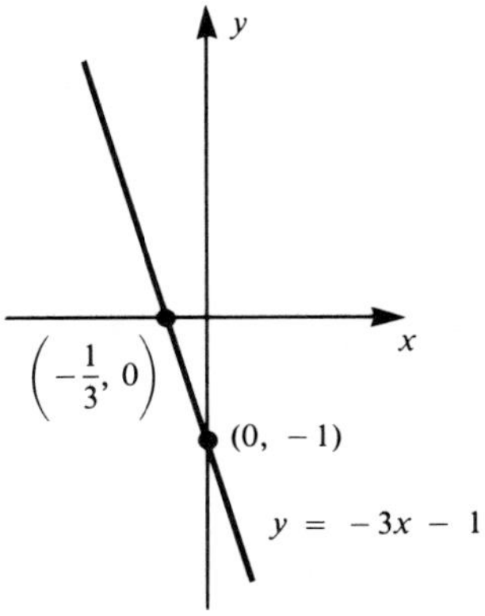

7. slope 7, y-intercept 22 **9.** slope $-\frac{2}{5}$, y-intercept 6 **11.** slope $-\frac{2}{5}$, y-intercept $\frac{3}{5}$
13. slope $\frac{1}{3}$, y-intercept 2 **15.** slope 2 **17.** no slope **19.** slope 0 **21.** slope 0 **23.** $x = 3$
25. $x = 10$ **27.** $y = 4x + 3$ **29.** $y = -x + 6$ **31.** $y = \frac{x}{2}$ **33.** $y - 1 = -4(x - 2)$
35. $y - 4 = \frac{1}{2}(x - 5)$ **37.** $y - 5 = 7(x - 1)$ **39.** $y - 6 = \frac{1}{5}(x - 9)$

41.

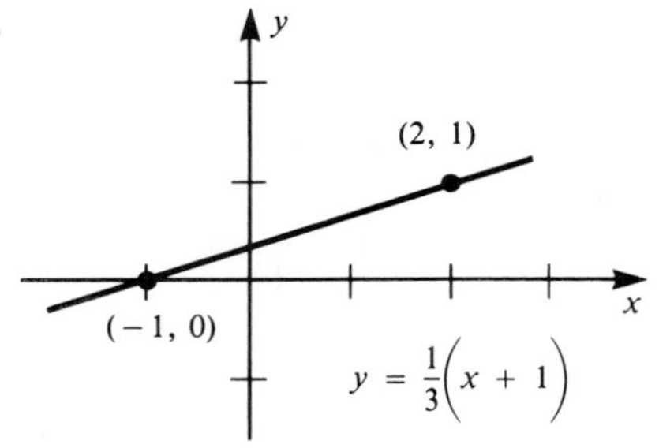

43.

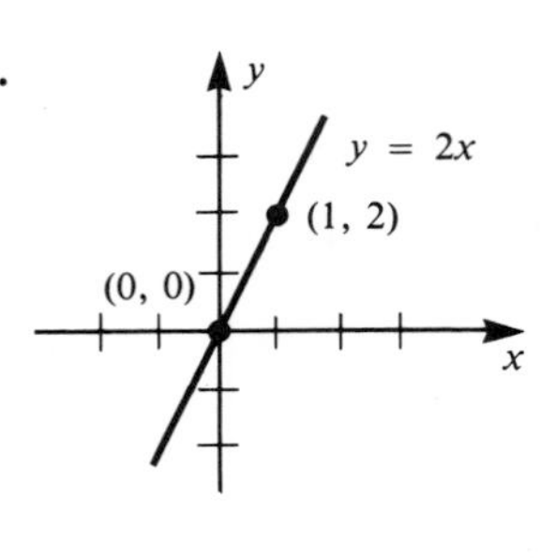

45.

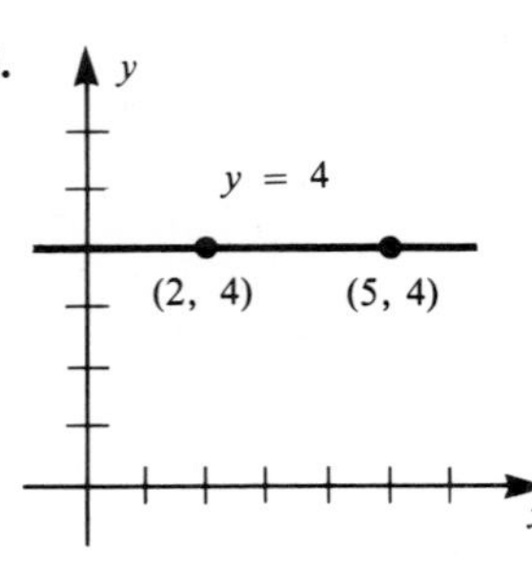

47. yes, $m = 1$ **49.** yes, $m = -\frac{9}{7}$ **51.** yes, $m = -\frac{3}{2}$ **53.** $m = \frac{3}{5}$, but only one line. **55.** $3x + 2y = 18$
57. $y - 5 = x + 4$ **59.** x = number of hamburgers sold each day, $C(x) = .67x + 480$ **61.** $C(x) = 1.4x + 640$
63. Let x = number of pounds gained per day, $f(x) = 3500x + 3000$
65. **(a)** y increases by 4 **(b)** y decreases by 3 **(c)** y increases by $\frac{2}{3}$ **(d)** y decreases by $\frac{1}{2}$ **(e)** y decreases by $\frac{2}{3}$ **(f)** y does not change
67. $c - 51.65 = \frac{15.9}{140}(x - 265)$

Section 1–3, page 63

1. **(a)** \$10,040 **(b)** 223 bicycles
3. **(a)** fixed cost = 400, unit cost = 3 **(b)** $x = 600$, $c = 2200$; $x = 1000$, $c = 3400$
5. **(a)** $R(x) = 32x$ **(b)** \$2496 **(c)** 21 pairs **7.** **(a)** $R(x) = 3.39x$ **(b)** \$2,827.26
9. 123,000 pages per month **11.** **(a)** $C(x) = .50x + 750$ **(b)** fixed cost = 750; unit cost = .50 **(c)** 1350
13. $C(x) = 4x + 500$; $C(800) = 3700$
15. **(a)** $C(x) = 649x + 1500$, x = number sold per week **(b)** $R(x) = 899x$ **(c)** $C(37) = \$25{,}513$ **(d)** $R(37) = \$33{,}263$ **(e)** 6 computers per week
17. \$70 **19.** **(a)** $BV = 425 - 50x$ **(b)** \$50 **(c)** \$275
21. **(a)** $BV = 9750 - 1575x$ **(b)** \$1575 **(c)** 2 years → 6600, 5 years → 1875 **23.** \$885 **25.** $3\frac{1}{2}$ years
27. $C(x) = 4.15x + 1850$ **29.** $R(x) = 38x$
31. **(a)** **(i)** $P(x) = 24x - 465$ **(ii)** 135 **(b)** **(i)** $P(x) = 80x - 1200$ **(ii)** 240 **(c)** **(i)** $P(x) = 80x - 5200$ **(ii)** 800

Section 1–4, page 81

1. $R(p) = (300 - \frac{1}{6}p)p = 300p - \frac{1}{6}p^2$ **3.** up; $5 > 0$ **5.** down; $-6 < 0$ **7.** $(3, -2)$ **9.** $(-2, -11)$

11.

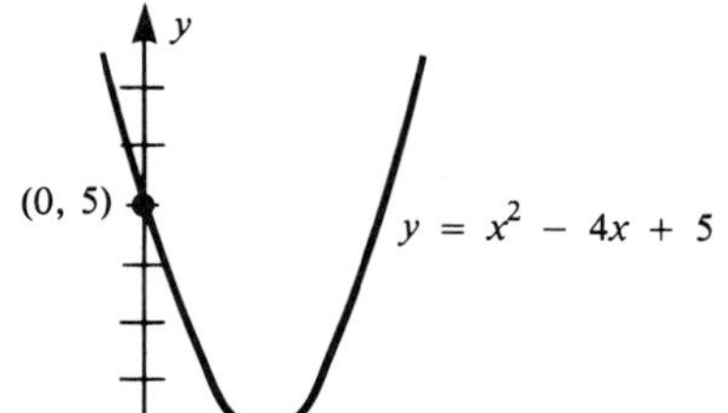

13.

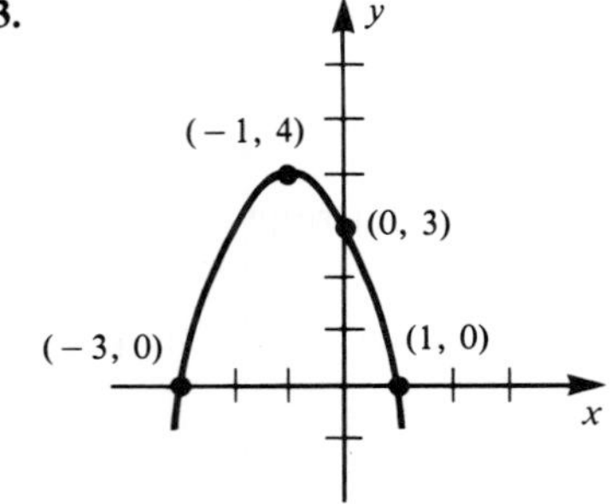

15. (1, 13) **17.** $\dfrac{-7 \pm \sqrt{49 + 480}}{12} = \dfrac{-7 \pm 23}{12} = \dfrac{-5}{2}$ and $\dfrac{4}{3}$

19. $\dfrac{-8 \pm \sqrt{64 + 48}}{-6} = \dfrac{-8 \pm 4\sqrt{7}}{-6} = \dfrac{4 + 2\sqrt{7}}{3}$ and $\dfrac{4 - 2\sqrt{7}}{3}$

21. no real zeros **23.**

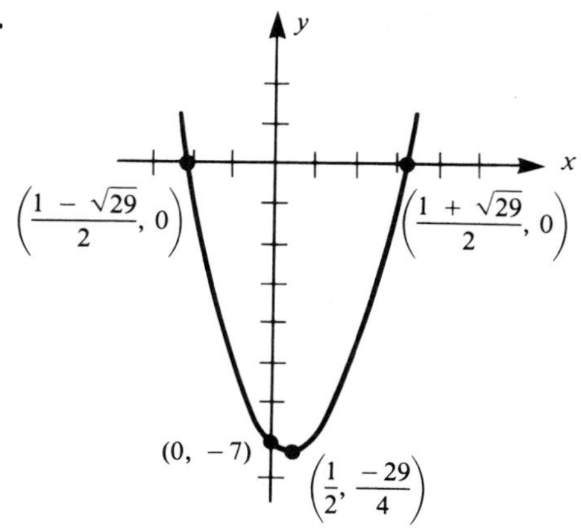

25.

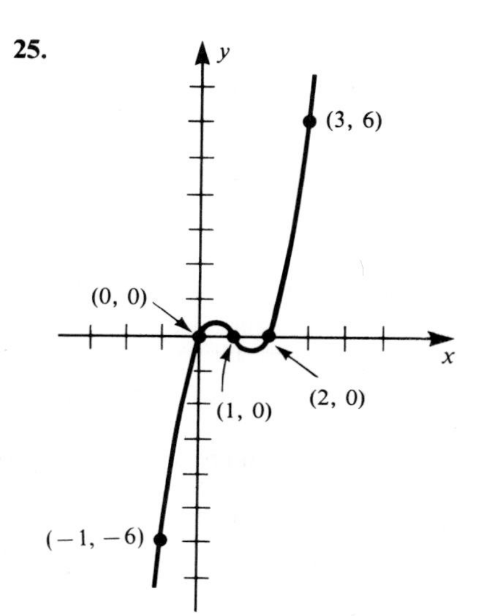

27.

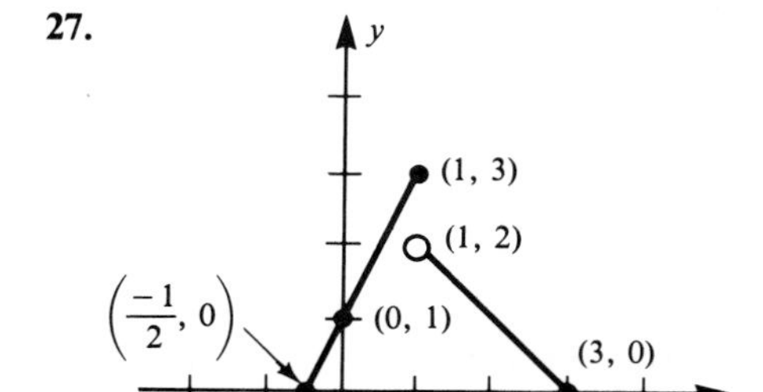

29.

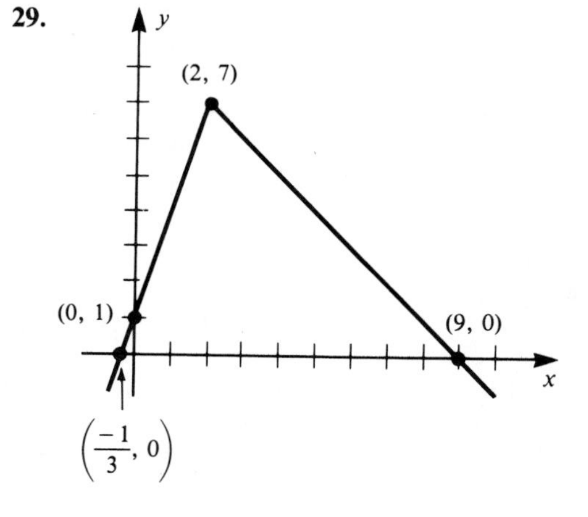

31.

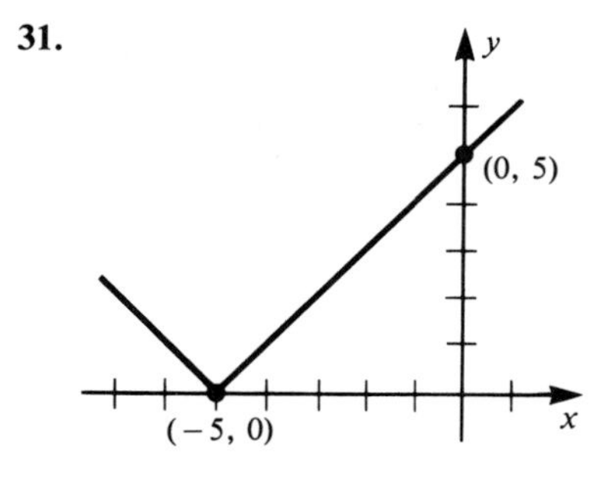

33.

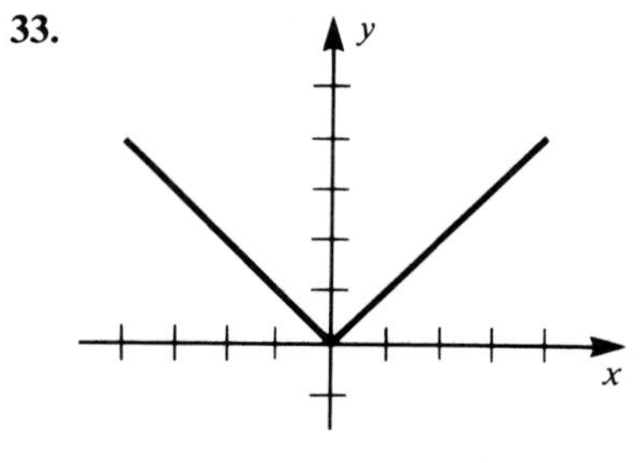

35. **(a)** $W(.3) = .16$; 16%
(b) $W(.9) = .84$; 16%
(c) $W(.5) = \frac{1}{3}$, $x = .5$; Lower 50% of families earn $33\frac{1}{3}$% of total income
37. $R(x) = x(1800 - 6x) = 1800x - 6x^2$

Section 1–5, page 88

1. $(-\infty, 4]$ **3.** $(-\infty, -2)$ **5.** $(3, \infty)$ **7.** $(-\infty, -2]$ **9.** $[-2, 3]$ **11.** $\left(-\infty, -\frac{1}{2}\right)$ and $(4, \infty)$
13. $(-\infty, -2)$ and $(2, \infty)$ **15.** $[-4, 2]$ **17.** $(-\infty, -5)$ and $\left[-\frac{14}{3}, \infty\right)$ **19.** $(-2, 2)$ and $[4, \infty)$
21. $[-4, 0)$ **23.** $[2 - 2\sqrt{3}, 2 + 2\sqrt{3}]$ **25.** $(-9, -6]$ **27.** $(-\infty, 6)$ **29.** all $x > 20$

Review Exercises, Chapter 1, page 89

1. 22 **2.** All real numbers **3.** **(a)** \$4.20 **(b)** 2.75 lbs **4.** **(a)** 2475 **(b)** 13
5. **(a)** $f(x) = (x + 5)^2$ **(b)** $f(x) = \sqrt{x - 4}$ **6.** $f(1 + t) = 8 + 3t$; $f(a + h) = 3a + 3h + 5$
7. $f(4 + D) = 2D^2 + 12D + 19$; $f(a + q) = 2a^2 + 4aq + 2q^2 - 4a - 4q + 3$
8. **(a)**

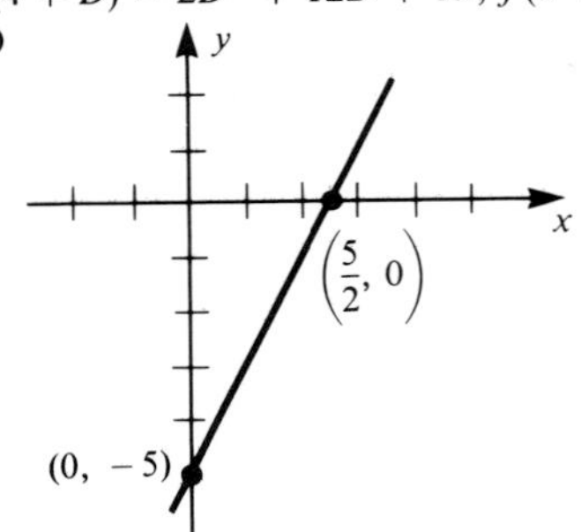

(b)

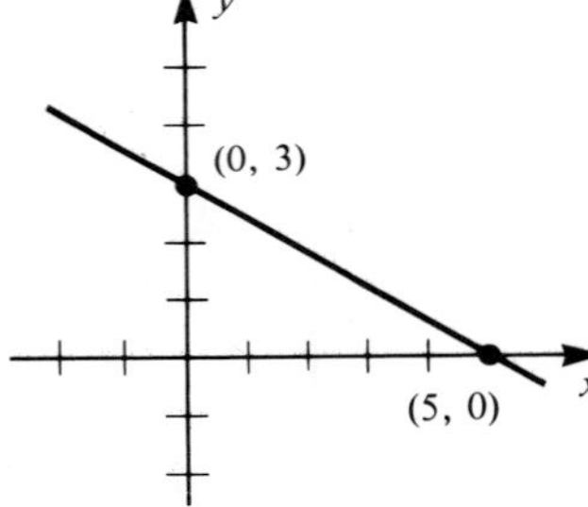

9. **(a)** slope = −2, y-intercept = 3 **(b)** slope = $\frac{5}{4}$, y-intercept = $\frac{3}{2}$ **(c)** slope = $-\frac{6}{7}$, y-intercept = $-\frac{5}{7}$
10. **(a)** $\frac{3}{5}$ **(b)** 0 **(c)** no slope
11. **(a)** $y = -\frac{3}{4}x + 5$ **(b)** $y = 8x - 3$ **(c)** $y + 1 = -2(x - 5)$ **(d)** $y = 6$ **(e)** $y - 3 = -\frac{1}{6}(x - 5)$
(f) $x = -2$ **(g)** $4x - 3y = -13$
12. **(a)** yes **(b)** no **13.** no, $\frac{12}{7} \neq \frac{8}{12}$ **14.** $C(x) = 36x + 12{,}800$
15. **(a)** \$4938 **(b)** impossible, fixed cost > \$2648
16. **(a)** $R(x) = 11x$ **(b)** $C(x) = 6.5x + 675$ **(c)** 150 shirts per week
17. \$607.50 **18.** \$1710 **19.** **(a)** $BV = 17{,}500 - 2075x$ **(b)** \$2075 **(c)** 7125
20. **(a)**

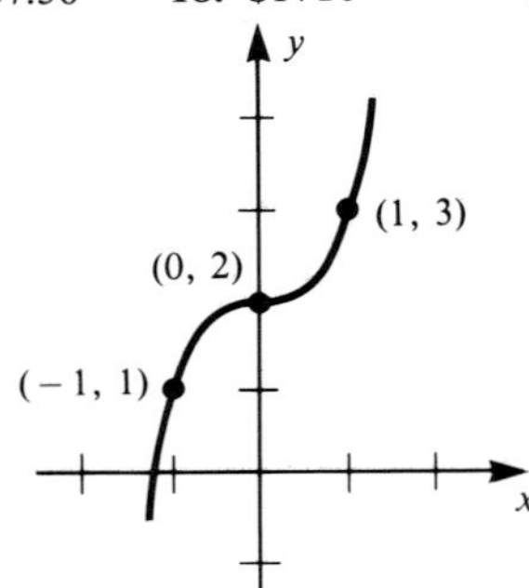

(b)

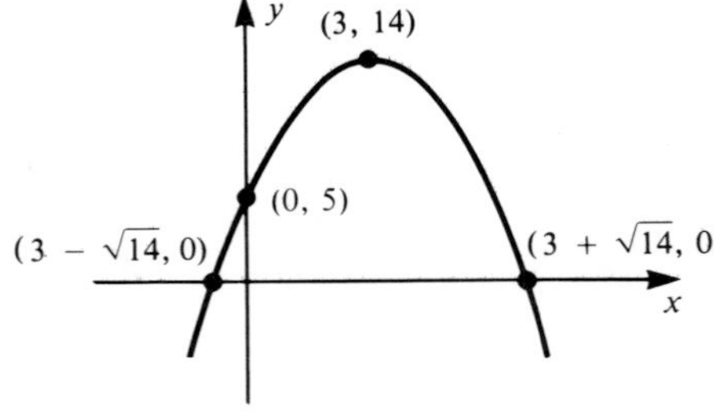

(c)

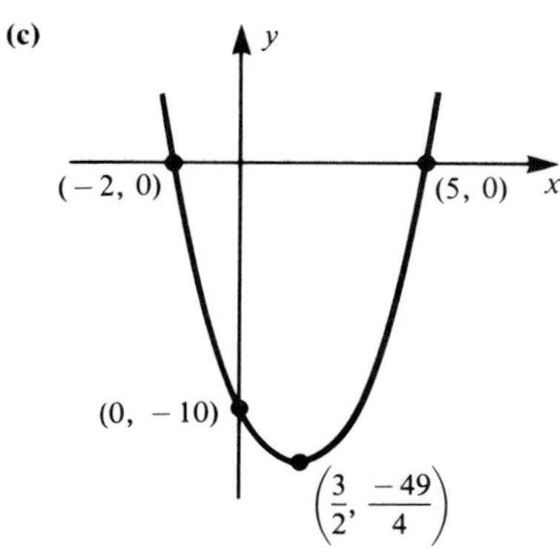

(d)

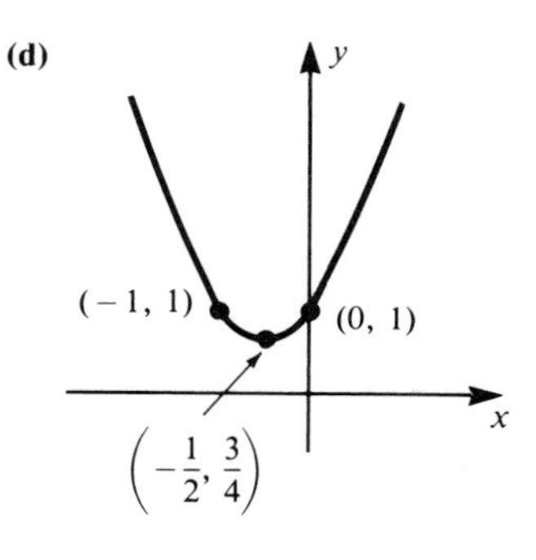

21. **(a)** $(-\infty, 0)$ and $[\frac{1}{5}, \infty)$ **(b)** $(-\infty, 2 - 2\sqrt{2}]$ and $[2 + \sqrt{2}, \infty)$ **(c)** $(0, 9)$ **(d)** $[-\frac{11}{2}, -1)$ and $(2, \infty)$

CHAPTER 2

Section 2–2, page 103

1. **(a)** -7 **(b)** -7 **3.** **(a)** $\frac{1}{2}$ **(b)** $\frac{1}{2}$ **5.** **(a)** 6 **(b)** 4 **7.** **(a)** 0 **(b)** 0 **9.** **(a)** $\frac{1}{9}$ **(b)** $\frac{1}{5}$
11. **(a)** -0.9 **(b)** -1 **(c)** -0.8 **13.** **(a)** $128 + 16\Delta t$ ft/sec **(b)** 128 ft/sec
15. **(a)** 390 ft **(b)** $390 + 250\Delta t + 40(\Delta t)^2$ ft **(c)** $250 + 40\Delta t$ ft/sec **(d)** 250 ft/sec **17.** 24 **19.** 2 **21.** 2
23. -1 **25.** $-\frac{3}{4}$ **27.** m **29.** 0 **31.** **(a)** 17 **(b)** 17 **(c)** 17

Section 2–3, page 118

1. 4 **3.** 6 **5.** -5 **7.** 4 **9.** 5 **11.** 4 **13.** -3 **15.** 1 **17.** does not exist **19.** 0
21. 30 **23.** **(a)** 8 **(b)** -7 **25.** **(a)** 3 **(b)** 81 **27.** 10 **29.** 36 **31.** 2 **33.** 5 **35.** 6 **37.** $\frac{1}{6}$
39. 10 **41.** does not exist **43.** 0 **45.** 0 **47.** does not exist **49.** 1

Section 2–4, page 129

1. **(a)** 1 **(b)** 3 **(c)** does not exist **(d)** 2 **(e)** no **3.** **(a)** 4 **(b)** 4 **(c)** 4 **(d)** 4 **(e)** yes
5. **(a)** 1 **(b)** 4 **(c)** does not exist **(d)** 4 **(e)** no **7.** **(a)** 5 **(b)** 6 **(c)** does not exist **(d)** 5 **(e)** no
9. **(a)** 5 **(b)** 7 **(c)** no **11.** yes $A = 8$ **13.** **(a)** -1 **(b)** 1 **(c)** does not exist **(d)** -1 **(e)** -1 **(f)** -1

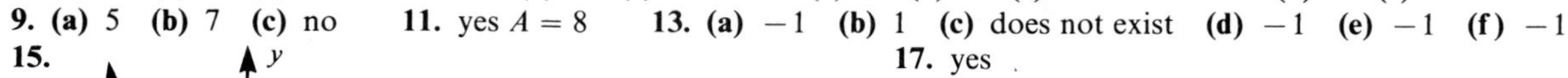

15.

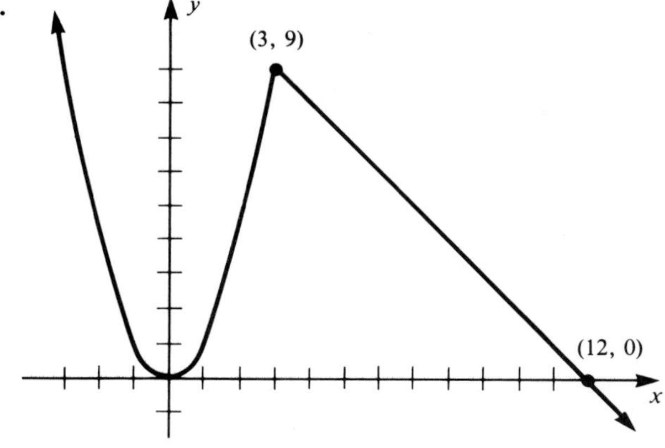

17. yes

Section 2–5, page 142

1. (a) $\frac{-1}{x^2}$ **(b)** $\frac{-1}{x^2}$ **(c)** $\frac{-1}{x^2}$ **(d)** $\frac{-1}{x^2}$ **3.** $y - 10 = 6(x - 3)$ **5.** $y - \frac{1}{10} = -\frac{1}{100}(x - 1)$

7. (a) $f'(x) = 2x - 8$ **(b)** $\frac{dy}{dx} = 2x - 8$ **9.** 4 **11.** $y + \frac{1}{2} = -\frac{1}{4}(x + 2)$ **13.** $y = 4(x - 3)$

15. (a) $-2, 2$, and 3 **(b)** $(2, 0)$ and $(3, -2)$ **(c)** 0 **(d)** 2 **(e)** 2 **(f)** 0

17. $\lim_{\Delta x \to 0} \frac{|3 + \Delta x - 3| - 0}{\Delta x} = \lim_{\Delta x \to 0} \frac{|\Delta x|}{\Delta x}$; does not exist

19. $\lim_{\Delta x \to 0} \frac{\sqrt{9 + \Delta x} - 3}{\Delta x} = \lim_{\Delta x \to 0} \frac{(\sqrt{9 + \Delta x} - 3)(\sqrt{9 + \Delta x} + 3)}{\Delta x(\sqrt{9 + \Delta x} + 3)} = \lim_{\Delta x \to 0} \frac{(9 + \Delta x) - 9}{\Delta x(\sqrt{9 + \Delta x} + 3)} = \lim_{\Delta x \to 0} \frac{\Delta x}{\Delta x(\sqrt{9 + \Delta x} + 3)}$

$= \lim_{\Delta x \to 0} \frac{1}{\sqrt{9 + \Delta x} + 3} = \frac{1}{\sqrt{9} + 3} = \frac{1}{6}$

21. $\lim_{\Delta x \to 0} \frac{(a + \Delta x)^2 - a^2}{\Delta x} = \lim_{\Delta x \to 0} \frac{2a\Delta x + (\Delta x)^2}{\Delta x} = \lim_{\Delta x \to 0} (2a + \Delta x) = 2a$

23. $\lim_{\Delta x \to 0} \frac{\frac{1}{a + \Delta x + 4} - \frac{1}{a + 4}}{\Delta x} = \lim_{\Delta x \to 0} \frac{a + 4 - (a + \Delta x + 4)}{(\Delta x)(a + \Delta x + 4)(a + 4)}$

$= \lim_{\Delta x \to 0} \frac{-\Delta x}{\Delta x(a + \Delta x + 4)(a + 4)} = \lim_{\Delta x \to 0} \frac{-1}{(a + \Delta x + 4)(a + 4)}$

$= \frac{-1}{(a + 4)^2}$

Review Exercises, Chapter 2, page 145

1. 0 **2.** Does not exist **3.** 6 **4.** $-\frac{20}{9}$ **5.** 2 **6.** 2 **7.** 0 **8.** 3 **9.** $f'(x) = 5x + 8$
10. $2x - 2$ **11. (a)** 0 **(b)** -2 **(c)** 3 **12.** $\frac{1}{10}$ hundred dollars per day = \$10 per day
13. (a) no **(b)** yes **14. (a)** yes **(b)** no **15. (a)** 3 **(b)** 4 **(c)** doesn't exist **16. (a)** 1 **(b)** 1 **(c)** 1
17. (a) 5 **(b)** 5 **(c)** 5 **18. (a)** 0 **(b)** -1 **(c)** does not exist **19. (a)** yes **(b)** no **(c)** no
20. (a) no **(b)** yes **(c)** no **21. (a)** yes **(b)** yes **(c)** yes **22. (a)** 5 **(b)** 0 **23. (a)** 1 **(b)** 2
24. (a) 2 **(b)** 0 **25. (a)** -1 **(b)** does not exist **26. (a)** 4 **(b)** does not exist
27. (a) 6 **(b)** 8 **(c)** 8 **(d)** no; $\lim_{x \to 4} C(x)$ does not exist.

28. yes; $\lim_{x \to 3^-} (2x + 1) = 7 = \lim_{x \to 3^+} (x^2 - 2)$, so $\lim_{x \to 3} f(x) = 7 = f(3)$ **29.** 3 **30.** 2 **31.** $-\frac{5}{2}$ **32.** 3

33. $\frac{1}{4}$ **34.** 12 **35.** $f'(1) = 6$ **36.** $\frac{1}{6}$ **37.** $f'(4) = \frac{1}{2}$ **38.** $-\frac{1}{8}$ **39.** $y - 6 = 3(x - 4)$

40. $y + 2 = 10(x - 7)$ **41.** $f'(a) = \frac{1}{2\sqrt{a}}$ **42.** $3a^2$ **43.** yes, $A = 15$ **44.** $B = 18$

CHAPTER 3

Section 3–1, page 158

1. $f'(x) = 5x^4$ **3.** $f(x) = -4x^{-5} = \frac{-4}{x^5}$ **5.** $f(x) = \frac{4}{3}x^{1/3}$ **7.** $f'(x) = \frac{5}{2}x^{3/2} = \frac{5}{2}\sqrt{x^3}$

9. $f'(x) = 40x^3$ **11.** $\frac{dy}{dx} = 6x$ **13.** $\frac{dy}{dx} = \frac{-9}{x^2} = -9x^{-2}$ **15.** $\frac{dy}{dx} = 30x^2 + \frac{5}{2}x^{-1/2} = 30x^2 + \frac{5}{2\sqrt{x}}$

17. $\frac{dy}{dx} = -30x^{-3} + 2x = \frac{-30}{x^3} + 2x$ **19.** $\frac{dy}{dx} = 20x - 4$ **21.** $f'(x) = 12x^2 + 5$ **23.** $f'(x) = 12x - 4$

25. $f'(x) = 27x^2 - 6x^{-4} = 27x^2 - \frac{6}{x^4}$ **27.** $f'(x) = 7x^{-1/2} - 48x^{-5}$ **29.** $\frac{dy}{dx} = 9x^2 - 5x^{-1/2} = 9x^2 - \frac{5}{\sqrt{x}}$

31. $\frac{dy}{dx} = 20x - 16x^{-3} = 20x - \frac{16}{x^3}$ **33.** $\frac{dy}{dx} = 4x^{-1/2} + 16 = \frac{4}{\sqrt{x}} + 16$

35. $\frac{dy}{dx} = \frac{9}{2}x^{-1/2} + 10x^{-3} = \frac{9}{2\sqrt{x}} + \frac{10}{x^3}$ **37.** $f'(x) = -\frac{2}{4}x^{-3} - \frac{1}{2}x^{-1/2} = \frac{-1}{2x^3} - \frac{1}{2\sqrt{x}}$

39. $f'(x) = \frac{-3}{5}x^{-2} - \frac{12}{7}x^{-3} = \frac{-3}{5x^2} - \frac{12}{7x^3}$ **41.** $g'(t) = \frac{-8}{5}t^{-3} - \frac{5}{3} = \frac{-8}{5t^3} - \frac{5}{3}$

43. $g'(x) = 9 - 2x^{-1/2} - \frac{9}{5}x^{-4} = 9 - \frac{2}{\sqrt{x}} - \frac{9}{5x^4}$ **45.** $f(x) = 3\sqrt{x}$; $f'(x) = \frac{3}{2\sqrt{x}}$

47. $f(x) = x^2 + 2 + \frac{1}{x^2}$; $f'(x) = 2x - \frac{2}{x^3}$ **49.** $g(t) = \frac{t^2}{3t} + \frac{6}{3t} = \frac{t}{3} + \frac{2}{t}$; $g'(t) = \frac{1}{3} - \frac{2}{t^2}$

51. $R'(x) = 2000 - \frac{1}{2}x$; $C'(x) = 720$; $P' = R' - C' = 1280 - \frac{1}{2}x$ **53.** $y - 8 = -6(x + 1)$ **55.** 24 ft/sec

---

Section 3–2, page 168

1. $f'(x) = 21x + 21x = 42x$ **3.** $f'(x) = 0 + 27x^2 = 27x^2$ **5.** $f'(x) = (2x - 14) + (2x + 3) = 4x - 11$

7. $f'(x) = \frac{-3x^2}{x^4} = \frac{-3}{x^2}$ **9.** $f'(x) = \frac{0 - 1}{x^2} = \frac{-1}{x^2}$ **11.** $f'(x) = \frac{16x}{64} = \frac{1}{4}x$

13. $f'(x) = (x^2 + 9)(3x^2 - 4) + (x^3 - 4x + 1)(2x)$ **15.** $f'(x) = \left(x - \frac{1}{x}\right)\left(1 - \frac{1}{x^2}\right) + \left(x + \frac{1}{x}\right)\left(1 + \frac{1}{x^2}\right)$

17. $f'(x) = \left(x^4 - 7x + \frac{1}{x}\right)(2x + 3) + (x^2 + 3x)\left(4x^3 - 7 - \frac{1}{x^2}\right)$ **19.** $f'(x) = \frac{-7}{(x - 2)^2}$

21. $f'(x) = \frac{(x^3 - 7x)(2x + 4) - (x^2 + 4x)(3x^2 - 7)}{(x^3 - 7x)^2}$ **23.** $f'(x) = \frac{(x^3 - 5)(6x - 9) - (3x^2 - 9x + 2)(3x^2)}{(x^3 - 5)^2}$

25. $f'(x) = \frac{(x^3 - 9)[(x + 2)(2x - 8) + (x^2 - 8x)(1)] - [(x^2 - 8x)(x + 2)](3x^2)}{(x^3 - 9)^2}$

27. $f'(x) = \frac{[(x + 2)(x^2 + 7)](3) - (3x + 7)[(x^2 + 7)(1) + (x + 2)(2x)]}{(x + 2)^2(x^2 + 7)^2}$

29. $A' = h \cdot w' + w \cdot h' = (16)(2) + (25)(3) = 107$; area is increasing at a rate of 107 sq in. per minute.

31. $A = \frac{1}{2}(h \cdot b' + b \cdot h') = \frac{1}{2}(9 \cdot 2 + 16 \cdot 6) = 57$; area is increasing at a rate of 57 sq in. per minute.

33. $(f \cdot g)'(2) = g(2) \cdot f'(2) + f(2)g'(2) = 6 \cdot 3 + 7 \cdot 4 = 46$

35. **(a)** $MC(x) = \frac{1}{10}x^2 + 2x + 100$ **(b)** $AC(x) = \frac{1}{30}x^2 + x + 100 + \frac{9500}{x}$

(c) $MAC(x) = (AC)'(x) = \frac{1}{15}x + 1 - \frac{9500}{x^2}$

37. $y + 6 = -3(x - 4)$

Section 3–3, page 177

1. **(i)** $f(g(x)) = 4x^2 + 10x + 4$ **(ii)** $g(f(x)) = 2x^2 + 6x + 1$

3. **(i)** $f(g(x)) = \sqrt{9x + 25}$ **(ii)** $g(f(x)) = 9\sqrt{x} + 25$ **5.** $\frac{dy}{dx} = 5(x^3 - 4x)^4(3x^2 - 4)$

7. $\frac{dy}{dx} = \frac{1}{2}(9x + 16)^{-1/2} \cdot 9 = \frac{9}{2\sqrt{9x + 16}}$ **9.** $\frac{dy}{dx} = 3(3x^2 + 8)^2 \cdot 6x = 18x(3x^2 + 8)^2$

11. $\frac{dy}{dx} = \frac{1}{2}(4x + 9)^{-1/2} \cdot 4 = \frac{2}{\sqrt{4x + 9}}$ **13.** $f'(x) = 3(x^2 - 9x + 5)^2(2x - 9)$

15. $f'(x) = \frac{1}{2}(16x - 4)^{-1/2} \cdot 16 = \frac{8}{\sqrt{16x - 4}}$ **17.** $\frac{dy}{dx} = -30(4x + 1)^{-4} \cdot 4 = \frac{-120}{(4x + 1)^4}$

19. $\frac{dy}{dx} = -12(3x + 1)^{-4} \cdot 3 = \frac{-36}{(3x + 1)^4}$ **21.** $f'(x) = \sqrt{4x + 9} + x \cdot \frac{1}{2}(4x + 9)^{-1/2} \cdot 4 = \frac{6x + 9}{\sqrt{4x + 9}}$

23. $f'(x) = (6x - 4)^2 3(7x - 8)^2 \cdot 7 + (7x - 8)^3 2(6x - 4) \cdot 6 = (6x - 4)(7x - 8)^2(210x - 180)$

25. $f'(x) = \frac{(8x + 10)[3(x^2 + 1)^2 \cdot 2x] - (x^2 + 1)^3(8)}{(8x + 10)^2} = \frac{(x^2 + 1)^2[40x^2 + 60x - 8]}{(8x + 10)^2}$

27. $f'(x) = \frac{(x^2 - 4)^2 3 - (3x + 1)[2(x^2 - 4)(2x)]}{(x^2 - 4)^4} = \frac{(x - 4)(-9x^2 - 4x - 12)}{(x^2 - 4)^4} = -\frac{9x^2 + 4x + 12}{(x^2 - 4)^3}$

29. $f'(x) = 5(3x + (2x + 1)^3)^4(3 + 3(2x + 1)^2 \cdot 2)$ **31.** $f'(x) = 3(7x + \sqrt{3x + 1})^2\left(7 + \frac{3}{2\sqrt{3x + 1}}\right)$

33. $y - 5 = \frac{1}{5}(x - 4)$

35. $R'(t) = \frac{2t + 6}{2\sqrt{t^2 + 6t}}$; $R'(2) = \frac{5}{4}$; revenue will be increasing at a rate of \$1,250,000 per year.

Section 3–4, page 181

1. $f'''(x) = 0$ **3.** $g'''(t) = 6$ **5.** $f'''(x) = \frac{-120}{x^5}$ **7.** $f'''(x) = 1620(3x + 1)^2$ **9.** $f'''(x) = 192(8x + 4)^{-5/2}$

11. $g'''(x) = \frac{-240}{(2x + 1)^4}$ **13.** $y'' = 6(x^2 + 1)^2 + 24x^2(x^2 + 1)$ **15.** $y'' = \frac{6}{\sqrt{6x^2 - 8}} - \frac{36x^2}{(\sqrt{6x^2 - 8})^3}$

17. $y'' = \frac{-18}{(x^2 + 1)^4} + \frac{144x^2}{(x^2 + 1)^5}$ **19.** $y'' = \frac{-120}{(3x^2 + 9)^6} + \frac{4320x^2}{(3x^2 + 9)^7}$

21. $f''(x) = \frac{2}{\sqrt{2x^2 + 1}} - \frac{4x^2}{(\sqrt{2x^2 + 1})^3}$; $f''(2) = \frac{2}{27}$ **23.** $(MC)'(x) = C''(x) = \frac{1}{10}x + 10$; $(MC)'(150) = 25$

25. **(a)** $V(t) = -32t$, velocity when $t = 3$ is -96 ft/sec (The minus sign indicates that the rock is dropping; its height is decreasing.)
(b) $a(t) = -32$; acceleration when $t = 3$ is -32 (ft/sec)/sec $= -32$ ft/sec^2

Section 3–5, page 187

1. explicitly **3.** implicitly **5.** implicitly **7.** $y = \frac{2}{3}(x + 5) + 1$ **9.** $y = \frac{3x - 7}{x + 1}$ **11.** $y = \frac{10}{x + 3}$

13. $y = \dfrac{-9x}{x^2 + x - 5}$ **15.** $y' = \dfrac{-x}{y}$ **17.** $y' = \dfrac{-5}{12y + 2}$ **19.** $y' = \dfrac{8x - 2xy}{x^2 + 3y^2}$ **21.** $y' = \dfrac{12xy - 6x}{2y - 6x^2}$

23. $y' = \dfrac{6y^2 + 9}{12y^2 - 12xy + 3}$ **25.** $y' = \dfrac{1 - y}{x + 3y^2}$ **27.** $y + 1 = 6(x - 2)$ **29.** $C = -77;\ y + 2 = \dfrac{69}{104}(x - 3)$

Section 3–6, page 193

1. $2x\dfrac{dx}{dt} + 2y\dfrac{dy}{dt} = 0$ **3.** $\dfrac{dx}{dt} + 2\dfrac{dy}{dt} = 0$ **5.** $6x\dfrac{dx}{dt} - \dfrac{dx}{dt} + 4y\dfrac{dy}{dt} = 0$ **7.** $\dfrac{dy}{dt} = \dfrac{1}{2}$ **9.** $\dfrac{dy}{dt} = -\dfrac{3}{14}$

11. $2x\dfrac{dx}{dt} + 2y\dfrac{dy}{dt} = 0;\ \dfrac{dy}{dt} = \dfrac{3}{2}$ **13.** $2\dfrac{dx}{dt} + 3\dfrac{dy}{dt} = 0;\ \dfrac{dy}{dt} = -\dfrac{8}{3}$ **15.** $2x\dfrac{dx}{dt} - \dfrac{1}{2\sqrt{y}}\dfrac{dy}{dt} = 0;\ \dfrac{dy}{dt} = 120$

17. $\dfrac{dy}{dt} = -15$ **19.** $\dfrac{dy}{dt} = 48$ **21.** 52 mph **23.** y is increasing at a rate of $27\frac{1}{2}$ rentals per week.

Section 3–7, page 203

1. $dy = 15x^2\,dx$ **3.** $dy = \dfrac{x}{\sqrt{x^2 + 1}}dx$ **5.** $dy = \dfrac{-1}{2(\sqrt{x})^3}dx$ **7.** $dy = (24x^2 - 4)\,dx$ **9.** $dy = 10$

11. $dy = \dfrac{24}{1000}$ **13.** $dy = 30$ **15.** $\Delta y \approx dy = 1$ **17.** $\Delta y \approx dy = \dfrac{1}{200}$ **19.** $\sqrt[3]{29} \approx 3\dfrac{2}{27}$

21. $\sqrt{23} \approx 4\dfrac{4}{5}$ **23.** $\sqrt{52} \approx 7\dfrac{3}{14}$ **25.** $\sqrt[3]{61} \approx 3\dfrac{15}{16}$ **27.** \$21 **29.** \$120 per hour

---

Review Exercises, Chapter 3, page 205

1. $f'(x) = \dfrac{x}{\sqrt{x^2 + 1}}$ **2.** $f'(x) = \dfrac{2x^2 + 1}{\sqrt{x^2 + 1}}$ **3.** $f'(x) = \dfrac{(x^3 - 4x)2x - (x^2 + 1)(3x^2 - 4)}{(x^3 - 4x)^2}$

4. $f'(x) = \left(8x^3 - 9x + \dfrac{6}{x^2}\right)\left(14x - \dfrac{3}{x^2}\right) + \left(7x^2 - 4 + \dfrac{3}{x}\right)\left(24x^2 - 9 - \dfrac{12}{x^3}\right)$

5. $f'(x) = (5x + 1)^3 2(7x + 4) \cdot 7 + (7x + 4)^2 \cdot 3(5x + 1)^2 \cdot 5$ **6.** $f'(x) = \dfrac{(9x^2 + 3x)(6) - (6x - 4)(18x + 3)}{(9x^2 + 3x)^2}$

7. $f'(x) = 4(3x + \sqrt{5x + 1})^3\left(3 + \dfrac{5}{2\sqrt{5x + 1}}\right)$ **8.** $f'(x) = 5[(3x + 1)^2 + (4x - 8)^2]^4[6(3x + 1) + 8(4x - 8)]$

9. $y'' = \dfrac{48}{x^5}$ **10.** $\dfrac{d^2y}{dx^2} = \dfrac{-9}{4}x^4(x^3 + 8)^{-3/2} + 3x(x^3 + 8)^{-1/2}$ **11.** $\dfrac{dy}{dx} = \dfrac{3 - y}{x + 2y}$ **12.** $\dfrac{dy}{dx} = \dfrac{4y^2 - 3x^2}{3y^2 - 8xy}$

13. $y - 3 = \dfrac{4}{3}(x + 4)$ **14.** $y + 3 = \dfrac{-15}{4}(x + 2)$ **15.** $\dfrac{dy}{dt} = -\dfrac{21}{20}$ **16.** $\dfrac{dx}{dt} = -7$ **17.** $dy = \dfrac{-37}{(2x - 5)^2}dx$

18. $dy = \dfrac{3x^2 - 8x}{2\sqrt{x^3 - 4x^2}}dx$ **19.** $\sqrt[3]{10} \approx 2\dfrac{1}{6}$ **20.** $6 - \dfrac{5}{12} = 5\dfrac{7}{12}$

21. $MP(x) = MR(x) - MC(x) = \sqrt{x^2 + 4x} - 6\sqrt{x} - 5$ **22.** increasing \$1775 per week

23. **(a)** $AC(x) = \dfrac{1}{1000}x + 12 + \dfrac{6200}{x}$ **(b)** $MAC(x) = AC'(x) = \dfrac{1}{1000} - \dfrac{6200}{x^2}$ **24.** $MC(100) = 20\dfrac{1}{3}$

CHAPTER 4

Section 4–1, page 217

1. **(a)** 4 **(b)** -2 **(c)** -1 and 3 **(d)** 2 **(e)** $-1, 0, 2, 3$ **(f)** $(-3, -1)$ and $(2, 3)$ **(g)** $(-1, 0)$, $(0, 2)$ and $(3, 5)$
3. $x = \frac{1}{3}$ **5.** $x = 2$ and 1 **7.** $(-\infty, \frac{4}{5})$ **9.** $(-\infty, -2)$ and $(4, \infty)$ **11.** $f(\frac{4}{5}) = -\frac{29}{5}$ is a rel max
13. $f(-2) = \frac{46}{3}$ is a rel max and $f(4) = -\frac{62}{3}$ is a rel min **15.** $x = \frac{-2}{3}$ and $x = 1$ **17.** $x = -4$
19. $(-\infty, -2)$ and $(0, 1)$ **21.** $(-\infty, 3)$
23. inc on $(2, \infty)$, dec on $(-\infty, 2)$

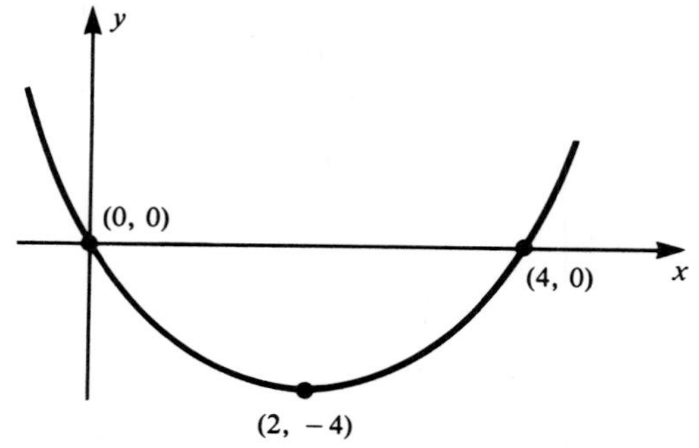

25. inc on $(-\infty, \frac{5}{2})$, dec on $(\frac{5}{2}, \infty)$

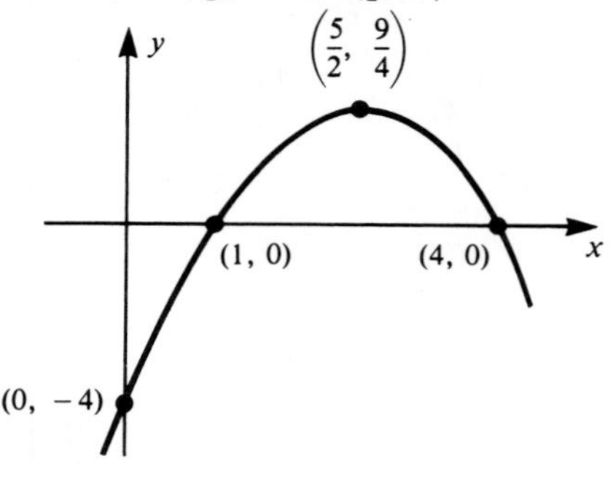

27. inc on $(-\frac{5}{2}, \infty)$, dec on $(-\infty, -\frac{5}{2})$; $f(\frac{-5}{2}) = -\frac{49}{4}$ is a rel min

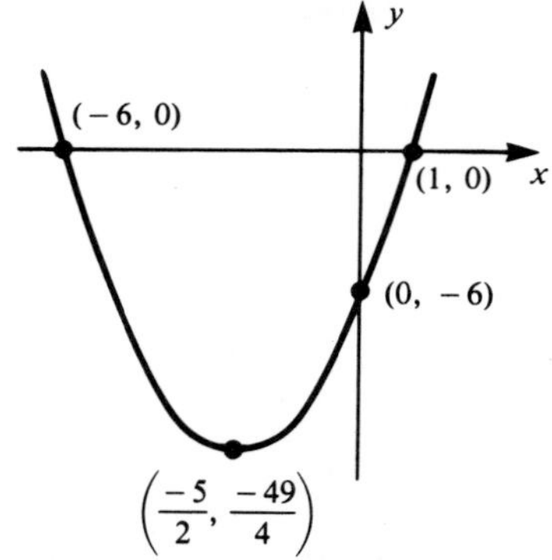

29. inc on $(-\infty, 0)$ and $(4, \infty)$, dec on $(0, 4)$; $f(0) = 2$ is a rel max; $f(4) = -30$ is a real min

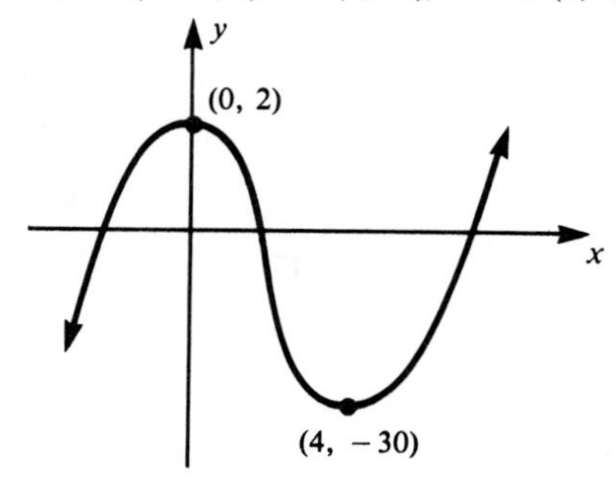

31. inc on $(-\infty, \infty)$; no relative extremum

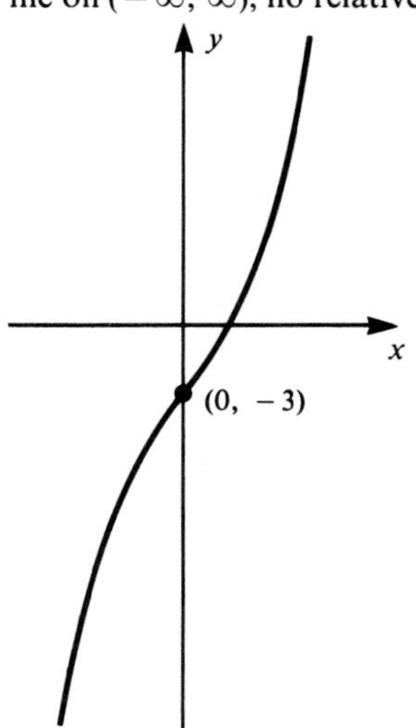

33. inc on $(-\infty, 8)$ and $(8, \infty)$; no relative extremum

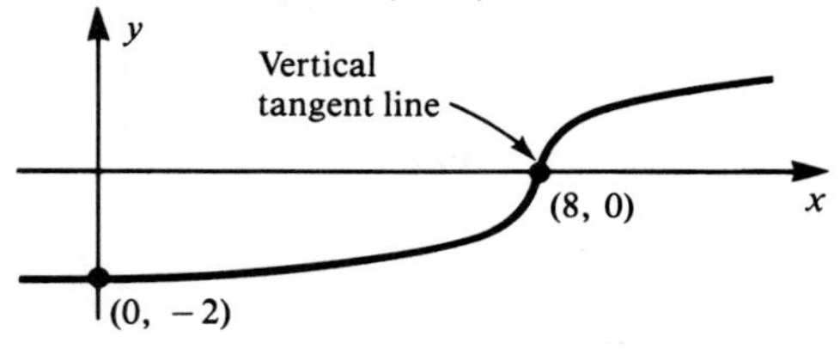

35. inc on $(-3, \infty)$, dec on $(-\infty, -3)$; $f(-3) = -28$ is a rel min

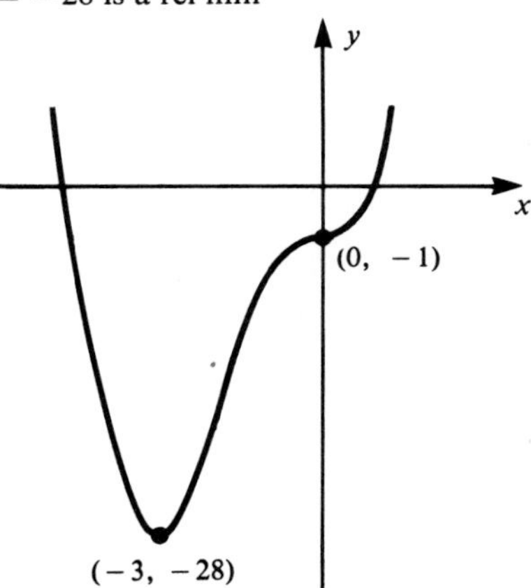

37. inc on $(-3, 0)$, dec on $(0, 3)$; $f(0) = 3$ is a rel max

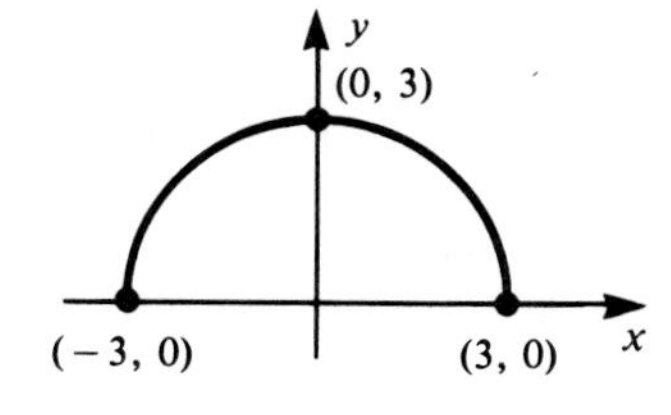

39. dec on $(-\infty, 0)$ and $(0, 2)$, inc on $(2, 4)$ and $(4, \infty)$; $f(2) = -4^{1/3}$ is a rel min

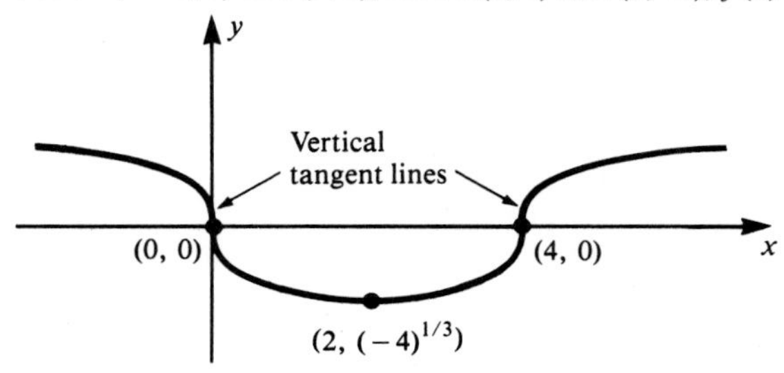

41. dec on $(-6, -4)$, inc on $(-4, \infty)$; $f(-4) = -8$ is a rel min

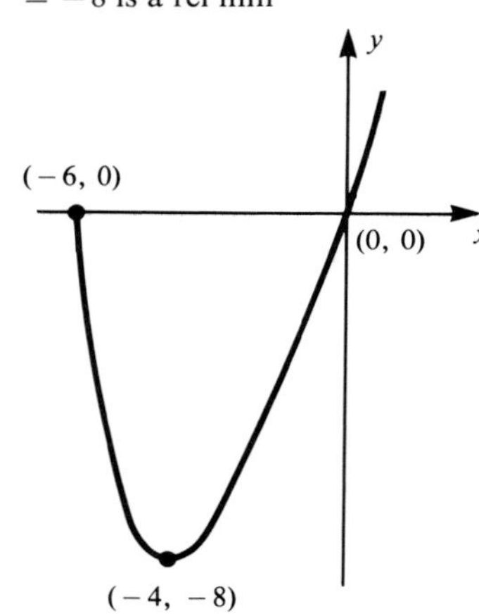

43. inc on $(-\infty, 2)$, dec on $(2, 3)$; $f(2) = 8$ is a rel max;

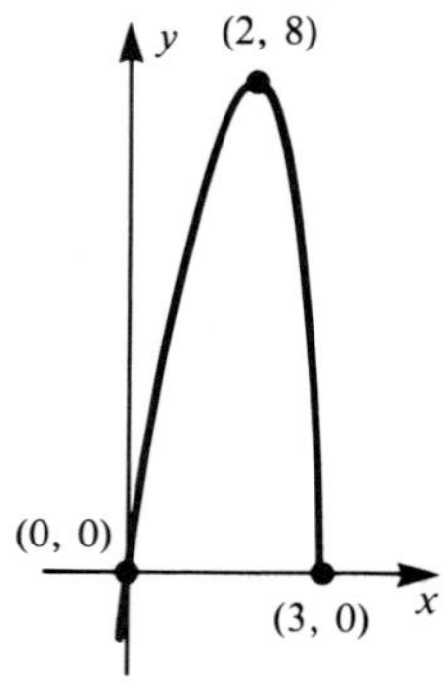

45. **(a)** $0 < t < 20$ **(b)** 240 **47.** **(a)** $0 < t < 3$ **(b)** 150 feet

---

Section 4–2, page 232

1. $(-5, -3)$, $(0, 2)$, and $(3, 5)$ **3.** The points $(-3, 2)$, $(0, 3)$, $(2, 2)$, and $(3, 1)$ **5.** rel min at $x = \frac{2}{7}$
7. rel max at $x = 3$ **9.** rel max at $x = -5$; rel min at $x = 2$
11. rel min at $x = -1$ and $x = 2$; rel max at $x = 0$ **13.** rel max at $x = -\frac{1}{3}$; rel min at $x = \frac{1}{3}$
15. Concave up on $(-\infty, \infty)$, no points of inflection.
17. Concave down on $(-\infty, 2)$, concave up on $(2, \infty)$; $(2, -2)$ is the point of inflection.
19. Concave up on $(-\infty, \frac{2}{3})$, concave down on $(\frac{2}{3}, \infty)$; $(\frac{2}{3}, -\frac{82}{27})$ is the point of inflection.
21. Concave up on $(-\infty, -3)$ and $(2, \infty)$, concave down on $(-3, 2)$. The points of inflection are $(-3, -325)$ and $(2, -100)$.
23. Concave up on $(-\infty, 0)$, concave down on $(0, \infty)$; no points of inflection.
25. Concave up on $(-\infty, -\frac{1}{2})$ and on $(0, \infty)$, concave down on $(-\frac{1}{2}, 0)$. Point of inflection is $(-\frac{1}{2}, 0)$.
27. Concave up on $(-\infty, 5)$, concave down on $(5, \infty)$; $(5, 2)$ is a point of inflection.
29. **(a)** increasing on $(-\infty, \infty)$ **(b)** concave up on $(4, \infty)$; concave down on $(-\infty, 4)$
(c)

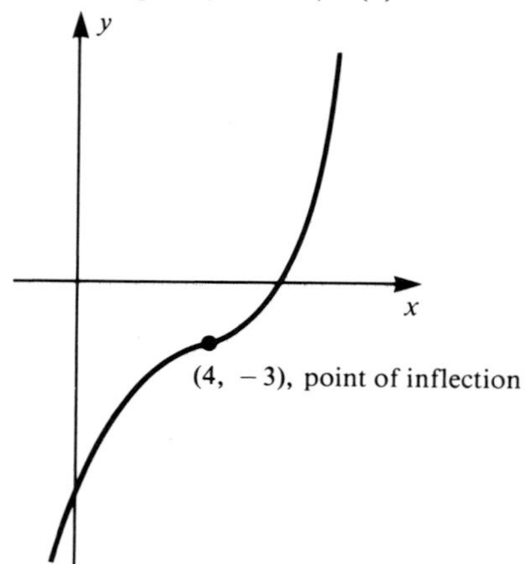

31. **(a)** increasing on $(-\infty, -3)$ and $(1, \infty)$, decreasing on $(-3, 1)$
(b) concave down on $(-\infty, -1)$, concave up on $(-1, \infty)$
(c)

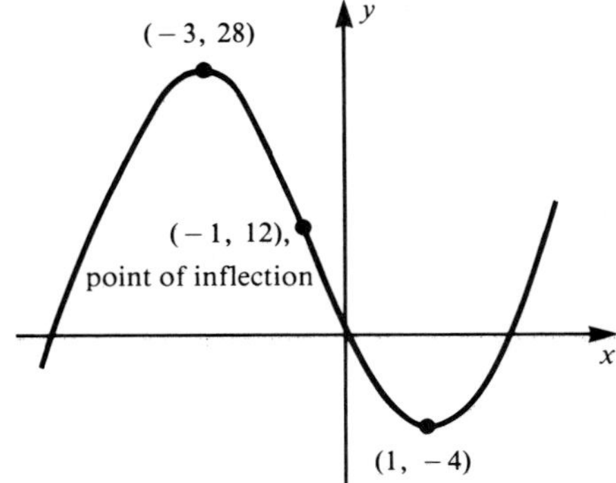

33. **(a)** dec on $(0, 1)$, inc on $(1, \infty)$ **(b)** concave up on $(0, \infty)$

(c)

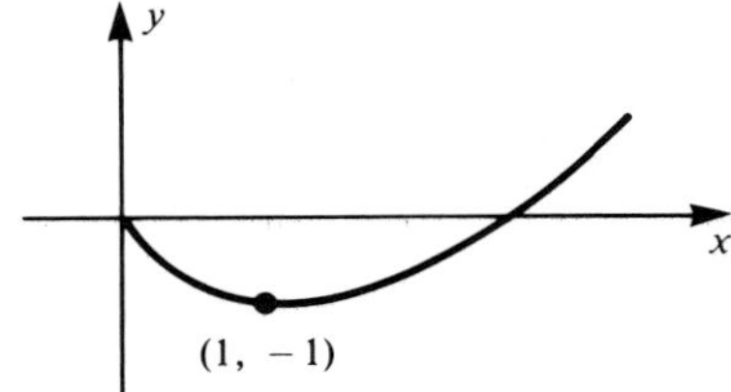

35. $x = 400$

Section 4–3, page 241

1. max is 22, min is -3 **3.** max is -2, min is -29 **5.** min is 6 **7.** min is 24
9. max is 22, min is -13 **11.** max is 0, min is -20 **13.** min is 6, max is 86 **15.** max is 4
17. max is -12 **19.** max is 1 **21.** max is 3, min is $(-9)^{1/3}$ **23.** min is 0, max is 9
25. min is 0, max is 8 **27.** min is 40, no max **29.** min is $\frac{1}{4}$, no max **31.** min is -2, no max
33. $x = 750$, $p = 15$ **35.** 40 mph **37.** 50 mph

Section 4–4, page 256

1. V.A. is the line $x = 6$; H.A. is the line $y = 1$.
3. V.A.'s are the lines $x = 3$ and $x = -3$; H.A. is the line $y = 0$.
5. V.A.'s are the lines $x = -3$ and $x = 4$; H.A. is the line $y = 0$.
7. V.A. is the line $x = -2$; H.A. is the line $y = \frac{4}{5}$. **9.** No V.A.; H.A. is the line $y = 0$.
11. ∞ **13.** $-\infty$ **15.** $\frac{3}{5}$ **17.** $\frac{1}{2}$
19. V.A. is the line $x = -5$; $\lim_{x \to -5^+} f(x) = -\infty$; $\lim_{x \to -5^-} f(x) = \infty$.
H.A. is the line $y = 3$; $\lim_{x \to \infty} f(x) = 3 = \lim_{x \to -\infty} f(x)$.

21. No V.A.; H.A. is $y = 0$, the x-axis. $\lim_{x\to\infty} f(x) = 0 = \lim_{x\to-\infty} f(x)$

23. $\lim_{x\to 1} f(x) = -1$; line $x = 2$ is a V.A. $\lim_{x\to 2^+} f(x) = \infty$.
$\lim_{x\to 2^-} f(x) = -\infty$; line $y = 0$ is H.A. $\lim_{x\to\infty} f(x) = \lim_{x\to-\infty} f(x) = 0$.

25. $\lim_{x\to 4} f(x) = \frac{3}{5}$; line $x = -1$ is a V.A. $\lim_{x\to -1^+} f(x) = -\infty$.
$\lim_{x\to -1^-} f(x) = \infty$; line $y = 1$ is H.A. $\lim_{x\to\infty} f(x) = \lim_{x\to-\infty} f(x) = 1$.

27. ∞ **29.** $-\infty$ **31.** ∞

33.

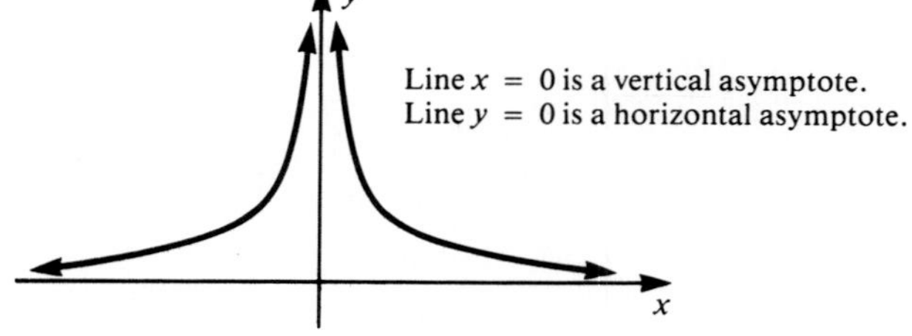

35.

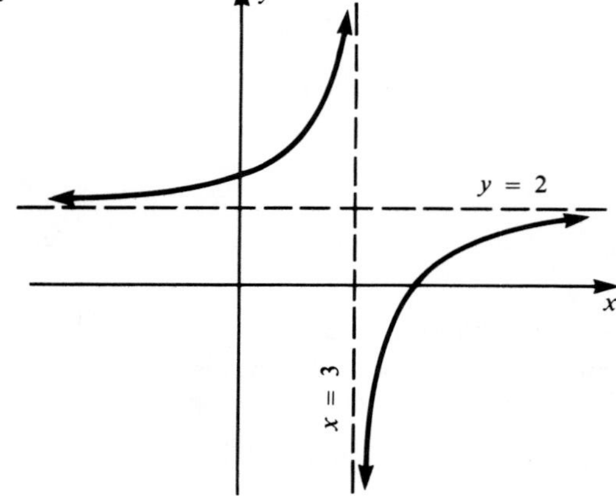

37.

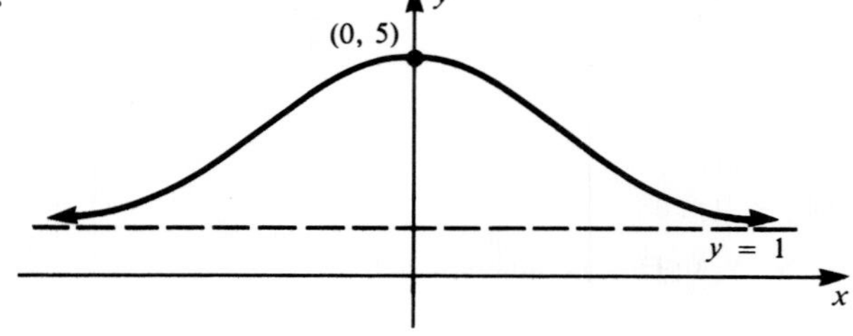

39.

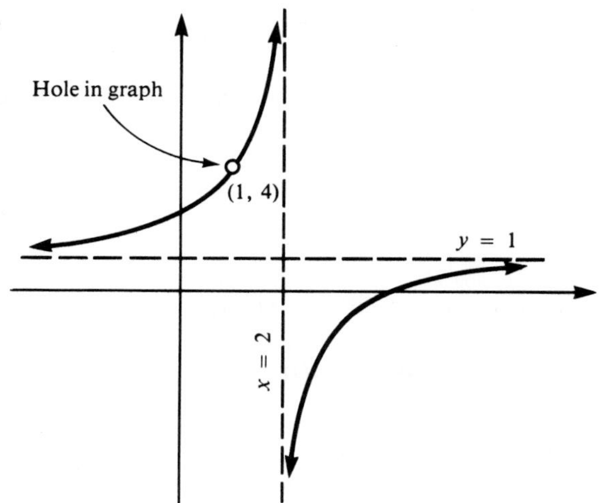

41.

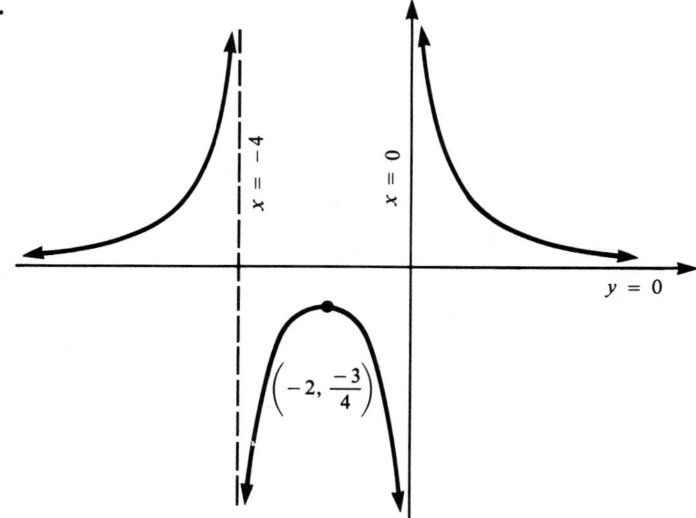

43. **(a)** Average cost when x is large is approx. 8.

(b)

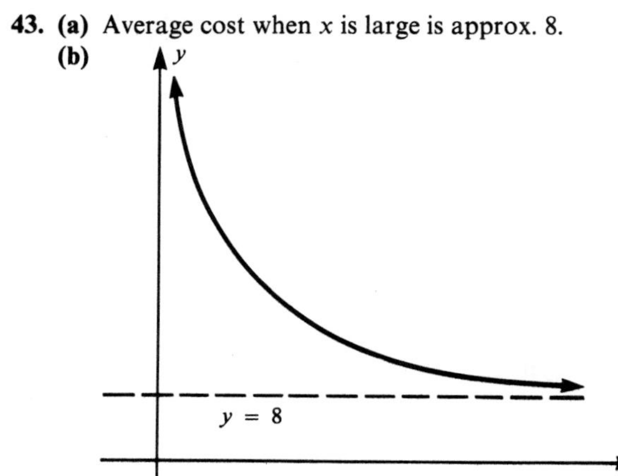

Section 4–5, page 264

1. 2,000,000 sq ft **3.** \$500 **5.** \$2.40 **7.** base is 15″ by 30″, height is 20″ **9.** a square 40′ on each side

Section 4–6, page 275

1. **(a)** $AC(x) = \frac{x^2}{20} - 15x + 3000 + \frac{25{,}600}{x}$ **(b)** $x = 160$ **(c)** avg cost is \$2040 **(d)** $MC(160) = 2040$

3. **(a)** $AC(x) = \frac{x^2}{4} - 49x + 4000 + \frac{10000}{x}$; $MC(x) = \frac{3x^2}{4} - 98x + 4000$ **(b)** $x = 100$
(c) $AC(100) = MC(100) = 1700$

5. $x = 300$; $P(300) = \$26{,}000$ **7.** Yes; increase production to 210 units per day.

9. Decrease to 300 per week. **11.** 250 sets per day **13.** **(a)** $x = 39$; **(b)** $x = 36$

15. When $x = 39$, $p = 1542$; after tax with $x = 36$, $p = 1548$.

Review Exercises, Chapter 4, page 279

1. Max is $\frac{1}{2}$; min is $\frac{1}{5}$; f is dec on (2, 5).

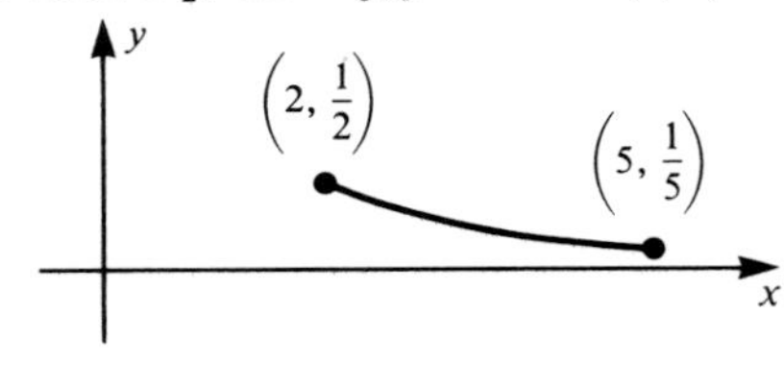

2. Max is $-\frac{1}{6}$; min is -2; f is dec on $(-6, -\frac{1}{2})$.

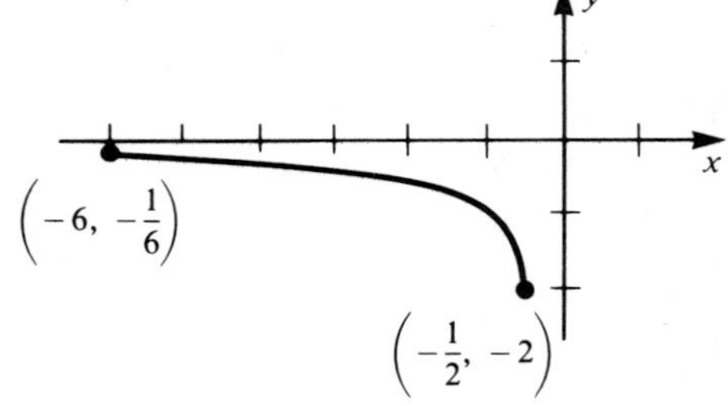

3. Max is 10; min is 1; f is inc on (2, 5).

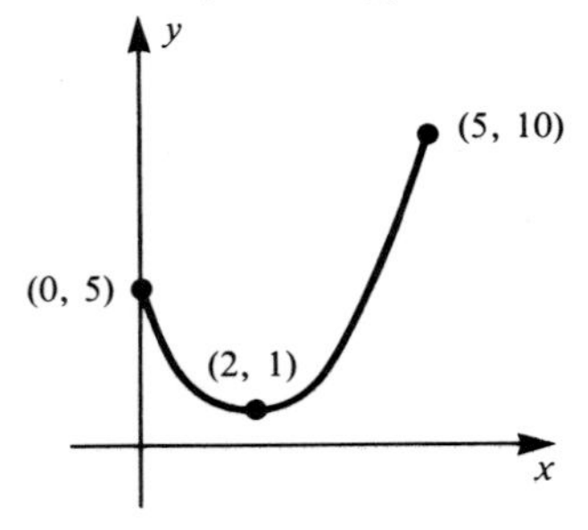

4. Max is 9; min is -39; f is dec on $(-1, 3)$.

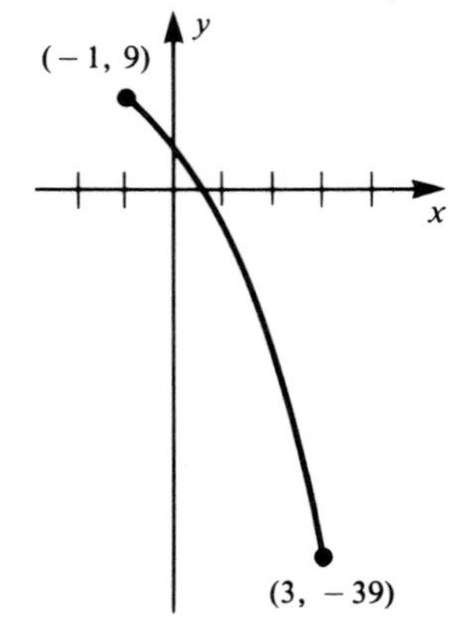

5. Max is 33; min is 6; f is inc on (4, 8).

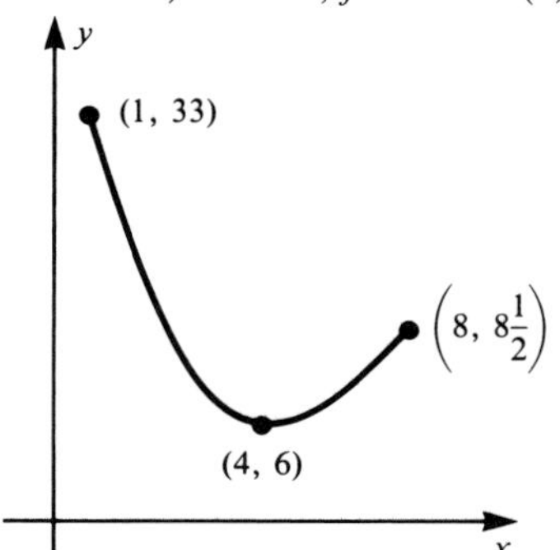

6. Max is 29; min is 9; f is dec on (1, 3), inc on (3, 6).

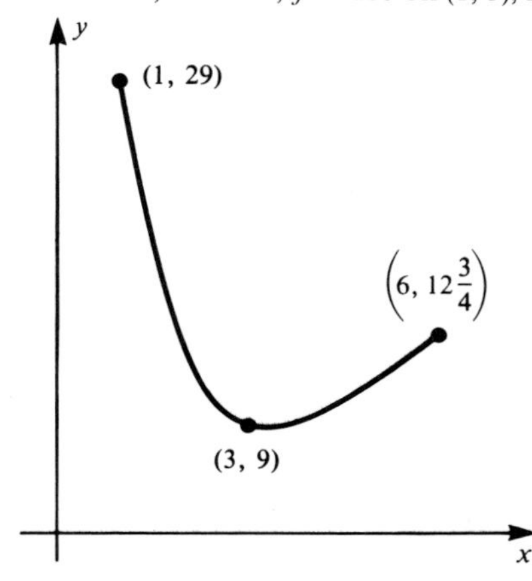

7. Max is 2; min is -254; f is inc on $(-1, 0)$ and (4, 5); $(3, -160)$ is a point of inflection.

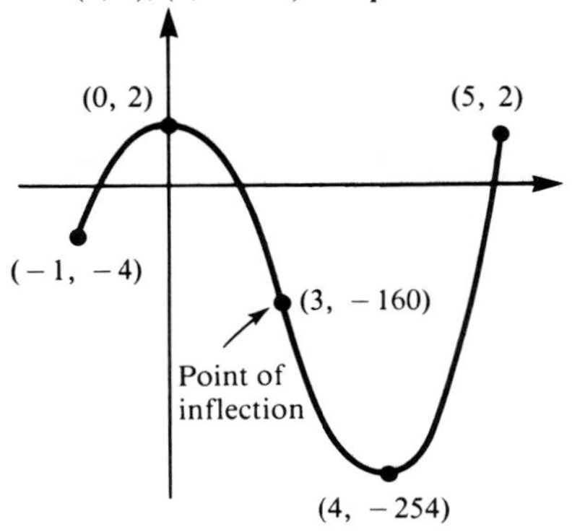

8. Max is 11; min is -21; f is inc on (3, 4), dec on $(-1, 3)$. Points of inflection are (0, 6) and $(2, -10)$.

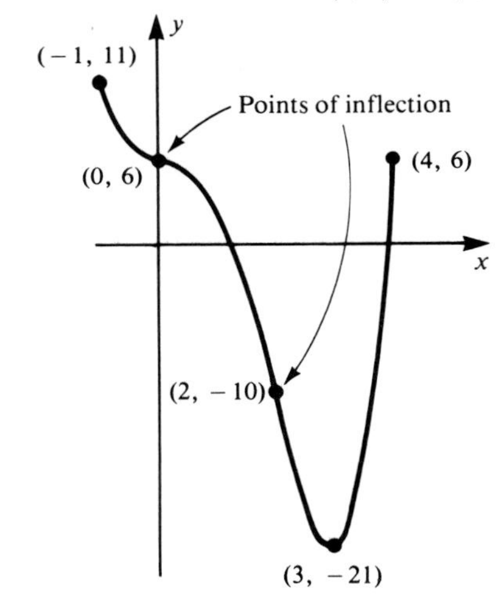

9. Max is 6; min is -6; f is inc on $(-6, 0)$.

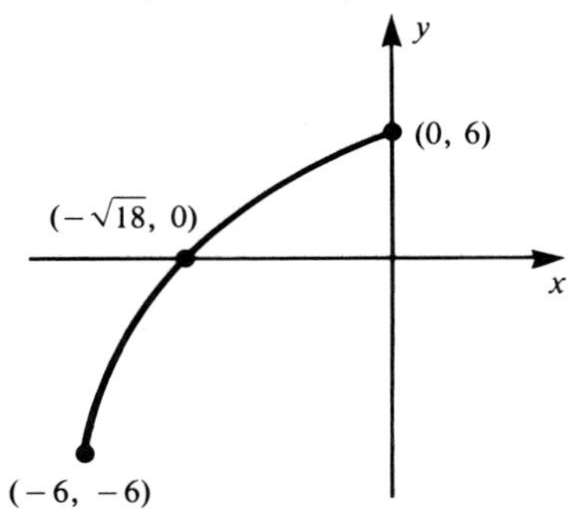

10. Max is $2\sqrt{18}$; min is 6; f is inc on $(0, \sqrt{18})$, dec on $(\sqrt{18}, 6)$.

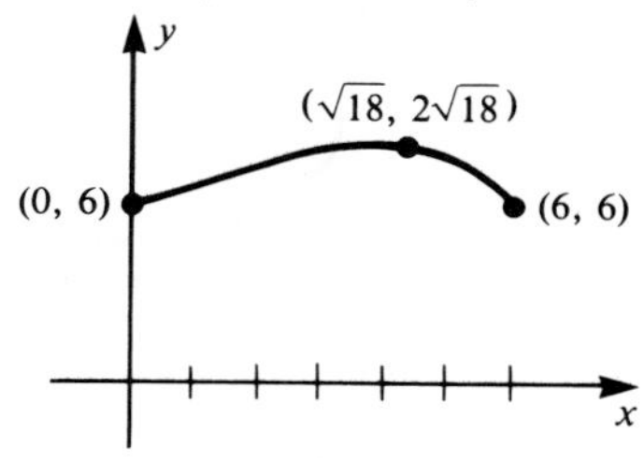

11. f is dec on $(-6, -\sqrt{18})$ and on $(\sqrt{18}, 6)$; $(0, 0)$ is a point of inflection.

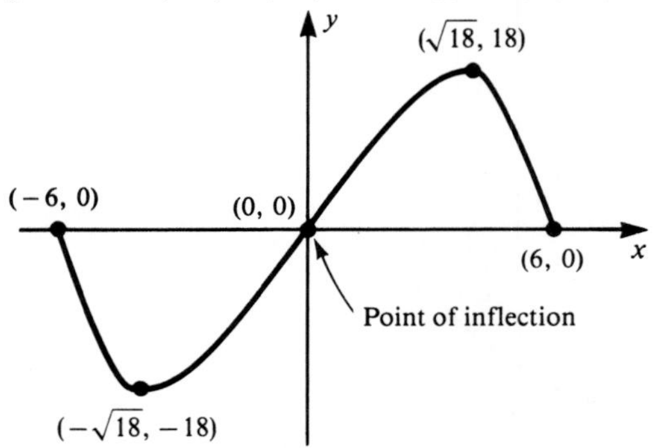

12. f is inc on $(-\infty, \infty)$; $(8, 0)$ is a point of inflection.

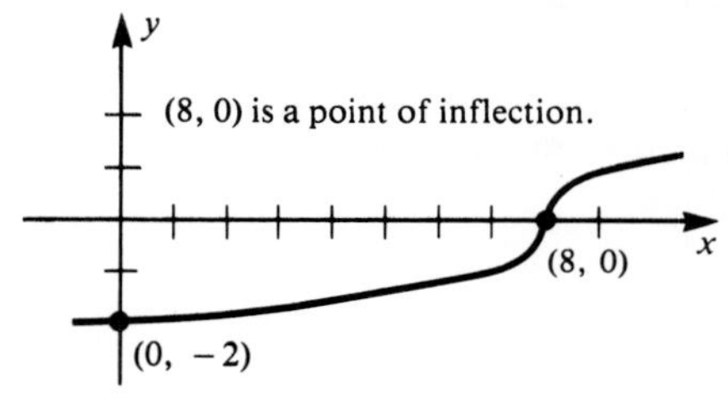

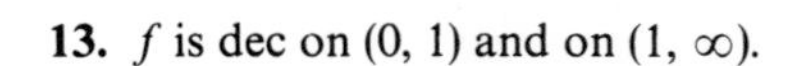

13. f is dec on $(0, 1)$ and on $(1, \infty)$.

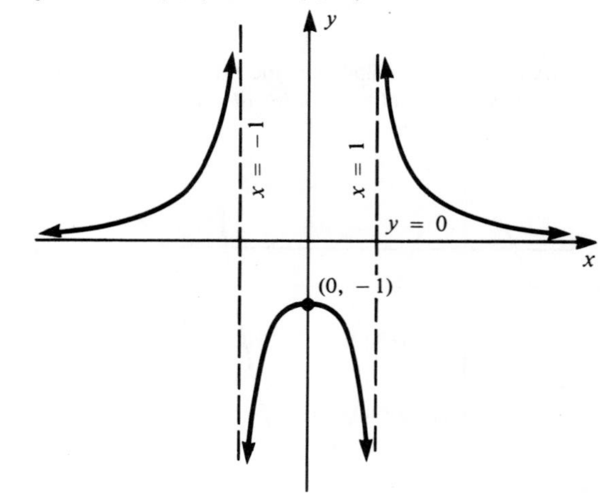

14. f is dec on $(-\infty, -1)$, $(-1, 1)$ and $(1, \infty)$; $(0, 0)$ is a point of inflection.

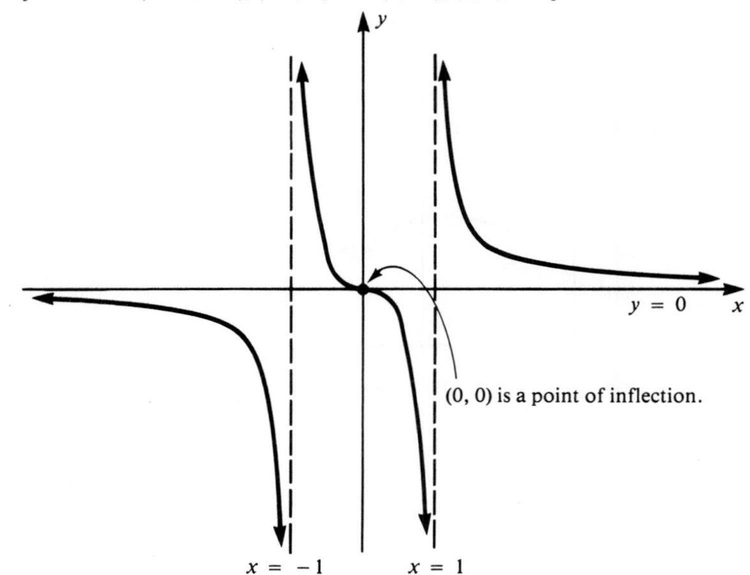

15. f is dec on (0, 1) and on (1, ∞).

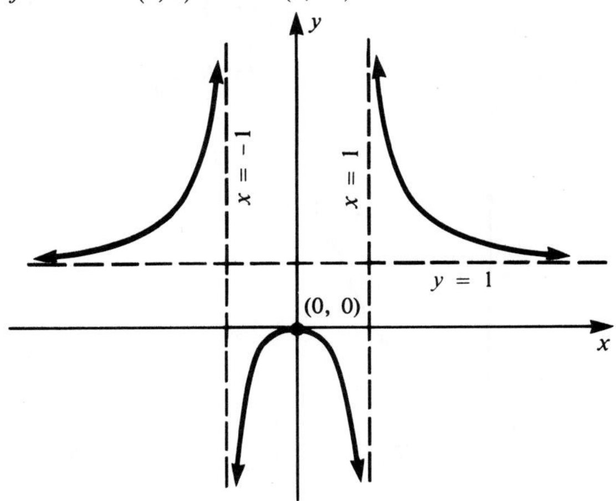

16. f is dec on (0, ∞); $(-\sqrt{2}, \frac{1}{8})$ and $(\sqrt{2}, \frac{1}{8})$ are points of inflection.

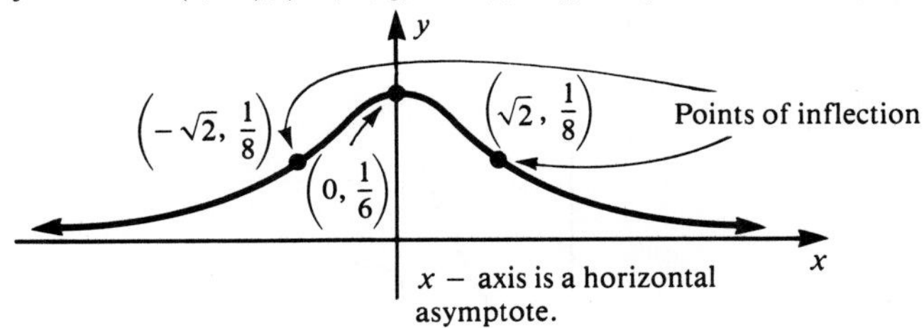

17.

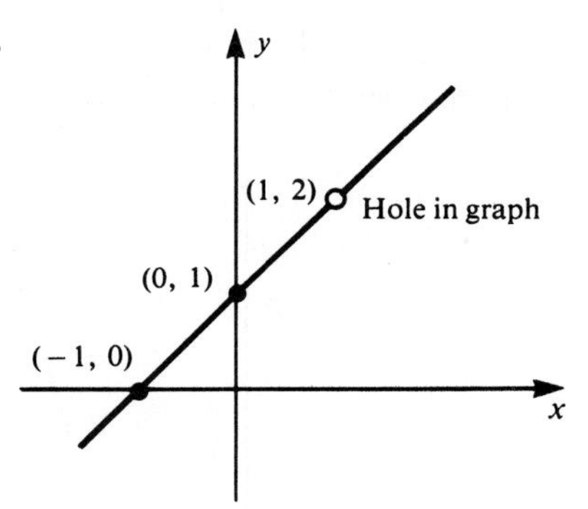

18. f is inc on $(-\infty, -3)$, and inc on $(-3, \infty)$.

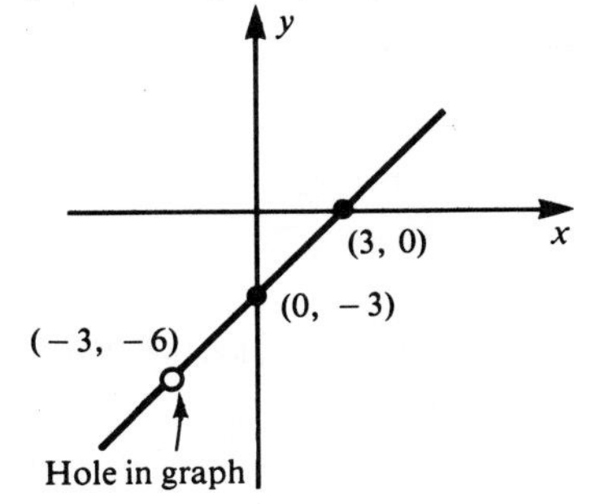

19. 64 **20.** 8 **21.** **(a)** min = $-\frac{3}{5}$ **(b)** max = $\frac{3}{5}$ **22.** $x = 200$ **23.** 69 ft **24.** $p = 20$ **25.** $x = 5$
26. $p = 20$ **27.** 180 refrigerators per week **28.** **(a)** $x = 100$ **(b)** $x = 95$

CHAPTER 5

Section 5–1, page 293

1. **(a)** \$1060 **(b)** \$1,123.60 **(c)** \$1,262.47 **3.** **(a)** \$1,259.71 **(b)** \$2,519.42 **(c)** \$12,597.10

5. **(a)** \$11,881 **(b)** \$11,948.31 **(c)** \$11,964.13 **7.** $f(1) = 4$ **9.** $g(3) = 8$ **11.** $f(-2) = \frac{1}{4}$ **13.** 3 years

15. $1000(1.03)^8$ **17.** two years **19.** $f(2) = 45$ **21.** $f(0) = 7$ **23.** $b = 25$ **25.** $b = \frac{1}{9}$ **27.** $c = 3$

29. $c = -1$ **31.** $k = 16$ **33.** $k = \frac{1}{9}$

35. **37.**

39. **41.**

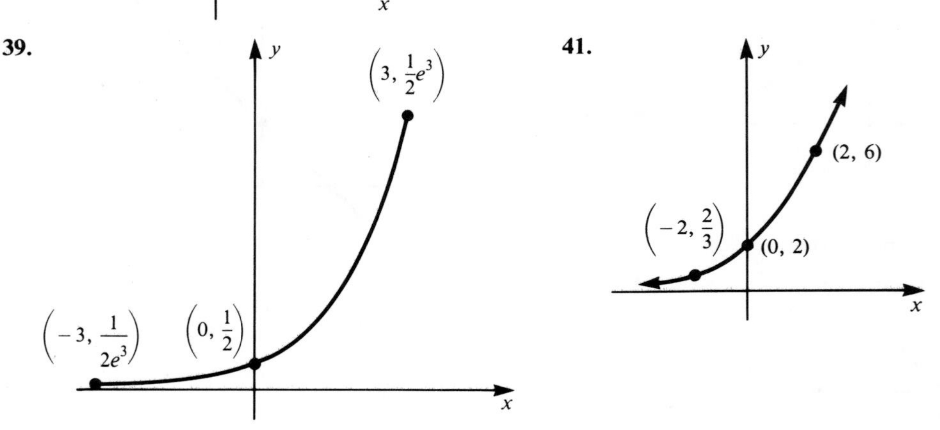

43. **(a)** \$1,808.72 **(b)** \$1,822.11

45. The king could only fill 39 squares. The 64th square *alone* would require over 9,000,000,000,000,000,000 grains of wheat!

Section 5–2, page 304

1. 3 **3.** -4 **5.** -2 **7.** 0 **9.** 4 **11.** 5 **13.** (7.4, 2) **15.** (1.4, 4) **17.** 9 **19.** 3 **21.** $\frac{1}{2}$

23. 6 **25.** 5 **27.** 3 **29.** 4 **31.** -4 **33.** 5 **35.** $\frac{5 \ln 16}{4}$ **37.** $50 \ln 4$ **39.** $\frac{\ln 18}{\ln 2}$

41. $\frac{\ln 105}{2 \ln 5}$ **43.** $\frac{\ln 4}{\ln 3}$ **45.** 1 **47.** in $\frac{50 \ln 3}{3}$ years, a little over 18 years and 3 months

49. $\frac{50 \ln 2}{7}$ years, slightly less than 5 years **51.** $\frac{\$20{,}000}{e^{1.5}} \approx \$4{,}462.60$ **53.** $50 \ln 2$, a little bit less than 35 years

Section 5–3, page 312

1. $4e^x$ **3.** $3x^2 - 2e^x$ **5.** $3x^2e^x + x^3e^x = e^x[x^3 + 3x^2]$ **7.** $18e^{3x}$ **9.** $4e^{x/2}$ **11.** $-\dfrac{x^3 - 3x^2 + 2}{e^x}$

13. $\dfrac{xe^x - 2e^x - 4}{x^3}$ **15.** $4(e^x - 7)^3e^x$ **17.** $\dfrac{e^x}{2\sqrt{e^x + 1}}$ **19.** $2x^2e^{5x}(5x + 3)$ **21.** $12(e^{4x} + 1)^2e^{4x}$

23. $6x + e^{-x} - xe^{-x}$ **25.** $\dfrac{x}{\sqrt{x^2 + 1}}e^{4x} + 4e^{4x}\sqrt{x^2 + 1}$ **27.** $y - e^2 = \dfrac{1}{3}e^2(x - 6)$

29. $f'(x) = (2x + 1)e^{(x^2 + x)}$; $f''(x) = 2e^{(x^2 + x)} + (2x + 1)^2e^{(x^2 + x)}$

31. $f'(x) = 2xe^{4x} + 4x^2e^{4x} = e^{4x}(4x^2 + 2x)$; $f''(x) = e^{4x}(16x^2 + 16x + 2)$ **33.** $f'(x) = \dfrac{4e^{4x} + 5}{2\sqrt{e^{4x} + 5x}}$

35. $f'(x) = 2xe^{\sqrt{3x+1}} + \dfrac{3(x^2 + 1)}{2\sqrt{3x + 1}}e^{\sqrt{3x+1}}$ **37.** $y - \dfrac{1}{4} = -\dfrac{1}{4}(x - \ln 4)$ **39.** $x = 0, x = 2$

41. on $(-\infty, 0)$ and $(0, 1)$ **43.** $\left(2, \dfrac{2}{e^2}\right)$ **45.** max value $= 4e^2$, min value $= \dfrac{-2}{e}$

47.

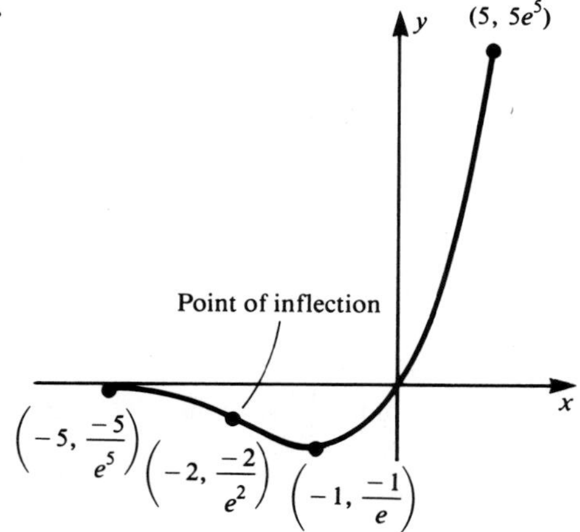

Section 5–4, page 319

1. $f'(x) = 2x + \dfrac{1}{x}$ **3.** $1 + \ln x$ **5.** $\dfrac{1}{x}$ **7.** $\dfrac{1}{x \ln 10}$ **9.** $e^x \ln x + \dfrac{e^x}{x}$ **11.** $\dfrac{1}{x}$ **13.** $\dfrac{1 - \ln x}{x^2}$

15. $\dfrac{2x}{x^2 + 1}$ **17.** $3[\ln(x + 1)]^2\left(\dfrac{1}{x + 1}\right)$ **19.** $\ln(3x - 8) + \dfrac{3x}{3x - 8}$ **21.** $\dfrac{12x^2 - 6(x^2 + 1)\ln(x^2 + 1)}{36x^2(x^2 + 1)}$

23. $\dfrac{4(\ln x)^3}{x}$ **25.** $\dfrac{1}{x}e^{5x} + 5(\ln x)e^{5x}$ **27.** $2(\ln 6)x$ **29.** $\dfrac{2x - 3}{x^2 - 3x + 1}$ **31.** $\dfrac{4}{4x + 8}$ **33.** 2 **35.** 1

37. $\dfrac{1}{2x}$ **39.** $\dfrac{12}{3x + 1}$ **41.** $[\ln(x^2 + 1)]^{-1/2}\left[\dfrac{x}{x^2 + 1}\right]$ **43.** $\dfrac{10\ln(5x + 1)}{5x + 1}$

45. $\dfrac{1}{e^{2x} + \sqrt{4x + 10}}\left[2e^{2x} + \dfrac{4}{2\sqrt{4x + 10}}\right]$ **47.** $3[\ln(x + e^{2x})]^2\left(\dfrac{1}{x + e^{2x}}\right)(1 + 2e^{2x})$

49. $f'(x) = 2x \ln x + x$; $f''(x) = 2\ln x + 3$ **51.** $f'(x) = \dfrac{1}{2}\ln(x + 1) + \dfrac{x}{2(x + 1)}$; $f''(x) = \dfrac{1}{2(x + 1)} + \dfrac{1}{2(x + 1)^2}$

Section 5–5, page 325

1. \$217,462 **3.** \$24,428 **5.** in 100 years, \$27,299; in 200 years, \$1,490,479
7. In *B*'s last year, budget grew about \$4.2 billion. Total growth under *A* was only \$9.8 billion.
9. about 1.5 billion tons **11.** about 3300 years old **13.** about 5000 years old

Section 5–6, page 331

1. $e^{x \ln 3}$ **3.** $e^{-x \ln 9}$ **5.** $e^{(x \ln 7)/2}$ **7.** $5e^{x \ln 8}$ **9.** $7e^{((\ln 4)/3)x}$ **11.** 4

13. $\frac{1}{3}$ **15.** -1 **17.** $\frac{\ln 36}{\ln 6} = 2$ **19.** $\frac{\ln 5}{\ln 25} = \frac{1}{2}$ **21.** $\frac{\ln 14}{\ln 10} \approx 1.146$ **23.** not defined **25.** not defined

27. $\frac{1}{x \ln 2}$ **29.** $2x \log_4 x + \frac{x}{\ln 4}$ **31.** $(\ln 10)10^x$ **33.** $\left(\frac{\ln 7}{2}\right)7^{x/2}$ **35.** $2x(\ln 16)16^{x^2}$

37. $\frac{4 \ln 10}{2\sqrt{4x+1}} 10^{\sqrt{4x+1}}$ **39.** $10(2x-2)(\ln 3)3^{x^2-2x}$ **41.** $(\ln 2)^2 2^{(2^x)} 2^x$ **43.** $\frac{4}{(\ln 3)(4x-2)}$

45. $\frac{12x}{(\ln 4)(2x^2-8)}$ **47.** $3[\log_4(x^2-9)]^2\left[\frac{2x}{(\ln 4)(x^2-9)}\right]$ **49.** $\frac{-1}{x(\ln x)^2}$

Review Exercises, Chapter 5, page 333

1. $f'(x) = 2xe^{3x} + 3x^2e^{3x}$ **2.** $f'(x) = e^x \ln x + \frac{e^x}{x}$ **3.** $f'(x) = \frac{2x}{x^2+1}$ **4.** $f'(x) = \frac{3}{6x+8}$ **5.** $\left(\frac{1}{e}, \infty\right)$

6. $(-\infty, 2)$ **7.** after 29 minutes

8. **(a)** $\frac{17 \ln 20}{\ln 2} \approx 73.5$ mins **(b)** $\frac{17 \ln 100}{\ln 2} \approx 112.9$ mins **(c)** $\frac{17 \ln 1000}{\ln 2} \approx 169.4$ mins **9.** about 1700 years old

10. **(a)** $\frac{100 \ln 4}{3} \approx 46.2$ years **(b)** $25e^3 \approx 502$ million **11.** **(a)** quarterly, \$22,080 **(b)** continuously, \$22,255

12. 7.25% **13.** about \$195 **14.** **(a)** 2.71828 $(\approx e^1 = e)$ **(b)** .367879 $\left(\approx e^{-1} = \frac{1}{e}\right)$

CHAPTER 6

Section 6–1, page 345

1. $f(x) = \frac{1}{4}x^4 - \frac{1}{2}x^2 + 5x + C$ **3.** $f(x) = \frac{-1}{4x^4} + C$ **5.** false **7.** false **9.** false

11. $\frac{1}{5}x^5 + C$ **13.** $\frac{3}{5}x^{5/3} + C$ **15.** $14x + C$ **17.** $\frac{3}{2}x^2 - 6x + C$ **19.** $\frac{3}{2}x^4 - 4x^2 + C$

21. $\frac{1}{4}x^4 - \frac{1}{2}x^2 + C$ **23.** $\frac{-1}{4x^4} + C$ **25.** $\frac{2}{3}x^{3/2} - 4\sqrt{x} + C$ **27.** $6x^{2/3} + C$ **29.** $\frac{-1}{8x^2} + \frac{4}{x} + C$

31. $\frac{2}{3}\sqrt{x} - \frac{x^2}{10} + C$ **33.** $\frac{1}{3}x^3 - 2x^2 + C$ **35.** $\frac{2}{5}x^{5/2} + 2x^{3/2} + C$ **37.** $x + \frac{9}{x} + C$

39. $\frac{1}{3}x^3 + 2x^2 + 3x + C$ **41.** $\frac{1}{2}x^2 - 5x + C$ **43.** $\frac{1}{2}x^2 + \frac{4}{3}x^{3/2} + x + C$

45. $R(x) = -3x^2 + 200x$ **47.** $C(x) = \frac{1}{6}x^3 - \frac{1}{2}x^2 + 2x + 11$

Section 6–2, page 355

1. $y = 7x^3 - 7x^2 + 8x + C$ **3.** $y = 4x^2 - \ln|x| + C$ **5.** $f(x) = \frac{-1}{3x} + 8\ln|x| + C$ **7.** true **9.** true

11. false **13.** true **15.** $f(x) = x^3 + 6x + 4$ **17.** $y = \frac{1}{2}x^2 + \frac{16}{x} - 32$ **19.** $y = 5x - \ln|x| + 13$

21. $y = 4e^{2x} + 10e^2$ **23.** $y = \frac{-1}{e^x} + 4$ **25.** $y = \frac{4}{3}x^3 + \ln|x| + \frac{20}{3}$ **27.** $f(x) = \frac{1}{3}x^3 + 2x - \frac{1}{x} + \frac{17}{3}$

29. $f(x) = \frac{2}{3}x^{3/2} + 3\ln|x| + \frac{13}{3}$ **31.** $y = \frac{1}{2}e^{2x} + 2e^x + x + C$ **33.** $f(x) = 7x^3 + 3x^2 + 4x + 7$

35. $y = 10x^2 - 36x + 40$ **37.** $h(t) = -16t^2 + 60t + 8$

39. $C(x) = \frac{x^3}{12} - 12x^2 + 350x + 2314$; fixed cost is \$2314 **41.** $x = -80p + 9400$

Section 6–3, page 365

1. 20 **3.** -30 **5.** 12 **7.** $\frac{14}{3}$ **9.** $\frac{3}{10}$ **11.** 15 **13.** $\frac{8}{2} + \frac{27}{2} = \frac{35}{2}$ **15.** 1 **17.** $-\ln 4$

19. $-\ln 4$ **21.** $\frac{-4}{e^3} + 4e^2$ **23.** $\frac{152}{3}$ **25.** 260 **27.** $\frac{1}{4}\ln 4$ **29.** $0 = \frac{56}{3} - \frac{56}{3}$ **31.** 3 **33.** $\frac{92}{3}$

35. $\frac{20}{3} + 4\ln 3$ **37.** 5 **39.** 14 **41.** 1 **43.** -8 **45.** 27

47. **(a)** increase by 9000 **(b)** decrease by 7200

49. **(a)** 192 ft

(b) 0 ft; same height

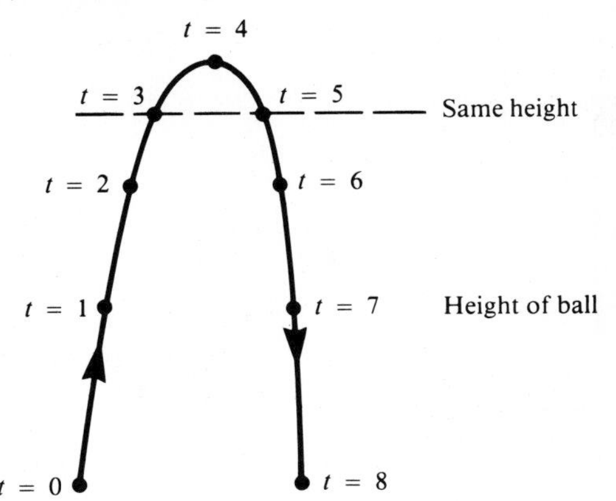

Section 6–4, page 383

1. $4 \cdot 1 = 4; \int_{-1}^{3} 1\,dx = 4$ **3.** $\left(\frac{1}{2}\right)(4)(6) = 12; \int_{-2}^{2} \left(\frac{3}{2}x + 3\right)dx = 12$ **5.** $\int_{-3}^{2} 4\,dx = 20$

7. area $= 2 \cdot 4 = 8 = \int_{-1}^{3} -(-2)\,dx$ **9.** area $= \frac{1}{2}(4)(2) = 4; \int_{-2}^{2} -\left(-\frac{1}{2}x - 1\right)dx = 4$

11. area $= 4 \cdot 2 + \frac{1}{2}(4)(4) = 16; \int_{1}^{5} (x + 1)\,dx = 16$ **13.** $\int_{2}^{7} (2x - 1)\,dx = 40$ **15.** $\int_{0}^{6} -(-4)\,dx = 24$

17. $\int_{2}^{6} -(5 - 3x)\,dx = 28$

19. $\int_{0}^{3} (x^2 - 2x + 1)\,dx = 3$

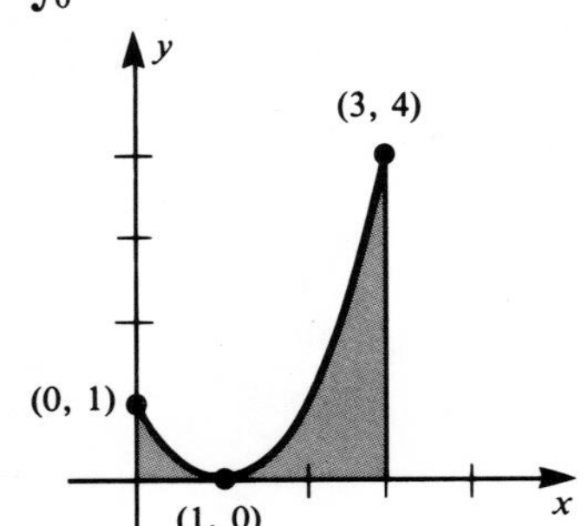

21. $\int_{2}^{10} \frac{2}{x}\,dx = 2 \ln 5$

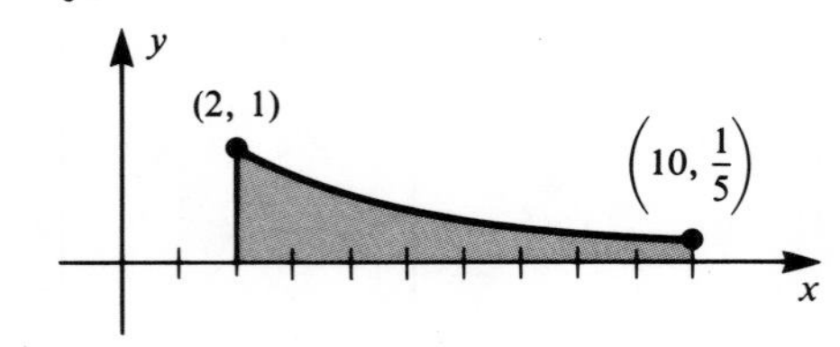

23. $\int_{1}^{3} [(x^2 + 1) - (x - 4)]\,dx = \frac{44}{3}$

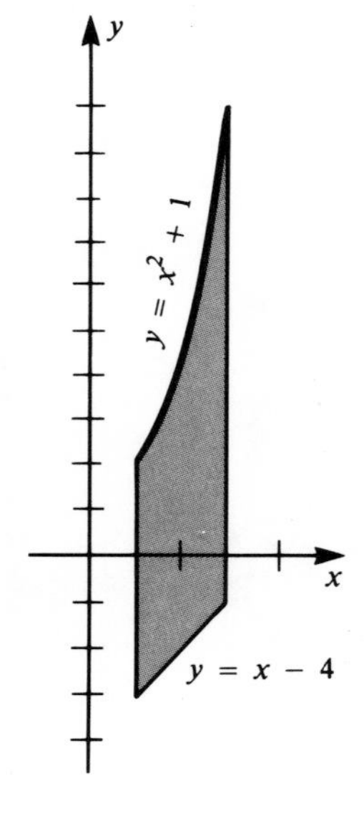

25. $\int_{0}^{2} [(x^3 + 2x^2 + 5x + 3) - (x + 2)]\,dx = \frac{58}{3}$

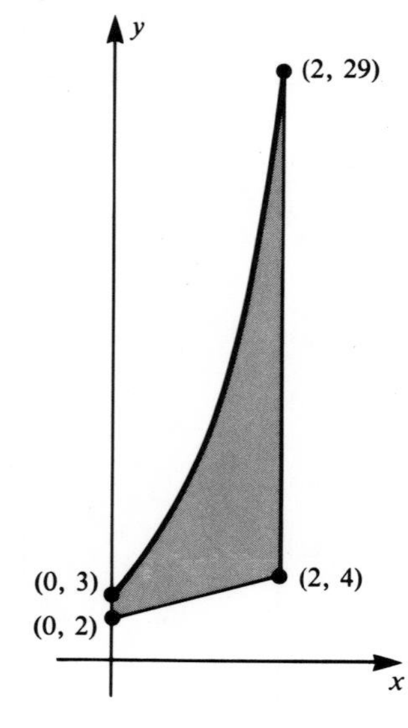

27. area $= \frac{1}{2}(2)(2) + \frac{1}{2}(3)(3) = \frac{13}{2}$; $\int_{-2}^{0} -x\,dx + \int_{0}^{3} x\,dx = \frac{4}{2} + \frac{9}{2} = \frac{13}{2}$

29. area $= (3)(6) + \frac{1}{2}(6)(2) = 24$; $\int_{-3}^{3} \left(\frac{1}{3}x + 2\right) - (-2)\,dx = 24$

31. $\int_{1}^{2} -(3x - 6)\,dx + \int_{2}^{4} (3x - 6)\,dx = \frac{15}{2}$

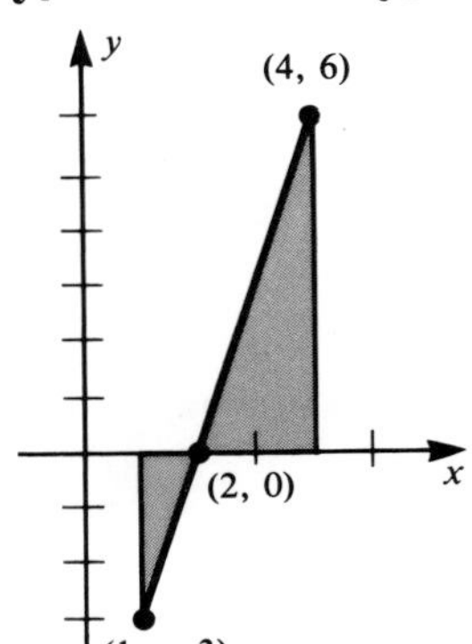

33. $\int_{-3}^{-1} [(x^2 + x + 2) - (3x + 5)]\,dx +$
$\int_{-1}^{0} (3x + 5) - (x^2 + x + 2)\,dx = \frac{32}{3} + \frac{5}{3} = \frac{37}{3}$

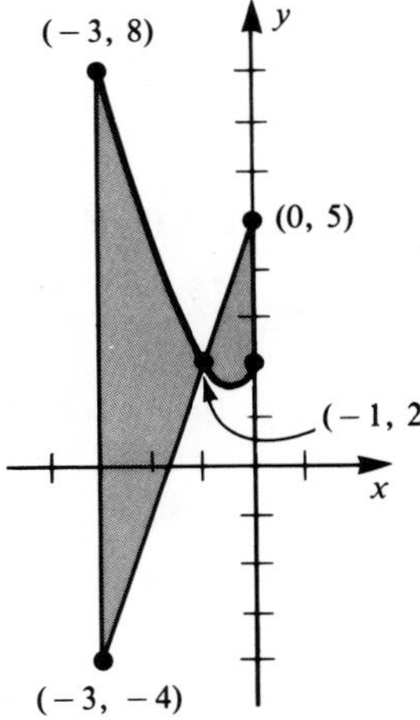

35. $\int_{-1}^{0} (x^3 + 3) - (x + 3)\,dx + \int_{0}^{1} (x + 3) - (x^3 + 3)\,dx = \frac{1}{4} + \frac{1}{4} = \frac{1}{2}$

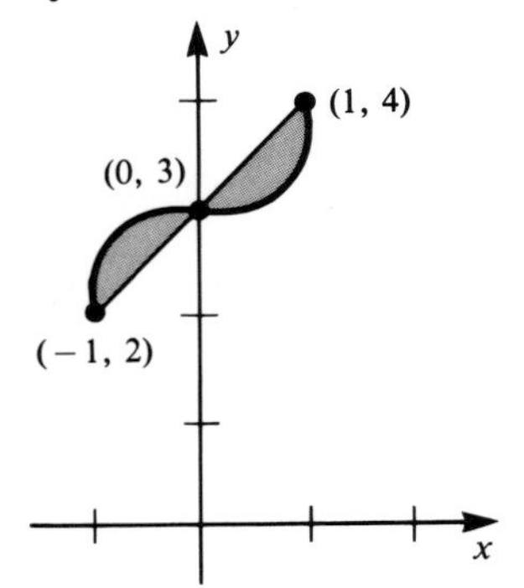

37. **(a)** $x = 12$ **(b)** $17\frac{1}{3} = \int_{10}^{12} MR(x) - MC(x)\,dx$

Section 6–5, page 397

1. $\frac{10}{3}$ **3.** $\frac{1}{4}\left(2-\frac{2}{e^2}\right)=\frac{1}{2}-\frac{1}{2e^2}$ **5.** $\frac{1}{4}\cdot\frac{1}{9}=\frac{1}{36}$ **7. (a)** 35 **(b)** 6650 **(c)** 10,325 **(d)** 3675
9. (a) 3000 **(b)** 630000 **(c)** 810000 **(d)** 180000
11. (a) $600(e^{6/20}-e^{3/20})$ thousand dollars $\approx$ \$112,800 **(b)** $600(e^{9/20}-e^{6/20})$ thousand dollars $\approx$ \$131,000
13. $120(1-e^{-2})$ thousand gallons $\approx$ 103,760 gallons **15. (a)** 0.55 **(b)** 0.63
17. (a) $\frac{1}{5}[245(e^{.1}-1]\approx 5.15$ (billion) **(b)** $\frac{1}{5}\cdot 245(e^{.2}-e^{.1})\approx 5.7$ (billion) **19. (a)** $45\frac{1}{3}$ ft/sec **(b)** $67\frac{1}{3}$ ft/sec

Section 6–6, page 411

1. (a) 8 **(b)** 8 **3. (a)** 9 **(b)** 9 **5. (a)** 8.5 **(b)** 8.625 **7.** $\int_1^3 (2x^2-3x)\,dx=\frac{16}{3}=5\frac{1}{3}$ **(a)** 5 **(b)** 5.25
9. $\int_1^5 \frac{1}{x^2}\,dx=\frac{4}{5}$ **(a)** 0.625 **(b)** 0.735 **11.** $\int_1^3 \frac{1}{x^3}\,dx=\frac{4}{9}=.444\ldots$ **(a)** 0.360 **(b)** 0.417
13. $\int_0^{60} x^2\,dx=72{,}000$, actual sum is 73,810 **15.** 288π **17. (a)** 1.0898 **(b)** 1.0963
19. (a) (i) 1.35 **(ii)** 1.377 **(b) (i)** 1.377 **(ii)** 1.384 **21.** 2.949 **23.** $\frac{64\pi}{3}$ **25.** $\frac{522\pi}{5}$ **27.** $\frac{4}{3}\pi R^3$
29. $\approx$ \$43,645 **31.** $\approx$ \$15,319 **33.** 3192 **35.** \$75,920,000

Review Exercises, Chapter 6, page 415

1. $2x^3-\frac{6}{x}+C$ **2.** $\frac{2}{3}x^{3/2}-2x^{1/2}+C$ **3.** $\frac{1}{7}x^7-\frac{2}{3}x^3-\frac{1}{x}+C$ **4.** $\frac{1}{3}x^3+2\ln|x|-\frac{1}{3x^3}+C$
5. $2x^2+8\ln|x|+C$ **6.** $6\ln|x|+\frac{4}{x}+C$ **7.** $\frac{1}{6}e^{6x}-\frac{1}{4}e^{-4x}+C$ **8.** $-\frac{1}{3}e^{-3x}+C$
9. $\frac{1}{2}e^{2x}+8e^x+16x+C$ **10.** $e^x-e^{-x}+C$ **11.** 204 **12.** 51 **13.** $\frac{1}{2}(e^8-17)$ **14.** 1 **15.** 5
16. 5 **17.** $4+2=6$ **18.** $-22+88=66$ **19.** $y=4x^4-6x^2+9x-3$ **20.** $y=4x^{3/2}-5x+1$
21. $y=2x^3-x^2+4x+9$ **22.** $y=9x^2-30x+34$ **23.** $h(t)=-4.9t^2+30t+2$
24. 6943.90 (careful with units) **25.** 12 **26.** 39 **27.** $\frac{125}{6}$ **28.** 30,000 **29.** 2 **30.** .40
31. (a) 1.57 **(b)** 1.60 **32.** \$84,160.79

CHAPTER 7

Section 7–1, page 430

1. $\frac{1}{20}(5x+1)^4+c$ **3.** $\frac{1}{5}(2x^2+3x)^5+c$ **5.** $(7u^2+3u-2)^3+c$ **7.** $\frac{1}{24}(4x^3-6x)^4+c$

9. $\ln|4x-3|+c$ **11.** $\frac{2}{3}e^{x^3}+c$ **13.** $4(x^2-x+5)^{1/2}+c$ **15.** $\ln|e^x-7|+c$ **17.** $-(1+\ln x)^{-1}+c$

19. $-\frac{1}{2}e^{1/x^2}+c$ **21.** $\int_5^{45}\frac{1}{2u}\,du=\frac{1}{2}(\ln 45-\ln 5)=\ln 3$ **23.** $\int_1^9\frac{1}{2}e^u\,du=\frac{1}{2}(e^9-e)$ **25.** $-\frac{1}{2}e^{-x^2}+c$

27. $\frac{2}{45}(3x-4)^{5/2}+\frac{8}{27}(3x-4)^{3/2}+c$ **29.** $\frac{1}{10}(x^2+5)^5-\frac{5}{8}(x^2+5)^4+c$ **31.** $25-\ln 26$

33. $c(x)=\frac{1}{3}(x^2+9)^{3/2}+141$ **35.** $\sqrt{65}-1$ **37.** $10(\ln 50-\ln 10)=10\ln 5$

Section 7–2, page 438

1. $\frac{1}{2}x^2\ln x-\frac{1}{4}x^2+c$ **3.** $\frac{1}{5}xe^{5x}-\frac{1}{25}e^{5x}+c$ **5.** $\frac{1}{2}x^2e^{2x}-\frac{1}{2}xe^{2x}+\frac{1}{4}e^{2x}+c$

7. $\frac{1}{4}x^4\ln x-\frac{1}{16}x^4+c$ **9.** $\frac{1}{3}x^3(\ln x)^2-\frac{2}{9}x^3\ln x+\frac{2}{27}x^3+c$ **11.** $\frac{1}{2}x^2e^{x^2}-\frac{1}{2}e^{x^2}+c$

13. $2e^{x^2}+c$ **15.** $-\frac{1}{x}\ln x-\frac{1}{x}+c$ **17.** $\frac{2}{e}-\frac{3}{e^2}$

Section 7–3, page 447

1. $\frac{1}{2}$ **3.** e^2 **5.** $\frac{1}{2}$ **7.** no finite area

9. e^{-4}; this is the area of the region under the curve $y=e^{-x}$, bounded below by the x-axis and bounded on the left by the line $x=4$.

11. $\frac{1}{3}$; this is the area of the region above the axis, under the curve $y=\frac{1}{x^4}$, and bounded on the left by the line $x=1$.

13. $\frac{1}{4}$; this is the area of the region between the curve $y=e^{4x}$ and the x-axis and that is bounded on the right by the y-axis.

15. This integral diverges; the region bounded on the left by the line $x=4$, and which lies between the graph of $y=\frac{1}{\sqrt{x}}$ and the x-axis does not have finite area.

17. diverges **19.** $\frac{1}{3}$ **21.** diverges **23.** 0.435 **25.** 18

27. approximately \$17 million (The first \$2 million is not *spent* in the state; the initial amount spent is \$1.8 million.)

Section 7–4, page 452

1. $\frac{1}{10}\ln\left|\frac{x-5}{x+5}\right|+c$ **3.** $\frac{-5}{12}\ln\left|\frac{x-6}{x+6}\right|+c$ **5.** $\frac{-2}{9(3x-2)}+\frac{1}{9}\ln|3x-2|+c$

7. $x\sqrt{x^2+25}+25\ln|x+\sqrt{x^2+25}|+c$ **9.** $\frac{3}{4}\ln\left|\frac{x}{8-2x}\right|+c$ **11.** $\frac{6}{5}\ln|x+\sqrt{x^2+4}|+c$

13. $\frac{x+2}{2}\sqrt{(x+2)^2+25}+\frac{25}{2}\ln|x+2+\sqrt{(x+2)^2+25}|+c$

15. $\frac{1}{2}\left[\frac{2x+1}{2}\right]\sqrt{(2x+1)^2+100}+25\ln|2x+1+\sqrt{(2x+1)^2+100}|+c$

17. $\frac{1}{3}\ln\left|\frac{x}{x+3}\right|+c$ **19.** $3\left[\frac{-2}{x-2}+\ln|x-2|\right]+c$ **21.** $10+\frac{9}{2}(\ln 9-\ln 3)$

23. $\frac{\sqrt{15}}{4}-\frac{\sqrt{3}}{2}$ **25.** $30-\frac{15}{2}\ln 45+\frac{15}{2}\ln 25$

Review Exercises, Chapter 7, page 454

1. $\frac{1}{3}e^{x^3}+C$ **2.** $\frac{2}{9}(3x+4)^{3/2}+C$ **3.** $\frac{1}{5}\ln|5x+8|+C$ **4.** $-xe^{-x}-e^{-x}+C$ **5.** $\ln|\ln x|+C$

6. $x^2e^{x^2}-e^{x^2}+C$ **7.** $\frac{x^2}{2}\ln(3x)-\frac{x^2}{4}+C$ **8.** $xe^{x+1}-e^{x+1}+C$ **9.** $\frac{1}{2}e^{(x^2+1)}+C$

10. $\frac{2}{3}x^{3/2}\ln x-\frac{4}{9}x^{3/2}+C$ **11.** $\frac{x^2}{2}\ln x^2-\frac{x^2}{2}+C$ **12.** $\frac{2}{7}(x+1)^{7/2}-\frac{4}{5}(x+1)^{5/2}+\frac{2}{3}(x+1)^{3/2}+C$

13. divergent **14.** 1 **15.** $\frac{1}{2e^2}+\frac{1}{e}$ **16.** 0 **17.** $\frac{1}{40}\ln\left|\frac{x-5}{x+5}\right|+C$ **18.** $\frac{1}{4}\ln\left|\frac{e^x-2}{e^x+2}\right|+C$

19. $\ln|(x+1)+\sqrt{x^2+2x+10}|+C$ **20.** $e^x-5\ln|e^x+5|+C$ **21.** $\frac{1}{6}\ln\left|\frac{x}{x+6}\right|+C$

CHAPTER 8

Section 8–1, page 461

1. 11 **3.** 6 **5.** $8e^3$

7.

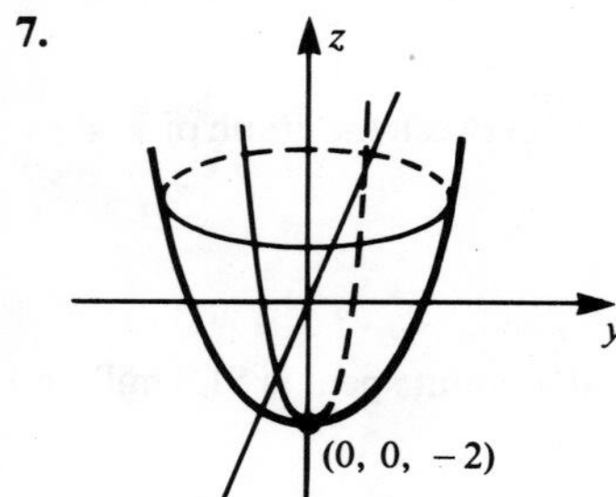

9.

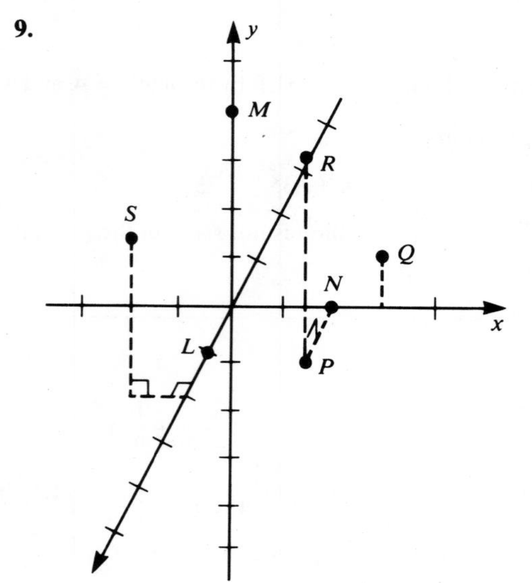

11.

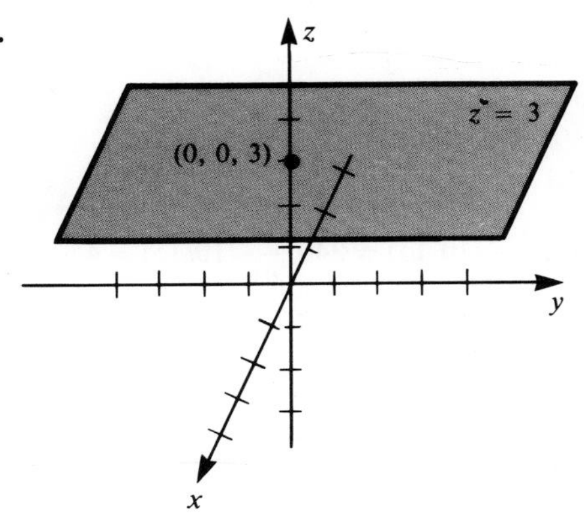

13. 207,200 **15.** 1760 **17.** 240

Section 8–2, page 476

1. $\dfrac{\partial f}{\partial x}(x, y) = 2x + 3y; \dfrac{\partial f}{\partial y}(x, y) = 3x$ **3.** $\dfrac{\partial z}{\partial y} = 10x^2y + 2; \dfrac{\partial z}{\partial x} = 10xy^2$

5. $f_x(x, y) = \dfrac{y^2 - x^2}{(x^2 + y^2)^2}; f_y(x, y) = \dfrac{-2xy}{(x^2 + y^2)^2}$ **7.** $\dfrac{\partial f}{\partial x}(x, y, z) = \dfrac{2xy}{z^3}; \dfrac{\partial f}{\partial y}(x, y, z) = \dfrac{x^2}{z^3}; \dfrac{\partial f}{\partial z}(x, y, z) = -\dfrac{3x^2y}{z^4}$

9. $\dfrac{\partial w}{\partial x} = \dfrac{y}{xy + yz^2}; \dfrac{\partial w}{\partial y} = \dfrac{x + z^2}{xy + yz^2}; \dfrac{\partial w}{\partial z} = \dfrac{2yz}{xy + yz^2}$

11. $\dfrac{\partial f}{\partial x}(x, y) = 6x^2y; \dfrac{\partial f}{\partial y}(x, y) = 2x^3; \dfrac{\partial^2 f}{\partial y \partial x}(x, y) = 6x^2; \dfrac{\partial^2 f}{\partial x \partial y}(x, y) = 6x^2; \dfrac{\partial^2 f}{\partial x^2}(x, y) = 12xy; \dfrac{\partial^2 f}{\partial y^2}(x, y) = 0$

13. $\dfrac{\partial f}{\partial x}(x, y) = 4xy - 3y^2; \dfrac{\partial f}{\partial y}(x, y) = 2x^2 - 6xy; \dfrac{\partial^2 f}{\partial y \partial x}(x, y) = 4x - 6y; \dfrac{\partial^2 f}{\partial x \partial y}(x, y) = 4x - 6y; \dfrac{\partial^2 f}{\partial x^2}(x, y) = 4y;$
$\dfrac{\partial^2 f}{\partial y^2}(x, y) = -6x$

15. $\dfrac{\partial f}{\partial x}(x, y) = 4xe^{x^2+y}; \dfrac{\partial f}{\partial y}(x, y) = 2e^{x^2+y}; \dfrac{\partial^2 f}{\partial y \partial x}(x, y) = 4xe^{x^2+y}; \dfrac{\partial^2 f}{\partial x \partial y}(x, y) = 4xe^{x^2+y};$
$\dfrac{\partial^2 f}{\partial x^2}(x, y) = 4e^{x^2+y} + 8x^2e^{x^2+y}; \dfrac{\partial^2 f}{\partial y^2}(x, y) = 2e^{x^2+y}$

17. $\dfrac{\partial f}{\partial x}(x, y) = 2 + 2xye^{x^2y}; \dfrac{\partial f}{\partial y}(x, y) = x^2e^{x^2y}; \dfrac{\partial^2 f}{\partial y \partial x}(x, y) = 2xe^{x^2y} + 2x^3ye^{x^2y}; \dfrac{\partial^2 f}{\partial x \partial y}(x, y) = 2xe^{x^2y} + 2x^3ye^{x^2y};$
$\dfrac{\partial^2 f}{\partial x^2}(x, y) = 2ye^{x^2y} + 4x^2y^2e^{x^2y}; \dfrac{\partial^2 f}{\partial y^2}(x, y) = x^4e^{x^2y}$

19. $f_x(2, 3) = 12; f_y(2, 3) = 4; f_{xy}(2, 3) = 4; f_{xx}(2, 3) = 6; f_{yy}(2, 3) = 0$

21. $f_x(-1, 0) = 0; f_y(-1, 0) = -1; f_{xy}(-1, 0) = 1; f_{xx}(-1, 0) = 0; f_{yy}(-1, 0) = 0$

23. $f_x(e, 1) = 2; f_y(e, 1) = e; f_{xy}(e, 1) = 1; f_{xx}(e, 1) = \dfrac{1}{e}; f_{yy}(e, 1) = -e$

25. (a) 2500

(b) The marginal productivity of labor $= f_x(x, y) = \dfrac{5}{2}x^{-3/4}y^{3/4}$.

The marginal productivity of capital $= f_y(x, y) = \dfrac{15}{2}x^{1/4}y^{-1/4}$.

(c) $f_x(16, 625) = \frac{1250}{32}$; $f_y(16, 625) = 3$

(d) Production will increase by approximately $\frac{1250}{32}$ units.

(e) Production will decrease by approximately 3 units.

27. **(a)** 10,750 **(b)** $\frac{\partial R}{\partial x}(x, y) = -6x + 400$; $\frac{\partial R}{\partial y}(x, y) = -4y + 500$ **(c)** $\frac{\partial R}{\partial x}(10, 15) = 340$; $\frac{\partial R}{\partial y}(10, 15) = 440$

(d) Revenue will increase by approximately 340 units.

(e) Revenue will decrease by approximately 440 units.

Section 8–3, page 484

1. (0, 0) **3.** (0, 0) **5.** (−1, 0) and (3, 0) **7.** f has a relative minimum at (2, −1).
9. f has a relative minimum at (−2, −2).

11. f has a relative minimum at $\left(\frac{27}{2}, 5\right)$ and a saddle point at $\left(\frac{3}{2}, 1\right)$.

13. f has a relative minimum at (1, 1) and a saddle point at $\left(-\frac{2}{3}, -\frac{2}{3}\right)$.

15. $x = 10, y = 5$ **17.** $x = 150, y = 100$ **19.** 4 replacements for x, 6 replacements for y

---

Section 8–4, page 493

1. $\frac{111}{5}$ **3.** 18 **5.** 203 **7.** 612 **9.** 21 **11.** 150 units of labor; 225 units of capital
13. 81 from X, 45 from Y **15.** 7 inspections at X, 15 at Y **17.** \$4040 on X, \$15,960 on Y
19. base is 20.7 feet by 20.7 feet, height is 14.5 feet (to nearest tenth)

Section 8–5, page 502

1. $6x^3y^2 - 6xy + f(x)$ **3.** $x\ln|y| + f(x)$ **5.** $63y + 15$ **7.** $9x^2 - 12x + 21$ **9.** 1 **11.** −1
13. 1600 **15.** 4 **17.** 0 **19.** 4 **21.** 32 **23.** 140 **25.** 4 **27.** 30

Review Exercises, Chapter 8, page 504

1. $f(1, -2) = -23$; $f_x(1, -2) = -14$; $f_y(1, -2) = 28$; $f_{xx}(1, -2) = 2$; $f_{yy}(1, -2) = -20$; $f_{xy}(1, -2) = 16$

2. $f(-1, 4) = 2$; $f_x(-1, 4) = 2$; $f_y(-1, 4) = \frac{1}{2}$; $f_{xx}(-1, 4) = 2$; $f_{yy}(-1, 4) = 0$; $f_{xy}(-1, 4) = \frac{1}{2}$

3. $f(0, 2) = 1$; $f_x(0, 2) = 4$; $f_y(0, 2) = 0$; $f_{xx}(0, 2) = 16$; $f_{yy}(0, 2) = 0$; $f_{xy}(0, 2) = 4$

4. $f(2, -1) = \ln 5$; $f_x(2, -1) = \frac{4}{5}$; $f_y(2, -1) = -\frac{4}{5}$; $f_{xx}(2, -1) = -\frac{6}{25}$; $f_{yy}(2, -1) = \frac{44}{25}$; $f_{xy}(2, -1) = \frac{16}{25}$

5. $f(-4, 2) = -2$; $f_x(-4, 2) = \frac{1}{2}$; $f_y(-4, 2) = 1$; $f_{xx}(-4, 2) = 0$; $f_{yy}(-4, 2) = -1$; $f_{xy}(-4, 2) = -\frac{1}{4}$

6. $f(1,2)=3$; $f_x(1,2)=\frac{1}{3}$; $f_y(1,2)=2$; $f_{xx}(1,2)=\frac{8}{27}$; $f_{yy}(1,2)=\frac{2}{3}$; $f_{xy}(1,2)=-\frac{6}{27}$

7. $f(2,4)=10$; $f_x(2,4)=5$; $f_y(2,4)=\frac{8}{5}$; $f_{xx}(2,4)=0$; $f_{yy}(2,4)=\frac{18}{125}$; $f_{xy}(2,4)=\frac{4}{5}$

8. $f(3,1)=e$; $f_x(3,1)=e$; $f_y(3,1)=-2e$; $f_{xx}(3,1)=e$; $f_{yy}(3,1)=4e$; $f_{xy}(3,1)=-2e$

9. $(3,-2)$ is a saddle point. **10.** $(0,0)$ is a saddle point; f has a relative minimum at $(18,6)$.

11. $(0,0)$ is a saddle point; f has a relative minimum at $(2,2)$. **12.** $(1,2)$ is a saddle point.

13. maximum is 50; minimum is -50 **14.** minimum is 8 **15.** 4 **16.** minimum is 18 **17.** $\frac{9}{4}$

18. 84 **19.** $2e-2$ **20.** $9\ln 9-12\ln 6+3\ln 3$ **21.** 36

Index

E

F

G

H

I

Q

R

S

T

U

V

W

X

Y

Z